Stimulant Drugs and ADHD
Basic and Clinical Neuroscience

Stimulant Drugs and ADHD

Basic and Clinical Neuroscience

Edited by

MARY V. SOLANTO, Ph.D.

AMY F. T. ARNSTEN, Ph.D.

F. XAVIER CASTELLANOS, M.D.

OXFORD
UNIVERSITY PRESS
2001

OXFORD
UNIVERSITY PRESS

Oxford New York
Athens Auckland Bangkok Bogotá Buenos Aires Calcutta
Cape Town Chennai Dar es Salaam Delhi Florence Hong Kong Istanbul
Karachi Kuala Lumpur Madrid Melbourne Mexico City Mumbai
Nairobi Paris São Paulo Shanghai Singapore Taipei Tokyo Toronto Warsaw

and associated companies in
Berlin Ibadan

Published by Oxford University Press, Inc.,
198 Madison Avenue, New York, New York 10016
http://www.oup-usa.org

Library of Congress Cataloging-in-Publication Data
Stimulant drugs and ADHD : basic and clinical neuroscience/
edited by Mary V. Solanto, Amy F. T. Arnsten, F. Xavier Castellanos.
p. cm. Includes bibliographical references and index.
ISBN 0-19-513371-4
1. Attention-deficit hyperactivity disorder—Chemotherapy.
2. Stimulants. 3. Neurobiology. 4. Psychopharmacology.
I. Solanto, Mary V. II. Arnsten, Amy Frances Torrance, 1954– III. Castellanos, F. Xavier.
RJ506.H9 S74 2001 618.92′8589061—dc21 00-025748

2 4 6 8 9 7 5 3 1

Printed in the United States of America
on acid-free paper

Preface

More than 60 years ago, in 1937, Charles Bradley, a physician in Providence, Rhode Island, made a serendipitous but astute observation that benzedrine (a racemic mixture of D- and L-amphetamine) had a distinct calming effect on the behavior of children at a residential treatment center. Since that time, thousands of studies have documented the effectiveness of stimulant drugs in the treatment of what is now called Attention-Deficit/Hyperactivity Disorder (ADHD). Many health professionals, educators, and parents have had occasion to observe first-hand the often dramatic improvements in attentiveness, impulse-control, academic productivity, and social adjustment which use of these medications—notably D-amphetamine (Dexedrine) and methylphenidate (Ritalin)—can effectuate.

Despite widespread clinical use of stimulant compounds and interest in their effects, the mechanisms of the therapeutic action of these drugs in the central nervous system remain poorly understood. Basic pharmacological research established that the stimulant drugs augment central norepinephrine and dopamine. Reasoning in part from these data, Paul Wender proposed, in an influential book entitled *Minimal Brain Dysfunction* published in 1971, that alterations in catecholamine neurotransmission mediate the therapeutic effects of the stimulants in ADHD. A later series of neurotransmitter-selective clinical drug trials in children with ADHD conducted at the National Institute of Mental Health (NIMH) substantiated the belief that *both* dopamine and norepinephrine are important to the therapeutic effects of stimulants. A major milestone in understanding stimulant effects concerned the long-standing belief that the response to stimulants in children with ADHD was "paradoxical" in that normal children and adults would respond with an *increase* in excitation and activity. Indeed, a reduction in hyperactive and inattentive symptoms following stimulant administration was often regarded as post-hoc confirmation of the ADHD diagnosis. The idea of a paradoxical effect was finally laid to rest in 1978 with the publication of a pivotal report published in *Science* by Judith Rapoport and colleagues at NIMH that normal children, normal adults, and children with ADHD *all* responded to D-amphetamine with an increase in attention and a reduction in impulsivity and hyperactivity. Since then there have been studies documenting efficacy across the life span and stimulant medications are used in patients of all ages.

Contrasting with a dearth of specific knowledge concerning the clinical mechanisms of action of stimulants are significant advances during the past 20 years in the basic, preclinical, and clinical neurosciences relevant to a comprehensive neuropsychopharmacology of catecholamines. Electron microscopy in conjunction with the use of labeled tracers has enabled an increasingly precise mapping of dopaminergic and noradrenergic circuitry. Single-unit electrophysiological recording and in vivo microdialysis now permit measurement of changes in neural activity and in neurotransmitter levels, respectively, in awake-behaving animals. Many recently developed receptor-selective agonists and antagonists are being used to elucidate functional relationships between catecholamine receptor subtypes and specific domains of behavior, including working memory, delayed responding, attentional control and locomotor activity. The development of neuroimaging technology, including functional magnetic resonance imaging (MRI) and positron emission tomography (PET), has opened a window on the functioning human and primate brain. Studies using these techniques during performance of cognitive tasks reveal the specific brain regions that are targeted by stimulants and that mediate their effects on specific neurocognitive functions. Use of parallel cognitive tasks in primates and humans has fostered cross-species generalization and integration of findings. In clinical populations, neuroimaging, neuropsychological, and genetic studies are providing important clues to the aberrant neurobiological substrates of ADHD that may be altered by therapeutic drugs. The purpose of this book is to systematically bring the new knowledge in these fields to bear on the central question of why and how stimulants exert their therapeutic effects on individuals with ADHD. A better understanding of these mechanisms is likely not only to lead to the development of better pharmacological treatments, but may also ultimately elucidate the neuropathology and pathophysiology of ADHD, and the neurobiology of self-regulatory processes generally.

Such an integration clearly would not be possible without the collaboration of accomplished scientists from an array of neuroscience disciplines. The request to contribute was an invitation to join a working group of scientists committed to bringing their individual spheres of knowledge and expertise to bear on the fundamental questions of the volume. To foster interdisciplinary dialogue, authors first completed and circulated drafts of their individual chapters to all other members of the group. Thanks to a generous educational grant from Medeva Pharmaceuticals, the contributors were able to come together for 2 ½ intensive days of extraordinary presentations and dialogue. Factual and conceptual integration of the material presented and discussed at the meeting shaped the subsequent revisions of the chapters and enhanced the cohesiveness of the final volume.

The book opens with two chapters that provide comprehensive descriptions of the clinical disorder of ADHD and the response to stimulants, respectively. Part II of the book is devoted to the basic and preclinical neuroscience of cortical and subcortical dopamine and norepinephrine systems, including their neuroanatomy, cellular neurochemistry, and functional neurophysiology with respect to arousal and attention, activity, impulsivity and reward-related behavior, and executive functions. The section concludes with a review and illustration of animal models of

ADHD. Part III presents the clinical neuroscience, including neuroimaging, pharmacokinetics and pharmacodynamics of methylphenidate in humans, as well as the comparative psychopharmacology of stimulants in adult volunteers and individuals with ADHD. Introductions to each section elucidate the goals of each chapter and its place on the emerging canvas of stimulant drug action. The goals of the final, challenging, integrative chapter in Part IV were multiple: to highlight the critical findings across the relevant disciplines, to identify the gaps in our knowledge, to propose alternate models of stimulant drug action that integrate current findings, and to chart a course for future translational research. The project, we are delighted to observe, has already led to some new proposals for research in these areas. Included as an Appendix is the now-classic but seldom reread 1937 article titled, "The Behavior of Children Receiving Benzedrine" by Charles Bradley. In hindsight, Bradley's observations concerning the effects of amphetamine treatment on primary symptoms, academic and social functioning, pharmacodynamics, and common side effects were remarkably astute.

In closing, we would like, first of all, to thank the other contributors for their hard work and commitment to the project. It was indeed a privilege to work with such outstanding scientists and to be the first to review such interesting, insightful, and provocative material. Thanks are also due to Fiona Stevens, Senior Editor at Oxford University Press, for guiding us, with unending patience and good grace, through the process of editing the book. I (MVS) would also like to recognize Jeffrey Newcorn, M.D., Chief of the Division of Child and Adolescent Psychiatry at the Mount Sinai School of Medicine, for his support, encouragement, and suggestions during a project that required more time and effort than either of us had imagined. Finally, we are grateful to Medeva Pharmaceuticals for the generous educational grant that allowed the contributors to travel from far-flung places to meet in what proved to be a highly productive and critically important conference.

New York, NY — M. V. S.
New Haven, CT — A. F. T. A.
Bethesda, MD — F. X. C.

Contents

Contributors

AMY F. T. ARNSTEN, PH.D.
Associate Professor
Section of Neurobiology
Yale University School of Medicine
New Haven, Connecticut

CRAIG W. BERRIDGE, PH.D.
Associate Professor
Psychology Department
University of Wisconsin
Madison, Wisconsin

F. XAVIER CASTELLANOS, M.D.
Chief, ADHD Research Unit
Child Psychiatry Branch
National Institute of Mental Health
Bethesda, Maryland

COLIN B. DENNEY, PH.D.
Postdoctoral Fellow
Department of Psychology
University of Hawaii
Honolulu, Hawaii

SHERRY A. FERGUSON, PH.D.
Research Psychologist
Division of Neurotoxicology
National Center for Toxicological Research/FDA
Jefferson, Arkansas

ANTHONY A. GRACE, PH.D.
Professor of Neuroscience and Psychiatry
Department of Neuroscience
University of Pittsburgh
Pittsburgh, Pennsylvania

LAURENCE L. GREENHILL, M.D.
Professor of Clinical Psychiatry
Division of Child and Adolescent Psychiatry
New York State Psychiatric Institute
New York, New York

J. DAVID JENTSCH, PH.D.
Postdoctoral Fellow
Department of Neuroscience
University of Pittsburgh
Pittsburgh, Pennsylvania

DAVID A. LEWIS, M.D.
Professor of Psychiatry and Neuroscience
University of Pittsburgh
Pittsburgh, Pennsylvania

MITUL A. MEHTA, PH.D.
MRC Cyclotron Unit
Hammersmith Hospital
Imperial College School of Medicine
London
United Kingdom

STEVEN R. PLISZKA, M.D.
Associate Professor and Chief
Division of Child and Adolescent Psychiatry
Department of Psychiatry
University of Texas Health Science Center
San Antonio, Texas

MARK D. RAPPORT, PH.D.
Professor of Clinical Psychology
Director of Clinical Training
Department of Psychology
University of Central Florida
Orlando, Florida

TREVOR W. ROBBINS, PH.D.
Professor of Cognitive Neuroscience
Department of Experimental Psychology
University of Cambridge
Cambridge
United Kingdom

TERJE SAGVOLDEN, PH.D.
Professor
Department of Physiology
University of Oslo
Norway

BARBARA J. SAHAKIAN, PH.D.
Reader in Cognitive Neuropsychology
Department of Psychiatry
University of Cambridge
Cambridge
United Kingdom

MARY V. SOLANTO, PH.D.
Associate Professor
Division of Child and Adolescent Psychiatry
Department of Psychiatry
The Mount Sinai School of Medicine
New York, New York

JAMES SWANSON, PH.D.
Professor
Department of Pediatrics
University of California at Irvine
Irvine, California

JANE R. TAYLOR, PH.D.
Associate Professor
Department of Psychiatry
Yale University School of Medicine
New Haven, Connecticut

NORA D. VOLKOW, M.D.
Associate Laboratory Director
Brookhaven National Laboratory
Upton, New York
Professor, Department of Psychiatry
SUNY, Stony Brook
Stony Brook, New York

I

PHENOMENOLOGY

1

Attention-Deficit/Hyperactivity Disorder: Clinical Features

MARY V. SOLANTO

The condition now referred to as Attention-Deficit/Hyperactivity Disorder (ADHD) was first described by George Still in 1901 (Still, 1902). In papers based upon a presentation to the Royal College of Physicians in London, Dr. Still enumerated problems of overactivity, inattention, and poor "inhibitory volition" that he observed in a case series of 20 children. Also noted were aggressiveness, defiance, resistance to discipline, "lawlessness," spitefulness, and dishonesty. Today, the latter would be diagnosed as symptoms of Oppositional Defiant Disorder and/or Conduct Disorder, which are often seen in conjunction with ADHD.

Diagnosis

The symptoms of inattention and overactivity have been recognized as core features of what we now refer to as ADHD since the publication of the second edition of the *Diagnostic and Statistical Manual of Mental Disorders* (DSM) in 1968 (American Psychiatric Association, 1968), which labeled the condition "Hyperkinetic Disorder of Childhood." The relative importance of these symptom dimensions and the combinations of criteria required for diagnosis have varied in subsequent editions, however, as reflected by changes in diagnostic nomenclature and delineation of subtypes.

The definition of this syndrome in the DSM-III (American Psychiatric Association, 1980) reflected a significant shift in conceptualization toward the primacy of the attention deficit in the criteria and the condition was renamed "Attention Deficit Disorder" (ADD). Whereas previously it had been proposed that the apparent attentional problems were secondary to the motor overactivity, the formulation in DSM-III, influenced by the work of Virginia Douglas (1972), suggested the reverse. DSM-III, furthermore, recognized for the first time the existence of two subtypes of ADD—with and without hyperactivity. Impulsivity was a necessary symptom feature for diagnosis of *both* subtypes.

The subtype distinction was abandoned in DSM-III-R (American Psychiatric Association, 1987) on the grounds of insufficient supporting evidence. The symptom of overactivity was restored to its original status as a core symptom, along with inattention and impulsivity, of "Attention Deficit Hyperactivity Disorder" (ADHD). Behavioral criteria were specified in a list of 14 symptoms spanning these areas, the presence of at least eight of which was required for diagnosis. Because this list included fewer than eight symptoms that primarily reflected inattention, an individual whose symptoms were primarily those of inattention was relatively unlikely to be diagnosed using this schema.

After publication of DSM-III-R, the results of research using factor analysis by Lahey and colleagues (1994) became available and clearly indicated that the behavioral symptoms of children with ADHD loaded onto two separable dimensions: one reflecting symptoms of inattention, and the second reflecting a *single* dimension of problems of hyperactivity and impulsivity. Furthermore, the DSM-IV field trials (Lahey et al., 1994) subsequently verified that these two clusters of symptoms can occur separately or together in individual children. The current edition, DSM-IV (American Psychiatric Association, 1994), thus recognizes three subtypes of Attention-Deficit/Hyperactivity Disorder (ADHD): "predominantly inattentive" (IN), requiring at least six of a list of nine attentional symptoms; "predominantly hyperactive-impulsive" (HI), requiring at least six of nine hyperactive and impulsive symptoms, and a "combined" (CB) subtype meeting *both* sets of symptom criteria (Table 1–1). For reasons that will be discussed, this book is largely concerned with the CB type.

ADHD has always been recognized as a disorder with childhood onset. Diagnostic criteria in DSM-III-R and the present edition, DSM-IV, specify an age of onset of symptoms before 7 years. DSM-IV further specifies that the onset of *functional impairment* due to the primary symptoms must also occur before the age of 7. The rationale for this feature has been that ADHD is a developmental disorder with a probable genetic cause and manifestation of symptoms at least by the time the child is in primary school and first experiences significant demands for attention and self-control. However, recent analysis of data from children in the DSM-IV field trials (Applegate et al., 1997) has indicated that this criterion may invalidly restrict the number of eligible cases, particularly with respect to the IN subtype, which appears to have a later age of onset (Applegate et al., 1997; McBurnett et al., 1999). These data suggest that more valid age-of-onset criteria may be 7 years for the CB and HI subtypes and 9 years for the IN subtype. For age of onset of impairment, more valid criteria may be 9 years for the CB subtype, 7 years for HI, and 9–14 years for the IN type.

The third criterion in DSM-IV for symptomatology in *two or more* settings (usually home, and school or work) ensures that the disorder will be *pervasive,* and not merely situational. Previous editions of the DSM did not specify whether symptoms were required to be present in more than one setting, resulting in considerable heterogeneity in subject samples across studies. The fourth criterion in DSM-IV emphasizes the requirement for *clinically significant impairment* in at least two settings. The criterion of duration of symptoms of at least 6 months has

Table 1–1. DSM-IV Criteria for Attention-Deficit/Hyperactivity Disorder

A. Either (1) or (2)

(1) Six (or more) of the following nine symptoms of *inattention* have persisted for at least 6 months to a degree that is maladaptive and inconsistent with developmental level:

Inattention

(a) Often does not give close attention to details or makes careless mistakes in schoolwork, work, or other activities
(b) Often has difficulty sustaining attention in tasks or play activities
(c) Often does not seem to listen when spoken to directly
(d) Often does not follow through on instruction and fails to finish schoolwork, chores, or duties (not due to oppositional behavior or failure to understand instructions)
(e) Often has difficulty organizing tasks and activities
(f) Often avoids, dislikes, or is reluctant to engage in tasks that require sustained mental effort (such as schoolwork or homework)
(g) Often loses things necessary for tasks or activities, such as toys, assignments, books, or tools
(h) Is often easily distracted by extraneous stimuli
(i) Is often forgetful in daily activities

(2) Six (or more) of the following nine symptoms of *hyperactivity–impulsivity* have persisted for at least 6 months to a degree that is maladaptive and inconsistent with developmental level:

Hyperactivity

(a) Often fidgets with hands or feet or squirms in seat
(b) Often leaves seat in classroom or in other situations in which remaining seated is expected
(c) Often runs about or climbs excessively in situations in which it is inappropriate (adolescents or adults may have feelings of restlessness)
(d) Often has difficulty playing or engaging in leisure activities quietly
(e) Is often "on the go" or often acts as if "driven by a motor"
(f) Often talks excessively

Impulsivity

(g) Often blurts out answers before questions are completed
(h) Often has difficulty awaiting turn
(i) Often interrupts or intrudes on others, such as butting into conversations or games

B. Some hyperactive-impulsive or inattentive symptoms that caused impairment were present before age 7 years.
C. Some impairment from the symptoms is present in two or more settings (e.g., at school or work, and at home).
D. There must be clear evidence of clinically significant impairment in social, academic, or occupational functioning.
E. The symptoms do not occur exclusively during the course of a Pervasive Developmental Disorder, Schizophrenia, or other Psychotic Disorder and are not better accounted for by another mental disorder (e.g., Mood Disorder, Anxiety Disorder, Dissociative Disorder, or a Personality Disorder).

314.01 **Attention-Deficit/Hyperactivity Disorder, Combined Type:** if both Criteria A1 and A2 are met for the past 6 months.
314.00 **Attention-Deficit/Hyperactivity Disorder, Predominantly Inattentive Type:** if Criterion A1 is met but Criterion A2 is not met for the past 6 months
314.01 **Attention-Deficit/Hyperactivity Disorder, Predominantly Hyperactive–Impulsive Type:** If Criterion A2 is met but Criterion A1 is not met for the past 6 months.

From: American Psychiatric Association (1994). *Diagnostic and statistical manual of mental disorders* (4th ed.). Washington, DC: Author.

been included beginning with the publication of DSM-III, and is necessary to distinguish a chronic developmental disorder such as ADHD from a more transient adjustment reaction—as to a move or birth of a sibling—and from response to an acute stressor. Finally, as in previous DSM editions, the following disorders are exclusionary: pervasive developmental disorder, schizophrenia, or other psychotic disorder if symptoms occur only during the course of the other disorder. With respect to diagnosis in the presence of mental retardation, accepted clinical practice is to render a second diagnosis of ADHD only if the child's inattentiveness, hyperactivity, and impulsivity are more extreme than what would be expected given the child's *mental age*.

The criteria for "Hyperkinetic Disorder" in the current edition of the *International Classification of Diseases* (ICD-10) (World Health Organization, 1990), used mainly in Europe, are similar to those in DSM-IV in that specific behavioral symptoms of the two dimensions of inattention and hyperactivity–impulsivity are delineated and both are required for diagnosis of the traditionally recognized disorder. The ICD-10, however, does not recognize predominantly inattentive and predominantly hyperactive–impulsive subtypes; children with these subsets of symptoms would be considered "subthreshold" cases in this classification schema. In addition, the ICD-10 is more stringent than DSM-IV in requiring *direct* observation of symptoms by the clinician—not simply report of symptoms by parents and teacher—and also in requiring onset before the age of 6 years. Duration of at least 6 months and an etiology not solely due to mental retardation (i.e., requirement of IQ above 50) are similar in both classification systems.

Presentation and Evaluation

Core Symptoms

As is true for most other psychiatric disorders, there is no single diagnostic test for ADHD. Instead, the diagnosis is based largely on careful elicitation and integration of reports by parents and teachers of past and current development and behavior, along with systematic exclusion of other conditions that may present with similar features, such as learning disorders, anxiety, depression, and post-traumatic stress disorder. It is also necessary to determine that the problem behaviors exceed what would be *expected*, given the child's age and gender. Symptom descriptions may be obtained from parents and teachers in open-ended clinical interview, or, more rigorously, by structured interview such as the comprehensive Diagnostic Interview Schedule for Children and Adolescents (DISC) (Shaffer et al., 1996) and age- and gender-normed behavioral questionnaires such as the Conners Parent and Teacher Rating Scales (Goyette et al., 1978) and the Child Behavior Checklist (Achenbach, 1991). Problems typically described by parents and teachers are enumerated in Table 1–1. It is noteworthy that parents and teachers of children with confirmed ADHD often report considerable situational variability in these symptoms. As will be elaborated later, in the section on Etiology, the central difficulty appears to be in the *volitional* control of attention and impulses during tasks and activities (e.g., homework) that are monotonous, rote, especially effortful, or lacking in inherent gratification or reward.

The core symptoms of ADHD have been studied using laboratory measures (see also Denney & Rapport, Chapter 12). In a key study conducted at NIMH (Porrino et al., 1983) activity counts of hyperactive and control participants were recorded continuously for 7 days using solid-state monitors with hourly "memory," worn around the waist. Activity counts for children with ADHD were significantly higher across a variety of situations, including academic subjects in school, outside play after school and on weekends, and during sleep. Activity counts for the children with ADHD were also higher, but not significantly so, during lunch/recess and physical education, when the activity level of the unaffected children also increased. Three additional studies, of a total of six reviewed (Corkum et al., 1998) have similarly found increased activity in ADHD children during sleep. These findings strongly indicate that overactivity in children with ADHD is a core symptom and not simply secondary to inattention in structured situations.

Impulsivity, defined as a greater propensity to make short-latency, inaccurate responses has been demonstrated on measures of signal detection, such as the continuous performance test (CPT) on which children with ADHD have been found to make more errors of commission (Corkum & Seigel, 1993). Impairments in inhibitory control have also been demonstrated on the Stop Signal (Oosterlaan et al., 1998; Schachar & Logan, 1990) and go/no-go Tasks (Casey et al., 1997), on which the individual is required to *withhold* a motor response in the presence of discrete stimuli. On the Delay Aversion Task, children with ADHD are significantly more likely to select response options associated with shorter intervals of delay, even when these entail forfeiture of larger rewards (Sonuga-Barke et al., 1992). Although some of these tasks, particularly the version of the CPT recently developed by Conners (Conners, 1996), are reported to have good sensitivity and specificity in distinguishing children with diagnosed ADHD from normal children, correlations between performance on these measures and teacher–parent ratings of behavior, as well as correlations among the tasks themselves, are disappointingly small, leading to concerns about their ecological validity (Barkley, 1991; Solanto et al., 2000). Differential expression of impulsivity in motor, cognitive, social, and emotional domains may account for some of the inconsistencies across measures (Kindlon et al., 1995).

Deficits in attention in ADHD have been the most difficult to isolate on clinical tests and to duplicate in laboratory paradigms. Children with ADHD make more errors of omission on a CPT, and have longer and more variable reaction times—features generally considered to index inattention (Corkum & Seigel, 1993; Losier et al., 1996). However, laboratory research has generally *not* revealed specific deficits in the encoding, memory search, or decision stages of information-processing in children with ADHD, or in orienting, divided attention, or attentional capacity (see reviews by Douglas, 1999; van der Meere, 1996). Most studies have similarly failed to find evidence for a sustained attention deficit when rigorously defined as a steeper slope of deterioration in performance *over time* in children with

ADHD and not merely worse *mean* performance (van der Meere, 1996). Finally, although children with ADHD certainly *appear* to be distractible, and this is explicitly delineated as a symptom in the DSM, this finding has been demonstrated inconsistently on laboratory tasks (Sergeant et al., 1999).

Associated Problems

The difficulties in the primary symptom areas of inattention and/or hyperactivity–impulsivity have secondary negative impact not only on the child's academic performance but also on relationships with parents and peers, and on the development of self-competence and self-worth.

Epidemiological studies have revealed that substantial proportions of children with the CB or IN subtype of ADHD are rated by their teachers as being impaired academically and socially (Baumgaertel et al., 1995; Gaub & Carlson, 1997b; Lahey et al., 1994; Wolraich et al., 1996). The latter findings are supported by sociometric ratings by classmates, on which hyperactive children were significantly more likely than others to be socially rejected (Henker and Whalen, 1999). This appears to be attributable to the high frequency and intensity of intrusive, unpredictable behaviors exhibited by children with ADHD, rather than to deficient prosocial behavior or faulty social information-processing.

The negative impact of ADHD behaviors on parent–child interaction has been amply documented. In an important study by Barkley, direct observations showed that parents of children with ADHD make significantly more critical and controlling statements to their children and provide less verbal reinforcement than do parents of nondiagnosed children (Barkley, 1989a). That these parental behaviors were the *result* rather than the *cause* of their children's behavior problems was indicated by the immediately ameliorating effect of subsequent administration of methylphenidate on *both* parental *and* child behavioral responses.

Prevalence

It is not a straightforward task to ascertain the population prevalence of ADHD. Reported figures vary considerably by methodology used to diagnose the disorder. In addition, the prevalence of the condition appears to vary systematically by diagnostic subtype, by age and sex of the individuals assessed, and by source of the sample (e.g., clinic-referred vs. school-based).

Earlier large-scale epidemiological studies, based on DSM-II and DSM-III criteria, reported prevalences varying from 1.2% to 14.3%, depending on whether criteria included parent and/or clinician ratings in addition to teacher ratings (Lambert et al., 1978; Szatmari et al., 1989; Trites et al., 1979).

More recently, Wolraich (Wolraich et al., 1996) surveyed the entire school-aged population in a county in Tennessee using teacher ratings of DSM-IV symptoms of ADHD. Prevalence rates for the IN, HI, and CB subtypes were 5.4%, 2.4%, and 3.6%, respectively, summing to 11.4%. Another large school-based study of 2,744 ethnically diverse children in Texas reported slightly lower rates of 4.5%, 1.7% and 1.9%, respectively, summing to 8.1% (Gaub & Carlson, 1997b). These rates are likely to be somewhat higher than the true population prevalence, however, because two important DSM-IV criteria—parent affirmation of symptoms at home, and significant impairment in both home and school settings—were not applied. Thus, a definitive population-based study of prevalence remains to be done. The DSM-IV conservatively estimates an overall population prevalence of 3% to 5% for ADHD (American Psychiatric Association, 1994).

Sex ratios among nonreferred children consistently favor boys, with ratios of approximately 3:1 and 2:1 reported for the CB and IN subtypes, respectively (Gaub & Carlson, 1997b; Wolraich et al., 1996). A meta-analysis of 18 studies (Gaub & Carlson, 1997a,b) concluded that among nonreferred children, but not among referred children with ADHD, boys had more severe symptoms. In general, qualitative differences between boys and girls meeting diagnostic criteria have *not* been observed (Biederman et al., 1999; Sharp et al., 1999).

Concern has been raised in the popular press about increasing rates of diagnosis of ADHD and prescription of methylphenidate. Swanson (Swanson et al., 1995) reported that in the period from 1990 to 1993 the number of patients diagnosed as having ADHD increased from 900,000 to 2 million. As highlighted by Goldman, representing the American Medical Association (Goldman et al., 1998), this increase is due to greater awareness of ADHD among professionals, knowledge that it may persist into adolescence and adulthood, and recognition of multiple subtypes. Safer (Safer et al., 1996) reported that during roughly the same period (1990–1995) there was a 2.5-fold increase in the prevalence of methylphenidate treatment of youth 5–18 years of age, such that the rate in mid-1995 was 2.8% (or 1.5 million). It is noteworthy that despite apparent increases, the percentage of children receiving methylphenidate is still below the estimated prevalence of ADHD of at least 3% to 5%. Indeed the school-based prevalence study by Wolraich found that only about a quarter of all children meeting criteria for ADHD were known by their teachers to have been so diagnosed and treated with stimulant medication. There are, however, wide regional differences in prevalence of drug treatment, as reflected in rates of 1.1% for all youth in the state of Michigan in 1992 (4.3% for boys 10–11 years of age) (Rappley et al., 1995), 6% of public elementary school students in Baltimore County, Maryland, in 1987 (Safer & Krager, 1988), and, most recently, 8% to 10% of children in grades 2–5 in two cities in southeastern Virginia (LeFever et al., 1999).

Comorbidity

Numerous studies have documented high rates of comorbidity between ADHD (CB subtype) and other disruptive behavior disorders. Prevalence rates reported across studies of 30%–50% for oppositional defiant disorder (ODD), 25% for comorbid anxiety, and 15%–75% for mood disorders (Biederman et al., 1991; Milberger et al., 1995; Newcorn and Halperin, 1994) are elevated compared to normal control groups. The most recent comorbidity figures, emerging from a large, multisite study of 579 children with ADHD and based on parent DISC interviews are: ODD, 40%; CD, 14%; anxiety disorder (excluding simple phobia), 35% (MTA Cooperative Group, 1999). Some evidence suggests that comorbid aggression may distinguish a biologically distinct subtype of ADHD (Halperin et al., 1997). Comorbid disruptive behavior disorders are found much less frequently with the IN subtype than the CB subtype (Eiraldi et al., 1997).

It is generally agreed that specific learning disabilities (LD) occur more commonly in children with ADHD than in those without. However, reported rates of comorbidity, reviewed by Semrud-Clikeman and colleagues (1992), vary enormously depending on LD criteria applied. When a moderate achievement–IQ discrepancy of at least 20 standard score points in both math and reading was imposed, 23% had a comorbid LD (Semrud-Clikeman et al., 1992).

Nonrandom associations between ADHD and other disorders admit of several possible explanations (Caron & Rutter, 1991). Among them are that (*1*) the primary and comorbid disorders share a common etiology or underlying vulnerability; (*2*) the presence of the primary disorder constitutes a risk factor for the development of the other disorder; (*3*) the combination of the disorders represents a hybrid that carries the unique characteristics of each of the disorders; (*4*) the combination of disorders represents a distinct subtype of the primary disorder; (*5*) the combination of the two disorders represents a third unique disorder. A review yields insufficient data as yet to distinguish among these possibilities with respect to ADHD (Jensen et al., 1997). However, a study of psychosocial, developmental and cognitive variables in children with ADHD, CD, and ADHD + CD yielded compelling support for the third hypothesis (Schachar & Tannock, 1995). The association between internalizing (affective and anxiety) disorders and ADHD has been most extensively studied by Biederman and colleagues who found higher rates of both affective and anxiety disorders in the first-degree relatives of 140 ADHD probands compared to relatives of 120 normal control subjects (Biederman et al., 1992). Patterns of familial segregation supported the authors' hypotheses (Faraone & Biederman, 1997) that ADHD and major depressive disorders share common familial (but probably nongenetic) risk factors. Anxiety disorders did not co-segregate, indicating they are independently transmitted in families. The results of a subsequent longitudinal follow-up study of 76 children with both depression and ADHD revealed that the two disorders had independent and distinct courses, challenging the idea that depression in children with ADHD and their relatives is simply demoralization consequent to living with ADHD (Biederman et al., 1998). The compelling companion hypothesis that comorbid anxiety in ADHD is secondary to chronic negative feedback and academic and social failures cannot be ruled out, however. Jensen et al. (1997) concluded that among all comorbidities with ADHD, evidence for a clinical subtype with distinctive age of onset, symptoms, pattern of inheritance, and clinical course and outcome is strongest for the combination of ADHD with CD.

Etiology

Genetic

Familial

There is now abundant evidence that genetic factors are principal in the etiology of ADHD. Numerous early studies, reviewed by Faraone and Biederman (1994), showed a significantly enhanced relative risk (odds ratios ranging from 2.1 to 7.6) for the disorder in siblings and parents of children with ADHD. Risk for ADHD was also shown to be higher in biological than in adoptive parents of probands. The pattern of genetic transmission in these studies suggested, in addition, that ADHD is *not* the product of either sex-linked or mendelian (single-gene) genetic transmission (Morrison & Stewart, 1973).

More recently, there have been eight twin studies, several of which have employed large epidemiological samples (see review by Tannock, 1998). The largest of these (Levy et al., 1997) enrolled 849 monozygotic (MZ) and 1089 dizygotic (DZ) twin pairs identified through the Australian National Health and Medical Research Council Twin Registry. Concordance rates for diagnosis of ADHD (DSM-III-R) for MZ and DZ twins were 82.7% and 37.9%, respectively. Similar concordance rates for parent-rated ADHD symptoms, at least twice as large for MZ as for DZ twins, were also reported in recent studies in the United States (Sherman et al., 1997; Silberg et al., 1996) and a large epidemiological sample in Norway (Gjone et al., 1996).

In order to ascertain the relative contributions of genetics and environment, twin data has been subjected to regression analysis that determines the relative magnitudes of three components: heritability (genetic determinants), shared environment (e.g., the impact of being raised in the same home with the same parents), and nonshared environment (i.e., events and environmental characteristics that affect only one twin—e.g., a head injury). Heritability in the study by Levy (Levy et al., 1997) was 0.91, corroborating other twin studies reporting genetic heritability estimates of 0.75 to 0.98. Notably, in addition, the contribution of shared environment in this study, as well as in nearly all the others, was *not significant.* Goodman and Stevenson (1989) reported that a combination of seven family adversity factors, including marital discord and criticism and coldness toward the child, accounted for only 10% of the variance in the occurrence of ADHD. These results provide compelling evidence that deficient parenting and family adversity do not *cause* ADHD. They further suggest that cross-sectional studies showing lower income (Scahill et al., 1999), a higher index of psychosocial adversity (Biederman et al., 1995a), and increased parental conflict (Biederman et al., 1995b) in families of ADHD probands have not identified *causes* of ADHD. Rather, the results were explainable by the *common* genetic vulnerability of the probands *and* their first-degree relatives to ADHD, learning disabilities, antisocial personality disorder, substance abuse, and depression, and the downward mobility and family adversity, dysfunction, and stress that *result* from these conditions. That socioeconomic status may be one variable that contributes to the manifestation of ADHD in genetically vulnerable individuals was suggested in a finding of lower SES in the *adoptive* families of hyperactive children when compared to the adoptive homes of children without ADHD (Cadoret & Stewart, 1991).

Levy and colleagues (1997) and Gjone and colleagues (1996) also addressed the long-standing issue of whether ADHD represents a distinct syndrome or the extreme end on a continuum of a trait of attention/activity. The result indicated that heritability was the same whether determined on the basis of the presence or absence of the diagnosis, or on the basis of a "trait" score anywhere along the continuum (DSM-III-R symptom score), thereby providing strong support for the trait interpretation. More recently, latent class analyses (Neuman et al., 1999) of DSM-IV symptom profiles obtained from three population-based data sets of children and adolescents have corroborated these conclusions. In addition, the findings of the latter study suggested that the inattentiveness and hyperactivity–impulsivity are genetically separable traits, corresponding to genetically distinct subtypes.

Molecular Genetics

Small-scale molecular studies in ADHD have provided intriguing results that have been marked by an unusual degree of convergence in a field where nonreplication has been otherwise uniform. Before discussing genetic findings in ADHD, a brief introduction to genetic terminology is useful.

Except for some genes carried on the sex chromosomes, all genes are inherited as pairs, one from each parent. Alternative forms of a given gene are called *alleles.* Some genetic variants are extremely rare, but others may be quite frequent. Allelic variations that are found in at least 1% of a given population are termed *polymorphisms.* Most polymorphisms have no detectable effect because they do not change the amino acid structure of any protein, or because they do not affect the rate at which such proteins are synthesized. Despite their apparent lack of *functional significance,* such polymorphisms can still function as *genetic markers* that are useful for genetic analyses. Such a marker may indicate the presence of nearby genes that may exert their functional effects by changing the protein structure of a component in a given pathway, or by altering the regulation of the production of such proteins.

The genetic influences on a particular disorder may be investigated in one of two ways. The *candidate gene approach* begins with a hypothesis about a specific gene based on the known pathophysiology of the disorder. There are two types of case-control designs that can be used to support or refute the involvement of a given candidate gene. Traditional case-control comparisons use a group of individuals who have the disorder of interest, and a group of control subjects who do not. Statistically significant association of a specific allele with the disorder could indicate that an allele conveys risk. However, because allele frequencies may vary widely in different populations, it is possible that significant case-control results may be ascribable to deficiencies in matching for ethnic origin.

The *family-based case-control design* has become the preferred manner of testing hypotheses about specific candidate genes. Under the null hypothesis that a particular allele does not convey additional risk for ADHD, parents who have one copy of the putative risk allele would be expected to transmit that allele randomly to 50% of their offspring. This type of analysis requires DNA from probands and their biological parents, and only those families in which at least one parent is heterozygous for the risk allele are informative. When such parent/child trios are analyzed for families in which the proband has the disorder of interest, such as ADHD, the number of transmissions of the risk allele should exceed the number of times the risk allele was not transmitted. The statistical test for this sort of analysis is called Transmission Disequilibrium Test (TDT), which has a chi-square distribution. Each parent is sorted into one of four classes: (*a*) high-risk allele transmitted, low-risk allele not-transmitted; (*b*) high-risk allele transmitted, high-risk allele not transmitted; (*c*) high-risk allele not transmitted, low-risk allele transmitted; (*d*) low-risk allele transmitted, low-risk allele not transmitted. If the number in each cell is represented by (a, b, c, and d), then $TDT = (a - c)^2/(a + c)$ with one degree of freedom. Thus, the control group for this test (and for the mathematically synonymous Haplotype Relative Risk) is composed of the parental alleles that were *not* transmitted, thereby eliminating any potential confounding due to ethnicity.

Because 30,000 to 50,000 genes are expressed in the human brain, there are theoretically 30,000 to 50,000 potential candidates for neuropsychiatric disorders. Thus, the *prior probability* that a particular candidate gene conveys risk is the most important determinant of the alpha level to be selected for candidate gene studies. We do not have compelling models of etiopathogenesis for any psychiatric disorder, thus the a priori probabilities for any given gene are low, and the likelihood that positive findings represent false positives are correspondingly high (Crowe, 1993). For this reason, an alternative approach has gained favor in psychiatric genetics (Genetics Workgroup of the National Institute of Mental Health, 1999).

In contrast to hypothesis-driven approaches, *positional cloning* refers to the attempt to find genetic factors that contribute to disease by genome-wide searches that are completely free of specific pathophysiological hypotheses. In single-gene disorders such as Huntington chorea or cystic fibrosis, the application of this approach to large families with multiple affected individuals has led to the discovery of the relevant genes and the specific mutations.

ADHD, like all psychiatric disorders, is not likely to be a single-gene disorder. Rather, ADHD falls into the category of complex genetic disorders that do not follow classic mendelian inheritance patterns and that are believed to result from the combined effects of several genes and interactions with the environment. The unit of study for positional cloning studies in ADHD is the family that contains at least a pair of affected siblings. Siblings, on average, share 50% of their genes with their co-sibling. A *genome-wide scan* is performed by studying a sufficiently large sample of affected sibling pairs for whom polymorphic markers that are evenly spread throughout the chromosomes are available. Findings of greater sharing by affected siblings than expected by chance indicate "hot spots" in the genome that may be worth further exploration. Ideally, such "hot spots" will contain some definite number of potential candidate genes that can then be tested individually.

To date, no group has reported a genome scan analysis in ADHD, but a number of candidate gene studies have been published. Possible involvement of the immune system in ADHD was suggested by an association between ADHD and a specific human leukocyte antigen (HLA) sequence in combination with a null gene for an immune system component (complement C4B) (Odell et al., 1997).

A number of dopaminergic candidates have been proposed in ADHD because of the consensus that dopaminergic dysregulation is likely to be central to the disorder. The first was an allelic variation in the dopamine transporter gene (DAT1) that was significantly associated with ADHD in a study of 53 families ($p = .006$) (Cook et al., 1995). This polymorphism is located in the untranscribed 3′ portion of the DAT1 gene, so that its functional significance remains unclear. Having previously confirmed the association between the 480 base pair DAT1 polymorphism and ADHD in a sample of 26 trios and 14 parent–proband pairs ($p = .01$) (Gill et al., 1997), the same group reported a replication with an additional sample of 58 trios and 5 parent–proband pairs ($p = .049$, one-tailed) (Daly et al., 1999). In the combined sample, they calculated the relative risk of the associated DAT1 allele as 1.20. Further supporting the possibility that DAT1 is associated with ADHD is another study that combined 117 families from clinics in Tucson, Arizona, and Atlanta, Georgia (Waldman et al., 1998). Instead of taking only well-characterized probands, the authors accepted a wide range of symptomatology and comorbidity and they relied exclusively on parent symptom ratings to derive continuous symptom scores of hyperactivity–impulsivity and inattention. Each symptom was rated 0–4, and scores of 1, 2, or 3 were interpreted as indicating presence of a symptom. Hyperactivity–impulsivity scores were significantly predicted by the number of DAT1 "high-risk" alleles ($p = .032$, one-tailed). This relationship did not hold for inattention scores, and was confirmed in secondary analyses with parent/proband trios and in 12 sets of siblings who were discordant for the number of DAT1 "high-risk" alleles. The authors estimated that the DAT1 allele explains 3.6% of the variance in hyperactivity–impulsivity symptoms, and only 1.1% of the variance in inattention symptoms.

Even if DAT1 were to be confirmed as the first gene to be associated with a psychiatric disorder, it is clear that it will not be the last. Another dopaminergic gene that has aroused interest is the dopamine 4 receptor (DRD4). DRD4 was first studied because clozapine is a highly effective dopamine 4 receptor antagonist. Once the gene was cloned, a large number of allelic variations were discovered, many of which are determined by the number of 48 base pair repeats in exon 3. The second most-common form of DRD4, which contains 7 repeats (DRD4*7R), is slightly less effective than the more common variants in inhibiting the second messenger cAMP (Asghari et al., 1995). It is of interest in this context that activation of cAMP/PKA second messenger systems in prefrontal cortex impairs prefrontal function (Taylor et al., 1999). The same DRD4 variant was associated with the psychological trait of Novelty Seeking in some samples (Benjamin et al., 1996; Ebstein et al., 1996) but not in others (Paterson et al., 1999). DRD4*7R was significantly associated with ADHD in a case-control study ($n = 39$ per group, $p = .01$) (LaHoste et al., 1996) that was followed up by an independent replication using within-family controls (52 trios, $p = .035$) (Swanson et al., 1998).

Despite one negative report (Castellanos et al., 1998), there have been several independent partial replications, all of which have been modestly significant. For example, in a study of 133 families, DRD4*7R was found to be associated with ADHD using one test (TDT, $p = .03$) but not with another (mean test of identity-by-descent sharing in affected siblings) (Smalley et al., 1998). In the Atlanta-Tucson study (Rowe et al., 1998), a higher proportion of questionnaire-diagnosed individuals with ADHD had DRD4*7R than controls ($p < .05$), and DRD4 genotypes were significantly correlated with continuous symptom counts ($r \sim .20$, $p < .01$, one-tailed), but the possibility of ethnic stratification could not be dismissed. Accordingly, within-family controls were obtained using two variations of the TDT. The categorical TDT, which requires that each proband be classified as affected or unaffected, did not yield any evidence in support of the hypothesis. Application of a quantitative version of the TDT resulted in correlations of 0.13 (not significant) for the total sample, and .18 for the patients ($p < .1$, one-tailed). Thus, under the most stringent conditions, these data provided minimal support for DRD4 as a susceptibility gene in ADHD.

Studies of the genetics of ADHD typically begin with child probands. In contrast, a study that began with 27 adult ADHD patients and their families (Faraone, 1999) found that the DRD4*7R allele was transmitted significantly more frequently than expected under the null hypothesis ($p = .03$ for multiallele TDT, $p = .007$ for a follow-up chi-square for the two most frequent alleles).

In summary, there is surprising consensus that two dopaminergic genes, DAT1 and DRD4, may be associated with susceptibility to ADHD. However, the effects of these genes are modest at best, and the possibility that these findings represent false positives must still be considered (Crowe, 1993). However, it is likely that the genetic contributors to the etiology of ADHD will be solidly established as the Human Genome Project approaches its goal of providing for the complete sequencing of human DNA. In addition, advances in technology and refinement of methodology are likely to yield increasingly precise information about genetic determinants of ADHD. It would be important to broaden the investigation of candidate genes by considering genes regulating norepinephrine as well as dopamine.

Neuroimaging and Electrophysiology

Neuroimaging studies, using PET, MRI, fMRI, and rCBF technology have revealed anomalies of volume as well as metabolic activity in the frontal cortex and basal ganglia of individuals with ADHD. These studies, as well as research on event-related EEG potential, are detailed in Chapter 10 by Castellanos.

Deficient Inhibitory Control

Various attempts have been made to describe ADHD in terms of fundamental neuropsychological or behavioral characteristics. Among the most influential of these is Barkley's (1997a, b) theory that a core deficit in inhibitory control impairs development of executive functions necessary for self-regulation of behavior, cognition, and emotions. In this schema, behavioral inhibition encompasses the abilities to inhibit a prepotent response, to stop an ongoing response, and to prevent interference by extraneous stimuli during the intervening interval. According to the theory, adequate inhibitory control is essential for the development of four types of executive function: (*1*) "working memory," that is the ability to retrieve, maintain, and mentally manipulate information (as, for example, when doing mental arithmetic); (*2*) self-regulation of arousal, affect, and motivation; (*3*) internalization of speech to mediate behavior via learning history, internalized "rules," and active problem solving; and (*4*) "reconstitution" or the ability to analyze and synthesize information in order to solve social and cognitive problems. Developmental deficits in these critical functions ultimately give rise to the primary and secondary symptom features of ADHD, CB type. In this model, inattentiveness, for example, can result from an inability to protect working memory processes from interference or to inhibit a response to an extraneous stimulus. The notable absence of disinhibited behavior in the IN subtype argues for a fundamentally different etiology in this subtype (Barkley, 1997b).

Although there are many additional untested hypotheses emerging from Barkley's model, the predicted deficits in working memory have been demonstrated for children with ADHD on verbal indices such as the Arithmetic, Coding, and Digit Span subtests of the Wechsler Intelligence Scale for Children (Barkley, 1997b) as well as on nonverbal measures of working memory such as the Tower of London (Klorman et al., 1999).

Previous researchers have also highlighted a critical deficit in inhibitory control in ADHD. Quay, drawing on seminal work by Jeffrey Gray (1982), proposed that the Behavioral Inhibitory System (BIS) is underactive in ADHD (Quay, 1988). The BIS is a septohippocampal frontal system that has been shown in animal studies to mediate passive and active avoidance responding to conditioned signals of punishment as well as extinction of previously reinforced responses (Gray, 1982). Although the hypothesis of an underactive BIS is consistent with behavioral observations of increased impulsivity and risk-taking in ADHD, as well as reduced inhibitory control on laboratory measures, it has been directly tested in only two studies, with inconsistent results. Pliszka and colleagues (1993) found that children with ADHD did not show smaller conditioned autonomic responses (skin conductance or heart rate) in response to an aversive stimulus than did normal controls. Iaboni and colleagues (1997), by contrast, reported that children with ADHD did show smaller skin conductance levels than controls in the extinction condition of performance on a repetitive motor task, as predicted by the theory. Douglas (1999) has pointed out that classical conditioning of autonomic responses, as in the study by Pliszka and colleagues (1993) are not encompassed by the BIS model. The BIS model is also challenged by the observation of comorbidity between ADHD and anxiety, which theoretically (Gray, 1982) result from an underactive BIS and an overactive BIS, respectively.

Reward-Seeking/Stimulation-Seeking

An alternative unifying theory is that children with ADHD have a dysfunctional reward system, such that more immediate, more frequent, or more intense reinforcers are necessary to maintain appropriate effortful and rule-governed behavior (Barkley, 1989b; Haenlein & Caul, 1987). This formulation is consistent not only with parental reports that children with ADHD appear to be immune to promise of reward and threat of punishment and with clinical research showing that systematically applied behavior therapy is effective in ameliorating core ADHD symptoms (Pelham & Murphy, 1986; Swanson, 1992), but also with the fact, well established in animal studies (Gardner, 1997), that stimulant drugs activate brain reward centers (see also Taylor & Jentsch, Chapter 4). Evidence for this hypothesis in ADHD was generated in early studies by Parry and Douglas (1983), who found that performance of children with ADHD on concept-learning and reaction time tasks was inferior to that of normal children under conditions of intermittent reinforcement, but was equivalent when reinforcement was continuous. Other studies by these investigators and by others, however, failed to confirm a differential response to reward in children with ADHD (see studies reviewed in Solanto, 1990). More recent investigations, although not finding a differential effect of reward manipulation on task performance of ADHD vs. normal children (Iaboni et al., 1995), have reported that the ADHD group displayed greater frustration during task performance on a partial reinforcement schedule (Douglas & Parry, 1994; Wigal et al., 1998). The related hypothesis—that stimulants act to reduce the reward threshold in ADHD—was supported by the report that methylphenidate increased the number of responses produced by children with ADHD to obtain monetary reinforcement on a progressive ratio schedule (Wilkison et al., 1995); this result suggested that the drug increased the reward value of the reinforcement, resulting in the subjects' expending greater effort to obtain it.

It has alternatively been proposed that children with ADHD are not explicitly reward-seeking but rather more broadly stimulation-seeking. As articulated by Zentall and Zentall (1983), the overactivity, impulsivity, and, secondarily, short attention span, as well as tendencies toward provocative or aggressive behavior, represent efforts to generate an optimal level of stimulation in individuals with abnormally low endogenous arousal. Despite the intuitive appeal of this model, it has generated only limited empirical investigation and support (Zentall, 1984, 1985; Zentall & Meyer, 1987).

Cognitive Dysregulation

Sergeant and colleagues (1999) and Douglas (1999) have argued that disinhibition is inadequate to account for the deficits in AD/HD and have postulated additional deficits of cognitive self-regulation with respect to availability and/or allocation of cognitive resources. Integrating work by Sanders (1983) concerning state factors, and by Sternberg (1969) delineating stages of information processing, Sergeant and colleagues (1999) have elaborated a *cognitive-energetic* model comprising *arousal* and *activation* pools, both of which are modulated and maintained by an *effort* pool. The arousal pool provides cognitive resources for phasic responding to stimuli, encompassing the processes of stimulus evaluation and encoding. The *activation* pool, which oversees the motor decision and response stages of information-processing and determines tonic preparedness to respond, is influenced by alertness and time on task. The *effort* pool regulates ability to adapt to task demands and is influenced by cognitive load and motivational variables. An overriding *evaluation* mechanism provides planning, monitoring, guidance, and self-correction.

Data relevant to tests of the cognitive–energetic model in ADHD have not yielded consistent support for a deficiency in the arousal, activation, or effort pools. As described, rigorous studies have failed to identify deficits in visual-spatial orienting to stimuli (Sergeant et al., 1999), or in the stimulus evaluation stages of information-processing (van der Meere, 1996). However, in more recent studies of EEG event-related potential (ERP) (reviewed in Solanto, 1998), children with ADHD had overly long P3b latencies, an index of duration of stimulus evaluation processes. Some studies (Carte et al., 1996; Hall et al., 1997; van der Meere et al., 1989), but not others (de Sonneville et al., 1994; Oosterlaan & Sergeant, 1995) have observed deficits in the response organization stage in children with ADHD. A meta-analysis of studies with the CPT in children with ADHD (Losier et al., 1996), showed a reduction, relative to normals, in d′, an index of the ability to differentiate target and nontarget stimuli that is considered to reflect arousal state, whereas no difference was found in Beta, the index of bias to respond, reflecting activation. A slow event rate, which may accentuate the effects of either underarousal or underactivation, exerted a differentially negative effect on the performance of children with ADHD in several studies (van der Meere et al., 1995; Zahn et al., 1991). Vigilance, an index of effortful maintenance of performance over time, has most often *not* differentiated the performance of children with ADHD and normal control subjects on the CPT (Sergeant et al., 1999). Inconsistencies such as these have led Douglas (1999) to argue that a critical deficit in ADHD may be better conceptualized as a reduced capacity for *adaptation* to changing situational and task demands. This is illustrated in recent studies in Douglas's laboratory showing deficits in response organization only when changes in response organization demands (compatibility-incompatibility) were presented unpredictably (Douglas, 1999). Previous work is congruous in showing that children with ADHD have greater difficulty speeding *or* slowing their performance in response to task instructions or subsequent to errors than do non-ADHD children (Sergeant et al., 1999).

It is of interest in this context that the effects of methylphenidate administration have *not* interacted with any stage of information-processing, excepting a single study showing a specific effect on response organization (de Sonneville et al., 1994). Methylphenidate improved d′, but not beta, in a meta-analysis of CPT studies (Losier et al., 1996), as well as in a subsequent CPT study (Solanto et al., 1997). However, an earlier study showed that D-amphetamine produced improvement on both indices (Sostek et al., 1980). Vigilance *has* consistently been improved by MPH (see review by Solanto, 1998). Thus, stimulants do not appear to target a cognitive processing deficit specific to ADHD. A more extended discussion of effects of stimulants on cognitive processes in ADHD is to be found in the chapter by Denney and Rapport in Chapter 12.

Hypo- versus Hypercatecholaminergic Status

Early investigators reasoned that, given the facilitative effect of stimulant drugs on catecholamines in pharmacological studies in animals, the key mechanism of action of these agents in ADHD was to compensate for a deficit in catecholaminergic neurotransmission (Wender, 1971). Data relevant to a model of deficient catecholaminergic transmission in ADHD have subsequently come from several sources. Several animal models of ADHD have been generated by neurotoxic *depletion* of dopamine (see Chapter 8, by Ferguson) or are characterized by genetically programmed dopamine hypofunction (see Sagvolden, Chapter 9). Dopamine *antagonists* have more generally impaired rather than improved cognitive functions (see Chapter 13, by Mehta, Sahakian, and Robbins). Results of recent PET studies of individuals with ADHD have been inconsistent, with one study showing decreased uptake of dopamine precursors in prefrontal cortex in adults (consistent with the hypothesis) and another showing *increased* levels in right midbrain in adolescents (Ernst et al., 1999). In addition, among children with ADHD, higher levels of a dopamine metabolite, homovanillic acid (HVA), were associated with *more severe* hyperactivity and a more pronounced positive response to stimulant than were lower levels, also suggesting that ADHD, at least in children, may be characterized by hyper-dopaminergic status (Castellanos et al., 1994, 1996).

In 1984, Solanto suggested that methylphenidate and D-amphetamine may act to stimulate inhibitory presynaptic autoreceptors, thereby *reducing* elevated levels of dopamine release in ADHD (Solanto, 1984, 1998). This hypothesis was indirectly supported by studies showing reduction of locomotor activity following low, presynaptically effective doses of dopamine agonists in animals, and reduction in activity level in children with ADHD after administration of a subclinical dose (0.1 mg/kg), in the range presumed to stimulate autoreceptors (Solanto, 1986). Also supportive were studies indicating that the beneficial effects of stimulants on higher-order cognitive processes in children were *not* mediated by the increases in stereotypy seen at higher, postsynaptically active, doses in animals (Solanto & Wender, 1989).

McCracken (1991) and Pliszka and colleagues (1996) proposed a similar model for noradrenergic neurotransmission in ADHD, specifically that noradrenergic neurons originating in the locus coeruleus and terminating in the cortex are overly active in children with ADHD, resulting in a reduced signal-to-noise ratio and consequent attentional deficiencies. Castellanos (1997) refined the model of dopaminergic pathophysiology in ADHD, limiting dopaminergic overactivity in ADHD to the nigrostriatal pathway. He proposed that stimulant drugs act on presynaptic dopaminergic autoreceptors in this pathway but stimulate *post*synaptic receptors of mesocorticolimbic dopamine projections, which are lacking in synthesis-modulating and impulse-modulating autoreceptors. In the model proposed recently by Seeman and Madras (1998), activation of autoreceptors at low stimulant doses is key to the divergent effects of low and high stimulant doses on activity level. In this formulation, the response of locomotor-controlling dopamine receptors is proportional to the ratio of phasic (acute) to tonic dopamine levels. Because of their greater relative effects on impulse-regulating autoreceptors, low doses produce a *net* decrease in the ratio of phasic to tonic dopamine levels and thus a decrease in locomotor activity. The controversial issues of pre- versus postsynaptic effects of stimulants and hyper- versus hypocatecholaminergic status of individuals with ADHD will be addressed in various chapters in this book and will be revisited in the final integrative summary.

Treatment

Pharmacological

The clinical usage and effects of psychostimulant drugs are comprehensively discussed in the chapter by Greenhill (see Chapter 2). We will simply note here that these drugs, which include D-amphetamine (Dexedrine), methylphenidate (Ritalin), and D-,L-amphetamine (Adderall), have been shown to significantly ameliorate the core ADHD symptoms of inattentiveness, hyperactivity, and impulsivity in 75% to 95% of treated individuals.

Psychosocial

In the past 15 years, behavior therapy programs for children with ADHD have been increasingly well developed; many have been formalized in treatment manuals. These programs aim to teach parents and teachers behavior management skills based on sound behavioral learning principles, particularly the systematic use of reinforcement contingencies. Among the most comprehensively described are parent and teacher training programs by Barkley (1990), a summer camp treatment program by Pelham and Waschbusch (1999), and experimental/treatment classrooms by Swanson (1992). An earlier review (Pelham & Murphy, 1986) of behavioral interventions in ADHD found that although these programs were more effective than placebo in enhancing attention and academic productivity and reducing disruptive behavior, they were less effective than stimulant drug treatment. The results of a recently concluded large-scale study further discussed by Greenhill (see Chapter 2) confirmed the superiority of a well-managed medication regimen (MTA Cooperative Group, 1999) for the treatment of ADHD.

Cognitive behavioral interventions in which the child with ADHD is explicitly taught *self*-monitoring and *self*-control techniques have not generalized beyond the immediate treatment context (Abikoff, 1987).

Multimodal Treatments

The efficacy of multimodal treatments that *combine* pharmacological and psychosocial treatments was studied in the large-scale Multimodal Treatment Study of ADHD (MTA Cooperative Group, 1999). The results showed that, for the sample as a whole, the combination of intensive psychosocial intervention and stimulant medication was not more effective in alleviating the primary symptoms of ADHD than was medication alone. Enhanced benefits with combination treatment were seen for some subgroups of children, however, as further discussed by Greenhill (see Chapter 2).

Outcome

There have been several well-controlled longitudinal follow-up studies that have tracked children diagnosed with ADHD into adulthood in order to ascertain persistence of the disorder, comorbid diagnoses at outcome, as well as ultimate academic and occupational achievement and social adaptation. Weiss and Hechtman (1993) followed and reassessed a sample of 103 children (DSM-II) at 5-, 10-, and 15-year intervals. In early adulthood (mean age 25 years), 66% of patients (vs. 4% of control subjects) experienced at least one disabling core symptom of ADHD and 23% (vs. 2.4%) met criteria for antisocial personality disorder (ASPD). Only 62% of the patients in the Weiss sample were interviewed at follow-up, however, resulting in exclusion of patients who were doing particularly well or particularly badly and biasing the results.

Mannuzza, Klein, and colleagues (Mannuzza et al., 1993) also followed into early adulthood a sample of 101 children given a clinical diagnosis of "hyperactivity" (DSM-II) (excluding cases with significant aggression or other conduct problems). However, they reported considerably lower rates of 8% for the full syndrome of ADD, and an additional 3% for those experiencing at least one disabling core symptom. Eighteen percent were diagnosed with antisocial personality disorder (ASPD), which was 7-fold more prevalent among those with persisting ADD. Although these researchers achieved an impressive follow-up rate of 88% of subjects initially enrolled, they based their diagnostic evaluation at follow-up exclusively on structured diagnostic interviews with the patients themselves, possibly resulting in an underestimate of psychiatric symptoms.

Of particular concern was a high rate of substance abuse disorder in the patients compared with control subjects (16% vs. 4%), mostly reflecting abuse of marijuana and cocaine (Mannuzza et al., 1993). Cases of substance abuse were much higher among those with ASPD (40%) than among those without (13%). Substance use outcome was assessed only with respect to alcohol use in the Weiss study and was found not to differ significantly between groups. A recent study of a large population-based sample of adolescents revealed that a diagnosis of ADHD did not independently increase the risk for substance use or abuse (Disney et al., 1999) over that conferred by conduct disorder, which is commonly a precursor of ASPD.

The studies by Mannuzza, Klein, and colleagues and by Weiss revealed worse educational and occupational status at outcome in the patient group, compared with the control group. Weiss reported a larger high school drop-out rate (19% vs. 4%) and lower frequency of attainment of bachelor's degrees (3% vs. 17%) (Weiss & Hechtman, 1993). Hyperactive subjects followed by Klein had a lower mean highest grade completed (12.1 vs. 14.6), as well as a lower rate of 4-year college completion (8% vs. 32%) (Mannuzza et al., 1993). More frequent job changes and inferior job status (Weiss & Hechtman, 1993), and lower occupational ranking (Mannuzza et al., 1993) characterized the hyperactive groups. Interestingly, Mannuzza, Gittelman-Klein, and colleagues also reported a *higher* rate of ownership of small businesses in the hyperactive group (16% vs. 5%) suggesting that hyperactives may function better in self-defined, self-paced roles than they do in more structured business environments.

General adaptive and social functioning were measurably worse in the hyperactive samples. This group had lower self-esteem and poorer social skills (Weiss & Hechtman, 1993), as well as worse automobile driving records in adolescence (Barkley et al., 1993) than the control group. Furthermore, subjects in the patient group had significantly higher rates of police arrest (39% vs. 20%) and incarceration (9% vs. 1%) (Mannuzza et al., 1993).

Barkley and colleagues (1990) and Biederman and colleagues (1996) have reported 8-year and 4-year follow-up data (respectively) of children diagnosed in childhood, and are continuing to follow these samples into adulthood.

Summary

ADHD is a chronic disorder characterized by developmentally inappropriate and impairing levels of inattention and/or hyperactivity–impulsivity. Estimated prevalence is at least 3% to 5% of school-aged children. The disorder is frequently comorbid with other disruptive behavior disorders, anxiety disorders, mood disorders, and specific learning disabilities. Outcome research indicates that individuals with ADHD are at greater risk for impaired social, academic, and occupational functioning in adulthood. Etiology is predominantly genetic, very likely mediated by catecholaminergic dysfunction in prefrontal cortex and basal ganglia brain regions, giving rise (at least in the combined subtype) to critical deficits in cognitive and motor self-regulation. Although behavior therapy may effectuate some improvement, stimulant medication remains the single most effective treatment.

Because the CB subtype of ADHD is by far the most widely recognized and best researched, and because differences between the CB and IN subtypes in age of onset, sex ratio, and comorbidity, as well as symptom features, suggest that the IN subtype has different neurobiological substrates, the CB subtype will be the primary focus of this book. Accordingly, ADHD will refer to the CB subtype unless otherwise specified.

References

The author gratefully acknowledges F. Xavier Castellanos for contributing the section on Molecular Genetics.

Abikoff, H (1987). An evaluation of cognitive behavior therapy for hyperactive children. In BB Lahey & AE Kazdin (Eds.), *Advances in clinical child psychology* (Vol. 10, pp. 171–216). New York: Plenum.

Achenbach, TM (1991). *Manual for the Child Behavior Checklist/4–18 & 1991 Profile.* Burlington, VT: University of Vermont, Department of Psychiatry.

American Psychiatric Association (1968). *Diagnostic and statistical manual of mental disorders* (2nd ed.). Washington, D.C.: Author.

American Psychiatric Association (1980). *Diagnostic and statistical manual of mental disorders* (3rd ed.). Washington, DC: Author.

American Psychiatric Association (1987). *Diagnostic and statistical manual of mental disorders* (3rd ed., rev ed.). Washington, DC: Author.

American Psychiatric Association (1994). *Diagnostic and statistical manual of mental disorders* (4th ed.). Washington, DC: Author.

Applegate, B, Lahey, BB, Hart, EL, Biederman, J, Hynd, GW, & Barkley, RA, et al. (1997). Validity of the age-of-onset criterion for ADHD: A report from the DSM-IV field trials. *J Am Acad Child Adolesc Psychiatry, 36,* 1211–1221.

Asghari, V, Sanyal, S, Buchwaldt, S, Paterson, A, Jovanovic, V, & Van Tol, HH (1995). Modulation of intracellular cyclic AMP levels by different human dopamine D4 receptor variants. *J Neurochem, 65,* 1157–1165.

Barkley, RA (1989a). Hyperactive girls and boys: stimulant drug effects on mother-child interactions. *J Child Psychol Psychiatry, 30,* 379–390.

Barkley, RA (1989b). The problem of stimulus control and rule-governed behavior in attention deficit disorder with hyperactivity. In LM Bloomingdale & J. Swanson (Eds.), *Attention deficit disorder: Current concepts and emerging trends in attentional and behavioral disorders of childhood* (pp. 203–232). Elmsford, NY: Pergamon Press.

Barkley, RA (1990). *Attention deficit hyperactivity disorder: A handbook for diagnosis and treatment.* New York: Guilford Press.

Barkley, RA (1991). The ecological validity of laboratory and analogue assessments of ADHD symptoms. *J Abnorm Child Psychol, 19,* 149–178.

Barkley, RA (1997a). *AD/HD and the nature of self-control.* New York: Guilford Press.

Barkley, RA (1997b). Behavioral inhibition, sustained attention, and executive functions: Constructing a unifying theory of ADHD. *Psychol Bull, 121,* 65–94.

Barkley, RA, Fischer, M, Edelbrock, CS, & Smallish, L (1990). The adolescent outcome of hyperactive children diagnosed by research criteria: I. An 8-year prospective follow-up study. *J Am Acad Child Adolesc Psychiatry, 29,* 546–557.

Barkley, RA, Guevremont, DC, & Anastopoulos, AD, (1993). Driving related risks and outcome of attention deficit hyperactivity disorder in adolescents and young adults. *Pediatrics, 92,* 212–218.

Baumgaertel, A, Wolraich, ML, & Dietrich, M (1995). Comparison of diagnostic criteria for attention deficit disorders in a German elementary school sample. *J Am Acad Child Adolesc Psychiatry, 34,* 629–638.

Benjamin, J, Li, L, Patterson, C, Greenberg, BD, Murphy, DL, & Hamer, DH (1996). Population and familial association between the D4 receptor gene and measures of Novelty Seeking. *Nat Genet, 12,* 81–84.

Biederman, J, Faraone, S, Milberger, S, Guite, J, Mick, E, & Chen, et al. (1996). A prospective 4-year follow-up study of attention-deficit hyperactivity and related disorders. *Arch Gen Psychiatry, 53,* 437–446.

Biederman, J, Faraone, SV, Keenan, K, Benjamin, J, Krifcher, B, Moore, C, Sprich, S, Ugaglia, K, Jellirek, MS, Steingaard, R, Spencer, T, Norman, B, Kolodny, R, Kraus, R, Perrin, J, Keller, MB, and Tsuang, MJ (1992). Further evidence for family-genetic risk factors in attention deficit hyperactivity disorder: Patterns of comorbidity in probands and relatives of psychiatrically and pediatrically referred samples. *Arch Gen Psychiatry, 49,* 728–738.

Biederman, J, Faraone, SV, Mick, E, Williamson, S, Wilens, TE, & Spencer, TJ, et al. (1999). Clinical correlates of ADHD in females: Findings from a large group of girls ascertained from pediatric and psychiatric referral sources. *J Am Acad Child Adolesc Psychiatry, 38,* 966–975.

Biederman, J, Mick, E, & Faraone, SV (1998). Depression in attention deficit hyperactivity disorder (ADHD) children: "True" depression or demoralization? *J Affect Disord, 47,* 113–122.

Biederman, J, Milberger, S, Faraone, SV, Kiely, K, Guite, J, & Mick, E, et al. (1995a). Family-environment risk factors for attention-deficit hyperactivity disorder: A test of Rutter's indicators of adversity. *Arch Gen Psychiatry, 52,* 464–470.

Biederman, J, Milberger, S, Faraone, SV, Kiely, K, Guite, J, & Mick, et al. (1995b). Impact of adversity on functioning and comorbidity in children with attention-deficit hyperactivity disorder. *J Am Acad Child Adolesc Psychiatry, 34,* 1495–1503.

Biederman, J. Newcorn, J, & Sprich, S (1991). Comorbidity of attention deficit hyperactivity disorder with conduct, depressive, anxiety, and other disorders. *Am J Psychiatry, 148,* 565–577.

Cadoret, RJ, & Stewart, MA (1991). An adoption study of attention deficit/hyperactivity/aggression and their relationship to adult antisocial personality. *Compr Psychiatry, 32,* 73–82.

Caron, C & Rutter, M (1991). Comorbidity in child psychopathology: Concepts, issues and research strategies. *J Child Psychol Psychiatry,* 1063–1080.

Carte, ET, Nigg, JT, & Hinshaw, SP (1996). Neuropsychological functioning, motor speed, and language processing in boys with and without ADHD. *J Abnorm Child Psychol, 24,* 481–498.

Casey, BJ, Castellanos, FX, Giedd, JN, Marsh, WL, Hamburger, SD, & Schubert, AB, et al. (1997). Implication of right frontostriatal circuitry in response inhibition and attention-deficit/hyperactivity disorder. *J Am Acad Child Adolesc Psychiatry, 36,* 374–383.

Castellanos, F (1997). Toward a pathophysiology of attention-deficit/hyperactivity disorder. *Clin Pediatr, 36,* 381–393.

Castellanos, FX, Elia, J, Kruesi, MJ, Gulotta, CS, Mefford, IN, & Potter, WZ, et al. (1994). Cerebrospinal fluid monoamine metabolites in boys with attention-deficit hyperactivity disorder. *Psychiatry Research, 52,* 305–316.

Castellanos, FX, Elia, J, Kruesi, MJP, Marsh, WL, Gulotta, CS, & Potter, WZ, et al. (1996). Cerebrospinal homovanillic acid predicts behavioral response to stimulants in 45 boys with attention-deficit/hyperactivity disorder. *Neuropsychopharmacology, 14,* 125–137.

Castellanos, FX, Lau, E, Tayebi, N, Lee, P, Long, RE, & Giedd, JN, et al. (1998). Lack of association between a dopamine-4 receptor polymorphism and attention-deficit/hyperactivity disorder: Genetic and brain morphometric analyses. *Mol Psychiatry, 3,* 431–434.

Conners, CK (1996). *Clinical use of the Continuous Performance Test: Diagnostic utility and drug sensitivity.* Paper presented at the American Academy of Child and Adolescent Psychiatry, Philadelphia, PA.

Cook, EH, Stein, MA, Krasowski, MD, Cox, W, Olkon, DM, & Kieffer, JE, et al. (1995). Association of attention-deficit disorder and the dopamine transporter gene. *Am J Hum Genet, 56,* 993–998.

Corkum, P, Tannock, R, & Moldofsky, H (1998). Sleep disturbances in children with attention-deficit/hyperactivity disorder. *J Am Acad Child Adolesc Psychiatry, 37,* 637–646.

Corkum, PV, & Seigel, LS (1993). Is the continuous performance task a valuable research tool for use with children with attention-deficit-hyperactivity disorder. *J Am Acad Child Adolesc Psychiatry, 34,* 1217–1239.

Crowe, RR (1993). Candidate genes in psychiatry: an epidemiological perspective. *Am J Med Genet, 48,* 74–77.

Daly, G, Hawi, Z, Fitzgerald, M, & Gill, M (1999). Mapping susceptibility loci in attention deficit hyperactivity disorder: Preferential transmission of parental alleles at DAT1, DBH, and DRD5 to affected children. *Mol Psychiatry, 4,* 192–196.

de Sonneville, LMJ, Njiokiktjien, C, & Bos, H (1994). Methylphenidate and information processing. Part I: Differentiation between responders and nonresponders; Part 2: Efficacy in responders. *J Clin Exp Neuropsychol, 16,* 877–897.

Disney, ER, Elkins, IJ, McGue, M, & Iacono, W (1999). Effects of ADHD, conduct disorder, and gender on substance use and abuse in adolescence. *Am J Psychiatry, 156,* 1515–1521.

Douglas, VI (1972). Stop, look and listen: The problem of sustained attention and impulse control in hyperactive and normal children. *Canadian Journal of Behavioral Science, 41,* 259–283.

Douglas, VI (1999). Cognitive control processes in Attention-Deficit/Hyperactivity Disorder. In H C Quay & A. E. Hogan (Eds.), *Handbook of disruptive behavior disorders* (pp. 105–138). New York: Kluwer Academic/Plenum.

Douglas, VI & Parry, PA (1994). Effects of reward and nonreward on frustration and attention in attention deficit disorder. *J Abnorm Child Psychol, 22,* 281–302.

Ebstein, RP, Novick, O, Umansky, R, Priel, B, Osher, Y, & Blaine, et al. (1996). Dopamine D4 receptor (D4DR) exon III polymorphism associated with the human personality trait of Novelty Seeking. *Nat Genet, 12,* 78–80.

Eiraldi, RB, Power, TJ, & Nezu, CM (1997). Patterns of comorbidity associated with subtypes of attention-deficit/hyperactivity disorder among 6- to 12- year-old children. *J Am Acad Child Adolesc Psychiatry, 36,* 503–514.

Ernst, M, Zametkin, AJ, Matochik, JA, Pascualvaca, D, Jons, PH, & Cohen, RM (1999). High midbrain 18F-DOPA accumulation in children with ADHD. *Am J Psychiatry, 156,* 1209–1215.

Faraone, SV & Biederman, J (1994). Is attention deficit hyperactivity disorder familial? *Harv Rev Psychiatry, 1,* 271–287.

Faraone, SV & Biederman, J (1997). Do attention deficit hyperactivity disorder and major depression share familial risk factors? *J Nerv Ment Dis, 185,* 533–541.

Faraone, SV, Biederman, J, Weiffenbach, B, Keith, T, Chu, MP, & Weaver, A, et al. (1999). Dopamine D4 gene 7-repeat allele and attention deficit hyperactivity disorder. *Am J Psychiatry, 156,* 768–770.

Fernandez, T, Castellanos, FX, Simonis, T, Jacobsen, LK, & Rapoport, JL (in press). Human leukocyte antigen alleles in attention-deficit hyperactivity disorder. *Mol Psychiatry.*

Gardner, EL (1997). Brain reward mechanisms. In JH Lowinson, P Ruiz, RB Millman, & JG Langrod (Eds.), *Substance abuse: A comprehensive textbook* (3rd ed., pp. 51–85). Baltimore: Williams & Wilkins.

Gaub, M, & Carlson, C (1997a). Gender differences in ADHD: A meta-analysis and critical review. *J Am Acad Child Adolesc Psychiatry, 36,* 1035–1045.

Gaub, M & Carlson, CL (1997b). Behavioral characteristics of DSM-IV ADHD subtypes in a school-based population. *J Abnorm Child Psychol, 25,* 103–111.

Genetics Workgroup of the National Institute of Mental Health (1999). Genetics and mental disorders. *Biol Psychiatry, 45,* 559–573.

Gill, M, Daly, G, Heron, S, Haaw, Z, & Fitzgerald, M (1997). Confirmation of association between attention deficit hyperactivity disorder and a dopamine transporter polymorphism. *Mol Psychiatry, 2,* 311–313.

Gjone, H, Stevenson, J, & Sundet, JM (1996). Genetic influence on parent-reported attention-related problems in a Norwegian general population twin sample. *J Am Acad Child Adolesc Psychiatry, 35,* 588–596.

Goldman, LS, Genel, M, Bezman, RJ, & Slanetz, PJ (1998). Diagnosis and treatment of Attention-Deficit/Hyperactivity Disorder in children and adolescents. *JAMA, 279,* 1100–1107.

Goodman, R & Stevenson, J (1989). A twin study of hyperactivity: II. The aetological role of genes, family relationships and perinatal adversity. *J Child Psychol Psychiatry, 30,* 691–709.

Goyette, CH, Conners, CK & Ulrich, RF (1978). Normative data on revised Conners Parent and Teacher Rating Scales. *J Abnorm Child Psychol, 6,* 221–236.

Gray, JA (1982). *The neuropsychology of anxiety.* New York: Oxford University Press.

Haenlein, M, & Caul, W (1987). Attention deficit disorder with hyperactivity: A specific hypothesis of reward dysfunction. *J Am Acad Child Adolesc Psychiatry, 26,* 356–362.

Hall, SJ, Halperin, JM, Schwartz, ST, & Newcorn, JH (1997). Behavioral and executive functions in children with Attention-Deficit Hyperactivity Disorder. *J Atten Disord, 1,* 235–244.

Halperin, JM, Newcorn, JH, Kopstein, I, McKay, KE, Schwartz, ST, Siever, LJ, & Sharma, V (1997). Serotonin, aggression and parental psychopathology in children with attention-deficit hyperactivity disorder. *J Am Acad Child Adolesc Psychiatry, 36,* 1391–1398.

Henker, B, & Whalen, CK (1999). The child with Attention-Deficit/Hyperactivity Disorder in school and peer settings. In HC Quay & AE Hogan (Eds.), *Handbook of disruptive behavior disorders* (pp. 157–178). New York: Kluwer Academic/Plenum.

Iaboni, F, Douglas, VI, & Baker, AG (1995). Effects of reward and response costs on inhibition in ADHD children. *J Abnorm Psychol, 104,* 232–240.

Iaboni, F, Douglas, VI, & Ditto, B (1997). Psychophysiological response of ADHD children to reward and extinction. *Psychophysiology, 34,* 116–123.

Jensen, PS, Martin, D, & Cantwell, DP (1997). Comorbidity in ADHD: Implications for research, practice, and DSM-V. *J Am Acad Child Adolesc Psychiatry, 36,* 1065–1079.

Kindlon, D, Mezzacappa, E, & Earls, F (1995). Psychometric properties of impulsivity measures: Temporal stability, validity and factor structure. *J Child Psychol Psychiatry, 36,* 645–661.

Klorman, R, Hazel-Fernandez, LA, Shaywitz, SE, Fletcher, JM, Marchione, KE, & Holahan, JM, et al. (1999). Executive functioning deficits in attention-deficit/hyperactivity disorder are independent of oppositional defiant or reading disorder. *J Am Acad Child Adolesc Psychiatry, 38,* 1148–1155.

Lahey, BB, Applegate, B, McBurnett, K, Biederman, J, Greenhill, L, Hynd, GW, et al. (1994). DSM-IV field trials for attention deficit hyperactivity disorder in children and adolescents. *Am J Psychiatry, 151,* 1673–1685.

Lahey, BB, Pelham, WE, Schaughency, EA, Atkins, MS, Murphy, HA, Hynd, G, & Russo, M, et al. (1988). Dimensions and types of attention deficit disorder. *J Am Acad Child Adolesc Psychiatry, 27,* 330–335.

LaHoste, GJ, Swanson, JM, Wigal, SB, Glabe, C, Wigal, T, & King, N, et al. (1996). Dopamine D4 receptor gene polymorphism is associated with attention deficit hyperactivity disorder. *Mol Psychiatry, 1,* 121–124.

Lambert, NM, Sandoval, J, & Sassone, D (1978). Prevalence of hyperactivity in elementary school children as a function of social system definers. *Am J Psychiatry, 48,* 446–463.

LeFever, GB, Dawson, KV, & Morrow, AL (1999). The extent of drug therapy for attention deficit-hyperactivity disorder among children in public schools. *Am J Public Health, 89,* 1359–1364.

Levy, F, Hay, DA, McStephen, M, Wood, C, & Waldman, I (1997). Attention-deficit hyperactivity disorder: a category or a continuum? *J Am Acad Child Adolesc Psychiatry, 36,* 737–744.

Losier, BJ, McGrath, PJ, & Klein, RM (1996). Error patterns on the continuous performance test in non-medicated and medicated samples of children with and without ADHD: A meta-analytic review. *J Child Psychol Psychiatry, 37,* 971–987.

Mannuzza, S, Klein, RG, Bessler, A, Molloy, P, & LaPadula, M (1993). Adult outcome of hyperactive boys: Educational achievement, occupational rank, and psychiatric status. *Arch Gen Psychiatry, 50,* 565–576.

McBurnett, K, Pfiffner, LJ, Willcutt, E, Tamm, L, Lerner, & M, Ottolini, et al. (1999). Experimental cross-validation of DSM-IV types of attention-deficit/hyperactivity disorder. *J Am Acad Child Adolesc Psychiatry, 38,* 17–24.

McCracken, JT (1991). A two part model of stimulant action on attention-deficit hyperactivity disorder in children. *J Neuropsychiatry Clin Neurosci, 3,* 201–209.

Milberger, S, Biederman, J, Faraone, SV, Murphy, J, & Tsuang, MT (1995). Attention deficit hyperactivity disorder and comorbid disorders: Issues of overlapping symptoms. *Am J Psychiatry, 152,* 1793–1799.

Morrison, JR, & Stewart, M (1973). The psychiatric status of the legal families of adopted hyperactive children. *Arch Gen Psychiatry, 28,* 888–891.

MTA Cooperative Group (1999). A 14-month randomized clinical trial of treatment strategies for Attention-Deficit/Hyperactivity Disorder. *Arch Gen Psychiatry, 56,* 1073–1086.

Neuman, RJ, Todd, RD, Heath, AC, Reich, W, Hudziak, JJ, & Bucholz, KK, et al. (1999). Evaluation of ADHD typology in three contrasting samples: A latent class approach. *J Am Acad Child Adolesc Psychiatry, 38,* 25–33.

Newcorn, JW, & Halperin, JM (1994). Comorbidity among disruptive behavior disorders. In L Greenhill (Ed.), *Child and adolescent psychiatric clinics of North America: Disruptive disorders* (Vol. 3,). Philadelphia: W.B. Saunders Co.

Odell, JD, Warren, RP, Warren, WL, Burger, RA, & Maciulis, A (1997). Association of genes within the major histocompatibility complex with attention deficit hyperactivity disorder. *Neuropsychobiology, 35,* 181–186.

Oosterlaan, J, Logan, GD, & Sergeant, JA (1998). Response inhibition in AD/HD, CD, comorbid AD/HD + CD, anxious, and control children: A meta-analysis of studies with the stop task. *J Child Psychol Psychiatry, 39,* 411–425.

Oosterlaan, J, & Sergeant, JA (1995). Response choice and inhibition in ADHD, anxious and aggressive children: The relationship between S-R compatibility and stop signal task. In JA Sergeant (Ed.), *European approaches to hyperkinetic disorder: Eunethydis* (pp. 225–240). Zurich: Fotorotar.

Parry, P, & Douglas, V (1983). The effects of reward on the performance of hyperactive children. *J Abnorm Child Psychol, 11,* 327–340.

Paterson, AD, Sunohara, GA, & Kennedy, JL (1999). Dopamine D4 receptor gene: Novelty or nonsense? *Neuropsychopharmacology, 21,* 3–16.

Pelham, WE, & Murphy, DA (1986). Behavioral and pharmacological treatment of hyperactivity and attention-deficit disorders. In M Herson & SE Bruening (Eds.),

Pharmacological and behavioral treatment: An integrative approach (pp. 108–147). New York: Wiley.

Pelham, WE, Jr., & Waschbusch, DA (1999). Behavioral intervention in Attention-Deficit/Hyperactivity Disorder. In HC Quay & AE Hogan (Eds.), *Handbook of disruptive behavior disorders* (pp. 255–278). New York: Kluwer Academic/Plenum.

Pliszka, SR, Hatch, P, Borcherding, BG, & Rogeness, GA (1993). Classical conditioning in children with attention deficit hyperactivity disorder (ADHD) and anxiety disorders: a test of Quay's model. *J Abnorm Child Psychol, 21,* 411–423.

Pliszka, SR, McCracken, JT, & Maas, JW (1996). Catecholamines in attention-deficit hyperactivity disorder: Current perspectives. *J Am Acad Child Adolesc Psychiatry, 35,* 264–272.

Porrino, LJ, Rapoport, JL, Behar, D, Sceery, W, Ismond, DR, & Bunney, WE (1983). A naturalistic assessment of the motor activity of hyperactive boys I. Comparison with normal controls. *Arch Gen Psychiatry, 40,* 681–687.

Quay, HC (1988). The behavioural reward and inhibition systems in childhood behavior disorder. In LM Bloomingdale (Ed.), *Attention deficit disorder: New research in attention, treatment and psychopharmacology* (Vol. 3, pp. 176–186). Oxford: Pergamon Press.

Rappley, MD, Gardiner, JC, Jetton, JR, & Houang, RT (1995). The use of methylphenidate in Michigan. *Arch Pediatr Adolesc Med, 149,* 675–679.

Rowe, DC, Stever, C, Giedenghagen, LN, Gard, JM, Cleveland, HH, & Terris, ST, et al. (1998). Dopamine DRD4 receptor polymorphism and attention deficit hyperactivity disorder. *Mol Psychiatry, 3,* 419–426.

Safer, DJ, & Krager, JM (1988). A survey of medication treatment for hyperactive/inattentive students. *N Engl J Med, 260,* 2256–2258.

Safer, DJ, Zito, JM, & Fine, EM (1996). Increased methylphenidate usage for attention deficit disorder in the 1990's. *Pediatrics, 98,* 1084–1088.

Sanders, AV (1983). Towards a model of stress and human performance. *Acta Psychol (Amst), 53,* 61–97.

Scahill, L, Schwab-Stone, M, Merikangas, KR, Leckman, JF, Zhang, H, & Kasl, S (1999). Psychosocial and clinical correlates of ADHD in a community sample of school-age children. *J Am Acad Child Adolesc Psychiatry, 38,* 876–984.

Schachar, R, & Logan, GD (1990). Impulsivity and inhibitory control in normal development and childhood psychopathology. *Dev Psychol, 26,* 710–720.

Schachar, R, & Tannock, R (1995). Test of four hypotheses for the comorbidity of attention-deficit hyperactivity disorder and conduct disorder. *J Am Acad Child Adolesc Psychiatry, 34,* 639–648.

Seeman, P & Madras, BK (1998). Anti-hyperactivity medication: Methylphenidate and amphetamine. *Mol Psychiatry,* 3, 386–396.

Semrud-Clikeman, M, Biederman, J, Sprich-Buckminster, S, Lehman, BK, Faraone, SV, & Norman, D (1992). Comorbidity between ADDH and Learning Disability: A review and report in a clinically referred sample. *J Am Acad Child Adolesc Psychiatry, 31,* 439–448.

Sergeant, JA, Oosterlaan, J, & van der Meere, J (1999). Information processing and energetic factors in Attention-deficit/Hyperactivity Disorder. In HC Quay & AE Hogan (Eds.), *Handbook of disruptive behavior disorders* (pp. 75–104). New York: Kluwer Academic/Plenum.

Shaffer, D, Fisher, P, Dulcan, M, Davies, M, Piacentini, J, & Schwab-Stone, et al. (1996). The NIMH Diagnostic Interview Schedule for Children Version 2.3 (DISC-2.3): Description, acceptability, prevalence rates, and performance in the MECA study. *J Am Acad Child Adolesc Psychiatry, 35,* 865–877.

Sharp, WS, Walter, JM, Marsh, WL, Ritchie, GF, Hamburger, SD, & Castellanos, FX (1999). ADHD in girls: Clinical comparability of a research sample. *J Am Acad Child Adolesc Psychiatry, 38,* 40–47.

Sherman, DK, Iacono, WG, & McGue, MK (1997). Attention-deficit hyperactivity disorder dimensions: A twin study of inattention and impulsivity-hyperactivity. *J Am Acad Child Adolesc Psychiatry, 36,* 745–753.

Silberg, J, Rutter, M, Meyer, J, Maes, H, Hewitt, J, & Simnoff, E (1996). Genetic and environmental influences on the covariation between hyperactivity and conduct disturbance in juvenile twins. *J Child Psychol Psychiatry, 37,* 803–816.

Smalley, SL, Bailey, JN, Palmer, CG, Cantwell, DP, McGough, JJ, & Del'Homme, MA, et al. (1998). Evidence that the dopamine D4 receptor is a susceptibility gene in attention deficit hyperactivity disorder. *Mol Psychiatry, 3,* 427–430.

Solanto, MV (1984). Neuropharmacological basis of stimulant drug action in attention deficit disorder with hyperactivity: A review and synthesis. *Psychol Bull, 95,* 387–409.

Solanto, MV (1986). Behavioral effects of low-dose methylphenidate in childhood attention deficit disorder: Implications for a mechanism of stimulant drug action. *J Am Acad Child Adolesc Psychiatry, 25,* 96–101.

Solanto, MV (1990). The effects of reinforcement and response-cost on a delayed response task in children with attention deficit hyperactivity disorder: a research note. *J Child Psychol Psychiatry, 31,* 803–808.

Solanto, MV (1998). Neuropsychopharmacological mechanisms of stimulant drug action in attention deficit/hyperactivity disorder: A review and integration. *Behav Brain Res, 94,* 127–152.

Solanto, MV, Abikoff, H, Sonuga-Barke, E, Schachar, R, Logan, G, Wigal, & T, Hechtman, et al. (2000). The ecological validity of delay aversion and response inhibition as measures of impulsivity in AD/HD: A supplement to the NIMH Multi-Modal Treatment Study of AD/HD. *J Abnorm Child Psychol,* (In press).

Solanto, MV, & Wender, EH (1989). Does methylphenidate constrict cognitive functioning? *J Am Acad Child Adolesc Psychiatry, 26,* 897–902.

Solanto, MV, Wender, EH, & Bartell, SS (1997). Effects of methylphenidate and behavioral contingencies on vigilance in AD/HD: A test of the reward dysfunction hypothesis. *J Child Adolesc Psychopharmacol 7,* 123–136.

Sonuga-Barke, EJS, Taylor, E, & Heptinstall, E (1992). Hyperactivity and delay aversion II: the effects of self-imposed and externally imposed presentation time on memory. *J Child Psychol Psychiatry, 33,* 399–409.

Sostek, AJ, Buchsbaum, MS, & Rapoport, JL (1980). Effects of amphetamine on vigilance performance in normal and hyperactive children. *J Abnorm Child Psychol, 8,* 491–500.

Sternberg, S (1969). Discovery of processing stages: Extensions of Donders' method. *Acta Psychol (Amst), 8,* 276–315.

Still, GF (1902). Some abnormal psychical conditions in children. *Lancet, 1,* 1008–1012, 1077–1082, 1163–1168.

Swanson, J (1992). *School-based assessments and interventions for ADD students.* Irvine, CA: K.C. Publications.

Swanson, JA, Sunohara, GA, Kennedy, JL, Regino, R, Fineberg, E, & Wigal, T, et al. (1998). Association of the dopamine receptor D4 (DRD4) gene with a refined phenotype of attention deficit hyperactivity disorder (ADHD): a family-based approach. *Mol Psychiatry, 3,* 38–41.

Swanson, JM, Lerner, M, & Williams, L (1995). More frequent diagnosis of attention deficit hyperactivity disorder. *N Engl J Med, 333,* 944.

Szatmari, P, Offord, DR, & Boyle, MH (1989). Ontario Child Health Study: Prevalence of Attention Deficit Disorder with Hyperactivity. *J Am Acad Child Adolesc Psychiatry, 30,* 219–230.

Tannock, R (1998). Attention deficit hyperactivity disorder: Advances in cognitive, neurobiological, and genetic research. *J Child Psychol Psychiatry, 39,* 65–99.

Taylor, JR, Birnbaum, SG, Ubriani, R, & Arnsten, AFT (1999). Activation of protein kinase A in prefrontal cortex impairs working memory performance. *Journal of Neuroscience (Online),* in press.

Trites, RL, Dugas, E, Lynch, G, & Ferguson, HB (1979). Prevalence of hyperactivity. *J Pediatr Psychol, 4,* 179–188.

van der Meere, J, Stemerdink, N, & B., G (1995). Effects of presentation rate of stimuli on response inhibition in ADHD children with and without tics. *Percept Mot Skills, 81,* 259–262.

van der Meere, JJ (1996). The role of attention. In S Sandberg (Ed.), *Monographs on child and adolescent psychiatry: Hyperactivity disorders* (pp. 109–145). Cambridge, UK.: Cambridge University Press.

van der Meere, JJ, van Baal, M, & Sergeant, JA (1989). The additive factor method: a differential diagnostic tool in hyperactive and learning disability. *J Abnorm Child Psychol, 33,* 409–422.

Waldman, ID, Rowe, DC, Abramowitz, A, Kozel, S, Mohr, J, & Sherman, SL, et al. (1998). Association and linkage of dopamine transporter gene (DAT1) and attention deficit hyperactivity disorder in children: Heterogeneity owing to diagnostic subtype and severity. *Am J Hum Genet, 63,* 1767–1776.

Weiss, G, & Hechtman, L (1993). *Hyperactive children grown up: Empirical findings and theoretical considerations* (2nd ed.). New York: Guilford.

Wender, PH (1971). *Minimal brain dysfunction.* New York: Wiley.

Wigal, T, Swanson, JM, Douglas, VI, Wigal, SB, Wippler, CM, & Cavoto, KF (1998). Effect of reinforcement on facial responsivity and persistence in children with attention-deficit hyperactivity disorder. *Behavior Modification, 22,* 143–166.

Wilkison, PC, Kircher, JC, McMahon, WM, & Sloane, HN (1995). Effects of methylphenidate on reward strength in boys with attention-deficit hyperactivity disorder. *J Am Acad Child Adolesc Psychiatry, 34,* 897–901.

Wolraich, ML, Hannah, JN, Pinnock, TY, Baumgaertel, A, & Brown, J (1996). Comparison of diagnostic criteria for attention-deficit hyperactivity disorder in a county-wide sample. *J Am Acad Child Adolesc Psychiatry, 35,* 319–324.

World Health Organization (1990). *International Classification of Diseases* (10th ed.). Geneva: Author.

Zahn, TP, Kruesi, MJP, & Rapoport, JL (1991). Reaction time indices of attention deficit in boys with disruptive behavior disorders. *J Abnorm Child Psychol, 19,* 233–252.

Zentall, SS (1984). Context effects in the behavioral ratings of hyperactivity. *J Abnorm Child Psychol, 12,* 345–352.

Zentall, SS (1985). A context for hyperactivity. *Advances in Learning and Behavioural Disabilities, 4,* 273–343.

Zentall, SS, & Meyer, MJ (1987). Self-regulation of stimulation for ADD-H children during reading and vigilance task performance. *J Abnorm Child Psychol, 15,* 519–536.

Zentall, SS, & Zentall, TR (1983). Optimal stimulation: A model of disordered activity and performance in normal and deviant children. *Psychol Bull, 94,* 446–471.

2

Clinical Effects of Stimulant Medication in ADHD

LAURENCE L. GREENHILL

A serendipitous observation made more than 60 years ago revealed that psychostimulants ameliorate symptoms of Attention-Deficit/Hyperactivity Disorder (ADHD) in children (Bradley, 1937, Appendix). Today, there are four immediate-release (IR) stimulant medications approved for use in children—methylphenidate (MPH), dextroamphetamine (DEX), Adderall, and pemoline (PEM)—approved for the treatment of school-aged children with ADHD. Although the Food and Drug Administration (FDA) approved these four medications more than 20 years ago, no entirely new psychostimulant has attained FDA approval and entered the market since then. The robust short-term beneficial effects of stimulants on the symptoms of ADHD have been documented by the largest treatment literature on any childhood-onset psychiatric disorder. By 1996, there were 161 randomized controlled trials (RCT) (Spencer et al., 1996a).

Increasing Use of Stimulants in Clinical Practice

Outpatient visits devoted to ADHD have increased from 1.6 to 4.2 million per year during the years 1990–1993 (Swanson et al., 1995); by 1995, these figures had climbed to 2 million visits and 6 million stimulant drug prescriptions (Jensen et al., 1999a). During those visits, 90% of the children were given prescriptions, 71% of which were for the stimulant methylphenidate (MPH). MPH production in the United States (U.S.) increased from 1,784 kg to 5,110 kg during the same time period, so that over 10 million prescriptions for MPH were written in 1996 (Vitiello & Jensen, 1997). It has been estimated that 2.8% of U.S. youth, ages 5 to 18 years, were prescribed stimulants in 1995 (Goldman et al., 1998). However, specific epidemiological surveys suggest that 12-month prescription rates for the school-age group—ages 6 to 12 years—may be higher, ranging between 6% urban (Safer et al., 1996) and 7% rural (Angold & Costello, in press) and extending up to 10% in some communities (LeFever et al., 1999).

Psychostimulant use has increased 5-fold over the past 12 years, and this has raised concerns at the U.S. Drug Enforcement Administration (DEA)—which regulates their production—about the risk of abuse and diversion. Production of MPH has tripled over a ten-year period, and 90% of U.S.-produced MPH is used in the U.S. Increased MPH use could mean increases in ADHD prevalence, a change in the ADHD diagnosis, improved recognition of ADHD by physicians, broadened indications for use, or an increase in drug diversion or prescription for profit or abuse (Goldman et al., 1998). Analyses of managed care data sets reveals a 2.5-fold increase in prescribing in the 1990–1995 time period, accounted for by longer durations of treatment, inclusion of girls and those with predominately inattention, and treatment of high school students (Safer et al., 1996). An epidemiologically based survey in four different communities found only one-eighth of diagnosed ADHD children received adequate stimulant treatment (Jensen et al., 1999b), while another survey in rural North Carolina found many school-aged children on stimulants did not meet DSM-IV criteria for ADHD (Angold & Costello, in press).

Controversy and Politics of Stimulant Treatments for Children with ADHD

Even though the psychostimulants are the most widely researched, clinically effective, and commonly prescribed treatments for ADHD, their use in children has become the focus of a major controversy. A 1998 Consensus Development Conference sponsored by the National Institutes of Health on ADHD concluded that stimulants were effective in reducing the defining symptoms of ADHD in children in the short term, but indicated that the controversy about their use demanded serious consideration (National Institute of Health Consensus Development Conference). They also noted the lack of evidence for their long-term benefit and safety; the considerable risks of treatment; the wide variation in prescribing practices among practitioners; and the absence of evidence "regarding the appropriate ADHD diagnostic threshold above which the benefits of psychostimulant therapy outweigh the risks."

The conclusions of the Consensus Development Conference were much less sanguine than a similar report published the same year by the Council on Scientific Affairs of the American Medical Association. After reviewing hundreds of trials involving thousands of patients, the council concluded that "the risk-benefit ratio of stimulant treatment in ADHD must be evaluated and monitored on an on-going basis in each case, but in general is highly favorable" (Goldman et al., 1998). Which point of view about the stimulants is more accurate?

Domains of Effectiveness

Stimulants ameliorate disruptive ADHD behaviors cross-situationally (classroom, lunchroom, playground, and home) when repeatedly administered throughout the day. In the classroom, stimulants decrease interrupting, fidgetiness, finger tapping, and increase on-task behavior (Abikoff & Gittelman, 1985); on the playground, stimulants reduce overt aggression (Gadow et al., 1990), covert aggression (Hinshaw et al., 1992), and signs of conduct disorder (Klein et al., 1997) and increase attention during baseball (Pelham et al., 1990a). At home, stimulants improve parent–child interactions, on-task behaviors, and compliance; in social settings, stimulants ameliorate peer nomination rankings (Whalen et al., 1989). Stimulants decrease response variability and impulsive responding on cognitive tasks (Tannock et al., 1995a); increase the accuracy of performance; and improve short-term memory, reaction time, seatwork computation, problem-solving games with peers (Hinshaw et al., 1989), and sustained attention. Studies of time-action stimulant effects show a different pattern of improvement for behavioral and for attentional symptoms, with behavior affected more than attention. For example, a controlled, analogue classroom trial ($n = 30$) of Adderall revealed rapid improvements on teacher ratings of behavior and math performance scores 1.5 hours after administration, with time of duration of effects on behavior specifically dependent on dose (Swanson et al., 1998a). While stimulant drugs show a large 0.8–1.0 effect size for behavioral measures, smaller 0.6–0.8 effect sizes are reported on cognitive measures (Spencer et al., 1996a).

Children with ADHD demonstrate low placebo response rates during clinical drug trials, between 3% (Spencer et al., 1996a) and 30% (Gillberg et al., 1997). The resulting large placebo-active drug differences enable small-N, single-site controlled trial designs to identify significant drug effects. The drugs' very short elimination half-lives make stimulants eligible for crossover designs (Swanson et al., 1998a), because pharmacokinetic carryover of active drug into placebo periods is unlikely.

The beneficial effects of stimulants on behavior and attention have also been shown in children with no mental disorder (Rapoport et al., 1980), so the response to stimulants is not diagnostically specific for ADHD. Stimulant medications continue to play a therapeutic role in other medical conditions, such as narcolepsy and depression (Goldman et al., 1998).

Cognitive Effects

Psychostimulants produce significant responses in ADHD children when compared with placebo using tasks that test vigilance and reaction time. These responses are task and dose dependent. The amount of effortful processing required is also a factor. The Continuous Performance Task (CPT) and the Gardner Steadiness Test (GST) show linear dose-response relationships, with improved performance as the MPH dose is increased. Early work on the CPT showed that ADHD children treated with MPH experienced a reduction in errors of commission (Weingartner et al., 1980), thought to relate to impulsivity, and occasionally, errors of omission, a measure of sustained attention. Later signal detection analysis (Bergman et al., 1991; Cornblatt, 1988) demonstrated that psychostimulant medication could change the sensitivity (d′) and style of responding (Log Beta) during these trials.

Effects on cognition could be measured outside the lab as well. Classroom arithmetic tasks (Carlson & Thomeer, 1991) showed medication-placebo effects, as did concentration during sports (Pelham et al., 1990a). Higher dosages of psychostimulants may interfere with complex memory tasks that involve set changes and other types of effortful processing. High MPH doses (1 mg/kg) produced less-than-expected reductions in errors on the most complex level of Sternberg memory task, although these high doses did better than lower doses in reducing abnormal behaviors in a classroom (Sprague & Sleator, 1977).

This dissociation of cognitive and social responses to stimulants was first reported by Sprague and Sleator in 1977. They studied ten hyperkinetic school-aged patients who were given a short-term memory paired associates task. The best percent correct score was reported on the lowest, 0.3 mg/kg, dose; increasing the dose to 1.0 mg/kg increased the error rate. In contrast, these same children obtained the best teacher global ratings of improved behavior on the higher, 1.0 mg/kg, dose. One might conclude, as did many clinicians, that the use of behavioral ratings for dosage adjustment may select doses that impair cognitive processes. This "window of response" can be modeled best as an inverted-U curvilinear dose-response curve (see Chapter 12 by Rapport for more details).

Sprague's findings have not been replicated in studies that followed. Group treatment study data have shown linear dose-response curves for MPH. A review of 11 studies, involving 180 children, that evaluated the stimulant dose-response relationship on cognitive tasks reported linear dose-response curves, with only two studies showing the decrease of cognitive abilities at higher doses (Rapport et al., 1989a). There is a great diversity among children in their dose-response curves, both between individuals and across domains of behavior and learning (Murphy et al., 1986; Solanto, 1999). Pelham believes optimal dosing of the individual ADHD child cannot be done a priori but must be carried out carefully on an empirical basis, adjusting the dose level to correct the child's most serious area of difficulty (Pelham & Milich, 1991).

Even with the lack of replication of Sprague's study, the impact on practice of this one study has been profound. Prior to this publication, investigators used higher MPH doses in their research (e.g., 51 mg/day—Gittelman-Klein et al., 1976), and each child's dose was titrated to maximal effect or toxicity. Since 1977, clinicians have adopted an increasingly conservative approach to MPH dose adjustments. The MTA study, to be discussed later in this chapter, and also in Chapter 15, found that, after a careful month-long titration, the average daily dose delivered on a three-times-per day dosage schedule was 32.1 ± 15.4 mg for the group on medication alone, and 28.9 ± 13.7 for the group on medication plus behavioral treatment.

Motor Effects

Psychostimulants reduce excessive motor behavior in children with ADHD. Early studies were limited to brief laboratory observations with stabilimetric chairs and mechanical wrist or ankle actometers (modified automatically winding wrist watches) (Barkley, 1977). Other approaches, such as counting grid crossings, did not measure restlessness and fidgetiness. Some postulated that motor activity per se did not differentiate ADHD children from normal children as much as the purposelessness of their activity.

Technological advances made it possible to obtain observational data in the child's natural settings, home and school. A portable actometer, developed at the National Institutes of Mental Health, consisted of an acceleration-sensitive device with a solid-state memory that stores data on the number of movements per unit time over a ten-day period (Porrino et al., 1983). Readings taken on 12 hyperactive boys were consistently higher than control values at all times, including during sleep and weekends. This argues against the concept that children's increased motor activity is a response to an increase in situational demands. This same research team treated male ADHD male participants with dextroamphetamine, 15 mg/day, in a double-blind ABAB design over 4 weeks. Motor activity was significantly decreased for 8 hours after dosing, followed by a slight increase in activity, which was the first objective demonstration of "rebound" (Porrino, 1983). Validating these reports were readouts from activity monitors that showed a significant stimulant-related reduction in motor activity during the school period and even during sleep (Porrino, 1983). These objective activity counts confirmed drug effects reported during global ratings and direct observations.

Borcherding et al. (1989) later replicated this work with 18 male participants with ADHD in a day hospital setting in an 11-week double-blind crossover trial comparing dextroamphetamine (0.2–0.6 mg/kg twice daily) and methylphenidate (0.45–1.25 mg/kg twice daily). Plasma drug concentrations failed to correlate with activity level changes, but high MPH and DEX plasma concentrations were reported by these investigators; this may have produced a ceiling effect, restricting the range of response, and reducing the correlation. MPH produced lower activity readings than did DEX.

Classroom Effects: Academic and Behavior

Many clinical treatment studies of hyperactive children have relied on global ratings from teachers. Despite their ubiquitous presence in the treatment literature, the psychometric properties of these rating instruments are often not well studied. Inter-rater reliability and even inter-test reliability of global rating forms are difficult to assess. The problem of validating hyperactive behaviors and their treatments has been greatly aided by using direct classroom observations with raters who are blind to student psychiatric and drug status. Often these studies have been conducted in summer school programs. An early study treated 23 hyperactive and 39 comparison boys in a complex design in which a number of conditions were systematically varied, including difficulty of work materials, self-paced versus other-paced activities, and single morning MPH dose (mean, 12.3 mg) (Whalen & Henker, 1976). Hyperactive boys on placebo were more often off-task, displayed more gross motor movement, more verbalizations, noise, and disruption, and had a more vigorous style of responding. These findings have been replicated, as shown by another study using higher mean MPH doses (41.5 mg daily) (Abikoff et al., 1985). Although the placebo-treated hyperactive children had much higher rates of off-task, out-of-seat, verbalization behaviors, they were indistinguishable from the normal children once on MPH.

Early reports suggested that the beneficial effects of psychostimulants on academic tasks were difficult to maintain over the months and years required for treatment of ADHD (Barkley & Cunningham, 1978; Charles et al., 1979; Heussy et al., 1973; Mendelson et al., 1971a; Minde et al., 1972; Sleator et al., 1974; Weiss et al., 1975). This was reflected by a MPH study, in which mean doses were 44.2 mg/day, of reading remediation in 61 children, half assigned to drug and half to placebo (Gittelman-Klein & Klein, 1976). No differences in reading skills between groups were found 6 months after treatment started. A 2-year follow-up of 72 children by Riddle and Rapoport (1976) treated with medication showed no significant advantage for those treated with MPH over those treated with other therapies. In fact, the 72 hyperkinetic children in this study did far worse on the WRAT reading and arithmetic than did the control group.

More recent short-term, controlled studies of ADHD children have shown that psychostimulants enhance arithmetic productivity, efficiency, and accuracy (Carlson & Thomeer, 1991; Douglas et al., 1986; Pelham, 1986). Children spent less time between problems (Carlson & Thomeer, 1991). These studies employed optimal experimental procedures, including standardized medication dosages and relevant dependent measures.

Aggression

Stimulant treatment has been associated with reductions in aggression ratings of children with ADHD in double-blind, placebo-controlled, crossover studies. Klorman and associates, in studies of 63 school-age ADHD patients—given MPH doses of 0.6 mg/kg—found that children with and without aggressive features show improvement in the form of reductions in ADHD behavior (Klorman et al., 1988). On the other hand, Hinshaw et al (1989) found only slight reductions in aggressivity in 24 ADHD school-aged children treated with MPH, 0.6 mg/kg (Hinshaw et al., 1989). Gadow and associates reported reductions in aggression in 11 ADHD boys, finding dose-related reductions in verbal aggression in the lunchroom and playground when the boys were treated with MPH (0.3 and 0.6 mg/kg) (Gadow et al., 1990). Aggression in nine severely aggressive adolescent inpatients with ADHD was reduced by MPH treatment, as measured by reduced ratings on a standardized nursing form, the Adolescent Antisocial Behavior Checklist (Kaplan et al., 1990).

Using a parallel-design, randomized, double-blind, placebo-controlled study of 62 school-aged ADHD children with comorbid conduct disorder, Klein and her colleagues reported that MPH treatment led to significant, dose-dependent reductions in aggressive behavior even when covaried for baseline severity of ADHD symptoms (Klein et al., 1997). Particularly in school-aged ADHD children, MPH reduces the aggression associated with conduct disorder.

Effects on Mother–Child Interactions

Maternal responses to MPH-treated children with ADHD show decreases of repetitive, harsh, command-like verbalizations on the part of the parent (Barkley et al., 1984). Rates of friction with siblings were also lower when the ADHD children were treated with MPH (Schachar et al., 1987). The medication-related change in style of interaction may be a result of increased compliance with mother's commands and decreased intensity on the part of the child. One study suggests a dose-dependent stimulant effect (Barkley et al., 1984).

Peer Relationships

Peer relationships have been a major problem area for ADHD children. Their impulsive interpersonal style leads them to be intrusive, abrasive, and loud and to relate to others with high intensity. They are apt to interrupt other children's games and conversations, to fail to wait their turn in games or to push ahead in line, and to miss social cues. Direct observations of ADHD boys in the classroom and playground or lunchroom have produced a picture of social interaction failure, aggressivity, and peer rejection. In a treatment study involving 25 ADHD boys, more frequent peer nominations were obtained on drug for being "cooperative," "fun to be with," and acting like "best friends," although their overall peer nomination ratings do not reach the levels given non-ADHD boys (Whalen et al., 1989). In a linear dose-response manner, higher-dose MPH (0.6 mg/kg) produced better peer ratings. MPH at that dosage reduced non-compliance—the number of episodes of verbal and physical aggression in both classroom and playground—to the level of the control children. MPH also reduced ADHD boys' verbal aggression in unstructured lunchroom environments (Gadow et al., 1990). However, MPH did not normalize the peer relationships of the ADHD children (Wilens & Biederman, 1992).

Short-Term Efficacy in Randomized Controlled Trials of Stimulants

Psychostimulants have consistently shown robust behavioral efficacy in randomized controlled studies (RCT). These effects were even evident in small-numbered, single-site, short-duration trials. By 1993, one review suggested that the stimulant drug treatment literature had grown so that there were 250 reviews (Swanson, 1993). Over 161 randomized controlled trials (RCTs) studies encompassing 5 preschool, 140 school-aged, 7 adolescent, and 9 adult RCTs (Spencer et al., 1996a) gave evidence that stimulants robustly reduced ADHD symptoms (Spencer et al., 1996a). Although early investigators suggested that stimulant medications uniquely and paradoxically calmed school-aged children with ADHD, these drugs have proven effective across the life span. Improvement was noted in 65% to 75% of the 5899 patients assigned to stimulant treatment versus only 4% to 30% of those assigned to placebo for MPH (133 trials), DEX (22 trials), and PEM (6 trials). Compared with placebo in short-term, double-blind trials, the stimulants demonstrate robust efficacy in improving the two prominent symptom clusters found in ADHD. Both inattentiveness and hyperactivity–impulsivity were reduced, as assessed by parent and teacher rating scales, direct observations in natural settings, and laboratory tests (Greenhill et al., 1999a; Jacobvitz et al., 1990). Three meta-analyses, conducted in the early 1980s, separately concluded that the efficacy for stimulants—in placebo-controlled, double-blind trials—was very strong, with effect sizes reaching 0.8 to 1.0 standard deviations (Kavale, 1982; Ottenbacher & Cooper, 1983; Thurber & Walker, 1983). Less powerful effects were found for laboratory measures for cognitive changes, in particular on the Continuous Performance Task, where effect sizes of these medications fall between 0.6 and 0.66 for omissions and commissions, respectively, in a within-subject design (Milich et al., 1989), and 0.6 and 1.8 in a between-subject study (Schechter & Keuezer, 1985).

Greatly improved methodology can be found in RCTs published over the past 15 years, as shown by their use of multiple-dose conditions with multiple stimulants (Elia et al., 1991), parallel designs (Wilens et al., 1995), multisite trials, and use of common definitions of response such as normalization (Abikoff et al., 1985; Rapport et al., 1994). These trials now address psychostimulant responses in adolescents (Klorman et al., 1990), adults (Spencer et al., 1995), mentally retarded subjects (Horn et al., 1991), ADHD subjects with anxiety disorders and internalizing disorders (Diamond et al., 1999), and ADHD subjects with tic disorders (Gadow et al., 1995; Castellanos et al., 1997). As shown in Table 2–1, 70% of ADHD subjects in large stimulant-randomized controlled trials since 1995 respond to stimulants and fewer than 10% to placebo.

Stimulant nonresponders have also been studied. A study conducted at NIMH in a day hospital found that the 32% non-response rate to a single psychostimulant dropped to less than 4% when two stimulants, DEX and MPH, were titrated sequentially in the same subject (Elia et al., 1991).

Use of Psychostimulants in Children Younger than 6 Years

The signs and symptoms of ADHD may be evident before age 3, particularly pronounced motor activity, excessive climbing, aggressivity, and destructiveness. These signs may be disruptive to family life and make nursery school attendance impossible. Campbell found that hyperactive preschoolers did not "grow out of" this behavior, but maintained their hyperkinesis when they went on to grade school (Campbell et al., 1977). Even though these ADHD signs resemble the behaviors among older ADHD patients, the diagnostic manuals give little guidance about the validity of the ADHD diagnosis in the preschool years. School-age norms gathered on standard teacher global rating forms, such as the Conners Teacher Questionnaire (CTQ), have not included preschoolers until recently (Conners, personal communication).

Unfortunately, there are only a handful of published treatment studies on preschoolers. No pharmacokinetic studies have been done in this age group to determine if younger children, with their larger liver-to-body size ratio, might show more

(*text continues on page 42*)

Table 2–1. Controlled Stimulant Studies (N = 1702) Since 1985 Showing Efficacy for ADHD Symptoms

Principal Investigator (Year)	Number	Age Range (years)	Design	Drug (Dose)	Duration	Response	Comment
Abikoff (1985)	28	6–12	ADHD Controls	MPH (PB, 41 mg)	8 weeks	80.9%	ADHD kids normalized
Abikoff (1998)	103	6–12	ADHD	MPH TID (33.7 mg)	2 years	100% 2.7 SD	Multisite, multimodal study. All children on MPH
Barkley (1989)	74	6–13	Crossover 37 aggressive 37 non-aggressive	MPH (PB, 0.3, 0.5)	4 weeks	80%	Aggression responsive to MPH
Barkley (1991)	40	6–12	Crossover 23 ADHD 17 ADHD-W	MPH BID (5,10,15 BID) PB BID	6 weeks	ADHD 95% ADHD-W 76%	Fewer ADHD-W respond, need low dose
Castellanos (1997)	20	6–13	Crossover	MPH 45 mg BID DEX 22.5 mg BID	9 weeks	ADHD + TS	Tics dose-related at high doses
Douglas (1988)	19	7–3	Crossover	MPH (PB, 0.15, 0.3, 0.6)	2 weeks	100%*	Linear D/R relationships
Douglas (1995)	17	6–11	Crossover	MPH (0.3, 0.6, 0.9) PB	4 weeks	behavior 70%	No cognitive toxicity at high doses, linear D/R curves
DuPaul (1993)	31	6–12	Crossover 31 ADHD 25 normals	MPH (20 mg) PB BID	6 weeks	behavior 78% attention 61%	MPH can normalize classroom behavior 25% of ADHD subjects didn't normalize academics
DuPaul (1994)	40	6–12	Crossover 12 high Anx 17 mid Anx 11 low Anx	MPH (5,10,15 mg) PB single dose	6 weeks	high 68% normal mid 70% normal low 82% normal	25% in internalizing group deteriorated on meds ADHD subjects with comorbid int disorders are less apt to be normalized or to respond to MPH
Elia (1991)	48	6–12	Crossover	MPH (0.5; 0.8, 1.5) PB BID DEX (0.25, (0.5,0.75)	6 weeks	MPH 79% DEX 86%	Response rate for two stimulants = 96%

(continued)

Table 2–1. Controlled Stimulant Studies (N = 1702) Since 1985 Showing Efficacy for ADHD Symptoms *Continued*

Principal Investigator (Year)	Number	Age Range (years)	Design	Drug (Dose)	Duration	Response	Comment
Gadow (1995)	34	6–12	Crossover ADHD+tic Disorder	MPH (0.1,0.3,0.5) PB BID	8 weeks	100%MPH	No nonresponders to behavior; MD's motor tic ratings show 2-min increases on drug; only shows effects of 8-wk treatment
Gillberg (1997)	62	6–12	Parallel	AMP (17mg) PB BID	60 weeks	70% respond 27–40% improve	No dropouts but only 25% of the placebo grp left at 15-mo. assessment.
Klein (1997)	84	6–15	Parallel	MPH BID (1.0)	5 weeks	MPH 59–78% Pb 9–29%	MPH reduced ratings of antisocial behaviors
Klorman (1990)	48	12–18	Crossover	MPH TID (0.26) PB BID	6 weeks	MPH 60%	Less medication benefits for adolescents
MTA (1999)	579	7–9	Parallel	MPH TID (<0.8)	4 weeks (14 mo)	MPH 77% DEX 10% None 13%	Titration trial for multisite multimodal study Full study data for 288 on 38.7 mg MPH
Musten (1997)	31	4–6	Crossover	MPH BID (0.3, 0.5)	3 weeks	MPH>NA	MPH improves attention, not compliance
Pelham (1990)	22	8–13	Crossover	MPH 10 BID; PB BID; DEX span 10 mg PEM 56.25 daily	24 days	stimulant 68%	Dex-Span, PEM best for behavior; 27% did best on DEX; 18% on SR; 18% on PEM; 5% on MPH BID.
Pelham (1995)	28	5–12	Crossover	PEM (18.75,37.5, 75, 112.5 mg) PB O.D.	7 weeks	PEM response 89% PB response 0%	PEM dose ≥ 37.5 mg/day active for a 2–7 hour period Efficacy and time course, PEM = MPH

Rapport (1988)	22	6–10	Crossover	MPH (PB, 5, 10, 15 mg)	5 weeks	72%	MPH response same in home and school
Rapport (1994)	76	6–12	Crossover	MPH (5,10,15, 20 mg) PB BID	5 weeks	94% behavior 53% attention	MPH normalizes behavior > academics Higher doses better, linear D/R curve
Schachar (1997)	91	6–12	Parallel	MPH (33.5 mg) PB BID	52 weeks	0.7 SD effect size	15% side effects: affective, over-focusing Led to dropouts
Sharp (1999)	42	6–12	Crossover	MPH (.4,.8,1.2) Dex (.2, .4, .6)	9 weeks		Same response in girls as boys.
Spencer (1995)	23	18–60	Crossover	MPH (1 mg/kg/d)	7 weeks	78% PB 4%	MPH at 1 mg/kg/d produces improvement in adults equivalent to that seen in kids
Swanson (1998)	29	7–14	Crossover	Adderall (5,10,15, 20, Pb, MPH)	7 weeks	100%*	Adderal peaks at 3 hr, MPH at 1.5 hr
Tannock (1995)	40 22	6–12	Crossover	MPH (0.3, 0.6) 17 ADHD-Anx	2 weeks	70%	Activity level better in both groups; working memory not improved in anxious kids
Tannock (1995)	28	6–12	Crossover	MPH (0.3, 0.6, 0.9) PB	2 weeks	70% 70%	Effects on behavior D/R curve linear, but effects on resp inhibition U-shaped suggest adjust dose on objective measures
Taylor (1987)	38	6–10	Crossover	MPH (PB, 0.2–1.4)	6 weeks	58%	Severe AD/HD symptoms, better response
Whalen (1989)	25	6.3–12	Crossover	MPH (PB, 0.3,0.5)	5 weeks	48–72%	MPH helps, not normalizes, peer status

Doses listed as mg/kg/dose, and medication is given twice daily unless otherwise stated.

Abbreviations: PB = placebo; X-over = crossover design, Anx = anxiety; MPH-methylphenidate; DEX = dextroamphetamine; PEM = pemoline; AMP = d,l-amphetamine; int = internalizing.

accelerated metabolism of psychostimulants than seen in older children. To date there are no studies that would validate or refute clinicians' concerns that preschoolers suffer more pronounced stimulant withdrawal effects.

Comprehensive reviews (Spencer et al., 1996a) identified only five controlled studies ($n = 144$) of methylphenidate in preschoolers (Barkley et al., 1984; Barkley, 1988; Conners, 1975; Mayes et al., 1994; Musten et al., 1997; Schleifer et al.; 1975). Schleifer and colleagues (1975) found that MPH produced improvements in structured situations, but not in free play, although there were more side effects reported, such as tearfulness and irritability. A single-case intensive study design noted that dextroamphetamine reduced temper outbursts in a toddler (Speltz et al., 1988). The MPH treatment studies have found that the mother–child interactions involving preschoolers are affected by MPH (Barkley, 1988). There appeared to be a linear dosage effect of MPH on improvements in the mother–child interaction, perhaps related to increasing child compliance and decreased symptomatic intensity in the child. No disabling side effects were noted. Four of these six studies showed strongly positive results, and two reported higher rates of side effects than would be expected in school-aged children (Barkley et al., 1990b). However, the six studies utilize very different inclusion / exclusion criteria and designs, so that a meta-analysis would not be appropriate. Current labeling warns against using MPH in children below age 6. Even so, there has been a 180% increase over the past 5 years in the prescriptions of MPH for preschoolers. Clearly, more controlled data are needed.

Continuing Stimulant Treatment in Patients with Tic Disorders

The Physicians' Desk Reference (1999) characterizes the presence of tics in a patient or in a family member as a contraindication for MPH—but not dextroamphetamine or Adderall—usage. Patients with Tourette's syndrome often have comorbid ADHD. If the two conditions co-occur, concern arises that stimulant treatment—because of its dopamine agonist actions—might unmask, trigger, accelerate, or precipitate more severe tics. This worry was expressed in an editorial that accompanied a retrospective chart review (Cohen & Leckman, 1989). The report included 17 ADHD patients with Tourette syndrome who had been previously treated with stimulants, and reported their tics had been exacerbated by stimulant therapy (Lowe et al., 1982). Another study (Erenberg et al., 1985) found that symptoms in 11 of 39 patients who had both ADHD and Tourette worsened when stimulants were restarted.

The two controlled prospective studies of stimulants in children comorbid for ADHD and Tourette have included long-term longitudinal follow-ups extending 2 to 4 years (Gadow et al., 1995; Castellanos et al., 1997). Conclusions derived from these studies must be tentative because they are based on a total of 56 subjects. However, in both studies, stimulants produced improvements in ADHD symptoms, and no long-term worsening in tics, even when tics worsened temporarily with high doses of MPH or DEX. Despite the variable rate of adverse events across the whole sample in the Castellanos study, up to one-third of subjects did not continue long-term treatment with stimulants because of worsening of tics. Thus, both sides of the argument have been validated. Tics are unacceptably worsened for a minority of comorbid children, although reversible, and the majority of such comorbid children can be treated cautiously with low-to-moderate doses of stimulants, particularly MPH (Comings & Comings, 1987).

Mental Retardation

Many stimulant treatment studies for school-aged ADHD children have excluded subjects with a full-scale IQ below 80. As a result, children with Full Scale Wechsler IQS below 50, termed moderate mental retardation (MR), are underrepresented in these studies. This is unfortunate, for the signs of Attention-Deficit/Hyperactivity Disorder are far more prevalent in children with MR than in those without (Biederman et al., 1991). In addition, pharmacotherapy reviews have reported that institutionalized subjects with severe MR show a poor response to stimulants (Lipman et al., 1978). More recent studies, involving community samples of children with MR, have suggested that these children do respond to MPH. Thirty children treated in a double-blind, crossover study employing MPH (0.4 mg/kg/day), thioridazine (1.75 mg/kg/day), and placebo showed consistent MPH-related improvement on teacher's global ratings of behavior (Aman et al., 1991). This same report found that three "cognitive maturity parameters" (breadth of attention, mental age above 4.5 years, and Full Scale IQ above 45) predicted a good MPH response on 21 outcome variables. In contrast, children with IQs below 45 or a mental age of less than 4.5 years had more adverse side effects or an indifferent response. Another controlled study (Handen et al., 1991), reported drug responses of 27 children with ADHD and IQs between 48 and 74, treated with MPH (0.3 mg/kg and 0.6 mg/kg). Although 67% of this group were deemed stimulant responders, six of the 27 children reported side effects of such severity that medication had to be discontinued. These included motor tics (three children) and social withdrawal (two children). Thus, it appears that ADHD children with mild and moderate MR may be more vulnerable to certain stimulant-related adverse side effects.

Children with fragile X syndrome present with mental retardation, avoidance of eye contact, perseverative behaviors, and severe ADHD. Stimulant treatment has proven effective in this population, as shown in a double-blind crossover study of 15 patients with fragile X syndrome, comparing MPH, dextroamphetamine, and placebo (Hagerman et al., 1988).

The APA Diagnostic Statistical Manual advises that a diagnosis of Pervasive Developmental Disorder (PDD) or Early Infantile Autism "preempts" the ADHD diagnosis, even if the patient meets all criteria for the ADHD disorder. Symptomatically, children with PDD may exhibit severe overactivity, inattentiveness, and impulsivity requiring treatment. Early studies cautioned that PDD children may be sensitive to stimulant side effects, and may show increased irritability, motor activity and stereotypies (Aman, 1982). Others have reported improvements in global parent and teacher rating scales for hyperactivity and inattentiveness in nine children with early infantile autism treated with open-label MPH at daily doses of 10 to 50 mg (Birmaher et al., 1988). These findings were based on an open study. The lack of confirmatory studies in the past 17 years has not been encouraging. Clinicians use stimulants in children with pervasive developmental disorder in low doses because of the absence of good alternatives, and because of the good safety record of the stimulants.

Does Comorbidity Affect the Indications for the Use of Stimulant Medication?

ADHD children who are referred by schools and other physicians typically have more disorders than just ADHD, but most stimulant treatment studies have selected patients with few comorbidities. Extensive literature reviews (see Chapter 1 by Solanto) report that ADHD co-occurs with conduct and oppositional defiant disorder in 30% to 50% of cases, with mood disorders in 15% to 75% of cases, with anxiety disorders in 25% of cases, and with learning disabilities in 15% to 30% of cases. Specific patterns of signs and symptoms may co-exist in the patients and their families, with groups of disorders and vulnerabilities being genetically transmitted from generation to generation (Biederman et al., 1986). Does the presence of other psychiatric signs and symptoms affect response to stimulants?

In anxious ADHD children, it has long been speculated that comorbid anxiety may influence the outcome of treatment for ADHD. For example, Pliszka compared children with non-anxious ADHD to children with comorbid ADHD and overanxious disorder and found that the combination of ADHD and ANX reduced the risk of comorbid conduct disorder, decreased impulsive responding on a continuous performance task (CPT), and produced sluggish reaction times on a Memory Scanning Test (Pliszka, 1989). Subjects who were comorbid for anxiety and ADHD also showed a poorer response to MPH and more side effects than subjects with ADHD alone (Pliszka, 1989). Although failing to replicate the results of Pliszka 1989) with respect to behavioral outcomes, Tannock, Schachar and colleagues later demonstrated an enhanced risk of tics and dysphoria in ADHD subjects comorbid for anxiety treated with methylphenidate and less improvement in memory in comorbid children given a single dose of MPH (Tannock et al., 1995b). Conversely, using a flexible clinically appropriate titration trial, Tannock and colleagues recently found no difference in short-term outcome or side effects in children with ADHD and comorbid ADHD/ANX (Diamond et al., 1999), implying that method variance may in part account for conflicting results in these studies.

Analyses of the six-site NIMH collaborative Multimodal Treatment of Children with ADHD (MTA) Study, which addresses a priori questions about the individual and combined effects of pharmacological and psychosocial (behavioral) treatment for children age 7 to 9 years with ADHD (Arnold et al., 1997), suggested that parent-reported child anxiety differentially moderated the outcome of treatment. In particular, parent-reported child anxiety as a moderator of treatment outcome generally favored the addition of psychosocial treatment for anxious ADHD children irrespective of the presence or absence of comorbid conduct problems. Contravening earlier studies, no adverse effect of parent-reported child anxiety on medication for core ADHD or other outcomes was demonstrated. For some domains of outcome, anxious ADHD children showed significantly greater gains than nonanxious children even with medication alone. On the other hand, exploratory analyses documented that child self-reported anxiety on the Multidimensional Anxiety Scale for Children (MASC) was only weakly correlated with parent- or teacher-reported symptomatology at baseline and changed little with treatment, raising questions about whether the parent DISC accurately trapped anxiety per se. In this regard, analyses of baseline data suggest that child anxiety as ascertained on the parent DISC 2.3 reflects a broad mix of affective and behavioral dysregulation, suggesting that parent-reported anxiety may reflect a broader construct, perhaps an admixture of anxiety and affectively driven disruptive behavior.

Limitations of Short-Term Stimulant Trials

When the four marketed stimulants were developed—more than 20 years ago—most Phase I (safety) and Phase II (dosing and early efficacy) data were collected in animals and adults. As a result, modern New Drug Application (NDA) Phase I studies (e.g., effect of fatty diet on absorption and basic pharmacokinetics in children), Phase II studies (e.g., combined pharmacokinetic/pharmacodynamic) or Phase III studies (e.g., population pharmacokinetics) were not carried out for those stimulants now in use for children with ADHD.

Without the demands of the New Drug Application (NDA) approval process, the stimulant trials published in the last decade did not consistently use design features or analytic approaches employed in multi-site RCTs submitted to the FDA for drug approval. For example, sample ethnicities were not mentioned nor were representative child patient samples included that would allow findings to be generalizable. Most studies lasted less than 12 weeks (Schachar & Tannock, 1993), so could not inform maintenance treatment of children with ADHD. Many employed relatively small sample sizes and lacked control participants treated intensively with a non-medication therapy. Most did not individually titrate medication doses to optimize each child's response. Many studies did not report comorbid disorders, which can alter the response to medication. Others did not report the fidelity with which the treatments were administered by staff, nor did they always list results of the compliance measures used. Parallel design methodology and intent-to-treat designs were not used (Richters et al., 1995). Other details were not listed by most studies, such as the exact method of randomization (including a central randomization conducted away from the site) and details of dropouts. Few, if any, of the previous stimulant RCTs used parallel designs to test for the emergence of placebo responses during the drug trial. Furthermore, these estimates apply to group effects, and do not inform the clinician about the individual patient.

Predictors of Drug Response

Predicting drug response in the individual ADHD child has been a difficult goal to achieve in clinical research studies. While pretreatment patient characteristics (young age, low rates of anxiety, high severity of disorder, and high IQ) may predict a good response to methylphenidate on global rating scales (Buitelaar et al., 1995), most research shows that no clinically useful neurological, physiological, or psychological measures of functioning have been identified that are reliable predictors of response to psychostimulants (Pelham and Milich, 1991; Zametkin & Rapoport, 1987). Using the Irvine Laboratory Classroom, Swanson and colleagues have been able to measure academic performance during stimulant studies. The carefully timed laboratory classroom variables—which use simple addition problems—are meant to serve as surrogate measures for the broader attentional and learning problems seen in the standard weekday community classroom (Denney & Rapport, 1999).

Tolerance

Other than one study (Gillberg et al., 1997), the treatment literature does not clearly support the maintenance of positive effects from stimulants lasting after they have been discontinued. In addition, concern has been raised that children can become tolerant over time even when taking full doses of medications. Even though the mild elevations of blood pressure and pulse from stimulants disappear within a month (Safer, 1991), the beneficial effects on behavior have been thought to be stable over years of therapy(Safer & Allen, 1989). However, children may require slight dosage increases over time to maintain clinical efficacy (Zametkin & Rapoport, 1986).

However, total loss of stimulant effect has been more controversial. The evidence for "loss of stimulant effect" comes from a handful of studies carried out before 1983 (Charles et al., 1979; Heussy et al., 1973; Mendelson et al., 1971b; Riddle et al., 1976; Satterfield, 1982). These six studies followed up children from 2 to 10 years. Attrition rates were high (30%–50%), except for one (Riddle & Rapoport, 1976) 2-year study. Only a small percentage of the children in the study (12% to 28%) remained on medication for the follow-up period. Those on medication showed no advantage over unmedicated subjects in emotional adjustment, delinquency, academic failure, ratings of hyperactivity, and family ratings.

The methodology of these early studies has been criticized (Jacobvitz et al., 1990) for the lack of drug compliance measures, the failure to randomly assign children to drug and no-drug conditions, the high rates of subject attrition, and the failure to include relevant control groups (Jacobvitz et al., 1990). In addition, the studies did not use normed outcome measures, but relied on teacher or parent global measures of academic performance or antisocial behavior.

Although methodological problems in the studies of children treated with MPH over months and years make it difficult to discover the reasons for their poor chronic academic performance, how would they appear if they had become "tolerant" or unresponsive to the stimulants? Tolerance is a phenomenon that can be modeled as a shift to the right in the dose-response curve (Pearl & Seiden, 1976). Tolerance studies have been performed in animals, suggesting that tolerance to stimulants generally develops over a much shorter time frame—between 6 and 8 weeks.

Animal tolerance studies utilize high methylphenidate doses, often 10 to 20 mg/kg dose, and the route of administration is typically intraperitoneal, or intravenous. The effect studied most often is anorexia, measured by amount of milk consumed under different experimental conditions. Rats treated chronically with high intraperitoneal (i.p.) doses of MPH, ranging from 15 mg/kg (Greenshaw et al., 1989) to 20 mg/kg (Pearl & Seiden, 1976), became tolerant to the drug's suppressive effect on appetite within 8 weeks.

Some, but not all, psychostimulant effects show tolerance. These include the loss of anorectic, hyperthermic, or cardiovascular effects in animals, over intervals that range from 1 week to 8 weeks (Cox, 1990, Greenshaw et al., 1989). In human studies, tolerance develops to dextroamphetamine's hyperthermic, appetite suppressant, mood elevating, and cardiovascular effects. Similarly, drug-related increases in the heart rate (Safer, 1991) and in growth hormone (Weizman et al., 1989) of acutely MPH-treated ADHD children disappear within a few weeks, the same time frame for psychostimulant tolerance in laboratory animals (Safer, 1991).

Other psychostimulant effects do not follow the several-weeks tolerance model. Adults with narcolepsy can remain on the same dose of psychostimulant for years, even though they have become tolerant to the appetite-suppressant effects (Gittelman-Klein, 1980a).

Reviews of the MPH treatment literature have clarified whether tolerance to the ADHD-symptom reduction effect of stimulants is a common phenomenon. Although Cantwell (Cantwell & Carlson, 1978) mentioned that tolerance to MPH is seen in clinical practice, no long-term treatment study has proved its existence (Barkley, 1982, 1977; DuPaul and Barkley, 1990; Gittelman-Klein, 1980b). One investigator reported that approximately 40% of their patients needed a small increase in MPH dose during the first year to maintain the drug's efficacy. Their careful follow-up studies (Satterfield et al., 1979, 1980) reported no tolerance effects. In another study (Gualtieri et al., 1981), ADHD children were maintained on fixed doses for 23 months with no loss of efficacy or change in plasma level concentrations (Gualtieri et al., 1981). Tolerance developed when 22 ADD girls and 86 ADD boys were treated for an average of 3.8 years; it was found that seven patients became nonresponders to MPH (Safer & Allen, 1989). Four turned out to be poor at taking their medications regularly; only 3 of 108 students might have developed tolerance. However, this study was retrospective, did not include MPH plasma level measures, and had a high attrition rate.

In other instances, apparent loss of stimulant effect may have been confounded with other changes in the child's life, such as the end of the school year (Kupietz et al., 1982).

Tachyphylaxis, which is the loss of drug effect within the first few doses on the same day, has been reported in a double-blind, stimulant study of children randomized to placebo, regular twice-daily MPH dosing, a flat MPH dosing, or ascending drug administration regimen (Swanson et al., 1999). The study revealed that an ascending MPH dosing profile was equivalent to twice-daily dosing, proving that an initial bolus was not needed for prolonged reduction of ADHD symptoms. Moreover, when the twice-daily doses were given close together, improvements in activity and attention were reduced, suggesting a tolerance effect. Although the authors of this study believe that the standard use of immediate-release MPH—with doses spread apart by 4 hours—does not induce tolerance, the widely observed ineffectiveness of long-duration preparations might be the result of a tachyphylactic effect.

Sensitization

Behavioral sensitization refers to the process whereby repeated, intermittent stimulant administration by a parenteral route produces a time-dependent, progressively greater or more rapid behavioral response to the same dose (Kalivas & Stewart, 1991). The same dose given repeatedly produces a larger effect. Sensitization has been most often invoked in adult addicts who abuse methamphetamine, cocaine, and 3,4-methylene-dioxymethamphetamine (MDMA) (Pierce, 1977), although progressive behavioral responses consistent with a sensitization responses were observed in a placebo-controlled study of 11 healthy adult volunteers taking two amphetamine doses (0.25 mg/kg) 48 hours apart (Strakowski et al., 1996). Subjects exhibited significant increases in ratings of elevated mood, in energy level, and drug liking compared to placebo. This human study—recently replicated—in normal volunteer adults as well as the vast animal literature on sensitization raises the question of whether sensitization may occur in children who are chronically treated with stimulants.

A voluminous literature in rodents has demonstrated that sensitization is a complex process affected by many factors including "route of administration, dose, duration of treatment, environmental context, conditioning, genetics, the nature of the parameter measured and its contingency to the experimental treatment" among others. Sensitization has not been detected in treatment trials of ADHD or in clinical experience within the time frames of days, weeks, or months, with motoric effects, nor with tics.

The concern about sensitization focuses on the question of whether chronic stimulant usage increases the risk of children with ADHD turning to illicit substances, particularly cocaine, in later adolescence. A naturalistic nonrandomized study has suggested that that long-term stimulant treatment may have enhanced the risks of substance abuse, especially for nicotine and cocaine (Lambert, 1988; Lambert & Hartsough, 1998). Contrary evidence has been found in a prospective longitudinal follow-up study in which pharmacotherapy was associated with an 85% reduction in risk for substance use disorders.

The association between ADHD and substance use has been shown repeatedly (Barkley et al., 1990a; Blouin et al., 1978; Gittelman et al., 1985; Milberger et al., 1997; Weiss & Hechtman, 1993) underlining the importance of substance use/abuse as a potential outcome for children with ADHD. This overlap between

ADHD and substance abuse risk factors points to possible mechanisms that may explain links between childhood ADHD and subsequent substance abuse. For example, previous studies of substance use/abuse (SA) by adolescents with ADHD histories have not systematically examined the potential adverse effects of psychostimulant treatment. Yet the recent literature concerning CNS effects of drugs of abuse (e.g., psychostimulants) suggests that the acute and long-term neural mechanisms involved in addiction and therapeutic response are similar (Hyman & Nestler, 1996), raising the possibility that treatment with therapeutic doses of CNS stimulants in ADHD individuals might increase the propensity to become addicted to drugs. Although several studies suggest that adolescents who use or used stimulant medications *are not* more likely to use other drugs (Beck et al., 1975; Henker et al., 1981), and that poor responders or nontreated adolescents are at increased risk for substance use (Blouin et al., 1978; Loney et al., 1981; Milberger et al., 1997), studies of stimulant medication effects on SA have not reliably and prospectively measured use of and responsivity to stimulant medications.

The rapid increase in use of psychostimulants has raised concerns at the U.S. Drug Enforcement Administration (DEA), which regulates their production, about the risk of diversion and subsequent abuse. While there might be an increase in drug diversion, prescription for profit, or abuse, the small changes in the reported rates of stimulant abuse do not begin to account for the increase in production quotas. National surveys indicate that snorting ground-up MPH tablets does occur among high school seniors, although this occurs far less frequently than marijuana or cocaine use; however, the lifetime nonmedical use has remained constant at 1% for years (Goldman et al., 1998). Analyses of annual school surveys of drug use and the Drug Abuse Warning network data on emergency room visit monitoring have not suggested growing abuse of MPH (Goldman et al., 1998). Reports of MPH abuse are limited to poly-substance-abusing adults who use intravenous methods (Haglund & Howerton, 1982). Because injectable MPH is not available, some addicts have attempted to solubilize the tablets before injecting them. Talc granulomatosis results, with severe pulmonary hypertension or ocular lesions developing in some patients.

Time-Response Effects

The short half-life of stimulant medications means that they must be administered several times a day for sustained benefit throughout the school schedule. Even so, their action can vary considerably over the schoolday, producing clinically meaningful time-response effects, as described more completely in Chapter 11 by Swanson and Volkow. Standard administration times for the immediate-release stimulant preparations are after breakfast and before school (8 a.m.); just after lunch (12 noon); and just before homework (about 3:30 p.m.). Dosing three times a day was used in the NIMH MTA Study, and has been shown not to increase adverse event rates having to do with appetite or with sleep (Kent et al., 1995). Standard doses and available preparations of the psychostimulants are shown in Table 2–2.

Table 2–2. Stimulant Drugs, Doses, and Pharmacodynamics

Variable	D-amphetamine (3 doses per day)	Adderall (2 doses per day)	Methylphenidate (3 doses per day)	Pemoline (2 doses per day)
Dosages available in brand medications	5 mg, 10 mg tablets 5, 10, 15 mg Spansule	5 mg, 10 mg, 20 mg, 30 mg capsules	5, 10, 20 mg tablets 10, 20 mg SR	18.75, 37.5, 75 mg tablet; 37.5 chewable
Package insert dose range	10–40 mg/day in split doses (3)	10–40 mg/day in split doses (3)	10–60 mg/day in split doses (3)	37.75–112.5 mg/day (2)
Administration	BID or TID	BID	TID	qAM or BID
Maximum effect	1–3 hours	1–3 hours	1–3 hours	2–4 hours
Length of effect	3–5 hours	3–5 hours	2–4 hours	7 hours

Pemoline effects on cognitive processing begins within the first 2 hours after administration. Unlike methylphenidate, the effects last up to 7 hours. The therapeutic effects of MPH and dextro-amphetamine are confined to the absorption phase, whereas pemoline has significant postabsorptive effects lasting into the post-distribution phase (Sallee et al., 1992). Disproving previous clinical suggestions that pemoline requires 3 to 6 weeks to work (Page et al., 1974), pemoline has been shown to be effective after the first dose (Collier et al., 1985, Pelham et al., 1995).

Group studies have consistently shown linear dose-response curves for the psychostimulants for their peak effects, although individual dose-response curves may differ across domains, exhibiting curvilinear or quadratic effects. Although an 1977 study suggested that optimal effects on cognitive functioning may be predictable using a weight-adjusted dose of 0.3 mg/kg (Sprague & Sleator, 1977), more recent work shows no consistent relationships between weight-adjusted doses and most response domains (Rapport et al., 1989b).

Do Long-Duration Stimulants Overcome Time-Response Limitations?

The main limitation of stimulant medications—their very short duration of action—has spurred the development of sustained-release preparations. Sustained-release preparations have been marketed for DEX and for MPH (Whitehouse et al., 1980). MPH's sustained-release preparation is available in 10 and 20 mg strengths, and DEX Spansules in 5, 10, and 15 mg strengths. Ritalin-SR and Metadate-ER use a wax-matrix vehicle for slow release; the dextroamphetamine Spansule is a capsule containing small medication particles. Both deliver a plasma bolus of the stimulant within hours of ingestion.

The efficacy of the long-acting preparations is controversial. A controlled study of nine children with ADHD reported that DEX Spansules and tablets produced identical duration of behavioral effects, despite the Spansule's high plasma levels over 12 hours (Brown et al., 1980). One center published an initial report that

MPH-SR20 was judged to be less effective than immediate-release MPH (MPH-IR) 10 mg BID for the treatment of ADHD children (N = 13) in a summer program (Pelham et al., 1989). In a later study, the same investigators reported that MPH-SR20, PEM and DEX Spansules were equally effective on a laboratory task of attention over a 9-hour period (Pelham et al., 1990b). In a 6-week, double-blind, crossover study of 34 children with ADHD, no differences in teacher's ratings or on performance on a CPT were detected 6 hours after ingestion of a single SR20 tablet or 10 mg of standard MPH, 8 a.m. and noon, yet all the children stopped responding to the MPH-SR within 6 months (Fried et al., 1987).

Clinicians in office practice find MPH-SR less effective, and combine it with MPH-IR, even though this combination was not found to be more effective than MPH-SR alone (Fitzpatrick et al., 1992). However, other controlled studies, carried out in standard clinic settings, also found that parents and teachers gave equal global ratings for MPH-SR and MPH-IR. A group of 42 children with ADHD were studied in a parallel-design, double-blind 8-week study comparing MPH 10mg BID, MPH SR-20 and placebo (Greenhill et al., 1994). Although there were differences in Cmax between MPH SR-20 and MPH 10 mg BID, no differences were seen in CPT performance, motor activity levels or teacher's reports for the two groups. Although 20 subjects per group may be sufficient to detect differences between a stimulant and placebo, the study lacked sufficient power to detect differences between the two forms of MPH. Clinicians, on the other hand, have found long-duration MPH-SR less effective over time than MPH-IR. This may explain why MPH-SR holds only a 14% market share of all stimulants sold. These reports are summarized in Table 2–3.

In response to clinician dissatisfaction with currently available long-duration preparations, particularly of MPH, several new preparations have been developed for testing. Alza Pharmaceuticals has announced the development of a new 12-hour, single dosing MPH preparation with a patterned delivery profile. Based on a proof-of-concept study, a caplet was designed that releases MPH at a predetermined rate that yields a gradually increasing plasma level concentration over a 12-hour period. Several other pharmaceutical companies are also designing full-day, single-dose delivery systems. At this time, however, only the Alza drug has obtained Food and Drug Administration approval for the treatment of ADHD.

Safety of Stimulants Used in Treatment of ADHD: Toxicology

Animal studies suggest that stimulants—particularly MPH—are among the safest medications used in the treatment of children, when given in the standard 0.3 to 1.2 mg/kg/day oral dose range. A 100 to 1 margin of safety exists between a single dose approximating a human clinical dose and one that produces lethality in two other animal species (Diener, 1991). In the animal lab, the median lethal dose (LD_{50}) for MPH is 48.3 mg/kg by intravenous route and 367 mg/kg by oral route. The LD_{50} for amphetamine in rats is 55 mg/kg. Ninety-day subchronic toxicity studies of MPH

Table 2–3. Sustained-Release Stimulant Preparations: Efficacy

Principal Investigator (year)	Number	Age (yr)	Long-Acting Drugs	Study Design	Daily Dose	Efficacy?
Brown (1988)	9	5–12	DEX Span	Crossover	0.5 mg/kg	Duration of action <3 hr
Whitehouse (1980)	30	5–12	MPH-SR	Parallel MPH-SR20 vs. MPH	MPH 10mg BID MPH-SR 20qAM	MPH 10 BID equivalent to MPH-SR 20
Pelham (1989)	13	7–11	MPH-SR	Crossover MPH-SR20 vs. placebo	MPH 10mg BID MPH-SR 20qAM	MPH-SR had delayed onset of action MPH 10 BID preferred by expert panel
Greenhill (1987)	38	7–12	MPH-SR	Crossover MPHSR20 vs. MPH vs. Placebo	MPH 10mg BID MPH-SR 20qAM	MPH 10 BID equal to MPH-SR 20 qAM
Birmaher (1989)	17	8–12	MPH-SR	Open study	MPH-SR20 MPH 1 mg/kg	MPH 10 BID gives high plasma level concentration >MPH-SR 20
Pelham (1990b)	22	8–13	MPH-SR,DEX PEM	Crossover Dex vs. MPH-SR PEM vs. MPH vs. Placebo	MPH-SR20 MPH 10 mg BID DEX Span 10 Pemoline 56.5	All sustained-release drugs lasted significantly longer than MPH and all more effective > PB
Greenhill (1991)	42	8–12	MPH-SR20	Parallel MPH-SR20 vs. Vs. MPH 10 mg BID vs. Placebo	MPH-SR20, 40 MPH 10 mg BID	MPH 10 BID equal to MPH SR20 qAM at 1 week and at 8 weeks
Fitzpatrick (1992)	19	6.8–11.5	MPH-SR20	Crossover MPH-SR20 vs. vs. MPH 10 BID vs. Combined SR+ Regular	MPH-SR20 MPH 10 mg BID Combination	All stimulants equally effective All better than PB
TOTAL	190					MPH–SR20 = MPH

Abbreviations: MPH = methylphenidate; DEX = dextroamphetamine; PEM = pemoline.

in rats (with doses up to 120 mg/kg/day) produced decreases in body weight, but no signs of growth inhibition, reproductive problems or carcinogenicity. A 120-day study of beagles treated with 10 mg/kg/day (10 times human dose) of MPH produced hyperactivity and hyperexcitability in the dogs, but there was no appetite suppression, growth suppression, convulsions, or changes in liver tissue.

Stimulant Adverse Events

Stimulant side effects are dose-dependent and range from mild to moderate in most children. Some stimulant side effects, such as headaches and delay in sleep onset, may be ascribed to adrenergic agonist activity; the anorexia associated with these drugs may be attributed to dopamine against activity in the hypothalamus. Based on 110 studies involving over 4200 children (Barkley, 1977; Barkley et al., 1984, 1990b) common adverse effects include insomnia, anorexia, irritability, weight loss, abdominal pain, and headaches. When side effect reports have been gathered on ADHD children taking placebo (DuPaul et al., 1990), anxiety, staring, disinterest, and sadness were reported; only parent reports of decreased appetite, stomach aches, and insomnia differentiated ADHD children on placebo from those on medication. Over several treatment studies, the percent of children that had to be taken off stimulants varies from 4% (Barkley et al., 1990b) to 10% (Schachar et al., 1997).

Staring, daydreaming, and irritability *decrease* with increasing stimulant dose, suggesting that these are symptoms of ADHD rather than adverse effects associated with taking the medication. No consistent relationship has been found in short-term controlled studies between stimulant dose and the less-frequently reported adverse events—behavioral rebound, motor tics, compulsive picking of nose or skin, emotional or cognitive constriction, and growth delays (Spencer et al., 1996a). One 4-month study reported persistent and worsening side effects—overfocusing and affective symptoms—that eventually led to stimulant discontinuation in 10% of those children who appeared to tolerate the drug during the first month of treatment, contradicting somewhat the popular notion that stimulant-associated adverse effects fade with time (Schachar et al., 1997).

The appearance of minor facial tics is a frequent reason for stopping stimulants in clinical practice, but a causal link between stimulant treatment and the continuation of these motor signs has not been established. A past history of involuntary muscle movements (tics) or a family history of Tourette disorder has become a contraindication to the use of methylphenidate as it may "unmask" or exacerbate Tourette's syndrome (Lowe et al., 1982). Switching to another stimulant (El-Defrawi & Greenhill, 1984), such as pemoline, may relieve or worsen the tic disorder. It is best to distinguish between minor motor tics that do not impair socially or academically—these do not require a change of medication—and other tics clearly affecting the child, either through subjective discomfort or through "cosmetic" factors. For these more serious tics, it is best to discontinue all psychostimulants and consider a trial of clonidine, guanfacine, or a tricyclic antidepressant.

In the MTA study, 289 patients were randomized to drug treatments for 14 months, and side effects were managed by reducing stimulants or switching to other psychotropics (Vitiello et al., in press). Approximately 85% were treated with stimulant medication, either MPH or DEX. At study end, 88 (35.9%) of the 245 families reporting noted that their children had no side effects, 122 (49.8%) had mild side effects, 28 (11.4%) had moderate side effects, and 7 (2.9%) had severe side effects. Because six of the rating scale side effects (measuring depression, worrying, or irritability) might have been associated with nonmedication factors, these data may overestimate the actual incidence of stimulant side effects.

Acute Withdrawal: "Rebound"

One adverse effect, commonly known as "behavioral rebound," appears when children experience psychostimulant withdrawal at the end of the school day. Children present with afternoon irritability, excessive talking, noncompliance, excitability, motor hyperactivity, and insomnia some 5 to 15 hours after the last dose. These behavioral signs are exacerbations of the presenting complaints. In a study involving 21 hyperkinetic boys (Porrino, et al., 1983) treated with DEX, three-quarters of the group on stimulant and none on placebo experienced rebound objectively quantified with actometers. A second study using a controlled crossover design of 21 ADHD boys taking MPH, at doses of 0.6 and 1.2 mg/kg/day or placebo, found no evidence of afternoon worsening based on parent ratings. It is not clear at this time if these contradictory reports represent differences between dextroamphetamine and MPH, and it is not known whether longer-acting stimulant preparations may be less likely to produce rebound. In any case, the controlled drug treatment literature does not report rebound effects, but this may be the result of the use of global parent rating forms that are not time-of-day specific. When symptoms consistent with rebound are reported, many clinicians add a small afternoon MPH dose. Treatment of rebound has involved adding a third, mid-afternoon dose of MPH added to the usual twice-daily regimen. One study of 10 boys with ADHD treated with three-times-daily versus twice-daily MPH dosing reported no additional sleep problems after the third daily dose (Kent et al., 1995). However, this was a small study in which sleep latency for all dose conditions averaged 45 minutes.

Do Psychostimulants Cause Cognitive Constriction?

The question of cognitive overfocusing has been a concern of clinicians, parents, and teachers, who may find a psychostimulant-treated child more perseverative and repetitive, and less spontaneous. This effect is a dose-related but subtle adverse reaction, which becomes more obvious at high doses of psychostimulants. As a result, MPH would be expected to interfere with cognitive tasks that require changes of set or divergent thinking. A team of investigators were unable to reproduce this cognitive constriction on the Wallach-Kogan test in which subjects are asked to name as many "alternate" uses an object can be put to (Solanto, 1989). MPH did not produce cognitive constriction relative to placebo in 19 ADHD males treated with three different doses of MPH (0.3, 0.6, and 1.0 mg/kg). Another group replicated these findings, finding that attention was enhanced by MPH, with no evidence for overfocusing even at higher dosages (Douglas et al., 1986).

Rare Side Effects

Psychosis

Stimulant-related toxic psychosis is a *rare* phenomenon in children, having been reported in less than 30 cases (DuPaul et al., 1990), including methylphenidate-induced mania or MPH-induced delusional disorders. Psychosis related to stimulant intake is well known in the adult literature, but only at higher-than-therapeutic doses of amphetamines, such as 300 mg of dextroamphetamine per day (Robinson et al., 1991). These reliably induce brief paranoid psychoses in adult volunteers (Angrist & Gershon, 1972). Even though the psychotomimetic effects are short-lived, this side effect can be very serious, and requires immediate cessation of stimulant treatment. Psychosis is a contraindication for psychostimulant usage (Dulcan, 1990). Physicians should avoid dispensing psychostimulants to agitated children with psychotic symptomatology, and consider neuroleptics instead.

The *Physicians' Desk Reference* lists a number of other adverse reactions, without stating frequency, that are not found in standard clinical practice or in published reports from treatment studies. These *rare* side effects of stimulants include alopecia and leukocytosis. The alopecia on stimulants is reversible and temporary, having to do with too many follicles going into dormant phase in synchrony (Castellanos, personal communication). There is no utility in performing screening blood counts, as the chance of detecting such rare leukocytosis on a random yearly CBC are impossibly low. Other rare adverse affects involve skin hypersensitivity (skin rash, urticaria, fever, arthralgia, exfoliative dermatitis, erythema multiforme with histopathological findings of necrotizing vasculitis, and thrombocytopenic purpura), angina, and cardiac arrhythmia.

Hepatotoxicity

Altered liver function tests and fulminant hepatotoxicity have been reported in 44 children treated with pemoline (Berkovitch et al., 1995), which is 4 to 17 times the expected rate, leading to a warning inserted in the drug's package insert. The manufacturer now recommends that pemoline not be used as a first-line medication to treat ADHD, and that biweekly liver function tests be obtained indefinitely from all patients maintained on the medication.

Hepatoblastomas in mice treated for 2 years with oral doses of 5 to 67 mg/kg of MPH (Dunnick & Hailey, 1995) have been reported. The incidence of these rare tumors was only significantly increased at the highest dose tested, and not in F344/N rats. Other studies show the same species-specific hepatic vulnerability to high MPH doses, particularly when given together with beta-adrenergic agonist drugs (Roberts et al., 1994). Hepatic tumors are species-specific and the MPH dose implicated in tumors is far higher than ever used in treatment. Furthermore, hepatic tumors have not been reported in children with ADHD treated with MPH, and they are exceedingly rare in preschool and school-aged children.

Seizure Disorders

A commonly held notion is that stimulants lower the seizure threshold. This may be related to toxicology studies carried out in animals. Convulsions are one of the terminal signs in animals treated with 300 mg/kg doses for the determination of psychostimulant lethality. The *Physicians' Desk Reference* repeats this notion of decreased seizure threshold in warning against the use of stimulants in children with pre-existing seizure disorders (Arky, 1994). ADHD and seizure disorders are both quite prevalent and may occur in the same individuals (Wilens et al., 1992), so the question of the risk of stimulant treatment may arise frequently. A naturalistic study suggests that stimulant treatment of ADHD children maintained on effective anticonvulsant doses produces no increase in seizure frequency, changes in the EEG waveforms, or difficulty in regulating anticonvulsant blood levels (Feldman et al., 1989). Current practice is to place ADHD children with epilepsy on a combination of anticonvulsant and MPH. Plasma levels of the anticonvulsant should be monitored to avoid toxicity resulting from MPH's competitive inhibition of metabolic pathways, boosting the plasma levels of both drugs. At the doses used to treat ADHD children, the psychostimulants have minimal effects on the seizure threshold.

Effects on Growth

Growth delays, particularly failure to attain weight at an expected rate during development (Gillberg et al., 1997), have been cited as a possible long-term risk of stimulant treatment. A later single-observation, cross-sectional study of adolescents with ADHD suggested that growth curves of untreated ADHD children may show significantly slowed growth in early years, and late catch-up, compared with non-affected children (Spencer et al., 1996b). These effects were thought to respond to short periods off drug ("drug holidays"), as one study with several hundred children showed less slow-down in rate of weight acquisition because children were taken off stimulants during summer vacation (Gittelman-Klein et al., 1988). This same group of investigators reported that prospective follow-up of ADHD children into adult life (Gittelman & Mannuzza, 1988) revealed no significant impairment of height attained. Long-term studies published or in press after 1996 showed decrements in weight gain during treatment up to 14 months of MPH treatment compared to those on nondrug treatments (Schachar et al., 1997; Greenhill & MTA Cooperative Group, 1999).

Long-Term Outcome: Randomized Controlled Trials of Stimulant Medication

Although short-duration stimulant studies have shown robust efficacy, their effects last only as long as the patient continues medication. ADHD is a chronic disorder lasting many years, and thus maintenance treatment constitutes the main component of care. Unfortunately, most stimulant RCTs last less than 3 months, with only 22 published studies lasting longer (Schachar & Tannock, 1993). Until recently, most of these long-duration studies were severely constrained by their retrospective methods, lack of nonstandard outcome measures, restrictive inclusion criteria that rejected patients with comorbid disorders, irregular prescribing patterns (Sherman, 1991), and the lack of compliance measures (Richters et al., 1995). More recently, long-duration trials have shown maintenance of stimulant medication effects over periods ranging from 12 months (Gillberg et al., 1997) to 24 months (Abikoff & Hechtman, 1998).

The NIMH Multimodal Treatment Study of Children with Attention Deficit Hyperactivity Disorder (MTA Study), was established to compare the relative effectiveness of four randomly assigned, 14-month treatments of children with ADHD: medication management (Medmgt), behavioral treatment (Beh), the combination (Comb), or a community comparison group (CC) getting treatment as usual (Arnold et al., 1997). The study hypotheses included: (*1*) Do Beh and MedMgt treatments result in comparable levels of improvement at study end? (*2*) Do participants assigned to Comb show greater improvement at study end than those assigned to either MedMgt or to Beh? (*3*) Do participants assigned to any of the 3 MTA intensive treatments show greater improvement at study end than those assigned to CC?

To have sufficient power to address these questions, 579 children with ADHD, Combined Subtype, aged 7.0 to 9.0 years, were recruited at multiple sites and assessed across six domains at four different points in time (Hinshaw et al., 1997). They were assigned to 14 months of (a) medication management, (b) intensive behavioral treatment (Beh—parent, school, and child components, with therapist involvement gradually faded over time), (c) the two combined, or (d) treatments by community providers). Outcomes were assessed in multiple domains before, during, and at treatment endpoint (with Comb and MedMgt subjects on medication at time of assessment).

Children randomized to medication management alone (MedMgt, n = 144) or to Combined Treatment (Comb, n = 145) in the MTA Study participated in a 28-day double-blind, placebo-controlled, titration study in order to identify each child's optimal dose of MPH. Three doses of methylphenidate and placebo were used in a three-times daily protocol (Greenhill et al., 1996). Medication conditions were switched each day, creating five repeats of the four drug states, to reduce error variance. Parents and teachers made daily ratings of ADHD symptoms and impairment.

During titration, a repeated-measures ANOVA revealed a main effect of MPH/placebo dose with greater effects on teacher's ratings—$F(3) = 100.6$, $n = 223$, $p = 0.0001$; effect sizes 0.8 to 1.3—than on parent's ratings—$F(3) = 55.61$, $n = 253$, $p = 0.00001$; effect sizes 0.4 to 0.6. Dose did not interact with between-subjects factors (period, dose order, comorbid diagnosis, site or treatment group). Drug-related adverse events were reported more often by parents than by teachers, who rated irritability highest on placebo. The distribution of the best MPH starting doses determined during titration (10–50 mg/day), response rate (77%), and adverse event profile across six sites and multiple subgroups suggest that MPH titration in office practice should explore MPH's full dosage range in 7- to 10-year-old children over 25 kg in weight. Parent weekend symptom ratings showed daytime drug effects, but not as clearly as weekday after-school ratings. Nonresponders to MPH were older, had less-severe symptoms, and were medication-naïve. The MTA controlled multi-site titration trial using daily dose switches replicated the published, single-site MPH trial effect sizes and adverse event rates (Spencer et al., 1996a).

Outcomes were analyzed through intent-to-treat, random-effects regression. When analyzed, all four groups showed sizeable reductions in symptoms over time, with significant differences among them in degrees of change. For most ADHD symptoms, children in the combined and medication management groups showed significantly greater improvement than those given intensive behavioral treatment or community care. Combined and medication management treatments did not differ significantly on any direct comparisons, but in several instances (internalizing symptoms, teacher-rated social skills, parent–child relations, and reading achievement) combined treatment proved significantly better at reducing symptom scores, compared with intensive behavioral treatment and/or community care, whereas medication management did not. Study medication strategies were superior to community care, despite the fact that two-thirds of community-treated subjects received medication during the study period. For ADHD symptoms, the MTA's medication management method was superior to behavioral treatment and to routine community care that included medication. The combined treatment did not yield significantly greater benefits than medication management for core ADHD symptoms, but may have provided modest additional advantages for non-ADHD-symptom and positive functioning outcomes.

The MTA is the largest and most methodologically sophisticated randomized multisite treatment study to date in ADHD children that includes mono-modal and combined treatments. These therapeutic methods that had been shown previously to be effective in simpler two-arm (active vs. placebo) controlled studies (Pelham & Murphy, 1986; Pelham, 1989). Because of its size, the MTA was able to address certain treatment questions. First, it shows that stimulant medication efficacy can be realized across diverse settings and patient groups, and can be maintained during chronic therapy lasting more than a year. The MTA's findings replicate those from smaller long-duration stimulant trials—the 102-child New York / Montreal study (Abikoff et al., 1998), the 91-child Toronto study (Schachar et al., 1997), and the 62-child Gillberg (Gillberg et al., 1997) studies (Table 2–4).

Collectively, these studies show a persistence of medication effects over time. Within-subject effect sizes reported after 12 to 24 months of MPH treatment resembled those previously reported in short duration studies (Elia et al., 1991; Thurber & Walker, 1983). Domain of greatest improvement differs among studies, with some (Gillberg et al., 1997) showing greater effects at home and another (Schachar

Table 2–4. Long Duration Stimulant Clinical Trials in Medication-Only Treatment Arm: Within-Subjects Effects

Study	Med-Only number (Study Total)	Design (Drug)	Duration Months	Compliance	Total mg Daily Dose (schedule)	Measure	Effect Size at Study End in SD*
Abikoff (1998)	33 (103)	Parallel RCT (MPH)	24	Saliva Levels	33.7 (TID)	Teacher CTRS	2.7
Gillberg (1997)	56 (62)	Double-blind Discontinuation RCT (DEX)	15	N/A	17 (BID) Amphetamine	Teacher CTRS Hyperactivity	0.27–0.40
Schachar (1997)	24 (91)	Parallel RCT (MPH)	12	Pill counts	33.5 (BID)	Telephone Interview Probe	0.7
MTA (1999)	133 (576)	Parallel RCT (MPH)	14	Saliva Levels pill counts	38.7 mg (TID)	SNAP Teacher Hyperactivity	1.5–1.8

SD = Standard deviations.

et al., 1997) showing greater effects at school. The total mean MPH daily doses reported by three long-duration studies ranged between 33 and 37.5 mg. The one DEX study reported a mean dose one and one half of this level, agreeing with the 1:2 mg ratio of DEX to MPH doses. Family-initiated treatment discontinuation was associated with persistent stimulant drug side effects or assignment to placebo treatment. Fortunately, attrition from placebo assignment was slow, allowing ample time for efficacy trials to be completed.

Limitations of Available Stimulants in Clinical Practice

Even with their robust efficacy and strong safety records, stimulant medication treatments may not be the treatment of choice for every child with ADHD. First of all, the stimulants have important limitations imposed by their very short pharmacokinetic and pharmacodynamic half-life (approximately 3 hours). This leads to instability of drug effects, to major problems with pill-taking compliance related to missed doses, to vulnerability to peer ridicule, and to possible diversion because of problems in monitoring its use outside the home (Swanson et al., 1998b). Time-action effects can produce drug wear-off during the late morning, mid-afternoon, or evening with return of ADHD symptoms. Although a small study suggested that TID dosing produces no delay in sleep onset (Kent et al., 1995), this last stimulant dose does not cover the evening homework period.

The long-duration stimulant preparations developed to counter these problems have not been widely accepted by clinicians (Greenhill et al., 1997). The efficacy of combination immediate and sustained-release stimulant preparations, though very popular in practice, has been tested in only one controlled study, which did not detect a difference from monotherapy (Fitzpatrick et al., 1992). Longer-acting pemoline treatment is not a realistic option because of concerns about hepatotoxicity appearing late during chronic treatment except in cases of severe comorbid stimulant abuse.

Second, not all children respond to the first stimulant given. Twenty percent of the 258 children titrated with MPH in the MTA Study were not assigned to MPH for treatment by blind raters on the basis of dose-response curves. Judging the dose-response curves, the raters assigned 9% to amphetamine because the responses to MPH left room for improvement. Another 11% of the dose-response curves showed equal responses to MPH and to placebo (Greenhill et al., in press). This latter group were followed on no medication during the 13-month maintenance period. Over 90% of that group showed a return of moderate to severe ADHD symptoms within 3 months of follow-up, so were restarted on MPH (Vitiello et al., 1999).

Third, for some children, stimulant side effects—including anorexia, weight loss, headaches, insomnia, and tics—may not improve over time as they do in animal tolerance experiments (Schachar et al., 1997). Fourth, many families will not enter their children into treatment programs using MPH and DEX because they are classified as schedule II medications with "high abuse potential with severe psychic or physical dependence liability." Fifth, these drugs show differential responses across settings and domains in the same child (Pelham & Milich, 1991), made all the more problematic because reliable predictors of drug response are not available (Jacobvitz et al., 1990).

Directions for Future Clinical Research

The NIMH Consensus Development Conference laid out a clear path for future research in stimulant treatments (NIH Consensus Statement, 1998). Future research needs to be directed towards developing more systematized treatment strategies. This includes studies of children with ADHD, predominately inattentive type, particularly because this group is seen more often in primary care, and might include higher proportions of girls than currently have been included in studies. Much longer-term prospective studies are needed to determine the long-term benefits and risks associated with stimulant treatment.

Also needed are studies to determine if combination of stimulants and behavioral modification can treat ADHD symptoms with reduced doses of stimulants. In the MTA Study, both the MedMgt and the Comb treatment showed equal efficacy in reducing ADHD symptoms, but the Comb group had a significantly lower end-of-treatment MPH total daily dose. This finding also needs to be corroborated in a controlled dismantling design. Because of increased usage in preschool children, studies of the safety and efficacy of methylphenidate in ADHD children under the age of 6 are needed.

As noted above, the MTA MedMgt medication strategy reduced the core symptoms of ADHD as effectively as any treatment used in the MTA Study. In fact, children randomized to the MTA MedMgt treatment arm did far better over the 14 months of treatment than did those ADHD children who were randomized to treatment as usual in the community where they received the same medications (68% of the CC group). This outcome suggests that the MTA medication strategy is more effective than community standard medication care. This finding needs to be examined in a prospective, controlled manner through the use of dismantling designs. The elements of MedMgt—intensive titration, three times daily stimulant dosing, monthly maintenance medication visits using regular parent and teacher assessment, the combination of supportive advice and bibliotherapy, and the algorithms for dose adjustment and new medications—need to be evaluated to determine which are essential for successful maintenance management. The report of 2 million office visits and the prescribing of stimulants to males with oppositional defiant disorder in rural U.S. areas (Angold et al., in press) suggests that more pharmacoepidemiological surveys are called for to monitor prescribing practices for children with disruptive behavior disorders (Jensen et al., 1999a). One report found that as many as 20% of white fifth grade boys in some communities are prescribed MPH, but this figure is probably too high for most schools (LeFever et al., 1999). There is a growing public concern about the increasing numbers of children being treated with stimulants (Safer et al., 1996), so improved prescribing practices are a public health necessity.

References

Abikoff, H & Gittelman, R (1985). The normalizing effects of methylphenidate on the classroom behavior of AD/HD children. *J Abnorm Child Psychol, 13*, 33–44.

Abikoff, H & Hechtman, L (1998). Multimodal treatment for children with AD/HD: Effects on AD/HD and social behavior and diagnostic status. (unpublished).

Aman, MG (1982). Stimulant drug effects in developmental disabilities and hyperactivity—toward a resolution of disparate findings. *J Autism Dev Disord, 12*, 385–398.

Aman, MG, Marks, RE, Turbott, SH, Wilsher, CP, Merry, SN (1991). Clinical effects of methylphenidate and thioridazine in intellectually subaverage children. *J Am Acad Child Adolesc Psychiatry 30*(2), 246–256.

Angold, A and Costello, E (in press). Stimulant medication: A general population perspective. *Psychopharmacol Bull.*

Angrist, BM & Gershon, S (1972). Psychiatric sequalae of amphetamine use. In RI Shader (Ed.), *Psychiatric complications of medical drugs.* New York: Raven Press. (pp. 175–199).

Arky, A (1994). *Physicians' Desk Reference.* Montvale, NJ: Medical Economics Data Production Company, Inc.

Arnold, L, Jensen, P, Richters, J, Abikoff, H, Conners, K, Greenhill, L, Hechtman, L, Hinshaw, S, Pelham, W, & Swanson, J (1997). The National Institute of Mental Health Collaborative Multisite Multimodal Treatment Study of Children with Attention-Deficit Hyperactivity Disorder (MTA): Design challenges and choices. *Arch Gen Psychiatry, 54*, 865–870.

Barkley, R, Fischer, M, Edelbrock, C, & Smallish, L (1990a). The adolescent outcome of hyperactive children diagnosed by research criteria: an 8-year prospective follow-up. *J Am Acad Child Adolesc Psychiatry 29*, 546–557.

Barkley, R, McMurray, M, Edelbrock, C, & Robbins, K (1990b). Side effects of MPH in children with attention deficit hyperactivity disorder: A systematic placebo-controlled evaluation. *Pediatrics, 86*, 184–192.

Barkley, RA (1977). A review of stimulant drug research with hyperactive children. *J Child Psychol Psychiatry, 18*, 137–165.

Barkley, RA (1982). *Hyperactive children: A handbook for diagnosis and treatment.* New York: Guilford.

Barkley, RA (1988). The effects of methylphenidate on the interactions of preschool AD/HD children with their mothers. *J Am Acad Child Adolesc Psychiatry, 27*, 336–341.

Barkley, RA, & Cunningham, CE (1978). Do stimulant drugs improve the academic performance of hyperkinetic children?: A review of outcome studies. *Journal of Clinical Pediatrics 17*(1), 85–92.

Barkley, RA, Karlsson, J, & Strzilecki, E (1984). Effects of age and Ritalin dosage on mother-child interactions of hyperactive children. *J Consul Clin Psychol 52*,750–758.

Beck, L, Langford, W, Mackay, W, & Sum, G (1975). Childhood chemotherapy and later drug abuse and growth curve: A follow-up study of 30 adolescents. *Am J Psychiatry, 132*, 436–438.

Bergman, A, Winters, L, & Cornblatt, B (1991). Methylphenidate: effects on sustained attention. In: Greenhill LL & Osman B (Eds), *Ritalin: Theory and patient management.* New York: Mary Ann Liebert, Inc. (pp 223–232)

Berkovitch, M, Pope, E, Phillips, J, & Koren, G (1995). Pemoline-associated fulminant liver failure: testing the evidence for causation. *Clin Pharmaco Therapy, 57*, 696–698.

Biederman, J, Munir, K, Knee, D, Habelow, W, Armentano, M, Autor, S, Hoge, SK, & Waternaux, C (1986). A controlled family study of patients with attention deficit disorder and normal controls. *J Psychiatry Res, 20*(4), 263–274.

Biederman, J, Newcorn, J, & Sprich, S (1991). Comorbidity of attention deficit hyperactivity disorder with conduct, depressive, anxiety, and other disorders. *Am J Psychiatry, 148* (5), 564–577.

Birmaher, B, Quintana, H, & Greenhill, LL (1988). Methylphenidate treatment of hyperactive autistic children. *J Am Acad Child Adolesc Psychiatry, 27*(2), 248–251.

Blouin, AG, Bornstein, R, & Trites, R (1978). Teenage alcohol use among hyperactive children: A five-year follow-up study. *J Pediatr Psychol, 3*, 188–194.

Borcherding, BG, Keysor, CS, Cooper, TB, Rapoport, JL (1989). Differential effects of methylphenidate and dextroamphetamine on the motor activity level of hyperactive children. *Neuropsychopharmacology, 2*, 255–263.

Bradley, C (1937). The behavior of children receiving benzedrine. *Am J Psychiatry, 94*, 577–585.

Brown, GL, Ebert, MH, Mikkelsen, EI, Hunt, RD (1980). Behavior and motor activity response in hyperactive children and plasma amphetamine levels following a sustained release preparation. *J Am Acad Child Adolesc Psychiatry, 19*, 225–239.

Buitelaar, J, Gary, R, Swaab-Barneveld, H, & Kuiper, M (1995). Prediction of clinical response to methylphenidate in children with attention deficit hyperactivity disorder. *J Am Acad Child Adolesc Psychiatry, 34*, 1025–1032.

Campbell, SB, Endman, M, & Bernfield, G (1977). A three-year follow-up of hyperactive preschoolers into elementary school. *J Child Psychol Psychiatry, 18*, 239–249.

Cantwell, DP, & Carlson, GA (1978), *Stimulants.* In Werry, JS (Ed.), *Pediatric psychopharmacology: The use of behavior modifying drugs in children.* New York: Brunner/ Mazel.

Carlson CL, Pelham WE, Swanson JM, & Wagner JL (in press). A divided attention analysis of the effects of methylphenidate on the arithmetic performance of children with attention deficit hyperactivity disorder. *J Child Psychol Psychiatry.*

Carlson, CL, & Thomeer, ML (1991). Effects of Ritalin on arithmetic tasks. In Greenhill LL & Osman B (Eds.), *Ritalin: Theory and patient management,* New York: Mary Ann Liebert, Inc., pp 195–202.

Castellanos, X, Giedd, J, Elia, J, Marsh, W, Ritchie, G, Hamburger, S, & Rapoport, J (1997). Controlled stimulant treatment of AD/HD and comorbid Tourette's Syndrome: Effects of stimulant and dose. *J Am Acad Child Adolesc Psychiatry, 36*, 589–596.

Charles, L, Schain, RJ, & Gutherie, D (1979). Long-term use and discontinuation of methylphenidate with hyperactive children. *Dev Med Child Neurol, 21*(6), 758–764.

Cohen, DJ, & Leckman, JF (1989). Commentary. *J Am Acad Child Adolesc Psychiatry, 28*(4): 580–583.

Collier, C, Soldin, S, Swanson, J, MacLeod, S, Weinberg, F, & Rochefort J, (1985). Pemoline pharmacokinetics and long term therapy in children with attention deficit disorder and hyperactivity. *Clin Pharmacokinetics, 10*, 269–277.

Comings, DE, & Comings, BG (1987). A controlled study of Tourette's syndrome. *Am J Hum Genet 41*, 701–741.

Conners, KC (1975). Controlled trial of methylphenidate in preschool children with minimal brain dysfunction. *Int J Mental Health, 4*, 61–74.

Cornblatt, B (1988). The continuous performance test, identical pairs version (CPT-IP): I. New findings about sustained attention in normal families. *Psychiatry Res, 26*, 223–238.

Cox, BM (1990). Drug tolerance and physical dependence. In Pratt WB & Taylor P (Eds.), *Principles of drug action: The basis of pharmacology*. New York: Churchill Livingstone. (pp. 639–690)

Curry, J & Murphy, L (1995). Comorbidity of anxiety disorders. In March J (Ed.), *Anxiety disorders in children and adolescents*. New York: Guilford Press. (pp. 301–317).

Denney, C & Rapport, M (1999). Predicting methylphenidate response in children with AD/HD: Theoretical, empirical and conceptual models. *J Am Acad Child Adolesc Psychiatry, 38*, 393–401.

Diagnosis and treatment of attention deficit hyperactivity disorder (AD/HD). NIH Consensus Statement Nov 16–18; *16*(2), 1–37.

Diamond, I, Tannock, R, & Schachar, R (1999). Response to methylphenidate in children with AD/HD and comorbid anxiety. *J Am Acad Child Adolesc Psychiatry, 38*, 402–409.

Diener, R.M. (1991). Toxicology of Ritalin. In Ritalin: Theory and Patient Management. Mary Ann Liebert. Larchmont, NY pp 35–45.

Douglas, VI, Barr, RG, O'Neill, ME, & Briton, BG (1986). Short-term effects of methylphenidate on the cognitive, learning and academic performance of children with attention deficit disorder in the laboratory and the classroom. *J Child Psychol Psychiatry, 27*, 191–211.

Dulcan, M (1990). Using psychostimulants to treat behavior disorders of children and adolescents. *J Child Adolesc Psychopharmacol, 1*, 7–20.

Dunnick, J & Hailey, J (1995). Experimental studies on the long-term effects of methylphenidate hydrochloride. *Toxicology, 103*, 77–84.

DuPaul, GJ, & Barkley, RA (1990). Medication therapy. In Barkley RA (Ed). *Attention deficit hyperactivity disorder: A handbook for diagnosis and treatment.* New York: Guilford Press. 2nd ed. (pp 573–612)

El-Defrawi, MH & Greenhill, L (1984). Substituting stimulants in treating behavior disorders. *Am J Psychiatry, 141*(4), 610.

Elia, J, Borcherding, B, Rapoport, J, & Keysor, C (1991). Methylphenidate and dextroamphetamine treatments of hyperactivity: Are there true non-responders? *Psychiatry Res, 36*, 141–155.

Erenberg, G, Cruse, RP, Rothner, AD (1985). Gilles del la Tourette's syndrome: Effects of stimulant drugs. *Neurology, 35*, 1346–1348.

Feldman, H, Crumrine, P, & Handen, BL (1989). Methylphenidate in children with seizures and attention-deficit disorder. *Am J Dis Child, 143*, 1081–1086.

Fitzpatrick, P, Klorman, R, Brumaghim, J, & Borgstedt, A (1992). Effects of sustained-release and standard preparations of methylphenidate on attention deficit disorder. *J. Am Acad Child Adolesc Psychiatry, 31*(2), 226–234.

Fried, J, Greenhill, LL, Torres, D, Martin, J, & Solomon, M (1987). Sustained-release methylphenidate: long-term clinical efficacy in ADDH males. *Am Acad Child Adolesc Psychiatry, Scientific Proceedings of the Annual Meeting 3*, 47.

Gadow, K, Sverd, J, Sprafkin, J, Nolan, E, & Ezor, S (1995). Efficacy of methylphenidate for attention deficit hyperactivity in children with tic disorder. *Arch Gen Psychiatry 52*, 444–455.

Gadow, KD, Nolan, EE, Sverd, J, Sprafkin, J, & Paolicelli, L (1990). Methylphenidate in aggressive-hyperactive boys: I. Effects on peer aggression in public school settings. *J Am Acad Child Adolesc Psychiatry, 29*(5), 710–718.

Gillberg, C, Melander, H, von Knorring, A, Janols, L, Thernlund, G, & Heggel, B, et al. (1997). Long-term central stimulant treatment of children with attention-deficit hyperactivity disorder: A randomized double-blind placebo-controlled trial. *Arch Gen Psychiatry, 54*, 857–864.

Gittelman-Klein, R (1980a). Diagnosis and drug treatment of childhood disorders: attention deficit disorder with hyperactivity. In DF Klein, R Gittelman-Klein, F Quitkin, A Rifkin, (Eds.), Diagnosis and drug treatment of psychiatric disorders: Adults and children. Baltimore: Williams & Wilkins.

Gittelman-Klein, R (1980b). The role of psychological tests for differential diagnosis in child psychiatry. *J Am Acad Child Adolesc Psychiatry, 19*, 413–437.

Gittelman-Klein, R, & Klein, DF (1976). Methylphenidate effects in learning disabilities. *Arch Gen Psychiatry, 33*, 655–664.

Gittelman-Klein, R, Klein, DF, Katz, S, Saraf, K, & Pollack, E (1976). Comparative effects of methylphenidate and thioridazine in hyperactive children. *Arch Gen Psychiatry, 33*, 1217–1231.

Gittelman-Klein, R, Landa, B, Mattes, JA, & Klein, DF (1988). Methylphenidate and growth in hyperactive children. *Arch Gen Psychiatry, 45*, 1127–1130.

Gittelman, R, & Mannuzza, S (1988). Hyperactive boys almost grown up: III. Methylphenidate effects on ultimate height. *Arch Gen Psychiatry, 45*, 1131–1134.

Gittelman, R, Mannuzza, S, Shenker, R, & Bonagura, N (1985). Hyperactive children almost grown up. I. Psychiatric status. *Arch Gen Psychiatry, 42*, 937–947.

Goldman, L, Genei, M, Bazman, R, & Stanetz, P (1998). Diagnosis and treatment of Attention-Deficit/Hyperactivity Disorder. *JAMA, 279,* 1100–1107 (abstract).

Greenhill, L, Abikoff, H, Conners, CK, Elliott, G, Hechtman, L, & Hinshaw, S, et al. (1996). Medication treatment strategies in the MTA: Relevance to clinicians and researchers. *J Am Acad Child Adolesc Psychiatry, 35*, 444–454.

Greenhill, L, Cornblatt, B, Cooper, T, Shapiro, B, Fried, J & Ramin, H (1994). Pharmacology and pharmacokinetics of sustained-release methylphenidate in the treatment of children with AD/HD: A replication. (Unpublished)

Greenhill, L, Halperin, J, & Abikoff, H (1999a). Simulant medications. *J Am Acad Child Adolesc Psychiatry, 35*, 1304–1313.

Greenhill, L, Halperin, J, & March, J (1997). Psychostimulants. In Tasman A, Kay J, & Lieberman J (Eds), *Psychiatry,* Philadelphia: W.B. Saunders. (pp. 1659–1682).

Greenhill, L, Swanson, J, Clevenger, W, Wu, M, Davies, M, & Rieppi, R, et al. (In press), Methylphenidate titration in the NIMH's multimodal treatment study of AD/HD. *J Am Acad Child Adolesc Psychiatry.* (Unpublished).

Greenhill, LL & MTA Cooperative Group (1999). Chronic stimulant treatment effects of weight acquisition rates of AD/HD children. *39*, 26–27. (Unpublished).

Greenshaw, AJ, Baker, GB, & Wishart, TB (1989). Dopamine receptor changes during chronic drug administration. In Goudie AJ & Emmett-Ogelsby MW (eds.), *Psychoactive drugs: Tolerance and sensitization* (pp. 353–390.) Clifton, NJ: Humana Press.

Gualtieri, CT, Kanoy, R, Koriath, U, Schroeder, S, Youngblood, W, Breese, GR, & Prange, AJ (1981). Growth hormone and prolactin secretion in adults and hyperactive children: relation to methylphenidate serum levels. *Psychoneuro, 6*(4), 331–339.

Hagerman, RJ, Murphy, MA, & Wittenberg, MD (1988). A controlled trial of stimulant medication in children with the fragile X syndrome. *Am J Med Genet, 30*, 377–392.

Haglund, RM & Howerton, LL (1982). Ritalin: Consequences of abuse in a clinical population. *Int J Addictions, 17*(2), 349–356.

Handen, BL, Feldman, H, Gosling, A, Breaux, AM, & McAuliffe, S (1991). Adverse side effects of methylphenidate among mentally retarded children with AD/HD. *J Am Acad Child Adolesc Psychiatry, 30*(2), 241–245.

Henker, B, Whalen C, Bugental, D, & Barker, C (1981). Licit and illicit substance use patterns in stimulant-treated children and their peers. In: K Gadow & J Loney (Eds.), *Psychosocial aspects of drug treatment for hyperactivity,* (pp 23–43) Boulder, CO: Westview Press.

Heussy, HR, Metoyer, M, & Townsend, M (1973). Eight-to-ten year follow-up of children treated in rural Vermont for behavior disorder. *Am J Orthopsychiatry, 43*, 236–238.

Hinshaw, S, Heller, T, & McHale, J (1992). Covert antisocial behavior in boys with attention-deficit hyperactivity disorder: External validation and effects of methylphenidate. *J Consult Clin Psychol, 60*, 274–281.

Hinshaw, S, Henker, B, Whalen, C, Ehrardy, D, & Dunnington, RE (1989). Aggressive, prosocial and nonsocial behavior in hyperactive boys: Dose effects of MPH in naturalistic settings. *J Consult Clin Psychology, 57*(4), 636–643.

Hinshaw, S, March, J, Abikoff, H, Arnold, L, Cantwell, D, & Conners, C, et al. (1997). Comprehensive assessment of childhood Attention-Deficit Hyperactivity Disorder in the context of a multisite, multimodal clinical trial. *Journal of Attention Disorders, 1*, 217–234.

Horn, WF, Islongo, NS, Pascoe, JM, Greenberg, G, Packard, T, Lopez, M, Wagner, A, & Puttler, L (1991). Additive effects of psychostimulants, parent training, and self-control therapy with AD/HD children. *J Am Acad Child Adolesc Psychiatry, 30*(2), 233–240.

Hyman, S & Nestler, E (1996). Initiation and adaptation: A paradigm for understanding psychotropic drug action. *Am J Psychiatry, 153*, 151–162.

Jacobvitz, D, Srouge, LA, Stewart, M, & Leffert, N (1990). Treatment of attentional and hyperactivity problems in children with sympathomimetic drugs: a comprehensive review. *J Am Acad Child Adolesc Psychiatry, 29*(5), 677–688.

Jensen, P, Bhatara, V, Vitiello, B, Hoagwood, K, Feil, M, & Burke, L (1999a). Psychoactive medication prescribing practices for US children: Gaps between research and clinical practice. *J Am Acad Child Adolesc Psychiatry, 38*, 557–565.

Jensen, P, Kettle, L, Roper, M, Sloan, M, Dulcan, M, & Hoven, C, et al. (1999b). Are stimulants overprescribed? Treatment of AD/HD in 4 US communities. *J Am Acad Child Adolesc Psychiatry*, *38*, 797–804.

Kalivas, P & Stewart, J (1991). Dopamine transmission in the initiation and expression of drug- and stress-induced sensitization of motor activity. *Brain Research, 16,* 223–244.

Kaplan, SL, Busner, J, Kupietz, S, Wasserman, E, & Segal, B (1990). Effects of methylphenidate on adolescents with aggressive conduct disorder and ADDH: A preliminary report. *J Am Acad Child Adolesc Psychiatry, 29*(5), 719–723.

Kavale, K (1982). The efficacy of stimulant drug treatment for hyperactivity: A meta-analysis. *J Learn Dis, 15* 280–289.

Kent, J, Blader, J, Koplewicz, H, Abikoff, H, Foley, C (1995). Effects of late-afternoon methylphenidate administration on behavior and sleep in attention-deficit hyperactivity disorder. *Pediatrics, 96*, 320–325.

Klein, R, Abikoff, H, Klass, E, Ganales, D, Seese, L, & Pollack, S (1997). Clinical efficacy of methylphenidate in conduct disorder with and without attention deficit hyperactivity disorder. *Arch Gen Psychiatry, 54*, 1073–1080.

Klorman, R, Brumagham, J, Fitzpatrick, P, & Burgstedt, A (1990). Clinical effects of a controlled trial of methylphenidate on adolescents with Attention Deficit Disorder. *J Am Acad Child Adolesc Psychiatry, 29*, 702–709.

Klorman, R, Brumaghim, JT, Salzman, LF, Strauss, J, Borgstedt, AD, McBride, MC, & Loeb, S (1988). Effects of methylphenidate on attention-deficit hyperactivity disorder with and without aggressive/noncompliant features. *J Abnorm Psychol, 97*(4), 413–422.

Kupietz, SS, Winsberg, BG, & Sverd, J (1982). Learning ability and methylphenidate (Ritalin[R]) plasma concentration in hyperkinetic children. *J Am Acad Child Adolesc Psychiatry, 21*(1), 27–30.

Lambert, NM (1988). Adolescent outcomes for hyperactive children: Perspectives on general and specific patterns of childhood risk for adolescent educational, social, and mental health problems. *Am Psychol, 43*, 786–799.

Lambert, NM & Hartsough, CS (1998). Prospective study of tobacco smoking and substance dependence among samples of AD/HD and non-AD/HD subjects. *J Learn Dis, 31*, 533–544.

LeFever, G, Sawson, KV, & Morrow, AL (1999). The extent of drug therapy for attention deficit-hyperactivity disorder among children in public schools. *Am J Public Health, 89*, 1359–1364.

Lipman, RS, DiMascio, A, Reatig, N, & Kirson, T (1978). Psychotropic drugs and mentally retarded children. In MA Lipton, A DiMascio, & KF Killam (Eds), *Psychopharmacology: A generation of progress* (pp. 1437–1449) New York: Raven Press.

Loney, J, Kramer, J, & Milich, R (1981). The hyperactive child grows up: Predictors of symptoms, delinquency and achievement at follow-up. In K Gadow & J Loney (Eds.), *Psychosocial aspects of drug treatment of hyperactivity* (pp. 381–415) Boulder, CO: Westview.

Lowe, TL, Cohen, DJ, Detlor, J, Kremenitzer, MW, & Shaywitz, BA (1982). Stimulant medications precipitate Tourette's syndrome. *JAMA, 26*, 1729–1731.

Mayes, S, Crites D, Bixler, E, Humphrey, B, & Mattison, R (1994). Methylphenidate and AD/HD: Influence of age, IQ and neurodevelopmental status. *Developmental Med Child Neurol, 36*, 1099–1107 (abstract).

Mendelson, W, Johnson, N, & Stewart, M (1971a). Hyperactive children as teenagers: A follow-up study. *J Nerv Ment Dis, 153*: 273–279

Mendelson, JH, Johnson, NE, & Stewart MA (1971b). Hyperactive children as teenagers: A follow-up study. *J Nerv Ment Dis, 153*: 273–279.

Milberger, S, Biederman, J, Faraone, S, Chen, L, & Jones, J (1997). AD/HD is associated with early initiation of cigarette smoking in children and adolescents. *J Am Acad Child Adolesc Psychiatry, 36*, 44.

Milich, R, Licht, B, & Murphy, D (1989). Attention-deficit hyperactivity disordered boys evaluations of and attributions for task performance on medication versus placebo. *J Abnorm Psychol, 98:* 280–284

Minde, K, Weiss, G, & Mendelson, M (1972). A five-year follow-up study of 91 hyperactive school children. *J Am Acad Child Adolesc Psychiatry 11*, 595–600.

Murphy, DA, Pelham, WE, & Lang, AR (1986). Methylphenidate effects on aggressiveness in ADD and ADD/CD children. Presented at the American Psychological Association, Washington, D.C., August 1986. (abstract)

Musten, L, Firestone, P, Pisterman, S, Bennett, S, & Mercer, J (1997). Effects of methylphenidate on preschool children with AD/HD: Cognitive and behavioral functions. *J Am Acad Child Adolesc Psychiatry, 36*, 1407–1415.

Ottenbacher, J, & Cooper, H (1983). Drug treatment of hyperactivity in children. *Dev Med Child Neurol, 25*, 358–366.

Page, JG, Bernstein, JE, Janicki, RS, & Michelli, FA (1974). A multicenter trial of pemoline (Cylert) in childhood hyperkinesis. In Conners, CK (Ed.), Clinical use of stimulant drugs in children. The Hague, Netherlands: Excerpta Medica.

Pearl, RG, & Seiden, LS (1976). The existence of tolerance and cross-tolerance between D-amphetamine and methylphenidate for their effect on milk consumption and on differential-reinforcement-of-low-rate performance in the rat. *J Pharmacol Exp Ther, 198*, 635–647.

Pelham, W & Murphy, H (1986). Behavioral and pharmacological treatment of hyperactivity and attention deficit disorders. In Hersen M & Breuning J, (Eds.), *Pharmacological and behavioral treatment: An integrative approach* (pp. 108–147). New York: Wiley.

Pelham, WE (1986), The effects of stimulant drugs on learning and achievement in hyperactive and learning-disabled children. In JK Torgesen, & B Wong, (Eds.), Psychological and educational perspective on learning disabilities. New York: Academic Press.

Pelham, WE (1989). Behavior therapy, behavioral assessment and psychostimulant medication in the treatment of attention deficit disorders: An interactive approach. In Swanson J & Bloomingdale L, (eds.), *Attention deficit disorder: 4. Emerging trends in the treatment of attention and behavioral problems in children* (pp. 169–195). London: Pergamon.

Pelham, WE, Greenslade, KE, Vodde-Hamilton, MA, Murphy, DA, Greenstein, JJ, Gnagy, EM, & Dahl, RE (1990b). Relative efficacy of long-acting stimulants on AD/HD children: A comparison of standard methylphenidate, Ritalin-SR, Dexedrine spansule, and pemoline. *Pediatrics 86*, 226–237.

Pelham, WE, McBurnett, K, Harper, G, Milich, R, Clinton, J, Thiele, C, & Murphy, DA (1990a). Methylphenidate and baseball playing in ADD children: Who's on first? *J Consult Clin Psychol, 22*, 131–135 (abstract).

Pelham, WE & Milich, R (1991). Individual differences in response to Ritalin in classwork and social behavior. In LL Greenhill & B Osman (Eds.), *Ritalin: Theory and patient Management* (pp. 203–222) New York: Mary Ann Liebert.

Pelham, WE, Hoza, B, Sturges, J, Schmidt, C, Bijlsma, J, & Moorer, S (1989). The effects of sustained release 20 and 10 mg Ritalin bid on cognitive and social behavior in children with attention deficit disorder. *Pediatrics, 80*, 491–501.

Pelham, WE, Swanson, J, Furman, M, & Schwindt, H (1995). Pemoline effects on children with AD/HD: A time-response by dose-response analysis on classroom measures. *J Am Acad Child Adolesc Psychiatry, 34*, 1504–1513.

Physicians' Desk Reference (1999). *Physicians' Desk Reference* (45th ed.). Oradell, NJ: Medical Economics Data.

Pierce, D (1977). Suicidal intent in self-injury. *Brit J Psychiatry, 130*, 377–385.

Pliszka, SR (1989). Effect of anxiety on cognition, behavior, and stimulant response in AD/HD. *J Am Acad Child Adolesc Psychiatry, 28*(6), 882–887.

Porrino, LJ, Rapoport, J, Behar, D, Sceery, W, Ismond, D, & Bunney, WE (1983). A naturalistic assessment of the motor activity of hyperactive boys, I: Comparison with normal controls. *Arch Gen Psychiatry 40*(6), 681–687.

Porrino, LJ, Rapoport, J, Behar, D, Sceery, W, Ismond, D, & Bunney WE (1983). A naturalistic assessment of the motor activity of hyperactive boys, II: Stimulant drug effects. *Arch Gen Psychiatry 40*(6), 688–697.

Rapoport, JL, Buchsbaum, MS, Weingartner, H, Zahn, P, Ludlow, C, Mikkelsen, EJ (1980). Dextroamphetamine: Cognitive and behavioral effects in normal and hyperactive boys and normal men. *Arch Gen Psychiatry 37*, 933–943.

Rapport, M, Denney, C, DuPaul, G, & Gardner, M (1994). Attention deficit disorder and methylphenidate: Normalization rates, clinical effectiveness and response prediction in 76 children. *J Am Acad Child Adolesc Psychiatry, 33*(6), 882–893.

Rapport, MD, DuPaul, GJ, & Kelly, KL (1989b). Attention deficit hyperactivity disorder and methylphenidate: the relationship between gross body weight and drug response in children. *Psychopharmacol Bull, 25*(2), 285–290.

Rapport, MD, Quinn, SO, DuPaul, GJ, Quinn, EP, Kelly, KL (1989a). Attention deficit disorder with hyperactivity and methylphenidate: The effects of dose and mastery level on children's learning performance. *J Abnorm Child Psychol, 17*(6), 669–689.

Richters, J, Arnold, L, Abikoff, H, Conners, C, Greenhill, L, Hechtman, L, Hinshaw, S, Pelham, W, & Swanson, J (1995). The National Institutes of Mental Health collaborative multisite multimodal treatment study of children with Attention-Deficit Hyperactivity Disorder (MTA): I. Background and rationale. *J Am Acad Child Adolesc Psychiatry, 34*, 987–1000.

Riddle, KD & Rapoport, JL (1976). A 2-year follow-up of 72 hyperactive boys. *J Nerv Ment Dis, 162*, 126–134.

Roberts, S., Harbison, R., Roth, L., & James, R (1994). Methylphenidate-induced hepatoxicity in mice and its potentiation by beta-adrenergic agonist drugs. *Life Sciences, 55*, 269–281.

Robinson, D, Jody, D, & Lieberman, J (1991). Provocative tests with methylphenidate in schizophrenia and schizophrenia spectrum disorders. In Osman B & Greenhill L, (Eds.) *Ritalin: Theory and patient management* (pp. 451–482) New York: Mary Ann Liebert.

Safer, D (1991). Cardiovascular effects of stimulants used to treat AD/HD. *J Am Acad Child Adoles Psychiatry, Scientific Proceedings 7*, 60–61 (abstract).

Safer, D, Zito, J, & Fine, E (1996). Increased methylphenidate usage for attention deficit hyperactivity disorder in the 1990s. *Pediatrics 98*, 1084–1088.

Safer, DJ & Allen, RP (1989). Absence of tolerance to the behavioral effects of methylphenidate and inattentive children. *J Pediatr 115*, 1003–1008.

Sallee, F, Stiller, R, & Perel, J (1992). Pharmacodynamics of pemoline in attention deficit disorder with hyperactivity. *J Am Acad Child Adolesc Psychiatry, 31*(2), 244–251.

Satterfield, JH, Cantwell, DP, & Satterfield, BT (1982). Prospective study of delinquency in 110 adolescent boys with attention deficit disorder and 88 normal adolescent boys. *Am J Psychiatry, 139* 797–798

Satterfield, JH, Cantwell, DP, & Satterfield, BT (1979). Multimodality treatment: A one-year follow-up of 84 hyperactive boys. *Arch Gen Psychiatry, 36*(8), 965–974.

Satterfield, JH, Satterfield, BT, & Cantwell, DP (1980). Multimodality treatment: A two-year evaluation of 61 hyperactive boys. *Arch Gen Psychiatry, 37*(8), 915–919.

Schachar, R, & Tannock, R (1993). Childhood hyperactivity and psychostimulants: A review of extended treatment studies. *J Child Adolesc Psychopharmacol, 3*, 81–97.

Schachar, R, Tannock, R, Cunningham, C, & Corkum, P (1997). Behavioral, situational, and temporal effects of treatment of AD/HD with methylphenidate. *J Am Acad Child Adolesc Psychiatry, 36*, 754–763.

Schachar, R, Taylor, E, Wieselberg, M, Thorley, G, & Rutter, M (1987). Changes in family function and relationships in children who respond to methylphenidate. *J Am Acad Child Adolesc Psychiatry, 26*(5), 728–732.

Schechter, M, & Keuezer, E (1985). Learning in hyperactive children: Are there stimulant-related and state-dependent effects? *J Clin Pharm, 25*, 276–280.

Schleifer, M, Weiss, G, Cohen, N, Elman, M, Cvejic, H, & Kruger, E (1975). Hyperactivity in preschoolers and the effect of methylphenidate. *American Journal of Orthopsychiatry, 45*, 38–50.

Sherman, M (1991). Prescribing practice of methylphenidate: The Suffolk County Study. In Osman B & Greenhill LL (Eds.), *Ritalin: Theory and patient management* (pp 401–420) New York: Mary Ann Liebert.

Sleator, E, von Neuman, A, & Sprague, R (1974). Hyperactive children: A continuous long-term placebo-controlled follow-up. *JAMA, 229*, 316–317.

Solanto, MV (1999). Dosage effects of Ritalin on cognition. In Osman B & Greenhill LL (Eds.), *Ritalin: Theory and patient management* (pp. 150–171) New York: Mary Ann Liebert.

Solanto, MV (1989). Does methylphenidate constrict cognitive functioning? *J Am Acad Child Adolesc Psychiatry, 28*(6), 897–902.

Speltz, ML, Varley, CK, Peterson, K, Beilke, RL (1988). Effects of dextroamphetamine and contingency management on a preschooler with AD/HD and oppositional defiant disorder. *J Am Acad Child Adolesc Psychiatry, 27*(2), 175–178.

Spencer, T, Biederman, J, Harding, M, Faraone, S, & Wilens, T (1996b). Growth deficits in AD/HD children revisited: Evidence for disorder related growth delays. *J Am Acad Child Adolesc Psychiatry, 35*, 1460–1467.

Spencer, T, Biederman, J, Wilens, T, Harding, M, O'Donnell, D, & Griffin, S (1996a). Pharmacotherapy of attention-deficit hyperactivity disorder across the life cycle. *J Am Acad Child Adolesc Psychiatry, 35*, 409–432

Spencer, T, Wilens, T, Biederman, J, Farone, S, Ablen, S, & Lapey, K (1995). A double-blind, crossover comparison of methylphenidate and placebo in adults with childhood onset AD/HD. *Arch Gen Psychiatry, 52*, 434–443

Sprague, RL & Sleator, EK (1977). Methylphenidate in hyperkinetic children: Differences in dose effects on learning and social behavior. *Science, 198*, 1274–1276.

Strakowski, S, Sax, K, Setters, M, & Keck, P (1996). Enhanced response to repeated D-amphetamine challenge: Evidence for behavioral sensitization in humans. *Biological Psychiatry, 40*, 872–880.

Swanson, J (1993). Effect of stimulant medication on hyperactive children: A review of reviews. *Except Child, 60*, 154–162.

Swanson, J, Gupta, S, Gupta, D, Flynn, D, Agler, D, & Lerner, M, et al., (1999). Acute tolerance to methylphenidate in the treatment of attention deficit hyperactivity disorder. *Clin Pharmacol & Ther, 66*, 295–305.

Swanson, J, Lerner, M, & Williams, L (1995). More frequent diagnosis of attention deficit-hyperactivity disorder. *New Engl J Med, 333*, 944.

Swanson, J, Sergeant, J, Sonuga-Barke, E, Jensen, P, & Cantwell, D P (1998b). Attention-deficit hyperactivity disorder and hyperkinetic disorder. *Lancet, 351*, 429–433.

Swanson, J, Wigal, S, Greenhill, L, Browne, R, Waslik, B, Lerner, M, Williams, L, Flynn, D, Agler, D, Crowley, K, Feinberg, B, Baren, M, & Cantwell, D (1998a). Analog classroom assessment of Adderall in children with AD/HD. *J Am Acad Child Adolesc Psychiatry, 37*, 1–8.

Tannock, R, Ickowicz, A, & Schachar, R (1995b). Differential effects of MPH on working memory in AD/HD children with and without comorbid anxiety. *J Am Acad Child Adolesc Psychiatry, 34*, 886–896.

Tannock, R, Schachar, R, & Logan, GD (1995a). Methylphenidate and cognitive flexibility: dissociated dose effects in hyperactive children. *Journal Abnorm Child Psychol, 23*, 235–267.

Thurber S, & Walker C (1983). Medication and hyperactivity: A meta-analysis. *J Gen Psychiatry, 108*, 79–86.

Vitiello, B & Jensen, P (1997). Medication development and testing in children and adolescents. *Arch Gen Psychiatry, 54*, 871–876.

Vitiello, B, Severe, J, Greenhill, L, Arnold, L, Abikoff H, & Jensen P, et al. Establishing the best maintenance dose of methylphenidate for children with Attention Deficit

Hyperactivity Disorder: The MTA Experience. *J Am Acad Child Adolesc Psychiatry.* (In Press).

Weingartner, H, Rapoport, JL, Buchsbaum, MS, Bunney, WE, Jr., Ebert, MH, Mikkelsen, EJ, & Caine, ED (1980). Cognitive processes in normal and hyperactive children and their response to amphetamine treatment. *J Abnorm Psychol, 89*, 25–35.

Weiss, G & Hechtman, L (1993). *Hyperactive children grown up.* New York: Guilford Press.

Weiss, G, Kruger, E, Danielson, V, & Elman, M (1975). Effect of long-term treatment of hyperactive children with methylphenidate. *Can Med Assoc J, 112*, 159–165.

Weizman, A, Mordechai, M, Gil-Ad, I, Tyrone, S, & Laron, Z (1989). Plasma cortisol, prolactin, growth hormone, and immunoreactive b-endorphin response to fenfluramine challenge in depressed patients. *Clin Neuropharmacol, 11*(3), 250–256.

Whalen, C, & Henker, B (1976). Psychostimulants and children: A review and analysis. *Psychopharmacol Bull, 83*, 1113–1130.

Whalen, C, Henker, B, Buhrmester, D, Hinshaw, S, Huber, A, & Laski, K (1989). Does stimulant medication improve the peer status of hyperactive children? *J Consult Clin Psychol, 57*, 545–549.

Whitehouse, D, Shah, U, & Palmer, FB (1980). Comparison of sustained-release and standard methylphenidate in the treatment of minimal brain dysfunction. *J Clin Psychiatry, 41*(8), 282–285.

Wilens, T, Biederman, J, & Spencer, T (1995). Venlafaxine for adult AD/HD. *American Journal of Psychiatry, 152*, 1099–1100.

Wilens, TE, & Biederman, J (1992). The stimulants. *Psychiatr Clin North Am 15*(1), 191–222.

Zametkin, AJ, & Rapoport, JL (1986). The pathophysiology of attention deficit disorder with hyperactivity: A review. In BB Lahey & AE Kazdin (Eds.), *Advances in clinical and child psychology* (pp. 177–216) New York: Plenum.

Zametkin, AJ, & Rapoport, JL (1987). Neurobiology of attention deficit disorder with hyperactivity: where have we come in 50 years? *J Am Acad Child Adolesc Psychiatry, 26*, 676–686.

II

BASIC NEUROSCIENCE

Introduction

AMY F. T. ARNSTEN

Stimulants are thought to ameliorate the symptoms of Attention-Deficit Hyperactivity Disorder (ADHD) through actions on dopamine and norepinephrine neurons in the central nervous system. Over the past 25 years, we have learned an enormous amount about the anatomical organization and modulatory properties of these catecholamines. Both norepinephrine and dopamine can have profound effects on cortical and subcortical functions. The chapters by Lewis, Taylor and Jentsch, Grace, Berridge, Arnsten, Sagvolden and Ferguson summarize much of this progress in relation to ADHD. Many of these chapters focus on catecholamine mechanisms in the cortical-striatal circuits that are likely substrates of stimulant actions in ADHD.

The first chapter, by Lewis, beautifully describes the catecholamine projections to the cortex in the nonhuman and human primate. Lewis and colleagues have shown that the dopamine and norepinephrine axons exhibit an intricate, heterogeneous innervation of the monkey cortex—the likely substrates for many stimulant actions. Impressively, the Lewis lab has performed parallel studies of the human brain, and have observed remarkably similar patterns in the monkey and human cortex. Lewis also describes developmental changes in the dopaminergic innervation of the monkey cortex. For example, in dorsolateral prefrontal cortex, dopamine markers increase in early childhood, peak in adolescence, and then decline to adult levels. Other cortical areas appear to show distinct developmental patterns in their catecholamine innervation. These developmental changes may provide the anatomical basis for age-related side effects and drug sensitivities in patients.

The chapter by Taylor and Jentsch provides a broad overview of dopamine anatomy and neurochemistry. Taylor and Jentsch offer an excellent review of the role of dopamine function within the ventral striatum and amygdala in motivational and incentive learning processes, and the role of frontal-striatal circuits in inhibitory control. They describe how chronic administration of high doses of stimulants may lead to neurochemical alterations of these circuits, and provide a model as to how changes in the prefrontal cortex, amygdala, and ventral striatum may lead to the impulsive behaviors so problematic in ADHD.

The chapter by Grace describes an elegant model whereby changes in tonic versus phasic dopamine release in the ventral striatum may lead to altered behavioral regulation. Tonic dopamine release is thought to reduce phasic dopamine release via presynaptic mechanisms. The model proposes that elevated phasic dopamine release results in increased impulsivity and stimulus control of behavior. Grace theorizes that low doses of stimulants may help to ameliorate ADHD symptomatology by increasing tonic dopamine release, thus decreasing phasic dopamine release, in the ventral striatum. As projections from the prefrontal cortex increase tonic dopamine release, low doses of stimulants may also optimize dopamine release in the ventral striatum by enhancing prefrontal cortical function. The model articulated by Grace succeeds in integrating and reconciling heretofore seemingly incompatible data concerning pre- versus postsynaptic effects of stimulants, and hypo- versus hyperdopaminergic status of children with ADHD.

The chapter by Berridge provides a comprehensive overview concerning norepinephrine: the distribution of noradrenergic axons, the activity of noradrenergic cells during different behavioral and sensory states, and the physiological and behavioral effects of noradrenergic modulation. Berridge describes the powerful effects of norepinephrine on attention, arousal and information-processing, and reveals a surprising degree of cohesion in norepinephrine's actions. Whether at the level of single cells, populations of cells, or the behavior of the organism, norepinephrine increases responses to relevant stimuli (signal) and/or suppresses responses to irrelevant stimuli (noise). Berridge emphasizes that stimulants have potent effects of norepinephrine release, although these are often overlooked by researchers focusing on dopamine. Indeed, recent research in animals suggests that low doses of methylphenidate preferentially increase norepinephrine release. He argues that noradrenergic mechanisms likely contribute to the therapeutic effects of stimulants because noradrenergic alpha-2 agonists and noradrenergic reuptake blockers are useful treatments for ADHD.

The chapter by Arnsten describes how catecholamines have critical modulatory influences on the functions of the prefrontal cortex in animals. Most researchers argue that dysfunction of the prefrontal cortex is fundamental to ADHD symptoms. Thus, changes in catecholamine mechanisms in the prefrontal cortex may contribute to the etiology of ADHD, and likely contribute to the beneficial actions of stimulants in this disorder. Current research in animals indicates that moderate levels of norepinephrine and dopamine are required for proper prefrontal cortical function, while very high levels of catecholamines (e.g., during stress) impair prefrontal cortical functions. Findings in humans have been similar, as described in the chapter by Mehta and colleagues in the clinical section of this book.

The chapters by Ferguson and Sagvolden illustrate how animal models may help us learn more about the etiology and treatment of ADHD. The chapter by Ferguson is a comprehensive review and critique of rodent animal models of ADHD and their response to stimulants. Ferguson reviews both older studies (e.g., dopamine depletion in developing rats) and recent research using genetically altered mice (e.g., dopamine transporter knock-out mice). She also describes her own research using animals with altered development of the cerebellum who exhibit locomotor hyperactivity. These studies may relate to recent imaging results showing reduced cerebellar volume in children with ADHD.

The chapter by Sagvolden focuses on the well-developed model of the spontaneously hypertensive rat (SHR), which reproduces clinical features of the disorder. This model, with the results of coordinated studies in children, suggests that a shorter and steeper delay-of-reinforcement gradient explains the hyperactive and impulsive behaviors characteristic of ADHD.

3

The Catecholamine Innervation of Primate Cerebral Cortex

DAVID A. LEWIS

The catecholamines dopamine (DA) and norepinephrine (NE) are known to exert critical modulatory influences over a number of cognitive and motor functions mediated by the neural circuitry of the frontal lobe. Understanding how alterations in cortical catecholamines may be involved in both the pathophysiology and treatment of ADHD requires knowledge of the normal regional, laminar, and cellular distribution of the axons that utilize DA or NE, of the receptors that are the targets of these neurotransmitters, and of the development of these systems during childhood and adolescence. In addition, the application of this information to human disorders such as ADHD requires that it be obtained in an animal model that most closely reflects the human brain. Consequently, this chapter will review the organization of the catecholamine innervation of the macaque monkey cerebral cortex with comparisons to the human brain where possible.

Regional and Laminar Distribution of Catecholamine Axons in Primate Neocortex

Background

Initial studies of the catecholamine innervation of monkey cerebral cortex utilized fluorescence histochemistry. Although this procedure labels all catecholamines, differences in the morphology of DA and NE axons were used to compare the distributions of these projections onto the cortical mantle (Levitt et al., 1984). Based on their morphological features in the rodent (Lindvall & Björklund, 1974), DA axons were identified as extremely fine and relatively smooth processes with few varicosities in their preterminal segments; in contrast, NE axons were recognized by their regularly spaced varicosities. Using these criteria, the largest numbers of DA-like axons in the monkey neocortex were observed in premotor and primary motor regions, prefrontal and anterior cingulate cortices, and the superior and inferior gyri of the temporal lobe. Fibers of the DA type were reported to be sparse in the parietal and occipital lobes. In contrast, NE axons were found in all cortical regions, with a particularly high density in the somatosensory areas.

The development of immunocytochemical techniques led to the more sensitive and specific detection of cortical DA and NE axons. The first studies were conducted with antibodies that recognized tyrosine hydroxylase (TH), the rate-limiting enzyme in the synthesis of all catecholamines, or dopamine-beta-hydroxylase (DBH), the enzyme that converts DA to NE. TH is expressed in all catecholaminergic neurons, whereas DBH is present only in NE and epinephrine-containing neurons. Because adrenergic fibers have not been detected in neocortex (Hökfelt et al., 1974), DBH immunoreactivity is a specific marker for NE cortical axons, whereas TH immunoreactivity would be expected to identify both DA and NE axons.

However, in several primate species, including humans, multiple lines of evidence indicate that anti-TH and anti-DBH antibodies predominantly label distinct populations of axons in primate neocortex, which appear to be DA- and NE-containing, respectively. For example, the different morphological features of TH-positive (fine with small varicosities) and DBH-positive (large, regularly spaced varicosities) cortical axons were similar to the differences between DA-like and NE-like axons observed in histofluorescent studies (Lindvall & Björklund, 1974; Levitt et al., 1984). These differences have been confirmed in studies using antibodies against DA (see what follows), where the majority of labeled axons were thin, nonmyelinated, and studded with multiple, small, spherical varicosities.

DBH-labeled axons were also present in cortical areas and layers where little or no TH immunoreactivity was found (Lewis et al., 1987; Campbell et al., 1987), indicating the existence of DBH-containing, NE axons in which TH was not detectable (Fig. 3–1). In contrast, TH-immunoreactive axons were very dense in other regions and layers of the monkey neocortex where DBH-containing processes were sparse or nonexistent, demonstrating that the TH antibodies were very sensitive and able to visualize a separate, non-NE population of cortical axons. In addition, dual-label studies demonstrated that anti-TH and anti-DBH antibodies labeled predominantly non-overlapping populations of axons in monkey and human cerebral cortex (Fig. 3–1) (Gaspar et al., 1989; Noack & Lewis, 1989; Akil & Lewis, 1993; Berger et al., 1993). Finally, in monkeys with histologically confirmed ablations of the NE projections from the locus coeruleus (Lewis et al., 1987, 1988a), cortical DBH immunoreactivity was eliminated or markedly reduced, but there was no apparent change in the density of TH-immunoreactive axons. These findings also suggest that the anti-TH antibody was not labeling cortical NE axons. However, it is important to note that the ability of anti-TH antibodies to identify cortical NE axons in primates, and consequently their selectivity for cortical DA axons, appears to differ somewhat across cortical regions and among anti-TH antibodies (Noack & Lewis, 1989; Akil & Lewis, 1993).

Although these TH antibodies fail to identify most, or in some cases virtually all, of the NE axons present in primate neocortex, these antibodies do clearly label the DBH-containing, NE cell bodies of the locus coeruleus (Lewis et al., 1987,1988a),

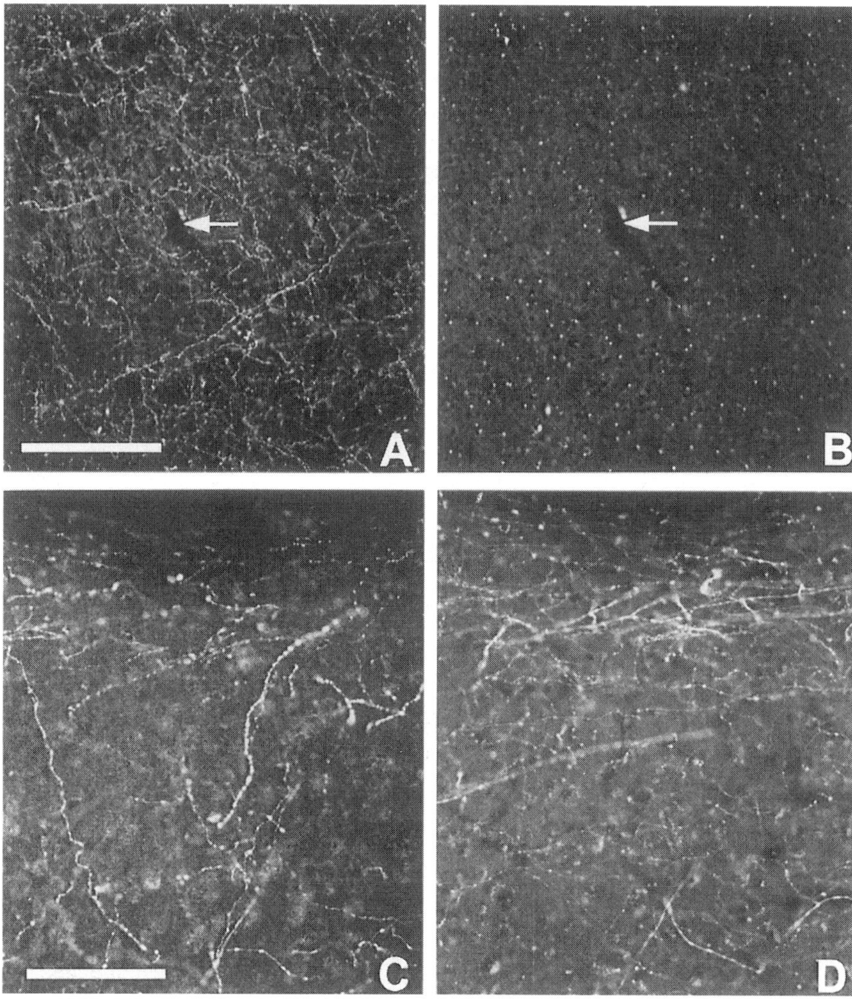

Figure 3–1. Fluorescent photomicrographs of immunoreactive axons visualized by rabbit anti-DBH (A,C) and mouse anti-TH (B,D) antibodies in area 3 (A,B) and in area 9 (C,D) of cynomolgus monkey neocortex. Note the numerous DBH-immunoreactive fibers that are not visualized by the mouse anti-TH antiserum. Arrows indicate the same blood vessel in A and B. Scale bar in A = 200 μm and applies to A and B; scale bar in C = 100 μm and applies to C and D. (Reproduced from Noack & Lewis, 1989.)

when used under the same conditions. These observations suggest that the selective labeling of DA cortical axons by some anti-TH antibodies might be due to differences in the affinities of the anti-TH antibodies, and consequently, in their ability to detect the lower concentrations of TH present in cortical axons than in cell bodies (Lewis et al., 1987; Noack & Lewis, 1989). For example, biochemical studies have demonstrated that cortical DA axons contain a much higher concentration of TH than do cortical NE axons (Emson & Koob, 1978; Schmidt & Bhatnagar, 1979). Alternatively, the differential labeling of DA and NE cortical axons by anti-TH antibodies might be attributable to differences in the TH molecule in DA and NE axons (Joh & Reis, 1974; Acheson et al., 1981). Although postranslational modifications of TH specific to one of these catecholamine systems cannot yet be excluded, the selective labeling of DA cortical axons by TH antibodies cannot be attributed to differences in the isoform of TH present in cortical DA and NE axons. For example, the primary transcript of the TH gene is alternatively spliced to give rise to two or four isoforms of TH in monkey (Ichikawa et al., 1990; Lewis et al., 1994) and human (Coker III et al., 1990; Lewis et al., 1993) brain, respectively. However, cell bodies of both DA neurons in the ventral mesencephalon and NE neurons in the locus coeruleus appear to contain all of the isoforms present in that species (Lewis et al., 1993,1994).

Although most investigations of the distribution of DA and NE axons in the monkey cortex have utilized TH and DBH antibodies, the findings of these studies, especially in the case of DA, have been confirmed through the use of complementary techniques. For example, dual-label studies (Lewis et al., 1998) have demonstrated that over 95% of all TH-positive axons in monkey neocortex were also labeled with an antibody specific for the DA transporter (DAT), a protein that appears to be a selective marker of DA axons. Similarly, the distribution pattern of TH-labeled axons in monkey neocortex has been shown to be virtually identical to that observed in studies that identified DA axons using autoradiographic techniques based on the specific uptake of tritiated DA (Berger et al., 1986, 1988) or antibodies directed against DA (Akil & Lewis, 1993; Williams & Goldman-Rakic, 1993). Although the relative regional densities and laminar distribution patterns of TH-, DA- and DAT-labeled axons were identical, within a given region of monkey neocortex, the density of TH-positive axons generally appeared greater than that of DAT-labeled axons, which exceeded the density of DA-positive fibers (Fig. 3–2). If these differences were due to the labeling of NE axons by TH and DAT antibodies, then one would also expect to see differences across antibodies in the laminar distribution of labeled axons, especially in areas with a high density of NE axons (Lewis et al., 1987). Thus, the apparent differences in axon density across antibodies probably reflect differences in antibody sensitivity, in the fixation requirements for each antigen, or the fact that each approach exploits different properties of DA axons. For example, DA uptake may primarily reflect the distribution of axon terminals, whereas TH immunoreactivity would also be likely to detect preterminal axons and axons of passage.

Innervation Patterns of DA Axons Across Monkey Neocortex

In studies using each of these approaches, the most striking finding was the widespread distribution, but substantial regional heterogeneity, of the DA innervation of neocortex in both Old and New World monkeys (Lewis et al., 1987,1988a; Campbell et al., 1987; Akil & Lewis, 1993). In contrast to the relatively restricted distribution of DA axons in rodent cerebral cortex (see Berger et al., 1991 for

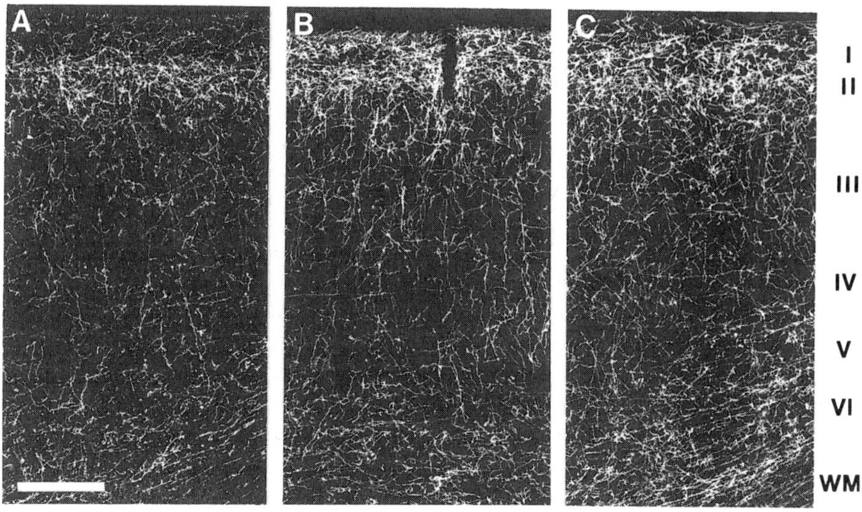

Figure 3–2. Dark-field photomicrographs of fibers labeled with anti-DA (A), anti-DAT (B), or anti-TH (C) antibodies in the inferior temporal cortex of macaque monkey. Note the similar bilaminar distribution of labeled axons with each antibody. Scale bar = 350 μm. (Reproduced from Lewis & Sesack, 1997.)

review), DA axons were present in every cortical region of each of these primate species. For example, the greatest density of DA axons was found in the premotor and primary motor cortices (Fig. 3–3). Rostrally, the anterior cingulate cortex and certain prefrontal cortical regions also contained a high density of labeled fibers, whereas other prefrontal cortical regions contained a lower density of labeled axons (see below). Caudally, the density of labeled axons decreased substantially across the central sulcus from motor to primary somatosensory cortex. However, the density of labeled axons then progressively increased again caudally, such that portions of the posterior parietal cortex contained a density of DA axons comparable to that seen in the prefrontal regions. Although not anticipated from the previous biochemical and fluorescence histochemistry studies, this dense DA innervation of the posterior parietal cortex has been confirmed by dual-label studies in which retrograde transport from the posterior parietal cortex revealed labeled neurons in the ventral mesencephalon that were also TH-immunoreactive (Lewis et al., 1988b).

From posterior parietal cortex, the density of DA axons progressively decreased in a caudal direction across the occipital lobe to area 17 (primary visual cortex), which contained the lowest density of labeled axons of any cortical region (Fig. 3–3). Within the lateral temporal lobe, caudal portions of the superior temporal gyrus (including the primary auditory cortex) contained a very low density of labeled axons, whereas the rostral portion of the superior temporal gyrus (auditory association cortex) had a much higher density of DA axons. By comparison, all regions of the inferior temporal gyrus (visual association cortex) contained an intermediate density of DA axons.

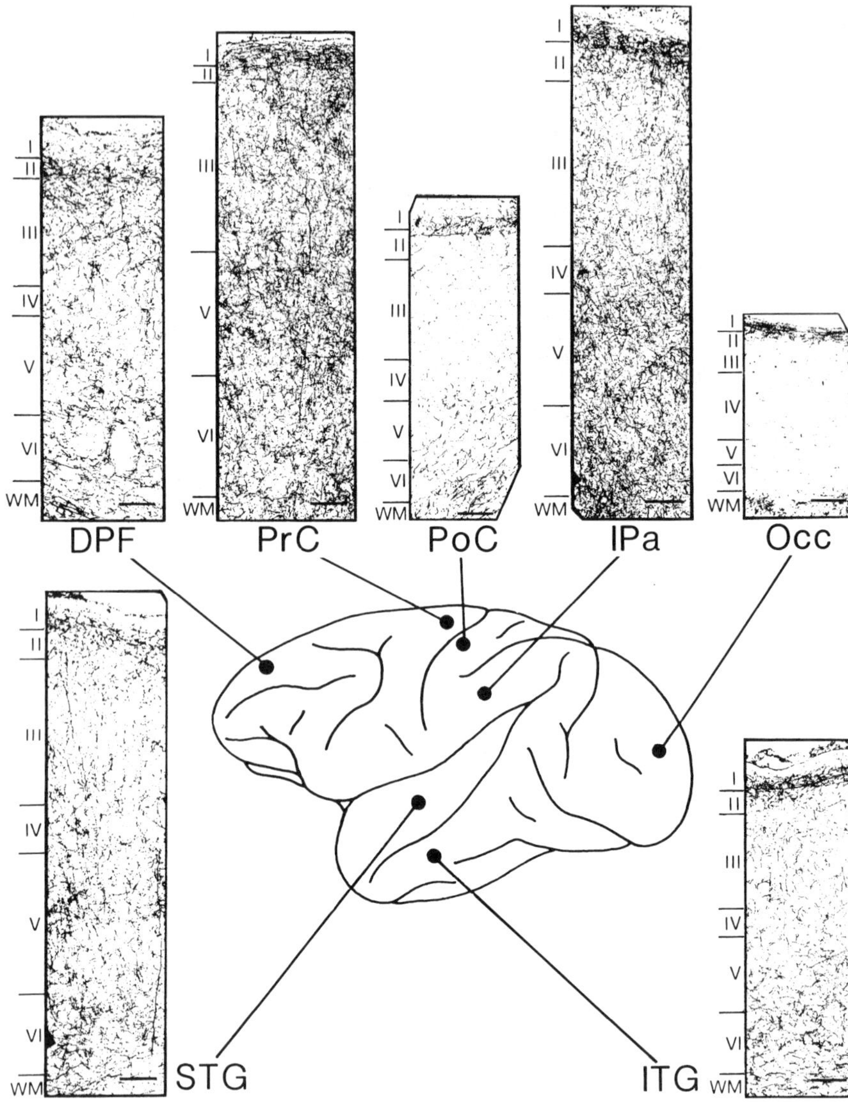

Figure 3–3. TH immunoreactivity in seven cytoarchitectonic areas of cynomolgus monkey neocortex. Note the extensive regional heterogeneity in the density and laminar distribution of labeled fibers. Photographs are reversed images of dark-field photomontages. DPF, dorsomedial prefrontal cortex (area 9); PrC, precentral (primary motor cortex); PoC, postcentral (primary somatosensory cortex); IPa, inferior parietal cortex (area 7); Occ, occipital (primary visual cortex); STG, rostral superior temporal gyrus (auditory association cortex); ITG, rostral inferior temporal gyrus (visual association cortex). Scale bars = 200 μm. (Reproduced from Lewis et al., 1987.)

These anatomical findings have been generally paralleled by the more limited number of biochemical studies of the distribution of DA in monkey cortex (Björklund et al., 1978; Brown et al., 1979). For example, the highest tissue concentrations of DA were found in the frontal lobe and the lowest in the occipital cortex. Most regions of the temporal cortex contained intermediate concentrations of DA. The apparent discrepancies between the anatomical and biochemical studies in some details may be due to the fact that the assessment of endogenous DA levels in the biochemical studies reflects the metabolic activity of the system, and not necessarily the density of innervation. Indeed, regional differences in in vivo synthesis rates may produce dissociations between the relative concentration of tissue DA and density of DA axons. Interestingly, though, recent studies using microdialysis techniques to assess basal extracellular levels of cortical DA in rhesus monkey frontal cortex in vivo (Moghaddam et al., 1993; Saunders et al., 1994) found the same regional patterns of DA innervation observed in the anatomical studies. For example, of the frontal cortical regions examined, extracellular levels of DA were highest in premotor cortex, intermediate in medial prefrontal cortex (PFC), and lowest in dorsolateral PFC. This relative ranking of extracellular DA levels corresponds to the relative density of DA axons across area 6 (premotor cortex), area 9 (dorsomedial PFC), and area 46 (dorsolateral PFC).

As demonstrated by all of these techniques for visualizing DA axons, the majority of cortical regions had a bilaminar distribution of labeled axons (Fig. 3–2). DA axons formed a dense band in layers 1 and 2 and the most superficial portion of layer 3, and a band of lower density in layers deep 5 and 6. In the densely innervated regions, such as the motor, premotor, and dorsomedial prefrontal regions, labeled axons were also present in high density in the middle cortical layers, forming a third distinctive band in deep layer 3. Finally, in more lightly innervated areas, such as primary visual cortex, labeled axons were primarily restricted to layer 1 (Fig. 3–3).

Rodents appear to have two separate mesocortical DA systems that differ in laminar distribution, as well as in fiber morphology, cells of origin, patterns of development, and neuropeptide colocalization (Berger et al., 1991). Observations of a selective loss of DA axons in the superficial cortical layers of patients with Parkinson's disease (Gaspar et al., 1991), and in the deep layers of subjects with schizophrenia (Akil et al., 1999), may be consistent with the existence of separate mesocortical DA systems in primates. However, other investigations have failed to reveal evidence for two such systems. For example, the light microscopic morphological features of DA cortical axons do not support the idea of two discrete classes of terminal fibers (Lewis et al., 1988a; Williams & Goldman-Rakic, 1993), and DA-containing axons have a homogeneous appearance at the ultrastructural level (Smiley & Goldman-Rakic, 1993). Finally, in contrast to rodents, neither neurotensin or cholecystokinin are present in DA cortical axons in primates (Gaspar et al., 1990; Oeth & Lewis, 1992), and consequently neuropeptide colocalization cannot be used to define subpopulations of DA axons in the primate neocortex.

Comparison to NE Innervation Patterns in Monkey Neocortex

Comparisons of the regional and laminar patterns of distribution of DA axons with those of NE axons in monkey and human cortex reveal interesting similarities and differences that may provide insight into their relative roles in cortical information processing. For example, the primary somatosensory cortex contains an extremely dense NE innervation, but a sparse number of DA axons, whereas primary motor cortex contains a high density of both types of axons (Morrison et al., 1982; Lewis et al., 1987; Berger et al., 1988) (Fig. 3–4). DA and NE afferents also exhibit different, and in some ways complementary, laminar innervation patterns. For exam-

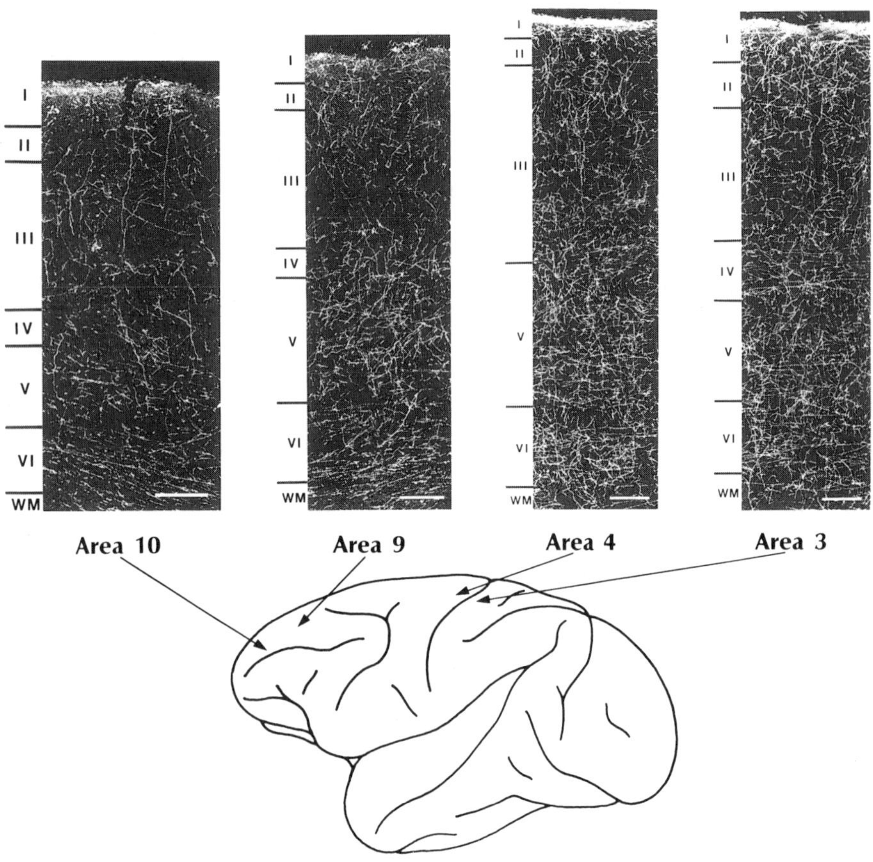

Figure 3–4. Dark-field photomicrographs of DBH-positive, noradrenergic axons in four regions of cynomolgus monkey neocortex. Note the progressive rostral to caudal increase in the density of labeled axons, including the high density in the postcentral, primary somatosensory cortex (Area 3). The regional and laminar distribution patterns of DBH-positive axons contrasts in many ways with those of TH-labeled axons (see Fig. 3–3). Scale bars = 200 μm. (Reproduced from Lewis and Sesack, 1997.)

ple, the density of NE axons is substantially greater in the deep cortical layers, especially layer 5, than in the more superficial cortical laminae. In particular, few NE axons are present in layer 1, which receives a dense DA innervation, and NE axons are heavily represented in layers deep 3 and 4, which in many cortical regions tend to receive a sparse DA innervation. Similar regional and laminar differences in innervation patterns between these two catecholamine systems have been observed in human neocortex (Gaspar et al., 1989).

These comparisons suggest that whether DA and NE axons share or have different target structures within a given cortical region, they may interact in regulating the output of that region. For example, of the pyramidal neurons that project to other cortical regions, those in the superficial layers may be preferentially influenced by DA, whereas those in the infragranular layers may receive both DA and NE inputs (Lewis, 1992). In contrast, infragranular pyramidal neurons, which project subcortically and which tend to have apical dendrites that extend to the superficial layers, may be regulated by DA inputs to their apical dendrites, and by NE inputs to their basilar dendrites.

Comparisons of the DA and NE Innervation of Monkey PFC

More detailed investigations of functionally linked cortical areas have provided additional evidence for the regional specificity of DA innervation patterns. For example, in the PFC of macaque monkeys (Fig. 3–5), the density of DA axons was greatest in area 9 (dorsal medial convexity). Within this region, the density of labeled axons progressively decreased in the medio-lateral direction from the dorsal bank of the cingulate sulcus to the lateral border of area 9 with area 46. A high density of labeled axons was also present on the medial cortical surface ventral to area 9. Among these areas, the density of DA axons was higher in area 24 (anterior cingulate cortex) than in the more rostral and ventral regions (areas 32 and 25). In contrast, the orbital PFC in general contained an intermediate density of labeled axons. Area 12 (lateral convexity) had a slightly greater density of labeled axons than did the more medial areas 11 and 13. The density of DA axons was lowest in prefrontal areas 46 (dorsal lateral surface) and 10 (frontal pole). Within area 46, fiber density was greatest on the dorsal surface, progressively decreased in both banks of the principal sulcus, and reached a nadir in fundus of this sulcus.

Differences in the density of DBH-labeled, NE axons were also evident both across and within many cytoarchitectonic regions of the monkey PFC (Fig. 3–6). For example, area 8b had the greatest density of DBH-immunoreactive axons; within this region, the medial surface had a greater density of labeled processes than the dorsal surface. Areas 9 and 24 also had a high density of labeled axons, areas 11–13 and 25 were of intermediate density, and portions of areas 10 and 46 had the lowest density of labeled axons. Despite these regional variations, the laminar distribution of DBH-immunoreactive axons was very similar throughout the PFC. Layer 1 contained a low density of labeled axons which were primarily tangential in orientation. The randomly oriented fibers in layers 2–4 were slightly higher in density. Fiber density increased substantially in layer 5 before declining again in layer 6.

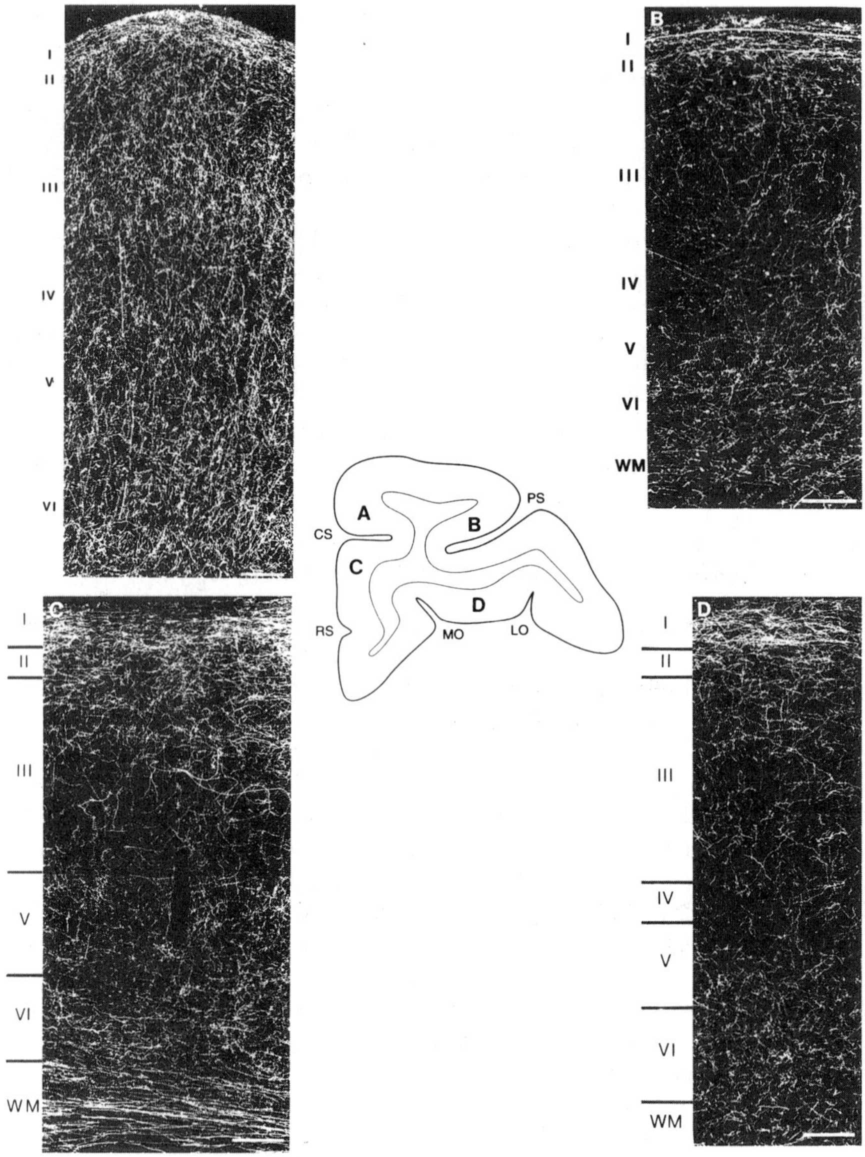

Figure 3–5. Dark-field photomicrographs of TH immunoreactivity in different regions of cynomolgus monkey prefrontal cortex. The drawing in the center depicts a standardized coronal section of prefrontal cortex. CS, cingulate sulcus; LO, lateral orbital sulcus; MO, medial orbital sulcus; PS, principal sulcus; RS, rostral sulcus. The letters A–D indicate the approximate location of the photomontage with the corresponding letter. Scale bars = 200 μm. (Reproduced from Lewis & Sesack, 1997.)

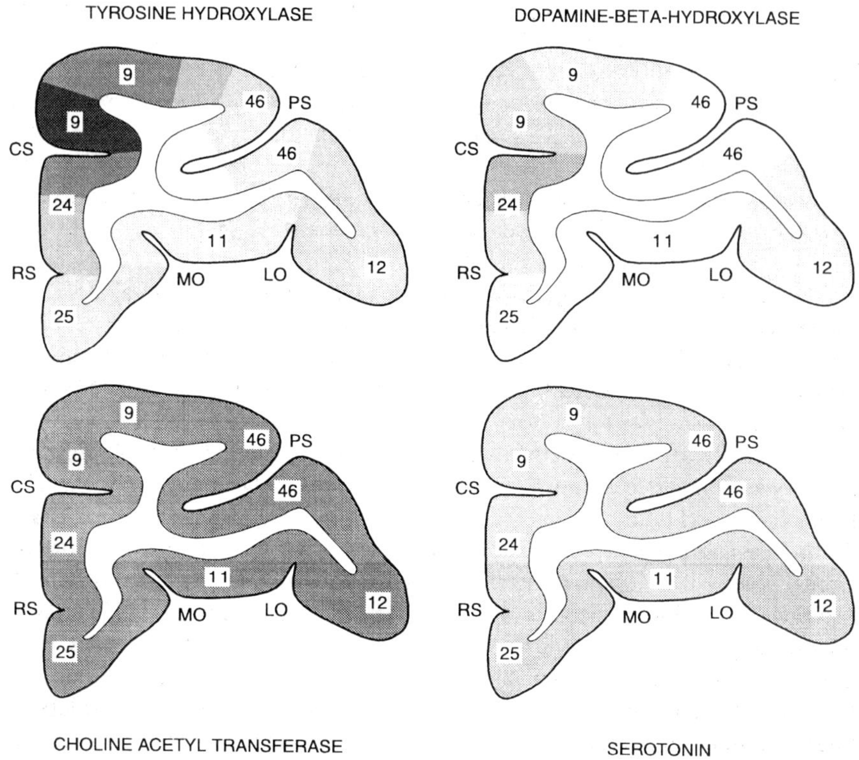

Figure 3–6. Schematic representation of coronal sections from macaque monkey prefrontal cortex illustrating the relative densities of tyrosine hydroxylase- (DA), dopamine-beta-hydroxylase-(norepinephrine), choline acetyltransferase-(acetylcholine), and serotonin-containing axons. CS, cingulate sulcus; LO, lateral orbital sulcus; MO, medial orbital sulcus; PS, principal sulcus; and RS, rostral sulcus. (Reproduced from Lewis, 1992.)

DA Innervation of the Human PFC

Due to the difficulties involved in obtaining human tissue under adequately controlled conditions, studies of catecholamine levels in postmortem human neocortex have been inconsistent and difficult to interpret. However, certain anatomical techniques, such as immunocytochemistry with antibodies against TH, have made it possible to reliably determine the regional and laminar innervation patterns of DA axons in human neocortex. Using this approach, Gaspar and colleagues (1989) found evidence of a DA innervation in every region of human neocortex examined. This expansion of the DA innervation to the entire cortical mantle is particularly impressive given that the human cortical surface is 400-fold greater than in rats, but the number of DA neurons is only increased by 10–20-fold (Berger et al., 1991). These investigators found that the density of TH-labeled axons was greatest in agranular areas, such as primary motor (area 4), premotor (area 6), and anterior cingulate (area 24) cortices. Similar to the homologous regions of monkey neocortex, labeled axons were distributed across all cortical layers in these regions of human neocortex. As in the monkey, the density of labeled axons was lowest in primary visual cortex (area 17), and intermediate in most association regions. In the majority of association regions, the laminar distribution of labeled axons was clearly bilaminar, with labeled axons forming dense bands in layers 1–2 and in layers 5–6. Similar regional and laminar patterns of innervation in the human cerebral cortex have also been observed with antibodies against DAT (Ciliax et al., 1999).

The cross-species similarities in the general distribution of cortical DA axons suggests that the organization of the mesocortical DA projections in monkey cortex accurately reflects its organization in the human brain, and that the nonhuman primate may be a very useful model system for investigating the role of cortical DA in the human cerebral cortex. This conclusion is supported by findings from more detailed regional comparisons across species. For example, both the macaque monkey and human PFC show very similar gradients in the density of TH-labeled fibers across and within regions (Lewis, 1992). On the dorsomedial convexity of the frontal lobes, fiber density increased in a rostral to caudal fashion from the frontal pole to the premotor and motor regions in both monkeys and humans. In addition, the medial cortical surface had a ventral to dorsal gradient of increasing fiber density in both species. The similarities across species in the distribution of TH-containing axons is further illustrated by comparison of Figures 3–5 and 3–7. For example, note that on these coronal sections of PFC, the density of labeled fibers was much greater in the dorsomedial regions and the anterior cingulate than in the lateral and orbital regions in both species.

Regional differences in the laminar distribution of labeled axons were also similar across species. For example, in both monkeys and humans the lightly innervated lateral and orbital regions had a bilaminar distribution of labeled fibers, whereas the more densely innervated dorsomedial region had axons distributed across all cortical layers (Figs. 3–5 and 3–7). These anatomical similarities suggest that the functional characteristics of DA cortical innervation derived from studies in nonhuman primates may be reasonably extrapolated to the function of that system in human PFC. This conclusion is further supported by the marked similarities in the distribution of DA axons across cytoarchitectonically matched regions of monkey and human entorhinal cortex (Akil & Lewis, 1993, 1994).

Functional Correlates of DA and NE Cortical Innervation Patterns

On a global level, the substantial regional differences in the density of DA axons present in monkey and human neocortex suggests a functional specialization of this afferent system rather than an organization based on a simple rostral-caudal gradient. For example, the premotor and primary motor cortices are more densely innervated than any of the more rostrally located prefrontal cortical regions, suggesting that DA plays an important role in the cortical regulation of movement.

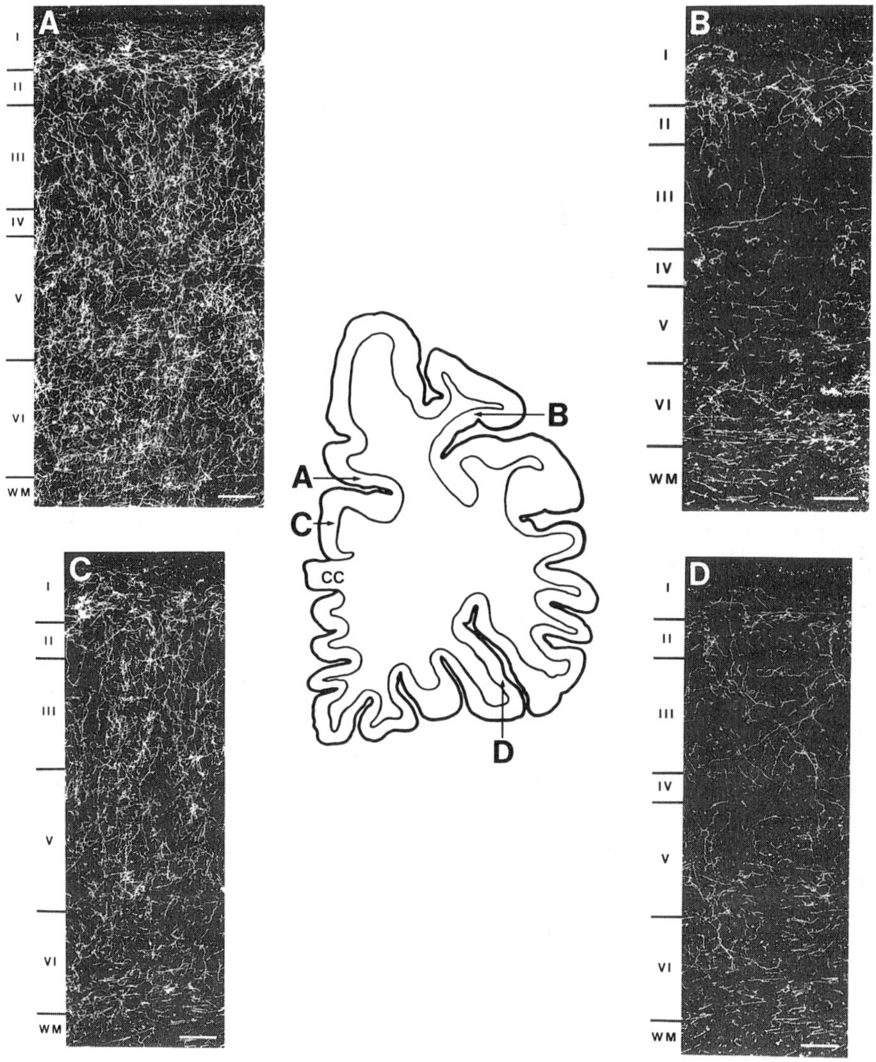

Figure 3–7. Dark-field photomontages of TH-immunoreactive fibers in human prefrontal cortex. The center panel is a camera lucida drawing of a section of human prefrontal cortex taken at the most anterior portion of the corpus callosum (cc). The letters A–D indicate the location of the surrounding photomontages. Note the differences in density and laminar distribution of the fibers present in medial prefrontal cortex (A) and anterior cingulate (C) from the dorsolateral (B) and orbital (D) prefrontal regions. Scale bars = 200 μm. (Reproduced from Lewis & Sesack, 1997.)

Indeed, neurons sensitive to iontophoretically applied DA have been detected in the primate motor cortex (Sawaguchi et al., 1986).

Although the density of DA axons decreases immediately caudal to motor cortex in the primary somatosensory areas, the density of DA axons increases again further caudally, revealing a preference for somatosensory association cortices over primary somatosensory cortex. Furthermore, within the rostral temporal cortex, auditory association regions in the superior temporal gyrus are more densely innervated than the visual association areas of the inferior temporal gyrus. Thus, DA axons preferentially innervate motor over sensory regions, sensory association over primary sensory regions, and auditory association over visual association areas.

The preferential innervation of sensory association regions over primary sensory areas is clearly evident not only in somatosensory systems, but in visual and auditory regions as well. In addition, within a certain range, the density of DA innervation appears to progressively increase through the regions that form a functional hierarchy within each of these systems. In the visual system, for example, the DA innervation of primary visual cortex (area V1) is quite sparse and restricted to layer 1. The density of DA axons increases in the immediately adjacent visual association area, V2, where labeled axons are present in both the superficial and deep cortical layers. From V2, visual information flows along both ventral (object recognition) and dorsal (object location) streams in the inferior temporal and posterior parietal cortices, respectively. In both streams, the density of DA axons is greater in the involved temporal and parietal cortices than in V2. These anatomical data suggest that while DA may play a role in every cortical region, this transmitter is particularly important in modulating the activity of regions involved in complex information processing.

Finally, although the presence of DA axons in other cortical layers varies across regions, DA axons are ubiquitous in layer 1. Given the rich connectivity of this layer, these observations suggest that DA may substantially influence the horizontal flow of information across the cortical mantle.

Distribution of DA and NE Receptors in Primate PFC

DA Receptors

DA receptors were initially categorized into two main types, D_1 and D_2 based on different biochemical and pharmacologic properties. D_1 receptor binding sites appear to be composed of the products of separate genes for D_1 and D_5 DA receptors, whereas the D_2, D_3 and D_4 DA receptors comprise the family of pharmacologically identified D_2 receptor binding sites. The mRNAs for all five (D_1–D_5) DA receptors have been detected in neurons located in layers 2–6 of monkey PFC, with the greatest density of labeled neurons present in layer 5 (Huntley et al., 1992; Brene et al., 1995; Lidow et al., 1998). In the frontal cortex of both human and monkey cortex, the mRNAs for D_2 and D_5 receptors are also expressed by neurons in layers 2–5, suggesting that many different classes of cortical neurons may receive DA input (Huntley et al., 1992).

Autoradiographic techniques with specific ligands have been used to examine the distribution of the pharmacologically identified classes of D_1 and D_2 receptors. Using [^{3}H] SCH23390 (in the presence of mianserin, which blocks $5HT_2$ and $5HT_{1c}$ sites), D_1 receptors were found to be highest in layers 1–3a, lowest in layers 3b–4 and intermediate in layers 5–6 in most regions of monkey neocortex (Richfield et al., 1989; Goldman-Rakic et al., 1990; Lidow et al., 1991a), a laminar pattern that, as described above, mirrors that of the DA axons innervating these regions. In all regions and layers examined, the density of D_1 receptors exceeds that of D_2 receptors (as assessed by [^{3}H] raclopride) binding (Richfield et al., 1989; Lidow et al., 1989) by a factor of 10–20 (Lidow et al., 1991a). In addition, in contrast to the predominant bilaminar distribution of D_1 receptors, D_2 receptors were preferentially localized to layer 5 (Lidow et al., 1991a). Receptor autoradiography studies in postmortem human brain have in general found similar patterns of DA receptor distribution to those observed in monkey (Dawson et al., 1987; Cortés et al., 1989; Camps et al., 1989).

More recent investigations have used comparisons of binding of new ligands to investigate the distributions of the different members of the D_2 family of receptors. For example, Lahti et al. (1995) determined that very low densities of D_2 and D_3 receptors were present in human cingulate, entorhinal, and temporal cortices compared to the densities of D_2 and D_3 receptors in caudate-putamen and nucleus accumbens, respectively. In contrast, D_4 receptors were present in relatively high densities in the cortical regions examined.

NE Receptors

Most studies of PFC NE receptors have employed ligand-binding techniques. Using ^{3}H prazosin as the ligand (Goldman-Rakic et al., 1990; Bigham & Lidow, 1995), alpha-1 NE receptors were found to be present in every layer of the macaque monkey PFC. Receptor density was highest in layers 1–3 and then decreased through the remaining layers, with the exception of a modest increase in layer 5 in some regions of the PFC. In general, this distribution paralleled that of DBH-immunoreactive axons. Similarly, alpha-1 NE receptors (using ^{3}H clonidine as ligand) were found to be highest in density in layers 1–3, with a general decrease in density from layer 1 to layer 6 (Goldman-Rakic et al., 1990; Bigham and Lidow, 1995). In contrast, beta NE receptors, identified by ^{3}H pindolol binding (Lavoie et al., 1989; Goldman-Rakic et al., 1990; Bigham & Lidow, 1995) have a much lower density, but the greatest number are present in the superficial layers.

Comparison of the distribution of DA receptors and those for NE show both overlapping and complementary patterns of distribution within the monkey PFC. For example, D_1 and alpha-1- and alpha-2-adrenergic receptors are all concentrated in layers 1–3a. In contrast, beta-adrenergic receptors are present in the middle cortical layers, laminae which contain relatively few D_1 or D_2 receptors.

Synaptic Targets of Cortical DA and NE Axons

DA System

In the monkey PFC, most DA synapses are of the symmetric type, although occasional asymmetric synapses are also formed (Goldman-Rakic et al., 1989; Smiley & Goldman-Rakic, 1993). DA axons target both the spines and distal dendritic shafts of cortical neurons, but rarely form synapses with cell soma or proximal dendrites. Axospinous synapses of DA axons are more prevalent in layers 4–6 than in layers 1–3, and the majority involve a "triadic" convergence of an unlabeled terminal forming an asymmetric synapse onto the same spine, supporting the view that DA modulates excitatory inputs to pyramidal cells (Van Eden et al., 1987; Goldman-Rakic et al., 1989).

The idea that DA principally targets cortical neurons is supported by the relatively prominent localization of immunoreactivity for the DA D_1 and D_5 receptors to pyramidal neurons in monkey PFC (Smiley et al., 1994; Bergson et al., 1995). The D_5 receptor appears to be more commonly distributed to dendritic shafts than is the D_1 receptor, which is preferentially localized to spines. Only a portion of the total population of dendritic processes are labeled for the D_1 or D_5 receptors. For example, in layer 3 of prefrontal area 46, approximately 20% of dendritic spines are immunoreactive for the D_1 receptor, and 5% contain the D_5 receptor. Some evidence (Smiley et al., 1994) suggests that the DA D_1 receptor has a nonsynaptic distribution. For example, D_1 receptor immunoreactivity is sometimes localized to spines that do not receive synaptic input from DA axons.

DA axons also synapse on dendrites with morphological features (varicose contours, received abundant synaptic input, and lacking spines) of cortical local circuit neurons (Smiley & Goldman-Rakic, 1993). The proportion of DA axon synapses that were made onto this type of dendritic shaft ranged from 7% in layer 5 to 39% in layer 2. These findings suggest that a considerable portion of DA axons in the PFC target interneurons in addition to pyramidal cells, particularly in the superficial layers where interneurons are more numerous. This interpretation has been confirmed by the direct demonstration that DA axons synapse on dendrites that contain immunoreactivity for GABA (Sesack et al., 1995b). The presence of both DA and unlabeled axons synapsing on common GABA dendrites suggests that convergence between DA and excitatory afferents (in the case of asymmetric synapses) occurs for local circuit neurons as well as for pyramidal cell spines. The relative incidence with which DA axons target GABA dendrites as opposed to unlabeled dendrites exhibits no quantitative difference between the monkey and rodent PFC, or between the monkey prefrontal and motor cortices (Sesack et al., 1995b). These findings suggest that the ultrastructural and synaptic features of DA varicosities are relatively consistent across cortical regions and species, including rodents, monkeys, and humans.

PFC GABA neurons comprise at least 12 subtypes, some of which can be distinguished by the differential expression of the calcium-binding proteins, calretinin, calbindin, or parvalbumin (Condé et al., 1994). Although DA axons have never been observed to synapse onto calretinin-containing local circuit neurons (Sesack et al., 1995a), they do synapse on the dendrites of interneurons that utilize parvalbumin (Sesack et al., 1998). These findings indicate that DA axons in the primate PFC innervate local circuit neurons such as the wide arbor and chandelier cells that provide potent inhibitory input to the soma and axon initial segments of pyramidal neurons, respectively, while tending not to innervate interneurons such as double bouquet cells that target more distal dendrites of both pyramidal and nonpyramidal neurons (Fig. 3–8). These observations are also consistent with the observation that the D_4 DA receptor is expressed by parvalbumin-containing cortical neurons

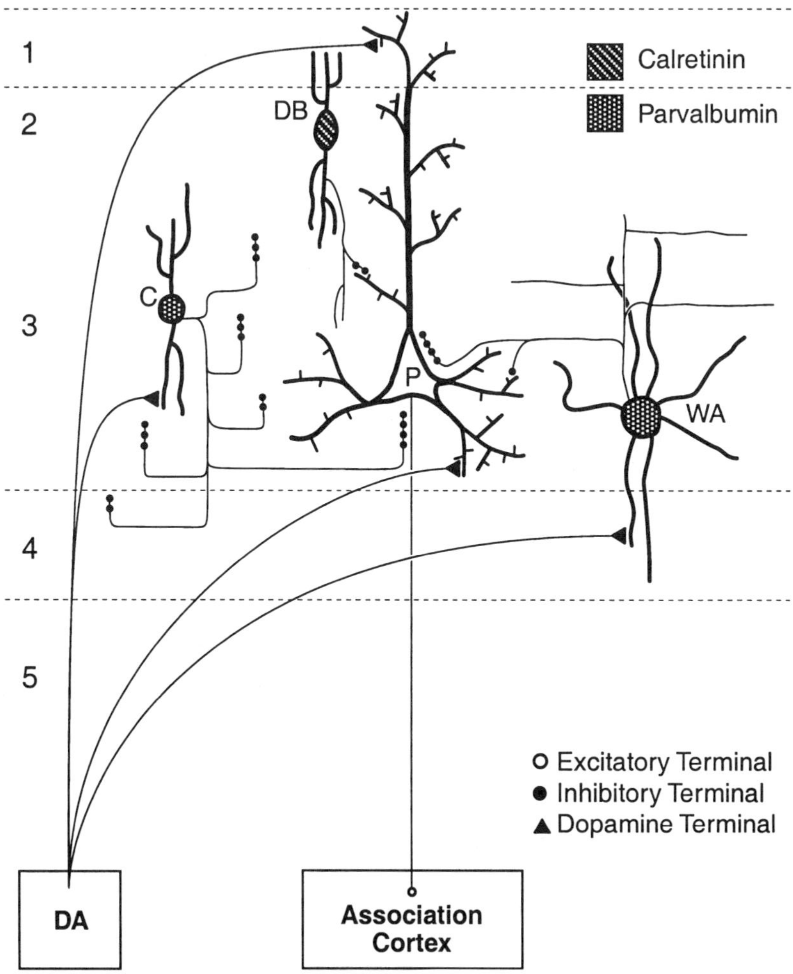

Figure 3–8. Schematic diagram illustrating the synaptic targets of DA projections to the monkey PFC. Note that DA axons innervate both pyramidal neurons and parvalbumin-containing local circuit neurons, but not calretinin-containing neurons.

(Mrzljak et al., 1996). Together, these observations suggest that DA does not exert a ubiquitous synaptic influence on all classes of cortical neurons, but rather has a more selective synaptic action on individual neuron types. The differential weighting of DA inputs to pyramidal and local circuit neurons is supported by recent observations that the density of putative inputs from TH-positive axons was twice as great for pyramidal neurons as for local circuit neurons (Krimer et al., 1997).

NE System

Similar to previous studies in rodents, DBH-immunoreactive terminals were found to form asymmetric synaptic junctions as well as symmetric synapses in the PFC of monkeys (Aoki et al., 1998). Although quantitative data regarding the postsynaptic targets for each type of synaptic specialization were not reported, DBH-immunoreactive terminals forming asymmetric synapses were shown to target dendritic spines, whereas those forming symmetric synapses targeted dendritic shafts. However, only 9% of these terminals examined in this study had morphologically identifiable synapses. Although this incidence is likely to be an underestimate, it may be that, like DA, DBH-immunoreactive terminals that do not form synaptic specializations may be involved in volume transmission.

In the PFC of macaque monkeys, immunocytochemical studies have indicated that the alpha2_A adrenergic receptor is the overwhelmingly predominant subtype (Aoki et al., 1998). Immunoreactivity for this receptor subtype was identified in both pyramidal and nonpyramidal neurons, and this receptor subtype was found at both pre- and postsynaptic locations. In postsynaptic locations, alpha2_A receptor immunoreactivity was found directly over the membranes of dendritic spines and shafts receiving asymmetric or symmetric synapses, respectively. However, synaptic localization was more prevalent for spines, which matches the synaptic pattern of NE axons. Thus, pyramidal neurons are the most likely target of NE axons in the PFC. alpha2_A receptor immunoreactivity was also associated with the presynaptic membrane and with both small clear and dense vesicles in axons forming synaptic junctions. These terminals predominantly formed asymmetric synapses. Some axons (including the preterminal portion) not forming any identifiable synapse were also found to be alpha2_A-immunoreactive. These terminals were often adjacent to other unlabeled axons and axon terminals. This axonal labeling may reflect the presence of autoreceptors and/or heteroreceptors.

Development of Cortical DA and NE Innervations

During the period of cortical neurogenesis and migration in primates, the first afferents to arrive in the cerebral wall appear to be the monoamines (Levitt, 1982; Berger et al., 1992), and these axons probably also form the first synaptic contacts in the marginal zone and subplate (Zecevic et al., 1989). In humans, monoaminergic axons penetrate the intermediate zone of the developing cortex at eight gestational weeks of age (Zecevic et al., 1991). At 11 gestational weeks, TH-labeled axons have invaded the subplate, and by 13 weeks, they are present in the cortical plate (Zecevic & Verney, 1995). In contrast, axons from the mediodorsal thalamic nucleus do not arrive in the subplate until gestational week 15 (Kostovic & Goldman-Rakic, 1983), and the afferent fibers from other cortical regions arrive still later (Goldman-Rakic, 1987). By gestational week 24, numerous presumptive DA axons can be identified in the cerebral wall, preferentially distributed in the deep cortical plate and upper subplate (Verney et al., 1993).

DA receptors also appear early in gestation, with D_1 receptors identified in the marginal zone as early as the second trimester (Lidow, 1995). In PFC, the highest densities of D_1 receptors are found in layers 1 and 5, with the adult pattern of bands of receptors in layers 1–3a and deep layers 5–6 present by E128. In general, similar developmental changes have been observed for D_2 receptors. The highest density of D_2 receptors is consistently found in layer 5. It is important to note that although the adult laminar pattern of receptor distribution was achieved early, as indicated below, the number of DA receptors within each layer undergoes substantial changes during postnatal development.

The early appearance of a DA innervation of the cortical plate has led to the suggestion that DA may play an important trophic role in the development of the cerebral cortex. Although comparable studies in primates are not available, early developmental lesions of DA neurons in rodents have been shown to produce a substantial decrease in cortical thickness (Kalsbeek et al., 1987), and in the total length of basal dendrites on layer 5 pyramids (Kalsbeek et al., 1989)

Substantial changes in the DA innervation of primate neocortex also occur postnatally (Rosenberg & Lewis, 1995), but these maturational patterns exhibit a high degree of regional and laminar specificity (Fig. 3–9). For example, in area 9 of monkey PFC, the density of TH-immunoreactive axons in layer 3 of remained relatively low during the first month of postnatal life, but then increased by a factor of three in animals 2 to 3 months of age. The density of labeled axons then continued to increase, reaching a peak value (6-fold greater than in neonates) in animals 2 to 3 years of age, the typical age of onset of puberty in this species. Axon density then rapidly declined to relatively stable adult levels by 5 years of age.

As indicated above, both dendritic spines and parvalbumin-containing local circuit neurons are synaptic targets of DA axon terminals in adult monkey PFC. Interestingly, synchronous changes occur in these three components of PFC circuitry during early postnatal development. However, the peripubertal increase in the density of DA axons appeared to begin prior to the decline in the densities of layer 3 pyramidal neuron dendritic spines and parvalbumin-immunoreactive chandelier neuron axon terminals in layer 3, and to persist until the adult levels of these markers of excitatory and inhibitory inputs were achieved (see Lewis, 1997a and 1997b for review). These patterns suggest that the neuromodulatory effects of DA may influence the adolescent refinement of excitatory and inhibitory inputs to layer 3 pyramidal neurons, and that DA may have a particularly strong influence on cortical information processing around the time of puberty. In this regard, it is interesting that the density of TH-positive axons in monkey cortex appears to change with exposure to ovarian hormones (Kritzer & Kohama, 1998).

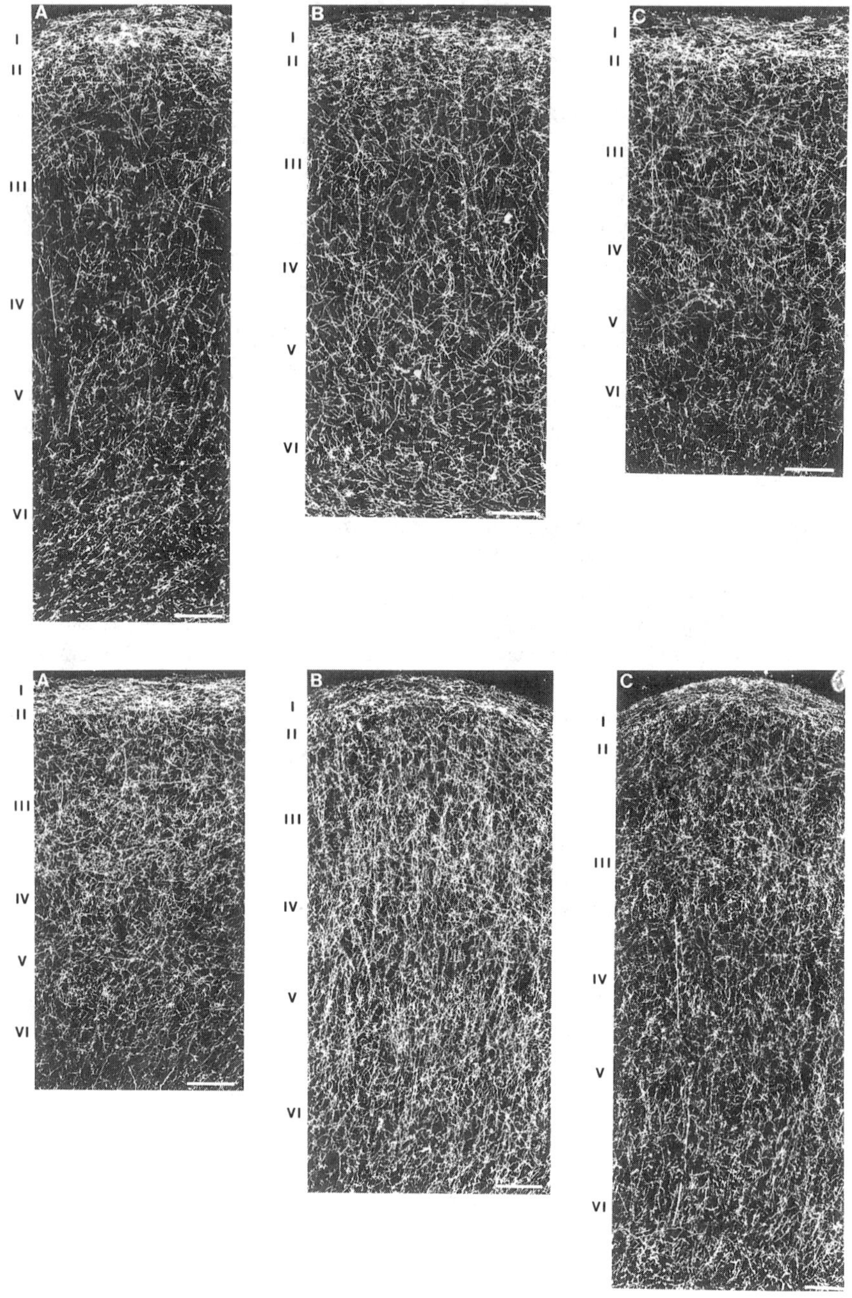

Figure 3–9. Dark-field photomicrographs of TH- immunoreactive axons in medial prefrontal cortex (area 9) from rhesus monkeys of different postnatal ages: A, 8 days; B, 37 days; C, 78 days; D, 1.0 year; E, 2.75 years; F, 5.7 years. Note the progressive, age-related increase in the density of labeled axons in the middle cortical layers, with maximal density achieved in the adolescent animal (E), before declining to adult levels (F). Scale bars = 200 μm. (Adapted from Rosenberg & Lewis, 1995.)

In contrast to these changes in layer 3, the densities of TH-labeled axons in layers 1 and 6 did not change appreciably over the entire period of postnatal development (Fig. 3–10A). Furthermore, in the rostral subdivision of monkey entorhinal cortex, the density of axons labeled with either antibodies against TH or the DAT showed yet a different pattern of developmental change (Erickson et al., 1998). In layer 3 of this region, the density of labeled axons increased from birth to peak at the middle of the first year of life, then declined over the next year to relatively stable levels, which were maintained through adulthood. In contrast to frontal cortical regions, DA axon density did not increase in any layer of the entorhinal cortex of animals 2 to 3 years of age.

Measures of DA tissue concentrations have also revealed regionally specific patterns of the postnatal development of cortical DA systems in monkeys (MacBrown & Goldman, 1977; Goldman-Rakic & Brown, 1982). For example, in the PFC, DA levels decreased during the first 6 months of postnatal life before gradually increasing to adult levels. This pattern of change contrasted with those observed in most other regions, where DA concentrations reached adult levels by 5 months of age and then remained stable through subsequent development. At all ages, however, the occipital cortex consistently contained the lowest concentration of DA.

Both D_1 and D_2 receptors, as measured by quantitative receptor autoradiography, also appeared to undergo distinctive patterns of postnatal development in monkey PFC (Lidow et al., 1991b; Lidow and Rakic, 1992). The densities of both receptor types were relatively low at birth, and then increased to peak values, about twice those present in adult animals, by 2 months of age. These elevated values appeared to be maintained until 4 months of age, and to then decline until adult levels were reached at the age of 3 years. D_1 binding sites were most dense in layers 1–3a and in layers 5–6, whereas D2 receptors (which had a much lower overall density) were most numerous in layer 5. These laminar patterns were consistently present throughout development and did not appear to change with the fluctuations in overall density of receptors. The changes in density of DA receptors did not correlate well with changes in cortical concentrations of DA (Goldman-Rakic & Brown, 1982) or with markers of the afferent axons (Rosenberg & Lewis, 1995), suggesting an independent time course for the development of pre- and postsynaptic markers of these systems (Lidow & Rakic, 1992).

Conclusion

The studies reviewed above indicate that the DA and NE projections to the monkey PFC exhibit regional, laminar, and target specificity, and that these patterns of innervation are likely to accurately reflect those present in the human brain. These findings provide insight into the anatomical substrate for possible alterations in PFC catecholamine systems in ADHD, and into the potential loci for the therapeutic action of catecholaminergic agents in this disorder. In addition, the substantial postnatal developmental and prolonged changes that occur in the DA innervation of the PFC may be important in considering both the therapeutic and adverse effects of pharmacological manipulations of catecholamine neurotransmission in children and adolescents.

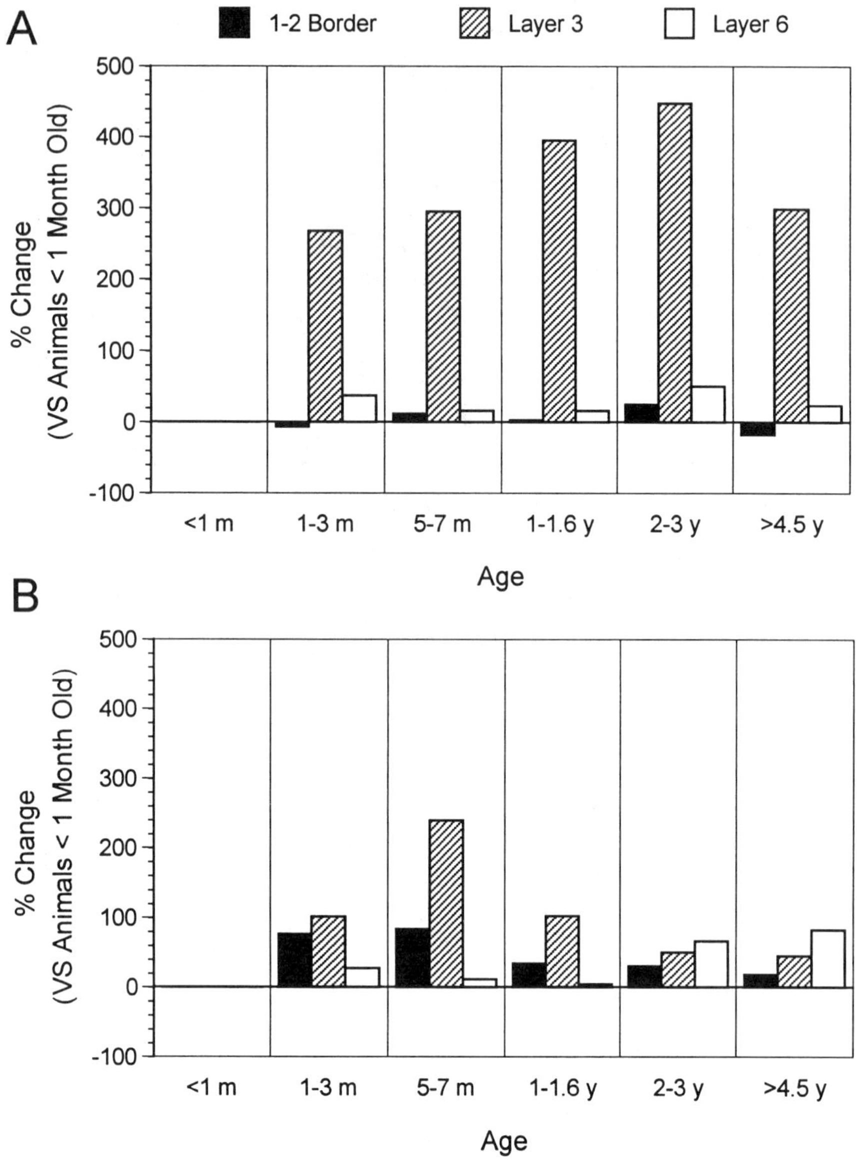

Figure 3–10. Postnatal developmental changes in the density of TH-immunoreactive axons, expressed as percent of the values in animals younger than one month of age, in layers 1, 3, and 6 of the prefrontal (*A*) and entorhinal (*B*) cortices of monkeys.

Work conducted by the author supported by USPHS grants MH00519, MH43784 and MH45156. The author thanks Mary Brady, Jenifer Hemphill, and Darlene Melchitzky for assistance in preparing this manuscript.

References

Acheson, AL, Kapatos, G, & Zigmond, MJ (1981). The effects of phosphorylating conditions on tyrosine hydroxylase activity are influenced by assay conditions and brain region. *Life Sci, 28,* 1407–1420.

Akil, M & Lewis, DA (1993). The dopaminergic innervation of monkey entorhinal cortex. *Cereb Cortex, 3*, 533–550.

Akil, M & Lewis, DA (1994). The distribution of tyrosine hydroxylase-immunoreactive fibers in the human entorhinal cortex. *Neuroscience, 60*, 857–874.

Akil, M, Pierri, JN, Whitehead, RE, Edgar, CL, Mohila, C, & Lewis, DA (1999). Lamina-specific alteration in the dopamine innervation of the prefrontal cortex in schizophrenic subjects. *Am J Psychiatry, 156*, 1580–1589.

Aoki, C, Venkatesan, C, Go, C-G, Forman, R, & Kurose, H (1998). Cellular and subcellular sites for noradrenergic action in the monkey dorsolateral prefrontal cortex as revealed by the immunocytochemical localization of noradrenergic receptors and axons. *Cereb Cortex, 8*, 269–277.

Berger, B, Alvarez, C, & Goldman-Rakic, PS (1993). Neurochemical development of the hippocampal region in the fetal rhesus monkey. I. Early appearance of peptides, calcium-binding proteins, DARPP-32, and monoamine innervation in the entorhinal cortex during the first half of gestation (E47 to E90). *Hippocampus, 3*, 279–305.

Berger, B, Gaspar, P, & Verney, C (1991). Dopaminergic innervation of the cerebral cortex: Unexpected differences between rodents and primates. *TINS, 14*, 21–27.

Berger, B, Trottier, S, Gaspar, P, Verney, C, & Alvarez, C (1986). Major dopamine innervation of the cortical motor areas in the Cynomolgus monkey. A radioautographic study with comparative assessment of serotonergic afferents. *Neurosci Lett, 72*, 121–127.

Berger, B, Trottier, S, Verney, C, Gaspar, P, & Alvarez, C (1988). Regional and laminar distribution of the dopamine and serotonin innervation in the macaque cerebral cortex: A radioautographic study. *J Comp Neurol, 273*, 99–119.

Berger, B, Verney, C, & Goldman-Rakic, PS (1992). Prenatal monaminergic innervation of the cerebral cortex: Differences between rodents and primates. In I Kostovic, S Knezevic, H Wisniewski, & G Spilich (Eds), *Neurodevelopment, aging and cognition* (pp. 18–36). Boston: Birkhauser.

Bergson, C, Mrzljak, L, Smiley, JF, Pappy, M, Levenson, R, & Goldman-Rakic, PS (1995). Regional, cellular, and subcellular variations in the distribution of D_1 and D_5 dopamine receptors in primate brain. *J Neurosci, 15*, 7821–7836.

Bigham, MH & Lidow, MS (1995). Adrenergic and serotonergic receptors in aged monkey neocortex. *Neurobiol Aging, 16*, 91–104.

Björklund, A, Divac, I, & Lindvall, O (1978). Regional distribution of catecholamines in monkey cerebral cortex, evidence for a dopaminergic innervation of the primate prefrontal cortex. *Neurosci Lett, 7*, 115–119.

Brene, S, Hall, H, Lindefors, N, Karlsson, P, Halldin, C, & Sedvall, G (1995). Distribution of messenger RNAs for D_1 dopamine receptors and DARPP-32 in striatum and cerebral cortex of the cynomolgus monkey: Relationship to D_1 dopamine receptors. *Neuroscience, 67*, 37–48.

Brown, RM, Crane, AM, & Goldman, PS (1979). Regional distribution of monoamines in the cerebral cortex and subcortical structures of the rhesus monkey: Concentrations and *in vivo* synthesis rates. *Brain Res, 168*, 133–150.

Campbell, MJ, Lewis, DA, Foote, SL, & Morrison, JH (1987). Distribution of choline acetyltransferase-, serotonin-, dopamine-beta-hydroxylase-, tyrosine hydroxylase-immunoreactive fibers in monkey primary auditory cortex. *J Comp Neurol, 261*, 209–220.

Camps, M, Cortés, R, Gueye, B, Probst, A, & Palacios, JM (1989). Dopamine receptors in human brain: Autoradiographic distribution of D_2 sites. *Neuroscience, 28*, 275–290.

Ciliax, BJ, Drash, GW, Staley, JK, Haber, S, Mobley, CJ, Miller, GW, Mufson, EJ, Mash, DC, & Levey, AI (1999). Immunocytochemical localization of the dopamine transporter in human brain. *J Comp Neurol, 409*, 38–56.

Coker III, GT, Studelska, D, Harmon, S, Burke, W, & O'Malley, KL (1990). Analysis of tyrosine hydroxylase and insulin transcripts in human neuroendocrine tissues. *Mol Brain Res, 8*, 93–98.

Condé, F, Lund, JS, Jacobowitz, DM, Baimbridge, KG, & Lewis, DA (1994). Local circuit neurons immunoreactive for calretinin, calbindin D-28k, or parvalbumin in monkey prefrontal cortex: Distribution and morphology. *J Comp Neurol, 341*, 95–116.

Cortés, R, Gueye, B, Pazos, A, Probst, A, & Palacios, JM (1989). Dopamine receptors in human brain: Autoradiographic distribution of D_1 sites. *Neuroscience, 28*, 263–273.

Dawson, TM, McCabe, RT, Stensaas, SS, & Wamsley, JK (1987). Autoradiographic evidence of [^{3}H]SCH 23390 binding sites in human prefrontal cortex (Brodmann's Area 9). *J Neurochem, 49*, 789–796.

Emson, PC, & Koob, GF (1978). The origin and distribution of dopamine-containing afferents to the rat frontal cortex. *Brain Res, 142*, 249–267.

Erickson, SL, Akil, M, Levey, AI, & Lewis, DA (1998). Postnatal development of tyrosine hydroxylase- and dopamine transporter-immunoreactive axons in monkey rostral entorhinal cortex. *Cereb Cortex, 8*, 415–427.

Gaspar, P, Berger, B, Fabvret, A, Vigny, A, & Henry, JP (1989). Catecholamine innervation of the human cerebral cortex as revealed by comparative immunohistochemistry of tyrosine hydroxylase and dopamine-beta-hydroxylase. *J Comp Neurol, 279*, 249–271.

Gaspar, P, Berger, B & Febvret, A (1990). Neurotensin innervation of the human cerebral cortex: Lack colocalization with catecholamines. *Brain Res, 530*, 181–195.

Gaspar, P, Duyckaerts, C, Alvarez, C, Javoy-Agid, F, & Berger, B (1991). Alterations of dopaminergic and noradrenergic innervations in motor cortex in Parkinson's Disease. *Ann Neurol, 30*, 365–374.

Goldman-Rakic, PS (1987). Development of cortical circuitry and cognitive function. *Child Devel, 58*, 601–622.

Goldman-Rakic, PS & Brown, RM (1982). Postnatal development of monoamine content and synthesis in the cerebral cortex of rhesus monkeys. *Dev Brain Res, 4*, 339–349.

Goldman-Rakic, PS, Leranth, C, Williams, SM, Mons, N, & Geffard, M (1989). Dopamine synaptic complex with pyramidal neurons in primate cerebral cortex. *Proc Natl Acad Sci USA, 86*, 9015–9019.

Goldman-Rakic, PS, Lidow, MS, & Gallagher, DW (1990). Overlap of dopaminergic, adrenergic, and serotoninergic receptors and complementarity of their subtypes in primate prefrontal cortex. *J Neurosci, 10*, 2125–2138.

Hökfelt, T, Fuxe, K, Goldstein, M, & Johansson, O (1974). Immunohistochemical evidence for the existence of adrenaline neurons in the rat brain. *Brain Res, 66*, 235–251.

Huntley, GW, Morrison, JH, Prikhozhan, A, & Sealfon, SC (1992). Localization of multiple dopamine receptor subtype mRNAs in human and monkey motor cortex and striatum. *Mol Brain Res, 15*, 181–188.

Ichikawa, S, Ichinose, H, & Nagatsu, T (1990). Multiple mRNAs of monkey tyrosine hydroxylase. *Biochem Biophys Res Commun, 173,* 1331–1336.

Joh, TH & Reis, DJ (1974). Different forms of tyrosine hydroxylase in noradrenergic and dopaminergic systems in brain [Abstract]. *Federation Proceedings, 33,* 535.

Kalsbeek, A, Buijs, RM, Hofman, MA, Matthijssen, MAH, Pool, CW, & Uylings, HBM (1987). Effects of neonatal thermal lesioning of the mesocortical dopaminergic projection on the development of the rat prefrontal cortex. *Dev Brain Res, 32,* 123–132.

Kalsbeek, A, Matthijssen, MAH, & Uylings, HBM (1989). Morphometric analysis of prefrontal and cortical development following neonatal lesioning of the dopaminergic mesocortical projection. *Exp Brain Res, 78,* 279–289.

Kostovic, I, & Goldman-Rakic, PS (1983). Transient cholinesterase staining in the mediodorsal nucleus of the thalamus and its connections in the developing human and monkey brain. *J Comp Neurol, 219,* 431–447.

Krimer, LS, Jakab, RL, & Goldman-Rakic, PS (1997). Quantitative three-dimensional analysis of the catecholaminergic innervation of identified neurons in the macaque prefrontal cortex. *J Neurosci, 17,* 7450–7461.

Kritzer, MF & Kohama, SG (1998). Ovarian hormones influence the morphology, distribution, and density of tyrosine hydroxylase immunoreactive axons in the dorsolateral prefrontal cortex of adult rhesus monkeys. *J Comp Neurol, 395,* 1–17.

Lahti, RA, Roberts, RC, & Tamminga, CA (1995). D_2-Family receptor distribution in human postmortem tissue: An autoradiographic study. *NeuroReport, 6,* 2505–2512.

Lavoie, B, Smith, Y, & Parent, A (1989). Dopaminergic innervation of the basal ganglia in the squirrel monkey as revealed by tyrosine hydroxylase immunohistochemistry. *J Comp Neurol, 289,* 36–52.

Levitt, P (1982). Central monoamine neuron systems: Their organization in the developing and mature primate brain and the genetic regulation of their terminal fields. In AJ Friedhoff & TN Chase (Eds), *Gilles de la Tourette Syndrome* (pp. 49–59). New York: Raven.

Levitt, P, Rakic, P, & Goldman-Rakic, P (1984). Region-specific distribution of catecholamine afferents in primate cerebral cortex: A fluorescence histochemical analysis. *J Comp Neurol, 227,* 23–36.

Lewis, DA (1992). The catecholaminergic innervation of primate prefrontal cortex. *J Neural Transm, 36,* 179–200.

Lewis, DA (1997a). Development of the prefrontal cortex during adolescence: Insights into vulnerable neural circuits in schizophrenia. *Neuropsychopharm, 16,* 385–398.

Lewis, DA (1997b). Development of the primate prefrontal cortex. In MS Keshavan & RM Murray (Eds), *Neurodevelopment & adult psychopathology* (pp. 12–30). Cambridge: Cambridge University Press.

Lewis, DA, Campbell, MJ, Foote, SL, Goldstein, M, & Morrison, JH (1987). The distribution of tyrosine hydroxylase-immunoreactive fibers in primate neocortex is widespread but regionally specific. *J Neurosci, 7,* 279–290.

Lewis, DA, Foote, SL, Goldstein, M, & Morrison, JH (1988a). The dopaminergic innervation of monkey prefrontal cortex: A tyrosine hydroxylase immunohistochemical study. *Brain Res, 449,* 225–243.

Lewis, DA, Morrison, JH, & Goldstein, M (1988b). Brainstem dopaminergic neurons project to monkey parietal cortex. *Neurosci Lett, 86,* 11–16.

Lewis, DA, Melchitzky, DS, & Haycock, JW (1993). Four isoforms of tyrosine hydroxylase are expressed in human brain. *Neuroscience, 54,* 477–492.

Lewis, DA, Melchitzky, DS, & Haycock, JW (1994). Expression and distribution of two isoforms of tyrosine hydroxylase in macaque monkey brain. *Brain Res, 656,* 1–13.

Lewis, DA, Sesack, SR, Levey, AI, & Rosenberg, DR (1998). Dopamine axons in primate prefrontal cortex: Specificity of distribution, synaptic targets, and development. In D Goldstein, G Eisenhofer, & R McCarty (Eds), *Advances in pharmacology: Catecholamines: bridging basic science with clinical medicine* (pp. 703–706). San Diego: Academic Press.

Lidow, MS (1995). D_1- and D_2 dopaminergic receptors in the developing cerebral cortex of macaque monkey: A film autoradiographic study. *Neuroscience, 65,* 439–452.

Lidow, MS, Goldman-Rakic, PS, Gallagher, DW, Geschwind, DH, & Rakic, P (1989). Distribution of major neurotransmitter receptors in the motor and somatosensory cortex of the rhesus monkey. *Neuroscience, 32,* 609–627.

Lidow, MS, Goldman-Rakic, PS, Gallagher, DW, & Rakic, P (1991a). Distribution of dopaminergic receptors In the primate cerebral cortex: Quantitative autoradiographic analysis using [^{3}H]raclopride, [^{3}H]spiperone and [^{3}H]SCH23390. *Neuroscience, 40,* 657–671.

Lidow, MS, Goldman-Rakic, PS, & Rakic, P (1991b). Synchronized overproduction of neurotransmitter receptors in diverse regions of the primate cerebral cortex. *Proc Natl Acad Sci USA, 88,* 10218–10221.

Lidow, MS & Rakic, P (1992). Scheduling of monoaminergic neurotransmitter receptor expression in the primate neocortex during postnatal development. *Cereb Cortex, 2,* 401–416.

Lidow, MS, Wang, F, Cao, Y, & Goldman-Rakic, PS (1998). Layer V neurons bear the majority of mRNAs encoding the five distinct dopamine receptor subtypes in the primate prefrontal cortex. *Synapse, 28,* 10–20.

Lindvall, O & Björklund, A (1974). The organization of the ascending catecholamine neuron systems in the rat brain as revealed by the glyoxylic acid fluorescence method. *Acta Physiol Scand, 412,* 1–48.

MacBrown, R & Goldman, PS (1977). Catecholamines in neocortex of rhesus monkeys: Regional distribution and ontogenetic development. *Brain Res, 124,* 576–580.

Moghaddam, B, Berridge, CW, Goldman-Rakic, PS, Bunney, BS, & Roth, RH (1993). *In vivo* assessment of basal and drug-induced dopamine release in cortical and subcortical regions of the anesthetized primate. *Synapse, 13,* 215–222.

Morrison, JH, Foote, SL, O'Connor, D, & Bloom, FE (1982). Laminar, tangential and regional organization of the noradrenergic innervation of monkey cortex: Dopamine-beta-hydroxylase immunohistochemistry. *Brain Res Bull, 9,* 309–319.

Mrzljak, L, Bergson, C, Pappy, M, Huff, R, Levenson, R, & Goldman-Rakic, PS (1996). Localization of dopamine D4 receptors in GABAergic neurons of the primate brain. *Nature, 381,* 245–248.

Noack, HJ & Lewis, DA (1989). Antibodies directed against tyrosine hydroxylase differentially recognize noradrenergic axons in monkey neocortex. *Brain Res, 500,* 313–324.

Oeth, KM & Lewis, DA (1992). Cholecystokinin- and dopamine- containing mesencephalic neurons provide distinct projections to monkey prefrontal cortex. *Neurosci Lett, 145,* 87–92.

Richfield, EK, Young, AB, & Penney, JB (1989). Comparative distributions of dopamine D-1 and D-2 receptors in the cerebral cortex of rats, cats, and monkeys. *J Comp Neurol, 286,* 409–426.

Rosenberg, DR & Lewis, DA (1995). Postnatal maturation of the dopaminergic innervation of monkey prefrontal and motor cortices: A tyrosine hydroxylase immunohistochemical analysis. *J Comp Neurol, 358,* 383–400.

Saunders, RC, Kolachana, BS, & Weinberger, DR (1994). Local pharmacological manipulation of extracellular dopamine levels in the dorsolateral prefrontal cortex and caudate nucleus in the rhesus monkey: An *in vivo* microdialysis study. *Exp Brain Res, 98,* 44–52.

Sawaguchi, T, Matsumura, M, & Kubota, K (1986). Catecholamine sensitivities of motor cortical neurons of the monkey. *Neurosci Lett, 66,* 135–140.

Schmidt, RH & Bhatnagar, RK (1979). Assessment of the effects of neonatal subcutaneous 6-hydroxydopamine on noradrenergic and dopaminergic innervation of the cerebral cortex. *Brain Res, 166,* 309–319.

Sesack, SR, Bressler, CN, & Lewis, DA (1995a). Ultrastructural associations between dopamine terminals and local circuit neurons in the monkey prefrontal cortex: A study of calretinin-immunoreactive cells. *Neurosci Lett, 200,* 9–12.

Sesack, SR, Snyder, CL, & Lewis, DA (1995b). Axon terminals immunolabeled for dopamine or tyrosine hydroxylase synapse on GABA-immunoreactive dendrites in rat and monkey cortex. *J Comp Neurol, 363,* 264–280.

Sesack, SR, Hawrylak, VA, Melchitzky, DS, & Lewis, DA (1998). Dopamine innervation of a subclass of local circuit neurons in monkey prefrontal cortex: Ultrastructural analysis of tyrosine hydroxylase and parvalbumin immunoreactive structures. *Cereb Cortex, 8,* 614–622.

Smiley, JF & Goldman-Rakic, PS (1993). Heterogeneous targets of dopamine synapses in monkey prefrontal cortex demonstrated by serial section electron microscopy: A laminar analysis using the silver-enhanced diaminobenzidine sulfide (SEDS) immunolabeling technique. *Cereb Cortex, 3,* 223–238.

Smiley, JF, Levey, AI, Ciliax, BJ, & Goldman-Rakic, PS (1994). D_1 dopamine receptor immunoreactivity in human and monkey cerebral cortex: Predominant and extrasynaptic localization in dendritic spines. *Proc Natl Acad Sci USA, 91,* 5720–5724.

Van Eden, CG, Hoorneman, EMD, Buijs, RM, Matthijssen, MAH, Geffard, M, & Uylings, HBM (1987). Immunocytochemical localization of dopamine in the prefrontal cortex of the rat at the light and electron microscopical level. *Neuroscience, 22,* 849–862.

Verney, C, Milosevic, A, Alvarez, C, & Berger, B (1993). Immunocytochemical evidence of well-developed dopaminergic and noradrenergic innervations in the frontal cerebral cortex of human fetuses at midgestation. *J Comp Neurol, 336,* 331–344.

Williams, SM & Goldman-Rakic, PS (1993). Characterization of the dopaminergic innervation of the primate frontal cortex using a dopamine-specific antibody. *Cereb Cortex, 3,* 199–222.

Zecevic, N, Bourgeois, J-P, & Rakic, P (1989). Changes in synaptic density in motor cortex of rhesus monkey during fetal and postnatal life. *Dev Brain Res, 50,* 11–32.

Zecevic, N, & Verney, C (1995). Development of the catecholamine neurons in human embryos and fetuses, with special emphasis on the innervation of the cerebral cortex. *J Comp Neurol, 351,* 509–535.

Zecevic, N, Verney, C, Milosevic, A, & Berger, B (1991). First description of the central catecholamine (CA) systems in 6–8 week-old human embryos [Abstract]. *Soc Neurosci Abstr, 17,* 745.

4

Stimulant Effects on Striatal and Cortical Dopamine Systems Involved in Reward-Related Behavior and Impulsivity

JANE R. TAYLOR AND J. DAVID JENTSCH

Attention deficit/hyperactivity disorder (ADHD) has recently been described as a disorder of the active inhibition of impulsive behavior (see for review, Barkley, 1997). This hypothesis predicts that the relative inability of ADHD patients to perform well in academic and social situations is a consequence of increased stimulus-control of behavior and reduced capacity for self-regulation at the cognitive level. Consequently, several researchers have suggested that dysfunction of the prefrontal cortex is fundamental to ADHD given the resemblance between ADHD patients and subjects with damage restricted to the frontal lobe. Interestingly, long-term intake of psychomotor stimulant drugs also results in a psychopathology that is strikingly similar to ADHD—augmented incentive motivational systems, hyperactivity, augmented stimulus control of behavior, and impaired inhibitory control. These findings seem paradoxical given that these very drugs are effective treatments for the classic clinical symptoms of ADHD.

In the following sections, we will outline the pharmacology and neuroanatomy of the subcortical dopamine (DA) systems (the mesolimbic and nigrostriatal DA system) because these neurons are the principal substrates of therapeutic action of stimulant drugs and because they mediate the unconditioned and conditioned responses toward stimuli with reinforcing value. The characteristic behavioral effects of acute and chronic psychomotor stimulant drugs will be reviewed, including locomotor activation, stereotypy, and conditioned reward and stimulus-reward learning. The evidence for impaired inhibitory control after chronic drug administration also will be discussed in terms of dysfunction of cortical-subcortical DA systems. It should be emphasized, however, that the majority of studies on psychomotor stimulant drugs have been performed in rodents and have used moderate to high doses of these drugs given by systemic injection. The implications of these studies for ADHD should be considered with this context.

Neuroanatomy of Subcortical DA Systems

DA-producing cells within the ventral mesencephalon project to widespread telencephalic regions, including the dorsal and ventral striatum, pallidum, septum, hippocampus and related structures, amygdala, and neocortex (Björklund & Lindvall, 1984). These cells can be parceled into a number of nuclear groups, including the ventral tegmental area (VTA, A10), substantia nigra pars compacta (A9), and retrorubral nuclei (A8). Moreover, these original anatomical subdivisions have themselves been subdivided into even more subnuclei; however, these cellular groups are not precisely defined, having considerable anatomical overlap, as described by Lewis (see Chapter 3).

Fallon and Moore (1978,1988) described several general topographic principles that appear to apply to the mesencephalic-to-telencephalon DA projections. First, there is rough medial-to-lateral topography, with more medial DA cells projecting to more medial striatal regions. Second, they described a rostral-to-caudal topography with more rostral DA neurons projecting to more rostral telencephalic structures. Finally, the most ventral DA neurons within the mesencephalon tend to project more dorsally within the forebrain, exhibiting an inverted dorsal-to-ventral topography.

It was classically thought that the A9 nucleus provided discrete input to the dorsolateral (sensorimotor) striatum, and the A10 nucleus provided the majority of the dopaminergic innervation of the ventral (limbic) striatum and allo- and neocortex (Björklund & Lindvall, 1984). Although these distinctions are generally valid, this view has been revised based on the finding that there is some overlap in the mesencephalic origins of the telencephalic DA innervation (Björklund & Lindvall, 1984; Lewis & Sesack, 1997; Williams & Goldman-Rakic, 1998). Moreover, single DA neurons can provide collateral projections to anatomically separate forebrain regions, such as the frontal cortex and the septum (Björklund & Lindvall, 1984). In general, the dorsolateral regions of the striatum are innervated by neurons from the ventrolateral regions of the A9 nucleus. These neurons have certain morphological and neurochemical characteristics that render them distinct from medial A10 and dorsal A9 neurons that provide a greater input to the nucleus accumbens (NAc) and neocortex. For instance, cells of the ventrolateral A9 nucleus are larger and more densely packed than in the A10 or dorsal A9 regions. Moreover, the ventral A9 neurons express higher levels of mRNA for the DA transporter (DAT) and for DA D2 'auto' receptors (Haber et al., 1995).

Within the striatum, DA terminals make direct synaptic contacts with the spines of striatal neurons. These synaptic contacts appear to provide the anatomical substrate for both compartmental (synaptic) and volume (extra-synaptic) transmission (see Fig. 4–1). The DAT is located intra- or peri-synaptically, suggesting that DA has a limited ability to escape the striatal synapse (Sesack et al., 1998). Moreover, DA, D2, and to a lesser extent D1, receptors are located postsynaptically to presumed DA terminals, suggesting compartmental transmission (An et al., 1998; Jansson et al., 1998). However, there is a significant cohort of D1 and D2 receptors that are not directly opposed to DA terminals, suggesting

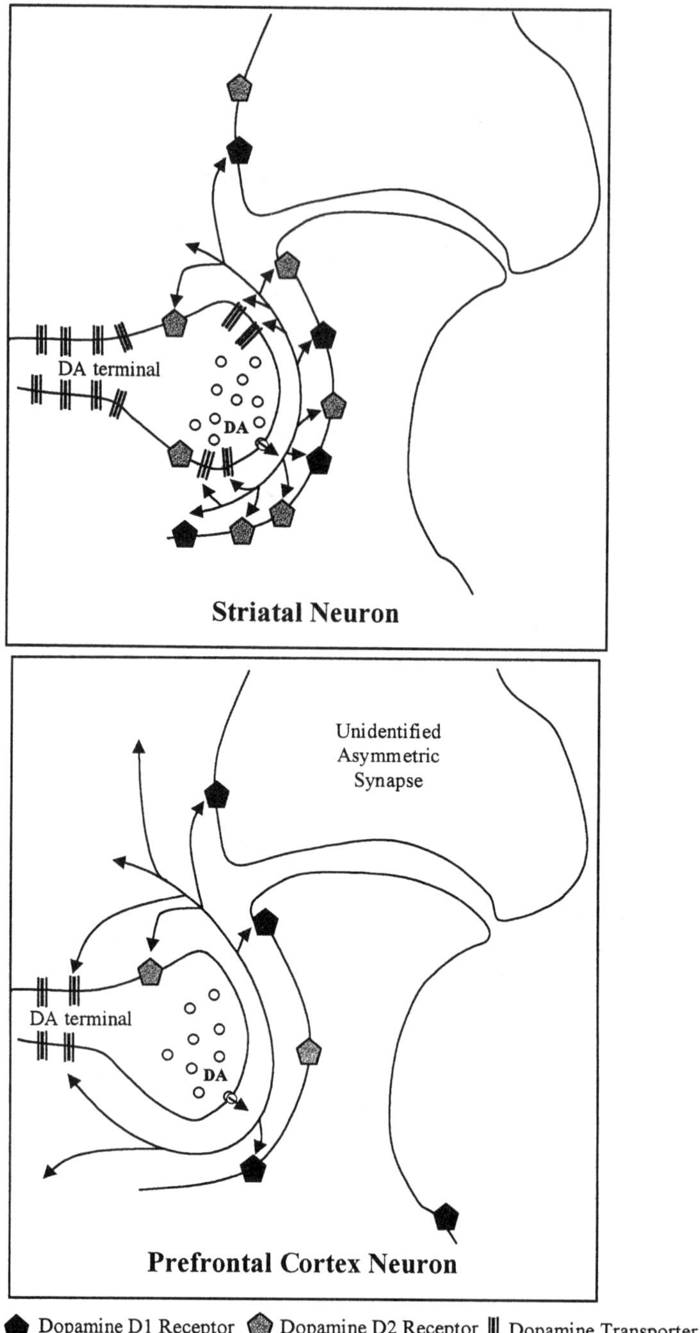
DA terminal
DA
Striatal Neuron
Unidentified
Asymmetric
Synapse
DA terminal
DA
Prefrontal Cortex Neuron
Dopamine D1 Receptor
Dopamine D2 Receptor
Dopamine Transporter

some component of volume transmission within the striatum (An et al., 1998). These data suggest that both tonic and phasic changes in DA transmission are critical in producing behavioral effects associated with the striatum.

In the cerebral cortex, volume transmission appears to be more critical in mediating the effects of DA (see Figure 1b). DAT density is reduced in the frontal cortex relative to the striatum, and it is primarily localized extra-synaptically (Sesack et al., 1998). Moreover, D1-immunoreactivity within frontal cortex is virtually never opposed to tyrosine hydroxylase-immunopositive terminals (An et al., 1998). These data suggest that DA's effects at the D1 receptor are mediated by volume transmission. This hypothesis has direct relevance to the neurochemistry of ADHD because cognitive functions known to be affected in this disorder, namely working memory and inhibitory control, are sensitive to manipulations of DA D1 receptor-mediated DA transmission (Broersen et al., 1994; Sawaguchi & Goldman-Rakic, 1991; Zahrt et al., 1997) and working memory is sensitive to depletion of DA levels within the prefrontal cortex (Brozoski et al., 1979; Bubser & Schmidt, 1990). These data suggest that the tonic component of DA transmission may be more critical for the behavioral functions of the frontal cortex.

Biochemical Pharmacology of Subcortical DA Systems

Transmission within the mesostriatal DA system can be pharmacologically modulated both at the somatodendritic and the terminal levels. Several classes of drugs modulate the firing rates or patterns of midbrain DA neurons by direct (monosynaptic) or indirect (polysynaptic) inputs to the cell bodies within the ventral mesencephalon, e.g., nicotine, opiates, benzodiazepines, cannabinoids, phencyclidine (Kalivas et al., 1998; Kalivas et al., 1993; White, 1996). In contrast, other drugs, including amphetamine, cocaine, and methylphenidate, act at the level of the DA terminal, interfering with normal processes of transmitter packaging,

Figure 4–1. Synaptology of dopaminergic transmission onto striatal and cortical neurons. Diagrams of striatal (top) and cortical (bottom) terminals showing DA transporter (DAT) sites and DA D1 and D2 receptor distributions. Figure is modified from Sesack et al. (1998). DA transmission within the striatum and prefrontal cortex is regulated at a variety of levels. *Top:* Synaptic DA in the striatum is subject to extensive reuptake by DAT localized on the presynaptic zone and preterminal regions, suggesting that DA has a limited ability to escape the synapse. DA D1 and D2 receptors are located postsynaptically as well as pre-synaptically, indicating both compartmental (synaptic) and volume transmission (extra-synaptic). *Bottom:* Because of the paucity of perisynaptic DAT in the frontal cortex, synaptically released DA is subject to less reuptake and thus greater extracellular diffusion. Released transmitter then stimulates postsynaptic and presynaptic DA receptors, producing post-synaptic potentials and feedback inhibition, respectively. Extrasynaptic DA D1 receptors are also stimulated by DA that escapes the synapse because of the low concentration of synaptic DAT.

release, reuptake, and metabolism (White, 1996). As the stimulant drugs that are known to be efficacious in ADHD largely act at the terminal level, we will focus on those mechanisms here.

Figure 4–2 depicts the biochemical processes that underlie the formation, release and subsequent disposition of DA within a classical mesostriatal DA synapse (see Cooper et al., 1996). DA is formed from the conversion of *L*-tyrosine to *L*-3-(3, 4-dihydroxy-o-phenylalanine) (DOPA) to DA by tyrosine hydroxylase (TH) and aromatic acid decarboxylase, respectively. TH is the rate-limiting step in the synthesis of DA, and this enzyme is allosterically regulated by a number of influences. Increases in impulse flow, cyclic AMP accumulation (that stimulates protein kinase A, PKA) and intracellular calcium (Ca^{++}) can lead to phosphorylation of TH, increasing its kinetic activity. In contrast, increases in nonvesicular DA concentrations can produce substrate ("end-product") inhibition of TH that is phosphorylation-dependent.

Following synthesis, vesicular DA is packaged into synaptic vesicles by the vesicular monoamine transporter (VMAT). These vesicles typically remain "docked" at the synaptic cleft until a local increase in the Ca^{++} concentration triggers fusion of the vesicle with the cell membrane and release of the amine into the synaptic space.

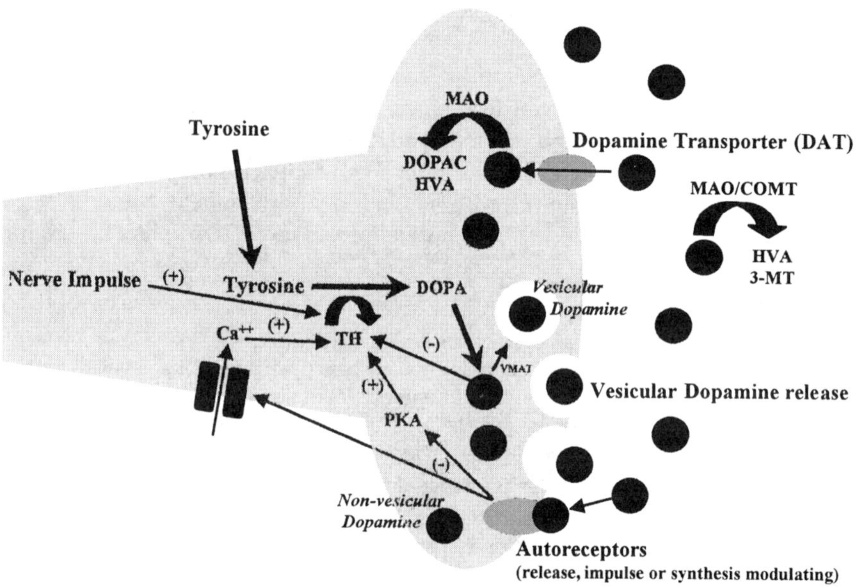

Figure 4–2. Mesostriatal dopaminergic nerve terminal. Schematized representation of a typical mesostriatal DA terminal. A number of pre- and intrasynaptic mechanisms govern the synthesis, storage, release and metabolism of DA. Principal sites of action of typical psychomotor stimulant drugs include the dopamine transporter (DAT) and monoamine oxidase (MAO).

This rapid ("phasic") increase in synaptic DA levels activates post-synaptic DA receptors.

After release, extracellular DA must be rapidly inactivated, and there appear to be three principal substrates to subserve this effect. First, DA can be metabolized within the synaptic cleft by the enzyme catechol-O-methyl transferase (COMT). The main product of this pathway is 3-methoxytyramine (3-MT), and this metabolite, though it exists in extremely low levels in the brain, has been argued to be the only true metabolite measure of DA release in vivo. Second, DA, itself, can stimulate presynaptic DA receptors, termed autoreceptors, and stimulation of these receptors results in decreases in DA synthesis (by inhibition of TH), DA release (by inhibition of Ca^{++} channels), as well as, impulse flow (by activation of K^+ channels). Third, DA can be transported back into the presynaptic DA terminal by the DAT. Intracellular DA is metabolized by the enzyme monoamine oxidase (MAO) to 3,4-dihydroxy-O-phenylacetic acid (DOPAC) and/or further catabolized by COMT to 4-hydroxy-3-methoxy-O-phenylacetic acid (homovanillic acid or HVA). Other monoamines, such as norepinephrine (NE) and serotonin (5-HT), utilize similar mechanisms.

Stimulant drugs have a number of different pharmacological effects on monoamine terminals that permit them to alter these normal mechanisms of transmitter production, release, and metabolism. Methylphenidate, like cocaine, largely acts by blocking reuptake of monoamines into the presynaptic terminal; this release is reserpine-insensitive because it does not involve vesicular- or impulse-dependent transmitter release. Methylphenidate administration produces an increase in the steady-state ("tonic") levels of monoamines within the synaptic space. Thus, DAT inhibitors, such as methylphenidate, increase extracellular levels of monoamines. In contrast, they decrease the concentrations of the monoamine metabolites that depend upon MAO (i.e., HVA and DOPAC), but not COMT, because reuptake by the transporter is required for the formation of these acidic monoamine metabolites (Nielsen et al., 1983). By stimulating presynaptic autoreceptors, methylphenidate-induced increases in tonic monoamine release can also (*1*) reduce monoamine synthesis, (*2*) inhibit monoamine neuron firing, and (*3*) reduce subsequent phasic transmitter release (Grace, 1991; Grace, 1995).

The pharmacology of amphetamine is considerably more complex. First, inhibition of monoamine reuptake alone is not sufficient to explain the effects of amphetamine. In fact, selective DA reuptake inhibitors block amphetamine-induced DA release because the pharmacological effects of amphetamine are exerted within the presynaptic DA terminal and uptake of amphetamine by the DAT is required for these processes. Data indicate that amphetamine gains access to the presynaptic terminal via the monoamine transporters and directly inhibits the activity of VMAT, resulting in an increase in cytosolic (non vesicular) levels of monoamine neurotransmitters. Eventually, increasing cytosolic concentrations of the neurotransmitters lead to reverse transport of the amines via the membrane-bound transporters (Sulzer et al., 1995; Sulzer et al., 1993). Thus, DAT inhibitors may block the actions of amphetamine by (*1*) preventing its entry into the presynaptic bouton and (*2*) preventing reverse transport.

In addition to increasing DA release, amphetamine has an additional, pharmacological effect that differentiates it from methylphenidate: it directly inhibits MAO. Amphetamine decreases formation of the acidic metabolites of DA, NE, and 5-HT in a dose-dependent manner by inhibiting the catabolism of the monoamines by MAO; even at low doses, it increases DA release while reducing DOPAC formation (Bowers & Hoffman, 1984; Elchisak et al., 1976; Imperato and Di Chiara, 1984; Karoum et al., 1994; Zetterstrom et al., 1988; Zetterstrom et al., 1983). As it does not inhibit COMT, amphetamine increases the accumulation of 3-MT and normetanephrine (a metabolite of NE) in the brain (Karoum et al., 1994).

Amphetamine administration can strongly evoke the release of both DA and 5-HT in the striatum; in contrast, there is little NE innervation of the striatum (except in the nucleus accumbens shell), and as such, the effects of amphetamine on this neurotransmitter system within the striatum are minimal. Alternatively, methylphenidate produces a substantial increase in DA, but not 5-HT release in the striatal complex (Kuczenski & Segal, 1997). These increases in release of DA from nigrostriatal dopamine neurons leads to strong feedback inhibition of the firing of DA neurons by both autoreceptor-mediated mechanisms as well as the striatonigral feedback pathways. Thus, mild increases in tonic DA release produced by these drugs could have important feedback consequences on subsequent phasic release.

Biochemical Pharmacology of Cortical DA Systems

Although DA terminals that innervate the cerebral cortex are similar to those in the striatum, there are several significant differences. First, although DA terminals that innervate the cortical regions contain release-regulating auto-receptors they lack synthesis- and impulse-regulating autoreceptors (Cooper et al., 1996). Second, DAT is expressed in considerably lesser amounts in the frontal cortex than in the striatum, and furthermore, DAT is located primarily *extrasynaptically* (Sesack et al., 1998), suggesting that it is only functional under situations of extremely high catecholamine release.

Because of the reduced levels of DAT in the frontal cortex, transporter inhibitors that have more selectivity for the DAT have somewhat weaker ability to increase DA efflux in the cortex than in the striatum (Tanda et al., 1997). Indeed, there is evidence to support the notion that the cortical NE transporter is a potential 'sink' for extracellular DA in the cerebral cortex (Tanda et al., 1997). Notable in this regard is the finding that DA actually has higher affinity for the NE transporter than does NE.

NE terminals utilize biochemical mechanisms that are very similar to those of dopaminergic boutons. Indeed, DA is the precursor for NE; the conversion of DA to NE is catalyzed by the enzyme dopamine-beta-hydroxylase (DBH). In these terminals, TH is still the rate-limiting step for NE synthesis {exception: under conditions of very high NE release, DBH becomes the rate-limiting step}. The catabolic pathways are identical (COMT {metabolite: nor-metanephrine} and MAO {metabolites: MHPG and DHPG}), and NE terminals are regulated by autoreceptors in a manner similar to that for DA.

Amphetamine administration produces a marked increase in cortical DA, NE, and 5-HT release that is impulse-independent. These effects are mediated by direct effects on cortical monoamine terminals, as described above. Methylphenidate can produce significant increases in DA and NA release, but its effects on cortical 5-HT release have not been studied. Given the paucity of impulse-regulating autoreceptors on mesocortical dopamine neurons and the lack of a direct feedback-inhibitory pathway to the DA cell bodies, drug-induced enhancements of DA release will have less capacity to produce feedback-inhibition of DA neuron firing; however, phasic release could still be modulated at the terminal level by increases in tonic levels of extracellular transmitters.

Acute Effects of Psychomotor Stimulants: Locomotion/Ambulation (Ventral Striatum, i.e., Nucleus Accumbens)

Both *d*-amphetamine and methylphenidate biphasically affect locomotor activity in rats and mice (Finn et al., 1990; Stromberg & Svensson, 1975). Low to moderate doses of both drugs linearly increase ambulation, while high doses can decrease locomotion because of drug-induced induction of motor stereotypy (Robbins & Sahakian, 1983). This represents an 'inverted-U' ("Yerkes-Dodson") dose-response curve for the effects of stimulant drugs on ambulation that has been repeatedly described. Nevertheless, it has been suggested that low doses of some psychomotor stimulant drugs in rodents can *reduce* locomotor activity, perhaps by releasing a small amount of DA that preferentially stimulates presynaptic DA autoreceptors. Seeman and Madras (1998) reviewed the evidence that doses of 0.5–1.0 mg/kg of *d*-amphetamine reduced locomotor activity in mice. For example, although *l*-amphetamine given at 'low' doses (2–4 mg/kg) reduced locomotor activity (Stromberg & Svensson, 1975), *d*-amphetamine (0.03–0.25 mg/kg) had no such effects, and doses of 0.5–8 mg/kg linearly increased locomotor activity. Nevertheless, similar effects have not been reported in rats, and several studies that found amphetamine- or methylphenidate-induced reductions in locomotor activity used animals that were hyperactive at baseline conditions (Seeman & Madras, 1998). In rats, over a wide range of doses, both methylphenidate (1.0–30 mg/kg) and *d*-amphetamine (0.1–3.0) have been found only to increase locomotor activity (Finn et al. 1990). Consequently, there is little evidence to suggest that methylphenidate or *d*-amphetamine decrease locomotor activity in rats even at low doses.

A series of studies by Iversen and colleagues demonstrated that dopaminergic mechanisms within the ventral striatum subserve the ability of stimulant drugs to increase locomotor activity. Destruction of DA terminals within the NAc blocked low-dose amphetamine-induced locomotor hyperactivity (e.g., Kelly & Iversen, 1976). Moreover, intra-NAc infusions of DA or *d*-amphetamine increase locomotion (Jackson et al., 1975; Staton & Solomon, 1984; Taylor & Robbins,

1984). This suggests that activation of DA release is important for the locomotor stimulant effects of amphetamine.

Recently, several investigators have hypothesized that the clinical effects of methylphenidate and amphetamine are mediated by the ability of small amounts of released transmitter to produce a net decrease in phasic release of monoamine transmitters by the selective activation of DA autoreceptors (Seeman & Madras, 1998; Solanto, 1998). As stated above, at low doses, both compounds could theoretically stimulate tonic, extracellular levels of the monoamines, and the small increase in steady-state levels would produce feedback inhibition of further release by stimulating presynaptic autoreceptors. While this mechanism is clearly an important one for the *normal* regulation of monoamine neurotransmission, as originally described by Grace (Grace, 1991), there is no direct evidence to support the notion that the doses used clinically to treat ADHD are low enough to have primarily presynaptic effects. However, alterations in phasic DA release could produce net reductions in DA release under putative altered/abnormal tonic dopaminergic conditions that might occur in ADHD and that might explain the beneficial effects of methylphenidate in ADHD (see Grace, this volume). Nevertheless, one should be cautious when extrapolating data-based dose-response curves in animal to human ones and from animal to human measures of behavior. Moreover, animal studies have not traditionally used oral routes of drug administration that are used clinically and that may produce a different biochemical and behavioral sequelae from that provoked by parenteral routes (see Swanson and Volkow, this volume).

It has been argued that the "decreased phasic release" hypothesis of stimulant drug action is supported by the finding that amphetamine decreases cerebrospinal concentrations of HVA, an acidic metabolite of DA, in ADHD children (Shetty & Chase, 1976) and that stimulant-induced decreases in HVA correlate with the degree of clinical improvement (Castellanos et al., 1996). As described above, amphetamine administration reduces HVA formation, independent of changes in DA release, because it directly inhibits MAO even at low doses (0.1–0.5 mg/kg; see Zetterstrom et al., 1986); thus, the interpretation of such data is confounded. Indeed, measurements of amphetamine- or methylphenidate-induced changes in acidic metabolite formation will remain uninterpretable because both drugs decrease acidic metabolite levels.

Nevertheless, there is general agreement that DA mediates key components of locomotor activity and that psychomotor stimulants affect activity by altering DA-dependent mechanisms. However, locomotor activity is not 'spontaneous' in that it is usually directed toward or away from a stimulus, or is motivated, for example, by appetitive behavior characterized by a heightened responsiveness to stimuli. Therefore, the involvement of DA in locomotor activity necessarily implicates DA in forms of locomotor activity that are collateral to the consummatory response, such as stimuli associated with appetitive behavior. It is thus important to consider locomotor activity within the context of "motivated" or exploratory behavior.

Acute Effects of Psychomotor Stimulants: Motor Stereotypy (Dorsal Striatum, i.e., Caudate Putamen)

At high doses, stimulant drugs produce motor stereotypy, including intense bouts of sniffing, biting, and chewing, in rodents. These effects appear to be modulated by DA-dependent mechanisms in the dorsal striatum, or caudate nucleus, of the rat. DA depletion from the dorsal striatum reduces amphetamine-induced stereotypy and augments hyperlocomotion produced by the drug (Creese & Iversen, 1975; Kelly & Iversen, 1976; Staton & Solomon, 1984). The motor stereotypy produced by hyperstimulation of caudate DA receptors appears to be relevant to certain forms of response perseveration and impairments in response switching that characterize several disorders typified by striatal DA dysfunction (Robbins, 1990; Robbins & Sahakian, 1983). Lyon and Robbins (Lyon & Robbins, 1975) hypothesized that amphetamine-induced motor-stereotypy results from decreases in the duration of pauses between individual behaviors and increases in the rapidity of behavioral execution. Behaviors begin to temporally overlap and longer, more complex, response chains are truncated by competition with simple, rapid behaviors. The result is a domination of behavior by response elements that are short in nature and that require little or no environmental feedback to be sustained.

Psychomotor Stimulants: Rate-Dependency and Behavioral Competition

The theory of rate-dependency first proposed by Dews (1958) predicts that the behavioral changes resulting from *d*-amphetamine administration are a function of the pre-drug or "control" rate of behavior generated by environmental context or reinforcement contingencies. Low doses act to increase low rates of behavior, leaving high rates virtually unaffected; conversely, high doses produced rate reductions under all schedules. Moreover, the pre-drug or control rate was suggested to be more important than the motivational state of the animal, the nature of the reinforcer, or the stimulus control exerted by the drug on behavior (Dews & Wenger, 1977).

Some findings confirm that control rate is one critical determinant for the effects of stimulant drugs (Bradshaw et al., 1981; Lucki & DeLong, 1983). Rate-dependency is in accord with large portions of available data, and it is surprisingly simple, since it is essentially derived from averaging the number of responses over time. Nevertheless, the theory ignores the possible significance of response distribution and topography and, most important, does not address the problem of or specify the causal factors by which these drugs exert their effect on behavior. In addition, in punishment paradigms, amphetamine does not increase low (suppressed) rates to the same extent as equivalent low (unsuppressed) rates (Geller & Seifter, 1960). Other data suggest that stimulants alter stimulus- or schedule-control of behavior by providing a strong interoceptive stimulus or by impairing sensory perception, thereby resulting in rate-dependent effects (Ksir, 1981; McKim, 1981). Robbins (1979, 1981) alternatively hypothesized that the rate-dependent effects of amphetamine stem from the propensity of the drug to enhance the control of responding by conditioned reinforcers, e.g., the salience of stimuli paired with food is enhanced by psychomotor stimulant drugs (see later).

With respect to stereotypies, the Lyon and Robbins hypothesis formulation, in essence, is that behavior under *d*-amphetamine increases in rate, with increasing dosage, within a decreasing number of response categories. Probabilistically, all existing categories of behavior with some minimal tendency to occur will increase in intensity and frequency. Therefore, competition arises for the expression of the behavioral sequence. Certain aspects of behavior are favored to survive, particularly those that can be performed most rapidly and repeatedly without losing their normal relationship with other events. The performance of the animal at very high doses consists of the perseveration of response sequences that eventually lose their relevance or position in normal behavioral sequences, and thus become stereotyped.

At a gross level, there is clearly response competition between locomotor activity and motor stereotypy and DA is clearly critical in these effects (Robbins & Sahakian, 1983). Behavioral competition predicts that low rates of operant behavior increase after d-amphetamine because of reductions in the duration of mediating behaviors. Indeed, reduced durations of these types of spontaneous behavior have been observed (Norton, 1973). Moreover, schedules such as the differential reinforcement of low rates of responding have been found to generate mediating behavior (e.g., tail-biting and grooming) in the interval between presentation of reinforcers (Weiss & Laties, 1964).

Sagvolden et al. (1989) and Heyman (1992) failed to find effects as predicted by rate-dependency after methylphenidate using fixed interval 60-second or various variable-interval schedules of reinforcement. Increases in low rates of reinforced responses were interpreted as a function of changes in reinforcement rate or efficacy, which yield rate-dependent changes in responses rate. Decreases in high rates of reinforced responding could be more parsimoniously explained by behavioral competition, as argued above. The effects of *d*-amphetamine and 6-OHDA lesions on, for example, schedule-induced polydipsia (Falk, 1971) have also failed to find support for rate-dependent effects of stimulant drugs on behavior (Nieto et al., 1979; Robbins et al., 1983a). Thus the apparent rate-reducing effects of stimulant drugs appear to depend more on DA and behavioral competition rather than being directly produced by the control rate of responding generated by the schedule.

Another major factor for determining what responses become activated by amphetamine-like drugs is their relation to reward or reinforcement. Given that alterations in response rate also affects reinforcement per se, the interpretation of the rate-dependent effects of psychomotor stimulant drugs on schedule-dependent responding can be confounded because these drugs can have appetite suppressant effects. Examination of the effects of DA manipulations using procedures that explicitly produce a behavioral bias or minimal tendency to respond may provide a means of disentangling the putative reward and motor effects of stimulant drugs. Moreover, explanations for the behavioral effects of psychomotor on pure rate-dependency principles may not adequately explain the clinical effects of psychomotor stimulants on ADHD (Swanson, 1988) and on more complex behaviors.

Acute Effects of Psychomotor Stimulants: Primary Reinforcing Effects (Ventral Striatum)

A substantial literature supports the notion that the ventral striatal DA system plays a critical role in reinforcement processes and that dopaminergic properties of stimulant drugs underlie their abuse potential in humans (e.g., Koob & LeMoal, 1998). The majority of studies suggest that stimulant drugs produce interoceptive effects that cause them to act as primary reinforcers or to support secondary reinforcement (Di Chiara & Imperato, 1988; Koob & LeMoal, 1998; Roberts et al., 1980; Taylor & Robbins, 1984) by increasing DA efflux and metabolism within the mesolimbic DA system (Di Chiara & Imperato, 1988; Hertel et al., 1996; Moghaddam & Bunney, 1989). Moreover, psychomotor stimulants are self-administered directly into the NAc (Hoebel et al., 1983) and lesions of the mesolimbic DA neurons or infusions of DA receptor antagonists into the NAc attenuates the reinforcing effects of self-administered psychostimulants (Maldonado et al., 1993; Roberts et al., 1980). Nevertheless, examining the reinforcing efficacy of drugs with self-administration procedures may be confounded by other acute effects of the drug that are not related to their rewarding effects. For example, rate-altering effects of stimulant drugs, as discussed above, often confound measures of reinforcement as assessed by self-administration behavior (Whitelaw et al., 1996).

Several findings indicate that the role of increased DA neurotransmission in the NAc in mediating reinforcement processes is complex. First, stimuli with aversive qualities (e.g., footshock, restraint stress, and anxiogenic drugs), as well as those with appetitive qualities, increase NAc DA release (Horger et al., 1995; Imperato et al., 1992; Kalivas & Duffy, 1995; McCullough & Salamone, 1992). Second, conditioned reward and punishment are both potentiated by intra-accumbens or systemic administration of amphetamine (Killcross et al., 1997a; Taylor, 1985; Taylor & Robbins, 1984). We, and others, have hypothesized that DA in the NAc is critical for gating, or modulating, behavior in the face of varying motivational valences and behavioral demands (Salamone et al., 1997; Taylor, 1985). Thus, increased DA release in the accumbens may affect the ability of stimuli with incentive qualities to affect behavior, regardless of whether the stimuli are associated with appetitive or aversive events. This phenomenon may be relevant to drug addiction, and to ADHD, because alterations in the sensitivity to stimuli with aversive or appetitive qualities may evoke intense or abnormal emotional responses. Thus there may be alterations in the modulation of reinforcement processes in ADHD (see Sagvolden, Chapter 9; Solanto, 1998), where stimuli with incentive motivational significance can abnormally control behavior.

Acute Effects of Psychomotor Stimulants: Conditioned Reward (Ventral Striatum/Amygdala)

Methods that utilize second-order schedules of reinforcement, in which responding is maintained by the presentation of stimuli paired with primary reward, may be useful for investigating the neurobiology of reinforcement mechanisms and reward-related behavior (Hill, 1970; Robbins, 1977; Taylor & Robbins, 1984; Taylor & Robbins, 1986). Arbitrary stimuli that are contingently associated with primary reinforcers can come, by associative processes, to act as conditioned reinforcers. Previous work, using second-order schedules of reinforcement, has demonstrated that the conditioned-reinforcing effects of stimuli associated with rewards are greatly enhanced by psychomotor stimulant drugs, including *d*-amphetamine, pipradrol, and methylphenidate (Beninger et al., 1980; Beninger et al., 1981; Hill, 1970; Robbins, 1975; Robbins, 1977; Robbins et al., 1983b).

The role for catecholamine neurotransmission in the NAc in conditioned reward (or reinforcement) was first demonstrated by the finding that intra-accumbens infusions of amphetamine enhanced responding for a conditioned reinforcer and that this effect could be selectively blocked by dopaminergic lesions of the NAc (Taylor & Robbins, 1984; Taylor & Robbins, 1986). Moreover, a selective DA-dependent mechanism in the NAc was implicated in the amphetamine potentiation of conditioned reward as direct infusions of DA, but not NE, enhanced responding for conditioned reinforcement (Cador et al., 1991). Additional studies suggesting a role of D1 and D2 DA receptor subtypes have focused on the use of D1 partial agonists (SKF38393) and mixed D2/3 agonists (quinpirole) (Chu & Kelley, 1992; Ranaldi & Beninger, 1995; Ranaldi et al., 1995; Wolterink et al., 1993). Recently, we have also found that intra-accumbens infusions of full D1 agonists (i.e., SKF81297) or of Sp-cAMPS (a cAMP analogue, that, like activation of D1 receptors, stimulates PKA) potentiate responding for conditioned reinforcement, thus implicating the PKA pathway in the NAc in the modulation of reward-related behaviors (unpublished observations).

The role of the NAc has, thus, been suggested to be that of a limbic-motor interface that gates or mediates incentive motivational response output (Mogenson et al., 1980; Parkinson et al., 1999; Robbins et al., 1989) and the incentive properties of motivationally relevant stimuli, including drugs of abuse (Koob, 1992; Le Moal & Simon, 1991; Markou et al., 1993). Another brain region that may play a significant role in conditioned reward is the amygdala.

Acute Effects of Psychomotor Stimulants: Stimulus-Reward Learning (Amygdala)

Several studies suggest that the amygdala plays an important role in mediating incentive learning and the incentive value of *conditioned* stimuli (Burns et al., 1993; Cador et al., 1989; Davis, 1992; Everitt & Robbins, 1992; Hiroi & White, 1991; Holland & Gallagher, 1999; Killcross et al., 1997b; LeDoux, 1992; Robbins et al., 1989). Importantly, the functions of the amygdala appear to apply equally to both rewarding and aversive stimuli.

At least two discrete nuclear complexes within the amygdala of the rat have been implicated in incentive processes. The central nucleus of the amygdala is primarily connected with hypothalamic and brainstem regions (LeDoux et al., 1988) that mediate autonomic and consummatory responses to stimuli with incentive value (i.e., approaching food or freezing at an aversive stimulus). Lesions to the central nucleus of amygdala largely prevent the development of autonomic responses to primary or conditioned stimuli and block amphetamine-induced potentiation of conditioned reinforcement (Killcross et al., 1997b; Robledo et al., 1996). Holland and Gallagher & (1999) have argued that these effects may represent a partial contribution of the central nucleus to "attentional" processing because lesions of the central nucleus block the development of 'orienting' responses to conditioned stimuli, indicating a possible failure to attend to such stimuli.

In contrast, the lateral-basal complex within the amygdala is primarily connected with nuclei involved in motor learning, as well as regions of limbic forebrain, such as the NAc (Kelley et al., 1982). Damage to the lateral-basal complex produces specific impairments in reinforcer devaluation and in the ability of conditioned stimuli to affect instrumental responding (Holland & Gallagher, 1999; Killcross et al., 1997b; Malkova et al., 1997). Thus, lesions of the lateral-basal complex fail to affect conditioned freezing behavior but do produce conditioned suppression of an instrumental response (Killcross et al., 1997b). Moreover, Hatfield et al. (1996) and Whitelaw et al. (1996) have demonstrated that lesions of the basolateral amygdala, but not the central nucleus, block second-order conditioning and impair reinforcer devaluation. Lesions of the basolateral amygdala also reduce the incentive value of an established conditioned reinforcer but fail to affect the potentiation of conditioned reinforcement by amphetamine (Burns et al., 1993). These data suggest that the basolateral nucleus of the amygdala is involved in creating the kinds of stimulus-reward associations that may be critical for the representation and transfer of information about the motivational value of conditioned stimuli to instrumental, motor learning and response output.

Several laboratories have demonstrated a clear role for catecholamine neurotransmission in the amygdala in stimulus-reward learning. Augmentation of dopaminergic function within the basolateral amygdala by intra-cerebral administration of amphetamine was found to augment acquisition of stimulus-reward learning, as measured by Pavlovian approach behavior (Hitchcott et al., 1997). Using one-trial inhibitory avoidance tasks, psychomotor stimulants have also been shown to facilitate memory consolidation (Cestari et al., 1996; Lee & Ma, 1995), and the effects of stimulant drugs on avoidance tasks were shown to be blocked by lesions of the amygdala (Cestari et al., 1996). Consequently, during situations of increased catecholamine release in the amygdala, associations between exteroceptive stimuli and reward may be enhanced. This is extremely relevant within the context of learning where stimuli with incentive value, such as psychomotor stimulants and stress, can produce increased DA release within the basolateral amygdala, as well as NAc and prefrontal cortex (Harmer et al., 1997; Hurd et al., 1997; Inglis & Moghaddam, 1999).

Chronic Effects of Psychomotor Stimulants

Repeated intermittent exposure to psychomotor stimulants can produce sensitization, where subsequent drug exposures produce augmented behavioral and neurochemical responses (Kalivas et al., 1998; Kalivas et al., 1993; Post & Rose, 1976; Robinson & Berridge, 1993). The ability of the drug, and ultimately of related stimuli, to elicit behavior may be augmented with repeated administrations or intake of the drug. This phenomenon may depend on sensitization of DA release and resulting neuroadaptations within the NAc produced by chronic drug intake (Kalivas et al., 1998; Kalivas et al., 1993; Kalivas & Stewart, 1991; Nestler & Aghajanian, 1998; Robinson & Berridge, 1993; Self & Nestler, 1998; White et al., 1995; Wolf, 1998). The phenomenon of sensitization appears to be most profound several weeks to months after repeated, intermittent injections of moderate doses of stimulant drugs, indicating possible interaction with withdrawal effects. In addition, the magnitude of sensitization can be affected by the environment, or context, in which the drug is administered. Interestingly, however, a recent report found that even a single exposure to 5 mg/kg amphetamine was sufficient to induce long-term context-independent behavioral and neurochemical sensitization in rats (Vanderschuren et al., 1999). Although the phenomenon of sensitization in humans has received relatively little attention, sensitization has been reported for several behavioral measures after a single low (0.25 mg/kg) oral dose of amphetamine in normal human volunteers (Strakowski et al., 1996). Nevertheless, there is a paucity of studies investigating sensitization to low doses of stimulant drugs, the effects of oral routes of administration, and potential differences between amphetamine- or methylphenidate-induced sensitization in experimental animals and humans.

Although the mesolimbic DA system has been implicated in both reward processes and behavioral sensitization, few studies have assessed whether there is a direct relationship between these two processes that extends beyond the dependence upon overlapping neural circuits (Wise & Leeb, 1993). Preexposure to cocaine, morphine or amphetamine has been shown to enhance the initial reinforcement value of cocaine as measured by self-administration or place-preference procedures (Horger et al., 1992; Horger et al., 1990; Shippenberg & Heidbreder, 1995). The result is progressively greater locomotion within that environment due to the enhanced incentive/motivational properties of contextual stimuli (Robinson & Berridge, 1993). Given the wealth of data that implicates the mesolimbic DA system in psychomotor stimulant sensitization and drug reinforcement processes (see above), this suggests that sensitization or augmentation of mesolimbic DA function could be one mechanism by which psychomotor stimulant drugs affect the control of behavior by reinforcing stimuli. These phenomena are clearly relevant to drug abuse and ADHD because they suggest that with repeated drug exposures, the incentive value of the drugs and related conditioned stimuli can be progressively enhanced. Moreover, in ADHD, alterations in the modulation (or regulation) of motivational processes have been implicated in the disorder (Barkley, 1997), and deficits in inhibitory control may depend, in part, on altered reinforcement processes (see later).

Our recent finding that cocaine sensitization facilitates responding for conditioned reinforcement (Taylor & Horger, 1999) is also consistent with the hypothesis that neuroadaptations in the mesolimbic system (Nestler, 1992; Nestler & Aghajanian, 1998; Self & Nestler, 1995) may be a critical consequence of long-term psychomotor stimulant exposure. The function of DA D1 receptors is also known to be altered after chronic cocaine treatment (D1 supersensitivity and autoreceptor subsensitivity; Henry & White, 1991; Henry & White, 1992; White, 1996), and these receptors are known to be involved in incentive motivation (Shippenberg & Heidbreder, 1995), as well as in sensitization.

Dopaminergic sensitization within the amygdala has also been found after repeated exposure to amphetamine and this can enhance appetitive and aversive learning even after drug cessation (Harmer et al., 1997; Harmer & Phillips, 1999a; Harmer & Phillips, 1999b; Hitchcott et al., 1997; Phillips et al., 1998). Specifically, facilitation of DA transmission in the amygdala *transiently* enhances (on the order of days) stimulus-reward learning. However, these changes in stimulus-reward learning may also be mediated by changes in signal transduction mechanisms in the amygdala. We have recently found that repeated administrations of cocaine or amphetamine can produce enduring enhancements (on the order of weeks to months) in stimulus-reward learning. Our preliminary data suggest that these effects may be mediated by neuroadaptations of the DA innervation of the amygdala or of the PKA pathway impinged upon by DA receptors (Jentsch & Taylor, 1999; Taylor et al., 1999b). These studies suggested that repeated inactivation of PKA in the amygdala may produce neuroadaptations consistent with a persistent, compensatory hyperactivation of the PKA pathway and an associated long-lasting enhancement of the incentive motivational qualities of conditioned reinforcers. Specifically, these putative neuroadaptations in adenylyl cyclase/PKA/CREB pathways in corticolimbic brain circuits may be relevant to the core dysfunctions of ADHD: poor inhibitory control (i.e., impulsivity) and alterations in the modulation of motivated behaviors (i.e., behavioral regulation, hyperactivity and altered stimulus control). Although the dose ranges and time courses and other parameters that can produce such behavioral effects are not known, these data may be relevant to the long-term treatment with psychomotor stimulants, as well as to the putative dysfunction of brain DA systems in ADHD.

Frontal Cortical Modulation of Subcortical DA

Parallel circuits connect discrete regions of the FC with specific subregions of the striatum, globus pallidus, and thalamus (e.g., Alexander et al., 1990; Cummings, 1995). These corticostriatal basal-ganglia "loops" have been hypothesized to mediate separable functions. One circuit mediates motor function, another eye movements and other circuits have been argued to mediate cognitive, emotional and motivational processes. These circuits are thought to integrate related information from widespread regions of the brain and subserve diverse behaviors. There are certain unique features of the "limbic circuit" in particular that may be especially relevant to the effects of psychomotor stimulant drugs and for ADHD. It receives substantial input from the amygdala at several points, including the ventral striatum, anterior cingulate cortex and limbic frontal cortex and has efferent connections from the ventral pallidum to other subcortical structures belonging to the limbic system including the VTA (Haber et al., 1995). Thus, this circuit engages a number of cortical and subcortical structures traditionally considered limbic. The ventral striatum receives projections from the cingulate and limbic frontal cortices; the ventral striatum then projects to the ventral pallidum. The loop is closed by thalamocortical projections from the nucleus medialis dorsalis to the frontal cortex. In addition, it has connections with the dorsolateral and orbitofrontal cortices. Importantly, the limbic frontal cortex receives DA input largely from the ventral tegmental area (Williams & Goldman-Rakic, 1998) and is the only cortical region which provides nearly exclusive input to the ventral striatum (Lynd-Balta & Haber, 1994).

Based on this circuitry, it has been suggested that the "limbic" loop is the primary substrate for motivational processes; however, of all the circuits examined using imaging and other techniques, it has been the least studied. Indeed, though multiple imaging studies have shown various alterations in dorsolateral prefrontal cortex activity and its projection zones (see Mehta et al., Chapter 13, Arnsten, Chapter 7, and Swanson & Volkow, Chapter 11), few studies have attempted to focus on the cortical-subcortical circuits that may be involved in inhibitory control of motivational states, and thus, ADHD.

Inhibition of Reward-Related Behavior and the Frontal Cortex

Damage to the frontal cortex in monkeys and humans leads to the inability to inhibit inappropriate responses, perseverative behavior and deficits in working memory (see Arnsten, Denney & Rapport, Mehta et al., Chapters 7, 12, 13, respectively). Frontal damage can cause disinhibition, such that behavior becomes largely guided by previously conditioned or 'prepotent' responses which are inappropriate in the current situation. Such deficits have been measured by go/no-go, object reversal, and other tasks after a variety of frontal cortical lesions. These experiments suggest that response inhibition is an active process for modulating stimulus-response associations and allowing animals to shift between old and new contingencies and to inhibit inappropriate responses to conditioned stimuli.

There is strong evidence for the beneficial effects of DA on inhibitory control and working memory, using manipulations that selectively deplete DA or block endogenous DA receptor stimulation. Depleting DA within the dorsolateral prefrontal cortex of monkeys (Brozoski et al., 1979; Goldman-Rakic, 1987) or prelimbic cortex of rats (Bubser & Schmidt, 1990; Simon, 1981; Stamm et al., 1989) produced profound deficits on delayed response or alternation tasks. A convincing body of data suggest that cortical DA, by actions on D1-like receptors and PKA (Sawaguchi & Goldman-Rakic, 1991; Taylor et al., 1999a; Zahrt et al., 1997), is critical in normal cognitive functioning in the primate and rodent (see also Arnsten, Chapter 7).

The cognitive/inhibitory modulation of reward-related behavior may also depend critically upon the corticostriatal projections from frontal cortex to the caudate nucleus and NAc core and shell. For example, lesions to the medial frontal cortex augment the hyperlocomotion or cognitive disruption produced by acute treatment with amphetamine (Jaskiw et al., 1990; Wilkinson et al., 1997). DA depletion within the prefrontal cortex can exacerbate or augment locomotor sensitization to amphetamine (Banks & Gratton, 1995) cocaine self-administration (Schenk et al., 1991), and increase subcortical DA in response to stress (Deutch & Roth, 1990). This has led to the hypothesis that cortical DA tonically inhibits subcortical DA systems (see Grace, Chapter 5; Deutch, 1992). Consequently, a relative hypoactivity of cortical DA function may result in vulnerability to self-administer psychomotor stimulant drugs or to develop locomotor sensitization with repeated exposures to the drug (Piazza et al., 1991; Prasad et al., 1999). Altered DA neurotransmission in the prefrontal cortex has also been confirmed after chronic cocaine (Sorg et al., 1997) and phencyclidine (Jentsch et al., 1997). Recently, Dalley et al. (1999) reported that lesions of the prefrontal cortex produced alterations in CREB phosphorylation in the NAc following amphetamine treatment. Thus, reductions in cortical DA function can lead to increased subcortical DA transmission and alterations in CREB regulation that may result in long-term alterations in gene expression. Recent studies have also suggested that chronic drug exposure can produce cognitive dysfunction, notably inhibitory control deficits (Jentsch & Taylor, 1999; McKetin & Mattick, 1998; O'Malley et al., 1992; Rogers et al., 1999). We have hypothesized that the result of chronic drug exposure may involve frontal cortical cognitive dysfunction, resulting in an inability to inhibit inappropriate unconditioned or conditioned responses elicited by drugs, by related stimuli, and by internal drive states, as outlined in Figure 4–3. Repeated psychomotor stimulant drugs are hypothesized to result in a continuous cycle of increases in the stimulus control of behavior and a fundamental impairment in the ability to gate or modulate behavioral responses to reward-related stimuli, coupled with impairments in the inhibitory response control capacities of the frontal cortex (Jentsch & Taylor, 1999). Such deficits may result from drug-induced neuroadaptations in the mesocorticolimbic DA neurotransmission produced by repeated intermittent exposures to moderate to high doses of stimulant drugs. As argued above, these behavioral deficits also appear to have striking similarities to the core deficits of ADHD.

Implications for ADHD

We have reviewed evidence that psychomotor stimulant drugs, after acute or chronic administration, increase locomotor activity, motor stereotypy and primary and conditioned reward (incentive motivation). Each of these processes depends upon increases in dopaminergic transmission within the striatum, and possibly, amygdala. Moreover, neurochemical actions within the frontal cortex

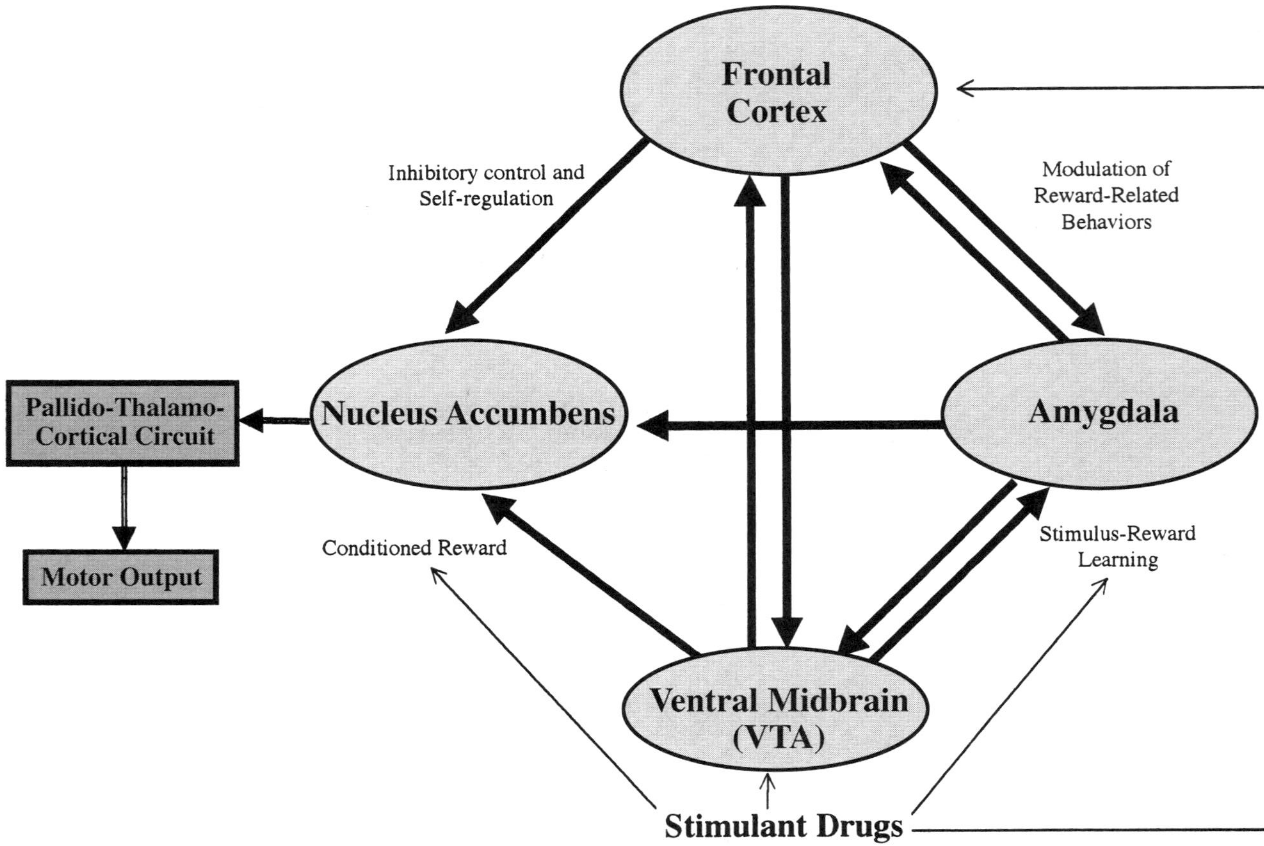
Frontal
Cortex
Inhibitory control and
Self-regulation
Modulation of
Reward-Related
Behaviors
Pallido-Thalamo-
Cortical Circuit
Nucleus Accumbens
Amygdala
Motor Output
Conditioned Reward
Stimulus-Reward
Learning
Ventral Midbrain
(VTA)
Stimulant Drugs

may contribute to the ability to modulate some of these basic motivational processes. These data are based primarily on studies in rodents given systemic injections of moderate to high doses of stimulants, and it is not known whether lower doses given by oral routes of administration would produce similar behavioral and neurochemical effects.

Nevertheless, when given acutely or chronically to animals, psychomotor stimulants appear to alter the neurochemistry of the striatum in such a way as to augment the control of behavior by conditioned or unconditioned stimuli associated with reinforcement processes. These effects may appear paradoxical given that psychomotor stimulant administration to children with ADHD increases performance in various academic and social conditions. However, they may be consistent with the suggestion that amphetamine or methylphenidate may exert some of its beneficial clinical effects by augmenting conditioned reinforcement and stimulus–reward associations that could enhance aspects of task performance (Robbins & Everitt, 1982; Robbins & Sahakian, 1983).

Stimulant-drug induced alterations in DA release in the striatum alone, based on studies in "normal" animals, may not be able to explain the clinical effects of stimulant drugs in ADHD. As discussed previously, a simple decrease in striatal DA transmission produced by even low clinical doses of the drugs according to current data seems an untenable hypothesis. However, if the tonic control of phasic release by DA is abnormally low in ADHD, as hypothesized by Grace (this volume), then a high phasic DA response may be associated with ADHD and treatment efficacy. Moreover, alterations in corticolimbic tonic/phasic regulation or balance may also provide an explanation of why chronic psychomotor stimulant drugs and ADHD produce a behavioral profile and response associated with subcortical DA neuronal hyperactivity and cortical hypoactivity, though issues

Figure 4–3. Corticolimbic-striatal circuitry and actions of psychomotor stimulant drugs. The nucleus accumbens and its GABAergic outputs to the pallido-thalamo-cortical motor circuits are modulated by a number of cortical and subcortical influences. Stimulant drugs are known to directly activate mesocorticolimbic DA neurotransmission. Chronic stimulant drugs are hypothesized to produce increases in drug-seeking behavior, in part, by augmenting dopaminergic transmission within the amygdala and nucleus accumbens and frontal cortex. These changes in dopaminergic transmission may be associated with augmented stimulus-reward learning dependent upon DA transmission in the amygdala and a potentiation of conditioned reward dependent upon DA transmission in the nucleus accumbens, respectively. Moreover, with increased stimulus-reward learning and conditioned reward, the inhibitory control functions of the frontocortical striatal system may be also impaired with chronic drug use. The loss of inhibitory control and self-regulation may thus be due to alterations in dopaminergic input to the frontal cortex and subsequent decreases in frontal cortical modulation of subcortical systems by descending corticostriatal projections. In addition, deficits in inhibitory control may further exacerbate impairments in the modulation of reward-related behaviors dependent on amygdala-accumbens interactions.

of dose and routes of administration should be considered. Nevertheless, it is tempting to speculate that the "neuroadaptive" processes that are responsible for the subsequent behavioral changes found after chronic psychomotor stimulants in "normal" individuals could be the result of similar underlying alterations in neuronal mechanisms in ADHD. For example, there are several potential alterations in corticolimbic brain circuits caused by long-term stimulant drug administration that have received little or no attention with respect to ADHD, such as alterations in adenylyl cyclase/PKA/CREB pathways, DAT uptake, VMAT uptake, calcium CaM-KII-dependent changes in release, and DA receptor internalization. Such changes in corticolimbic brain circuits could in ADHD result from underlying genetic alterations.

Another theory for the action of stimulant drugs in ADHD involves effects on nonstriatal monoamine systems. Frontal cortical DA, NE, and 5-HT are clearly important in cognitive functioning and impulse control (see previous discussion and Arnsten, Chapter 7, Mehta et al., Chapter 13, and Pliszka, Chapter 14), and, as we have discussed, these neurotransmitters may, by actions in the frontal cortex, directly modulate reward-related behaviors associated with the striatal DA system. Moreover, drug actions on the neurochemistry of the amygdala may contribute to enhancements in basic mechanisms of associative learning and recall (Cahill & McGaugh, 1998). Thus, the net behavioral effects of psychomotor stimulant drugs may promote changes in reward-motivated behaviors and impulsivity, as well as enhancement of neuromodulatory systems critical for inhibitory control, working memory, and incentive learning.

These data clearly stress the need for investigating the effects of psychomotor stimulant drugs on nonstriatal functions relevant to ADHD, namely inhibitory control (associated with the frontal cortex) and incentive learning (associated with the amygdala), as well as on models of ADHD associated with fundamental differences in DA regulation. Modulatory actions of stimulant drugs, at these levels, could provide a mechanism for the regulation of the potential subcortical impulsivity of ADHD and enhance our understanding of the pathophysiology of the disorder.

Preparation of this review was supported by USPHS grants DA11026 and DA11717 to J.R.T.

References

Alexander, GE, Crutcher, MD, & DeLong, Jr (1990). Basal gangliathalamocortical circuits: Parallel substrates for motor, oculomotor, "prefrontal" and "limbic" function. In HBM Uylings, CG Van Edit, JPC DeBruit, MA Corner, MGP Feenstra (Eds.), Progress in Brain Research, Vol. 85 (pp. 119–146). New York: Elsevier Science.

An, X, Lewis, DA, & Sesack, SR (1998). Ultrastructural localization of dopamine D1 or D2 receptor immunoreactivity in structures post-synaptic to tyrosine hydroxylase-positive terminals in the monkey striatum. *Soc Neurosci Abstr, 24,* 857.

Banks, KE & Gratton, A (1995). Possible involvement of medial prefrontal cortex in amphetamine-induced sensitization of mesolimbic dopamine function. *Eur J Pharmacol, 282,* 157–167.

Barkley, RA (1997). Behavioral inhibition, sustained attention, and executive functions: Constructing a unifying theory of ADHD. *Psychol Bull, 121,* 65–94.

Beninger, RJ, Hansen, DR, & Phillips, AG (1980). The effects of pipradrol on the acquisition of responding with conditioned reinforcement: a role for sensory preconditioning. *Psychopharmacology, 69,* 235–242.

Beninger, RJ, Hansen, DR, & Phillips, AG (1981). The acquisition of responding with conditioned reinforcement: effects of cocaine, (+)-amphetamine, and pipradol. *Br J Pharmacol, 74,* 149–154.

Björklund, A, & Lindvall, O (1984). Dopamine-containing systems in the CNS. In A Björklund & T. Hökfelt (Eds.). *Handbook of chemical neuroanatomy: Classical transmitters in the CNS, Part 1, Vol. 2* (pp. 55–122). Amsterdam: Elsevier Science B.V.

Bowers, JMB & Hoffman, FJ (1984). Homovanillic acid in rat caudate and prefrontal cortex following phencyclidine and amphetamine. *Psychopharmacology, 84,* 137–138.

Bradshaw, CM, Ruddle, HV, & Szabadi, E (1981). Relationship between response rate and reinforcement frequency in variable-interval schedules. *J Exp Anal Behav, 35,* 263–269.

Broersen, L, Heinsbroek, R, de Bruin, J, Joosten, R, van Hest, A, & Olivier, B (1994). Effects of local application of dopaminergic drugs into the dorsal part of the medial prefrontal cortex of rats in a delayed matching to position task: comparison with local cholinergic blockade. *Brain Res, 645,* 113–22.

Brozoski, TJ, Brown, RM, Rosvold, HE, & Goldman, PS (1979). Cognitive deficit caused by regional depletion of dopamine in prefrontal cortex of rhesus monkey. *Science, 205,* 929–931.

Bubser, M & Schmidt, WJ (1990). 6-OHDA lesions of the rat prefrontal cortex increases locomotor activity, impairs acquisition of delayed alternation tasks, but does not affect uninterrupted tasks in the radial maze. *Behav Brain Res, 37,* 157–168.

Burns, LH, Everitt, BJ, & Robbins, TW (1993). Differential effects of excitotoxic lesions of the basolateral amygdala, ventral subiculum and medial prefrontal cortex on responding with conditioned reinforcement and locomotor activity potentiated by intra-accumbens infusions of d-amphetamine. *Behav Brain Res, 55,* 167–183.

Cador, M, Robbins, TW & Everitt, BJ (1989). Involvement of the amygdala in stimulus-reward associations: Interaction with the ventral striatum. *Neuroscience, 30,* 77–86.

Cador, M, Taylor, JR, & Robbins, TW (1991) Potentiation of effects of reward-related stimuli by dopaminergic-dependent mechanisms in the nucleus accumbens. *Psychopharmacology, 104,* 377–385.

Cahill, L, & McGaugh, JL (1998). Mechanisms of emotional arousal and lasting declarative memory. *Trends in Neurosciences, 21*(7), 294–299.

Castellanos, XF, Elia, J, Kruesi, MJP, Marsh, WL, Gulotta, CS, Potter, WZ, Ritchie, GF, Hamburger, SD, & Rapoport, JL (1996). Cerebrospinal fluid homovanillic acid predicts behavioral response to stimulants in 45 boys with Attention Deficit/Hyperactivity Disorder. *Neuropsychopharmacology, 14,* 125–137.

Cestari, V, Mele, A, Oliverio, A, & Castellano, C (1996). Amygdala lesions block the effect of cocaine on memory in mice. *Brain Res, 713,* 286–289.

Chu, B & Kelley, AE (1992). Potentiation of reward related responding by psychostimulant infusion into nucleus accumbens: Role of dopamine receptor subtypes. *Psychobiology, 20,* 153–162.

Cooper, JR, Bloom, FE, & Roth, RH (1996). *The Biochemical Basis of Neuropharmacology.* New York, Oxford: Oxford University Press.

Creese, I, & Iversen, SD (1975). The pharmacological and anatomical substrates of the amphetamine response in the rat. *Brain Res, 83,* 419–436.

Cummings, JL (1995). Anatomic and behavioral aspects of frontal-subcortical circuits. *Annals NY Acad Sci, 769,* 1–3.

Dalley, JW, Thomas Howes, SR, Tsai, TH, Aparicio-Legarza, MI, Reynolds, GP, Everitt, BJ, & Robbins, TW (1999). Effects of excitotoxic lesions of the rat prefrontal cortex on CREB regulation and presynaptic markers of dopamine and amino acid function in the nucleus accumbens. *Eur J Neurosci, 11,* 1265–1274.

Davis, M (1992). The role of the amygdala in conditioned fear. In J P Aggleton (Ed.) *The Amygdala,* (pp. 255–306). New York: Wiley-Liss.

Deutch, AY (1992). The regulation of subcortical dopamine systems by the prefrontal cortex: Interactions of central dopamine systems and the pathogenesis of schizophrenia. *J Neural Transm [Gen Sect], 36,* 61–89.

Deutch, AY & Roth, RH (1990). Determinants of stress-induced activation of the prefrontal cortical dopamine system. *Prog Brain Res, 85,* 367–403.

Dews, PB (1958). Studies on behavior. IV. *JPET, 122,* 137–147.

Dews, PB & Wenger, AR (1977) Rate-dependency of the behavioral effects of amphetamine. In T Thompson & PB Dews (Eds.), *Advances in Behavioral Pharmacology, Vol. 1* (pp. 22–56). New York: Academic Press.

Di Chiara, G & Imperato, A (1988). Drugs of abuse preferentially stimulate dopamine release in the mesolimbic system of freely-moving rats. *Proc Natl Acad Sci USA, 85,* 5274–5278.

Elchisak, MA, Murrin, LC, Roth, RH, & Maas, JW (1976). Free and conjugated dihydroxyphenylacetic acid: effect of alterations in impulse flow in rat neostriatum and frontal cortex. *Psychopharmacology Communications, 2,* 411–420.

Everitt, BJ & Robbins, TW (1992). Amygdala-ventral striatal interactions and reward-related processes. In JP Aggleton (Ed.), *The Amygdala* (pp. 401–430). New York: Wiley-Liss.

Falk, JL (1971). The nature and determinants of adjunctive behavior. *Physiol Behav, 6,* 577–588.

Fallon, JH (1988). Topographic organization of ascending dopaminergic projections. *Ann NY Acad Sci, 537,* 1–9.

Fallon, JH & Moore, RY (1978). Catecholamine innervation of the basal forebrain. IV. Topography of the dopamine projection to the basal forebrain and neostriatum. *J Comp Neurol, 180,* 545–580.

Finn, IB, Iuvone, PM, & Holtzman, SG (1990). Depletion of catecholamines in the brain of rats differentially affects stimulation of locomotor activity by caffeine, *d*-amphetamine, and methylphenidate. *Neuropharmacology, 29,* 625–631.

Geller, T & Seifter, J (1960). The effects of meprobromate, barbiturates, d-amphetamine and promazine on experimentally induced conflict in the rat. *Psychopharmacologica (Berl), 1,* 482–492.

Goldman-Rakic, PS (1987). Circuitry of the frontal cortex and the regulation of behavior by representational knowledge. In F Plum & VM Mountcastle (Eds.), *Handbook of Physiology, the Nervous System Vol. 5,* (pp. 373–417). Bethesda, MD: American Physiological Society.

Grace, AA (1991). Phasic versus tonic dopamine release and the modulation of dopamine system responsivity: A hypothesis for the etiology of schizophrenia. *Neuroscience, 41,* 1–24.

Grace, AA (1995). The tonic/phasic model of dopamine system regulation: Its relevance for understanding how stimulant abuse can alter basal ganglia function. *Drug and Alcohol Dependence, 37*:111–129

Haber, SN, Kunishio, K, Mizobuchi, M, & Lynd-Balta, E (1995). The orbital and medial prefrontal circuit through the primate basal ganglia. *J Neurosci, 15,* 4851–4867.

Harmer, CJ, Hitchcott, PK, Morutto, SL, & Phillips, GD (1997). Repeated d-amphetamine enhances stimulated mesoamygdaloid dopamine transmission. *Psychopharmacology, 132,* 247–254.

Harmer, CJ & Phillips, GD (1999a). Enhanced conditioned inhibition following repeated pretreatment with d-amphetamine. *Psychopharmacology, 142,* 120–131.

Harmer, CJ & Phillips, GD (1999b). Enhanced dopamine efflux in the amygdala by a predictive, but not non-predictive stimulus: facilitation by prior repeated d-amphetamine. *Neuroscience, 90,* 119–130.

Hatfield, T, Han, JS, Conley, M, Gallagher, M & Holland, P (1996). Neurotoxic lesions of the basolateral, but not central, amygdala interfere with Pavlovian second-order conditioning and reinforcer devaluation effects. *J Neurosci, 16,* 5256–5265.

Henry, DJ & White, FJ (1991). Repeated cocaine administration causes persistent enhancement of D-1 dopamine receptor sensitivity within the rat nucleus accumbens. *JPET,* 258, 882–890.

Henry, DJ & White, FJ (1992). Electrophysiological correlates of psychomotor stimulant-induced sensitization. *Ann NY Acad Sci, 654,* 88–100.

Hertel, P, Mathe, JM, Nomikos, GG, Iurlo, M, Mathe, AA, & Svensson, TH (1996). Effects of d-amphetamine and phencyclidine on behavior and extracellular concentrations of neurotensin and dopamine in the ventral striatum and medial prefrontal cortex of the rat. *Behav Brain Res, 72,* 103–114.

Heyman, GM (1992). Effects of methylphenidate on response rate and measures of motor performance and reinforcement efficacy. *Psychopharmacology, 109,* 145–152.

Hill, RT (1970). Facilitation of conditioned reinforcement as a mechanism of psychomotor stimulation. In Costa, E & S. Garattini (Eds.) *Amphetamines and related compounds,* pp. 781–795. New York: Raven Press.

Hiroi, N & White, NM (1991). The amphetamine conditioned place preference: Differential involvement of dopamine receptor subtypes and two dopaminergic terminal areas. *Brain Res, 552,* 141–152.

Hitchcott, PK, Harmer, CJ, & Phillips, GD (1997). Enhanced acquisition of discriminative approach following intraamygdala d-amphetamine. *Psychopharmacology, 132,* 237–246.

Hoebel, BG, Monaco, AP, Hernandez, L, Aulisi, EF, Stanley, BG & Lenard, L (1983). Self-injection of amphetamine directly into the brain. *Psychopharmacology, 81,* 158–163.

Holland, PC & Gallagher, M (1999). Amygdala circuitry in attentional and representational processes. *Trends Cog Sci, 3,* 65–73.

Horger, BA, Elsworth, JD & Roth, RH (1995). Selective increase in dopamine utilization in the shell subdivision of the nucleus accumbens by the benzodiazepine inverse agonist FG7142. *J Neurochem, 65,* 770–774.

Horger, BA, Giles, MK & Schenk, S (1992). Preexposure to amphetamine and nicotine predisposes rats to self-administer a low dose of cocaine. *Psychopharmacology, 107,* 271–276.

Horger, BA, Shelton, K & Schenk, S (1990). Preexposure sensitizes rats to the rewarding effects of cocaine. *Pharmacol Biochem Behav, 37,* 707–711.

Hurd, YL, McGregor, A & Ponten, M (1997). In vivo amygdala dopamine levels modulate cocaine self-administration behaviour in the rat: D1 dopamine receptor involvement. *Eur J Neurosci, 9,* 2541–2548.

Imperato, A, Angelucci, L, Casolini, P, Zocchi, A & Puglisi-Allegra, S (1992). Repeated stressful experiences differently affect limbic dopamine release during and following stress. *Brain Res, 577,* 194–199.

Imperato, A & Di Chiara, G (1984). Trans-striatal dialysis coupled to reverse phase high performance liquid chromatography with electrochemical detection: a new method for the study of the in vivo release of endogenous dopamine and metabolites. *J Neurosci, 4,* 966–977.

Inglis, FM & Moghaddam, B (1999). Dopamine innervation of the amygdala is highly responsive to stress. *J Neurochem, 72,* 1088–1094.

Jackson, DM, Anden, N-E & Dahlstrom, A (1975). A functional effect of dopamine in the nucleus accumbens and some other dopamine-rich parts of the brain. *Psychopharmacologia (Berl), 45,* 139–149.

Jansson, A, Goldstein, M, Tinner, B, Zoli, M, Meador-Woodruff, JH, & Lew, JY, et al. (1998). On the distribution patterns of D1, D2, tyrosine hydroxylase and dopamine transporter immunoreactivities in the ventral striatum of the rat: Indications for volume transmission compartments. *Soc Neurosci Abstr, 24,* 857.

Jaskiw, GE, Karoum, G, Freed, WJ, Phillips, I, Kleinman, JE & Weinberger, DR (1990). Effect of ibotenic acid lesions of the medial prefrontal cortex on amphetamine-induced locomotion and regional brain catecholamine concentrations in the rat. *Brain Res, 534,* 263–272.

Jentsch, JD, Redmond Jr, DE, Elsworth, JD, Taylor, JR, Youngren, KD & Roth, RH (1997). Enduring cognitive deficits and cortical dopamine dysfunction in monkeys after long-term administration of phencyclidine. *Science, 277,* 953–955.

Jentsch, JD & Taylor, JR (1999). Impulsivity resulting from frontostriatal dysfunction in drug abuse: Implications for the control over behavior by reward-related stimuli. Review. *Psychopharmacology, 146,* 373–390.

Kalivas, PW & Duffy, P (1995). Selective activation of dopamine transmission in the shell of the nucleus accumbens by stress. *Brain Res, 675,* 325–328.

Kalivas, PW, Pierce, RC, Cornish, J & Sorg, BA (1998). A role for sensitization in craving and relapse in cocaine addiction. *J Psychopharmacol, 12,* 49–53.

Kalivas, PW, Sorg, BA & Hooks, MS (1993). The pharmacology and neural circuitry of sensitization to psychostimulants. *Behav Pharmacol, 4,* 315–334.

Kalivas, PW & Stewart, J (1991). Dopamine transmission in the initiation and expression of drug- and stress-induced sensitization of motor activity. *Brain Res Rev, 16,* 223–244.

Karoum, F, Chrapusta, SJ, & Egan, MF (1994). 3-Methoxytyramine, a major metabolite of released dopamine in the rat frontal cortex: Assessment of the effects of antipsychotics on the dynamics of dopamine release and metabolism in the frontal cortex, nucleus accumbens and striatum by a simple two pool model. *J Neurochem, 63,* 972–979.

Kelley, AE, Domesick, VB, & Nauta, WJH (1982). The amygdalostriatal projection in the rat: An anatomical study by anterograde and retrograde tracing methods. *Neuroscience, 7,* 615–630.

Kelly, PH & Iversen, SD (1976). Selective 6-OHDA induced destruction of mesolimbic dopamine neurons: Abolition of psychostimulant induced locomotor activity in rats. *Eur J Pharmacol,* 40, 45–56.

Killcross, AS, Everitt, BJ, Robbins, TW (1997a). Symmetrical effects of amphetamine and alpha-flupenthixol on conditioned punishment and conditioned reinforcement: Contrasts with midazolam. *Psychopharmacol, 129,* 141–152.

Killcross, S, Robbins, TW, & Everitt, BJ (1997b). Different types of fear-conditioned behavior mediated by separate nuclei within the amygdala. *Nature, 388,* 377–380.

Koob, GF (1992). Neuronal mechanisms of drug reinforcement. *Ann NY Acad Sci, 654,* 171–191.

Koob, GF & LeMoal, M (1998). Drug abuse: Hedonic homeostatic dysregulation. *Science, 278,* 52–58.

Ksir, C (1981). Rate-convergent effects of drugs. In T Thompson, PB Dews, & WA McKim (Eds.), *Advances in Behavioral Pharmacology, Vol. 3* (pp. 39–59). New York: Academic Press.

Kuczenski, R & Segal, DS (1997). Effects of methylphenidate on extracellular dopamine, serotonin, and norepinephrine: Comparison with amphetamine. *J Neurochem, 68,* 2032–2037.

Le Moal, M & Simon, H (1991). Mesocorticolimbic dopaminergic network: functional and regulatory roles. *Physiol Rev, 71,* 55–234.

LeDoux, JE (1992). Emotion and the amygdala. In JP Aggleton (Ed). *The Amygdala* pp. 339–351. New York: Wiley-Liss.

LeDoux, JE, Iwata, J, Cicchetti, P & Reid, DJ (1988). Different projections of the central amygdaloid nucleus mediate autonomic and behavioral correlates of conditioned fear. *J Neurosci, 8,* 2517–2529.

Lee, EH & Ma, YL (1995). Amphetamine enhances memory retention and facilitates norepinephrine release from the hippocampus in rats. *Brain Res Bull, 37,* 411–416.

Lewis, DA and Sesack, SA (1997) Dopamine systems in the primate brain. In FE Bloom, A Björklund & T Hökfelt (Eds.), *Handbook of Chemical Neuroanatomy. The primate nervous system., Part 1, Vol. 13* (pp. 263–375). Elsevier Science B.V.

Lucki, I & DeLong, M (1983). Control rate of response or reinforcement and amphetamine's effect on behavior. *J Exp Anal Behav, 40,* 123–132.

Lynd-Balta, E & Haber, SN (1994). The organization of midbrain projections to the striatum in the primate: Sensorimotor-related striatum versus ventral striatum. *Neuroscience, 59,* 625–640.

Lyon, M & Robbins, TW (1975). The action of central nervous system stimulant drugs: A general theory concerning amphetamine effects. In WB Essman & L Valzelli (Eds.), *Current developments in psychopharmacology Vol. 2* (pp. 79–163). New York: Spectrum.

Maldonado, R, Robledo, P, Chover, AJ, Caine, SB, & Koob, GF (1993). D1 dopamine receptors in the nucleus accumbens modulate cocaine self-administration in the rat. *Pharmacol Biochem Behav, 45,* 239–242.

Malkova, L, Gaffan, D, & Murray, EA (1997). Excitotoxic lesions of the amygdala fail to produce impairment in visual learning for auditory secondary reinforcement but interfere with reinforcer devaluation effects in rhesus monkeys. *J Neurosci, 17,* 6011–6020.

Markou, A, Weiss, F, Gold, LH, Caine, SB, Schulteis, G, & Koob, GF (1993). Animal models of drug craving. *Psychopharmacology, 112,* 163–182.

McCullough, LD & Salamone, JD (1992). Anxiogenic drugs beta-CCE and FG 7142 increase extracellular dopamine levels in nucleus accumbens. *Psychopharmacology, 109,* 379–382.

McKetin, R & Mattick, RP (1998). Attention and memory in illicit amphetamine users: Comparison with non-drug-using controls. *Drug Alcohol Dep, 50,* 181–184.

McKim, WA (1981). Rate-dependency a non-specific behavioral effect of drugs. In T Thompson, PB Dews, & WA McKim (Eds.), *Advances in behavioral pharmacology. Vol. 3* (pp. 61–73). New York: Academic Press.

Mogenson, GJ, Jones, DJ, & Yim, CY (1980). From motivation to actions: Functional interface between the limbic system and the motor system. *Prog Neurobiol, 14,* 69–97.

Moghaddam, B & Bunney, BS (1989). Differential effect of cocaine on extracellular dopamine levels in rat medial prefrontal cortex and nucleus accumbens: Comparison to amphetamine. *Synapse, 4,* 156–161.

Nestler, EJ (1992). Molecular mechanisms of drug addiction. *J Neurosci, 12,* 2439–2450.

Nestler, EJ & Aghajanian, GK (1998). Molecular and cellular basis of addiction. *Science, 278,* 58–63.

Nielsen, EB, Nielsen, M & Braestrup, C (1983). Reduction of 3H-spiroperidol binding in rat striatum and frontal cortex by chronic amphetamine: dose response, time course and role of sustained dopamine release. *Psychopharmacology, 81,* 81–85.

Nieto, J, Makhlouf, C & Rodriguez, R (1979). D-amphetamines effects on behavior produced by periodic food delivery in the rat. *Pharm Biochem Behav, 11,* 423–430.

Norton, S (1973). Amphetamine as a model for hyperactivity in the rat. *Physiol Behav, 11,* 181–180.

O'Malley, S, Adamse, M, Heaton, RK, & Gawin, FH (1992). Neuropsychological impairment in chronic cocaine abusers. *Am J Drug Alcohol Abuse, 18,* 131–144.

Parkinson, JA, Olmstead, MC, Burns, LH, Robbins, TW, & Everitt, BJ (1999). Dissociation of effects of lesions of nucleus accumbens core and shell in appetitive Pavlovian approach behavior and the potentiation of conditioned reinforcement and locomotor activity by D-amphetamine. *J Neurosci, 19,* 2401–2411.

Phillips, GD, Hitchcott, PK, & Harmer, CJ (1998). Potential role for the mesoamygdaloid dopamine projection in 'incentive-sensitization'. *Soc Neurosci Abstr, 24,* 1481.

Piazza, PV, Rouge-Pont, F, Deminiere, JM, Kharoubi, M, LeMoal, M & Simon, H (1991). Dopaminergic activity is reduced in the prefrontal cortex and increased in the nucleus accumbens of rats predisposed to develop amphetamine sensitization. *Brain Res, 567,* 169–174.

Post, RM & Rose, H (1976). Increasing effects of repetitive cocaine administration in the rat. *Nature,* 260, 731–732.

Prasad, BM, Hochstatter, T, & Sorg, BA (1999). Expression of cocaine sensitization: Regulation by the medial prefrontal cortex. *Neuroscience, 88,* 765–774.

Ranaldi, R & Benninger, RJ (1995). Bromocriptine enhancement of responding for conditioned reward depends on intact D1 receptor function. *Psychopharmacology, 118,* 437–443.

Ranaldi, R, Pantalony, D & Beninger, RJ (1995). The D1 agonist SKF 38393 attenuates amphetamine-produced enhancement of responding for conditioned reward in rats. *Pharmacol Biochem Behav, 52,* 131–137.

Robbins, T & Sahakian, BJ (1979). "Paradoxical" effects of psychomotor stimulant drugs in hyperactive children from the standpoint of behavioural pharmacology. *Neuropharmacology, 18,* 931–950.

Robbins, TW (1975). The potentiation of conditioned reinforcement by psychomotor stimulant drugs: A test of Hill's hypothesis. *Psychopharmacology (Berl), 45,* 103–114.

Robbins, TW (1977). Relationship between reward-enhancing and stereotypical effects of psychomotor stimulant drugs. *Nature, 264,* 57–59.

Robbins, TW (1990). The case for frontostriatal dysfunction in schizophrenia. *Schizophr Bull, 16,* 391–402.

Robbins, TW, Cador, M, Taylor, JR & Everitt, BJ (1989) Limbicstriatal interactions in reward-related processes. *Neurosci Biobehav Rev, 13,* 155–162.

Robbins, TW & Everitt, BJ (1982). Functional studies of the central catecholamines. *Intl Rev Neurobiol, 23,* 303–365.

Robbins, TW, Roberts, DCS, Koob, GF (1983a). Effects of d-amphetamine and apomorphine upon operant behaviour and schedule-induced licking in rats with 6-hydroxydopamine lesions of the nucleus accumbens. *JPET, 224,* 662–672.

Robbins, TW & Sahakian, BJ (1983). *Behavioral effects of psychomotor stimulant drugs: Clinical and neuropsychological implications.* New York: Raven Press.

Robbins, TW, Watson, BA, Gaskin, M, & Ennis, C (1983b). Contrasting interactions of pipradrol, d-amphetamine, cocaine, cocaine analogues, apomorphine and other drugs with conditioned reinforcement. *Psychopharmacology, 80,* 113–119.

Roberts, DCS, Koob, GF, Klonoff, P, & Fibiger, HC (1980). Extinction and recovery of cocaine self-administration following 6-hydroxydopamine lesions of the nucleus accumbens. *Pharmacol Biochem Behav, 12,* 781–787.

Robinson, TE & Berridge, KC (1993). The neural basis of craving: An incentive-sensitization theory of addiction. *Brain Res Rev, 18,* 247–291.

Robledo, P, Robbins, TW, & Everitt, BJ (1996). Effects of excitotoxic lesions of the central amygdaloid nucleus on the potentiation of reward-related stimuli by intra-accumbens amphetamine. *Behav Neurosci, 110,* 981–990.

Rogers, RD, Everitt, BJ, Baldacchino, A, Blackshaw, AJ, Swainson, R, & Wynne, K, et al. (1999). Dissociable deficits in the decision-making cognition of chronic amphetamine abusers, opiate abusers, patients with focal damage to the prefrontal cortex, and tryptophan-depleted normal volunteers: evidence for monoaminergic mechanisms. *Neuropsychopharmacology, 20,* 322–339.

Sagvolden, T, Wultz, B, Moser, E, Moser, M, & Morkrid, L (1989). Results from a comparative neuropsychological research program indicate altered reinforcement mechanisms in children with ADD. In T. Sagvolden & T. Archer, *Attention deficit disorder.* Hillsdale, NJ: Erlbaum.

Salamone, JD, Cousins, MS, & Snyder, BJ (1997). Behavioral functions of nucleus accumbens dopamine: Empirical and conceptual problems with the anhedonia hypothesis. *Neurosci Biobehav Rev, 21,* 341–359.

Sawaguchi, T & Goldman-Rakic, PS (1991). D1 dopamine receptors in prefrontal cortex: Involvement in working memory. *Science, 251,* 947–950.

Schenk, S, Horger, BA, Peltier, R, & Shelton, K (1991). Supersensitivity to the reinforcing effects of cocaine following 6-hydroxydopamine lesions to the medial prefrontal cortex in rats. *Brain Res, 543,* 25–36.

Seeman, P & Madras, BK (1998). Anti-hyperactivity medication: methylphenidiate and amphetamine. *Mol Psychiatry, 3,* 386–396.

Self, DW & Nestler, EJ (1995). Molecular mechanisms of drug reinforcement and addiction. *Ann Rev Neurosci, 18,* 463–495.

Self, DW & Nestler, EJ (1998). Relapse to drug-seeking: Neural and molecular mechanisms. *Drug Alcohol Dep, 51,* 49–60.

Sesack, SR, Hawrylak, VA, Matus, C, Guido, MA & Levey, AI (1998). Dopamine axon varicosities in the prelimbic division of the rat prefrontal cortex exhibit sparse immunoreactivity for the dopamine transporter. *J Neurosci, 18,* 2697–2708.

Shetty, T & Chase, TN (1976). Central monoamines and hyperkinase of childhood. *Neurology, 26,* 1000–1002.

Shippenberg, TS & Heidbreder, C (1995). Sensitization to the conditioned rewarding effects of cocaine: Pharmacological and temporal characteristics. *J Pharmacol Exp Ther, 273,* 808–815.

Simon, H (1981). Dopaminergic A10 neurons and the frontal system. *J de Physiol, 77,* 81–95.

Solanto, MV (1998). Neuropsychopharmacological mechanisms of stimulant drug action in attention-deficit hyperactivity disorder: a review and integration. *Behav Brain Res, 94,* 127–152.

Sorg, BA, Davidson, DL, Kalivas, PW, & Prasad, BM (1997). Repeated daily cocaine alters subsequent cocaine-induced increase of extracellular dopamine in the medial prefrontal cortex. *J Pharmacol Exp Ther, 281,* 54–61.

Stamm, CJ, de Bruin, JP, van Haelst, AM, van der Gegten, J, & Kalsbeek, A (1989). Influence of the mesocortical dopaminergic system on activity, food hoarding, social agonistic behavior, and spatial delayed alternation in male rats. *Behav Neurosci, 103,* 24–35.

Staton, DM, Solomon, PR (1984). Microinjection of d-amphetamine into the nucleus accumbens and caudate-putamen differentially affect stereotypy and locomotion in the rat. *Physiol Psychol, 12,* 159–162.

Strakowski, SM, Sax, KW, Setters, MJ, & Keck, PE, Jr. (1996). Enhanced response to repeated d-amphetamine challenge: Evidence for behavioral sensitization in humans. *Biol Psychiatry, 40,* 872–880.

Stromberg, U & Svensson, TH (1975). Differences between (+)- and (–)-amphetamine in effects on locomotor activity and L-Dopa potentiating action in mice. *Naunyn-Schmiedeberg's Arch. Pharmacol, 278,* 171–179.

Sulzer, D, Chen, TK, Y Y, L, Kristensen, H, Rayport, S and Ewing, A (1995). Amphetamine redistributes dopamine from synaptic vesicles to the cytosol and promotes reverse transport. *J Neurosci, 15,* 4102–4108.

Sulzer, D, Maidment, NT, & Rayport, S (1993). Amphetamine and other weak bases act to promote reverse transport of dopamine in ventral midbrain neurons. *J Neurochem, 60,* 527–535.

Swanson, JM (1988). What do psychoparmacological studies tell us about information processing deficits in ADD/Hyperactive children. In: J. Sergeant & L Bloomingdale (Eds.), *Attention: Criteria, cognition, intervention. Vol. V,* (pp. 97–115). New York: Pergamon Press.

Tanda, G, Pontieri, FE, Frau, R, & Di Chiara, G (1997). Contribution of blockade of the nonadrenaline carrier to the increase of extracellular dopamine in the rat prefrontal cortex by amphetamine and cocaine. *Eur J Neurosci, 9,* 2077–2085.

Taylor, JR (1985). Neural mechanisms of the potentiation of conditioned reinforcement by psychomotor stimulant drugs. In *Department of Experimental Psychology* (pp. 1–315). Cambridge, U.K: University of Cambridge.

Taylor, JR, Birnbaum, S, Ubriani, R & Arnsten, AFT (1999a). Activation of cAMP-dependent protein kinase A in prefrontal cortex impairs working memory performance. *J Neurosci, 19,* 1–5.

Taylor, JR & Horger, BA (1999). Enhanced responding for conditioned reward produced by intra-accumbens amphetamine is potentiated after cocaine sensitization. *Psychopharmacology, 142,* 31–40.

Taylor, JR, Nestler, EJ, Shaw, TZ, & Jentsch, JD (1999b). Acquisition of Pavlovian approach behavior is facilitated by repeated administration of d-amphetamine or cocaine. *Soc Neurosci Abtr, 25,* 2154.

Taylor, JR & Robbins, TW (1984). Enhanced behavioral control by conditioned reinforcers following microinjections of d-amphetamine into the nucleus accumbens. *Psychopharmacology, 84,* 405–412.

Taylor, JR & Robbins, TW (1986). 6-Hydroxydopamine lesions of the nucleus accumbens, but not of the caudate nucleus, attenuate enhanced responding with reward-related stimuli produced by intra-accumbens d-amphetamine. *Psychopharmacology 90,* 390–397.

Vanderschuren, LJMJ, Donne Schmidt, E, De Vires, TJ, Van Moorsel, CAP, Tilders, FJH, & Schoffelmeer, ANM (1999). A single exposure to amphetamine is sufficient to induce long-term behavioral, neuroendocrine, and neurochemical sensitization in rats. *J Neurosci, 19,* 9579–9586.

Weiss, B & Laties, VG (1964). Effects of amphetamine, chlorpromazine, pentobarbital and ethanol on operant response duration. *JPET, 144,* 17–23.

White, FJ (1996). Synaptic regulation of mesocorticolimbic dopamine neurons. *Ann Rev Neurosci, 19,* 405–436.

White, FJ, Hu, X-T, Zhan, gX-F, & Wolf, ME (1995). Repeated administration of cocaine or amphetamine alters neuronal responses to glutamate in the mesoaccumbens dopamine system. *JPET, 273,* 445–454.

Whitelaw, RB, Markou, A, Robbins, TW, & Everitt, BJ (1996). Excitotoxic lesions of the basolateral amygdala impair the acquisition of cocaine-seeking behaviour under a second-order schedule of reinforcement. *Psychopharmacology, 127,* 213–224.

Wilkinson, LS, Humby, T, Killcross, AS, Torres, EM, Everitt, BJ & Robbins, TW (1997). Dissociations in dopamine release in medial prefrontal cortex and ventral striatum during the acquisition and extinction of classical aversive conditioning in the rat. *Eur J Neurosci, 10,* 1019–1026.

Williams, SM & Goldman-Rakic, PS (1998). Widespread origin of the primate mesofrontal dopamine system. *Cerebral Cortex, 8*(4), 321–45.

Wise, RA & Leeb, K (1993). Psychomotor stimulant sensitization: A unitary phenomenon? *Behav Pharmacol, 4,* 339–349.

Wolf, ME (1998). The role of excitatory amino acids in behavioral sensitization to psychomotor stimulants. *Prog Neurobiol, 54,* 679–720.

Wolterink, G, Phillips, G, Cador, M, Donselaar-Wolterink, I, Robbins, TW & Everitt, BJ (1993). Relative roles of ventral striatal D1 and D2 dopamine receptors in responding with conditioned reinforcement. *Psychopharmacology, 110,* 355–364.

Zahrt, J, Taylor, JR, Mathew, RG & Arnsten, AFT (1997). Supranormal stimulation of D1 dopamine receptors in the rodent prefrontal cortex impairs spatial working memory performance. *J Neurosci, 17,* 8528–8538.

Zetterstrom, T, Sharp, T, Collin, AK, & Ungerstedt, U (1988). In vivo measurement of extracellular dopamine and DOPAC in rat striatum after various dopamine-releasing drugs: Implications for the origin of extracellular DOPAC. *Eur J Pharmacol, 148,* 327–34.

Zetterstrom, T, Sharp, T, Marsden, CA & Ungerstedt, U (1983). In vivo measurement of dopamine and its metabolites by intracerebral dialysis: Changes after d-amphetamine. *J Neurochem, 41,* 1769–1773.

Zetterstrom, T, Sharp, T, & Ungerstedt, U (1986). Further evaluation of the mechanism by which amphetamine reduces striatal dopamine metabolism: A brain microdialysis study. *Eur J Pharmacol, 132,* 1–9.

5

Psychostimulant Actions on Dopamine and Limbic System Function: Relevance to the Pathophysiology and Treatment of ADHD

ANTHONY A. GRACE

Understanding the mechanism of action of psychostimulants must begin with a comprehension of their neurochemical action at the site of the nerve terminal in the central nervous system. It is the catecholamine terminal at which drugs such as methylphenidate have their primary site of action. Moreover, the action of psychostimulants on the dopamine (DA) neurotransmitter system is believed to mediate the rewarding properties of these drugs (Colle & Wise, 1988; Gratton et al., 1988; Olds, 1990). However, to understand the effects of such compounds on normal behavior and their therapeutic actions in disorders such as Attention-Deficit/Hyperactivity Disorder (ADHD), psychostimulant effects must be placed in the context of multiple interacting systems. The focus of this chapter is the limbic system, a loosely defined system of central nuclei involved in motivation and emotion, which is regulated by higher cortical centers such as the prefrontal cortex; the limbic system appears to have a major role in the pathophysiology and pharmacotherapy of ADHD. Psychostimulant effects on the prefrontal cortex are discussed in detail in Chapter 7 by Arnsten.

Basic Pharmacology of Psychostimulant Action

Catecholamines, such as DA and norepinephrine (NE), are released from synaptic terminals primarily by a spike-dependent process, achieved via action potential discharge in DA and NE cell bodies. This action potential then propagates to the axon terminal to trigger calcium-dependent release of transmitter into the synaptic cleft. The release of neurotransmitter at high concentrations into the synapse then triggers postsynaptic receptors to elicit a response. This process is terminated via a high-capacity reuptake system that rapidly removes the transmitter from the synaptic cleft by taking it back up into the terminal that released the catecholamine.

Once taken back into the terminal, the transmitter can then undergo metabolism by exposure to the enzyme monoamine oxidase, or can be repackaged into vesicles for subsequent storage and release.

Psychostimulants such as amphetamine or methylphenidate are known to exert their actions via interference with the uptake process. In the case of amphetamine, this is achieved via mimicking the transmitter, allowing the drug to be taken up into the terminal. As a consequence, amphetamine both interferes with the uptake of the catecholamine by competing for the uptake carrier, as well as causing its release from the terminal via reversal of this transport process (Kuczenski, 1983; Reith et al., 1986; Heikkila et al., 1975). The catecholamine pool that is released is derived primarily from the newly synthesized pool in the cytoplasm of the synaptic terminal; therefore, release is not dependent on spike activity. Moreover, this amphetamine-induced release can be prevented by drugs that interfere with catecholamine synthesis, such as alpha-methyl-para-tyrosine, which is a competitive inhibitor of the rate-limiting enzyme in catecholamine synthesis, tyrosine hydroxylase. In addition, because the release process itself occurs via reversal of the transport system rather than via spike activity, this release is not calcium-dependent (Cooper et al., 1996; see Chapter 4 by Taylor & Jentsch). Because of its disruption of the catecholamine uptake process, amphetamine causes the release and accumulation of catecholamines, particularly NE and DA, in the synaptic cleft and extrasynaptic space. However, although amphetamine-induced *release* of catecholamines is not dependent on spike activity, this does not mean that the actions of the drug are independent of spike-dependent transmitter release. Thus, it should be noted that, in addition to causing a release of catecholamines, these drugs also would be expected to potentiate catecholamines released by spike activity as well, since the uptake process is the primary means by which catecholamines released by spike-dependent processes are removed from the synaptic cleft.

In contrast, drugs such as cocaine and methylphenidate are not taken up into the terminal by the uptake system. Instead, these drugs block the transporter, and thereby prevent the uptake of transmitters such as DA (Cooper et al., 1996; see Chapter 4 by Taylor & Jentsch). Since these drugs do not directly cause the release of catecholamines, they are dependent on spike activity for their action. Therefore, unlike amphetamine, all of the actions of these drugs on catecholamine systems are calcium-dependent, for they require action potential–triggered entry of calcium into the synaptic terminal in order to release the transmitter. Psychostimulants have actions on several transmitter systems. Amphetamine will primarily cause the release of the catecholamines, with most actions in the CNS mediated via NE and DA, with relatively little action on serotonin. Cocaine, on the other hand, is efficacious in blocking the reuptake of all of the monoamines. Methylphenidate blocks reuptake of DA more effectively than of NE and much more than serotonin (Gatley et al., 1996). The orientation of this chapter will be primarily on the DA system; potential involvements of the other monoamine systems in psychostimulant actions and AD/HD will be covered in other chapters (see Chapter 6 by Berridge and Chapter 14 by Pliszka).

Psychostimulant Effects on Limbic System Interactions

What is the effect of increases in DA system activity with respect to limbic system function? This may be derived from a consideration of the circuitry involved, and how it is affected by amphetamine administration. There are several structures within the limbic system that are relevant to the actions of the psychostimulants and all drugs of abuse. These are the prefrontal cortex, hippocampus subiculum, and the nucleus accumbens. The nucleus accumbens is the site to which each of the other systems sends projections. The accumbens, in turn, projects to the ventral pallidum and on to the thalamus, where it can influence thalamocortical activity (Fig. 5–1). The accumbens itself is comprised primarily of medium-spiny neurons, which are approximately 20 to 30 μm in diameter, with dendritic fields that extend approximately 200 μm (O'Donnell & Grace, 1993; Chronister et al., 1981). Despite their comparatively small size, these spine-covered dendrites are estimated to receive between 10,000 and 15,000 inputs from excitatory glutamatergic afferents, with the DA innervation overlapping with nearly half of these terminals (Kincaid et al., 1998; Doucet et al., 1986). Given this density of innervation, one may view these neurons as surrounded by a potential "cloud" of neurotransmitter

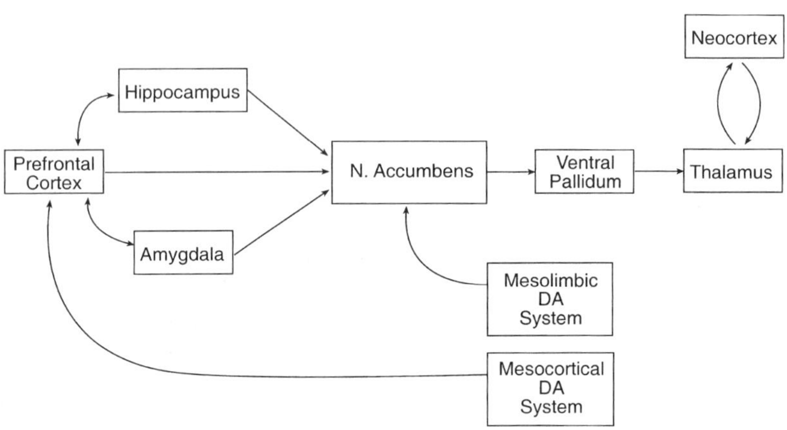

Figure 5–1. The nucleus accumbens receives input from a number of limbic-related cortical structures, including the prefrontal cortex, hippocampus, and amygdala. In particular, the hippocampus and amygdala strongly influence the ability of the prefrontal cortex to activate accumbens cell firing. As a result, these systems modulate the flow of information from the prefrontal cortex through the nucleus accumbens, where it can ultimately influence thalamocortical function via projections through the ventral pallidum. The DA system plays a significant role in the modulation of information integration in this system. Thus, the mesocortical DA system is known to provide a potent, direct influence over activity within the prefrontal cortex. Moreover, the mesolimbic DA system is positioned to modulate information integration within the nucleus accumbens via both presynaptic effects (on prefrontal cortical afferents; O'Donnell & Grace, 1994) and on the responsivity of accumbens neurons to these inputs (O'Donnell & Grace, 1996b).

due to spillover from the synaptic cleft, in addition to their discrete synaptic inputs. One of the prominent inputs is from the prefrontal cortex (Christie et al., 1987; Sesack et al., 1989). Stimulation of this afferent elicits a short-latency, glutamatergic excitatory postsynaptic potential (EPSP) in accumbens neurons (O'Donnell & Grace, 1993). Moreover, this EPSP is potently modulated by DA. Thus, low doses of D2 agonists will attenuate this EPSP, making it much less effective at evoking spike activity (O'Donnell & Grace, 1994). Indeed, the fact that D2 antagonists will increase this EPSP amplitude suggests that the EPSP itself is tonically down-modulated by DA stimulation, since this effect is lost when the rat is first depleted of endogenous DA by pretreatment with reserpine and alpha-methyl-para-tyrosine (O'Donnell & Grace, 1994). This effect of D2 agonists can also be mimicked by amphetamine (Grace et al., 1998). Thus, amphetamine, acting via the DA system, can diminish the influence of the prefrontal cortex over the accumbens when administered at a behaviorally activating dose.

There are several other systems that, *unlike* the DA system, serve to facilitate prefrontal cortical influence over the accumbens. In particular, the subiculum of the hippocampus provides a potent gating influence over this region, in that it can regulate information throughput arising from the prefrontal cortex. Thus, in the normal animal, accumbens neurons exist in a bistable state, with their membrane potential alternating between a hyperpolarized, nonfiring state and a depolarized plateau lasting for several hundred milliseconds, during which spike activity is generated. Our data show that this bistable state is driven by inputs from the hippocampus (O'Donnell & Grace, 1995).

What is the function of this bistable state? It appears that the depolarizing plateau associated with this state is required to facilitate inputs from other structures. Thus, stimulation of the prefrontal cortical afferents in vivo will rarely elicit spike activity; in the absence of spike firing, the accumbens fails to provide information throughput to its target structures. In contrast, if the accumbens neuron is first driven into the depolarized state, prefrontal cortex stimulation will readily activate spike discharge in the accumbens (O'Donnell & Grace, 1995). Spike activity in the accumbens is then transferred to the ventral pallidum and on to the thalamus (O'Donnell et al., 1997; Lavin & Grace, 1994, figure 1), where it can affect thalamocortical activity and higher levels of information processing. Therefore, in essence, the subiculum gates the ability of information from the prefrontal cortex to affect thalamocortical information processing.

This system may have a behaviorally relevant function as well. Studies have suggested that the subiculum of the hippocampus is involved in a type of context-dependency (Salzmann et al., 1993; Phillips & LeDoux, 1992), or the ability of an organism to stay "on task" and not be distracted by extraneous stimuli. Thus, stimulation of the ventral subiculum will increase investigatory behavior (Yang & Mogenson, 1987; Brudzynski & Gibson, 1997) in a DA-dependent manner (Wu & Brudzynski, 1995; Brenner & Bardgett, 1998). Alternately, lesions of the hippocampal subiculum projections to the accumbens interfere with the investigation of novel stimuli, disrupt the ability of the organism to ignore irrelevant stimuli, and increase the distractibility of the organism (Gray et al., 1991; Weiner, 1990). The ability of the accumbens to provide this gating influence is potently disrupted by two drugs known to increase distractibility and decrease focus in human subjects, and which also have been found to disrupt hippocampal inputs to the accumbens—phencyclidine (PCP; O'Donnell & Grace, 1998a) and DA agonists, including cocaine (O'Donnell & Grace, 1996; Cooper and White, 1998). Therefore, the hippocampus can provide a context-dependent gating influence over the ability of the prefrontal cortex to control thalamocortical information processing, enabling the organism to stay on-task and focused. On the other hand, this system can be disrupted by behaviorally activating doses of drugs which, in addition to altering hippocampal drive of accumbens gating, also are associated with deficits in context-dependent processes, leading to distractibility and confusional states in humans. Moreover, by interfering with normal investigatory behavior, dysfunctions in this system can also precipitate the type of misdirected hyperactivity that occurs in ADHD.

Another limbic input that exerts a potent modulatory influence over the accumbens is the amygdala (McDonald, 1991; O'Donnell & Grace, 1995). As with the subiculum, stimulation of the amygdala input facilitates the ability of the prefrontal cortex to cause spiking in the accumbens. However, in contrast to the long-duration gating influence provided by the subiculum, the amygdala will only gate prefrontal stimuli that arrive within a narrow time window after amygdala stimulation; that is, between 30–40 msec (Moore & Grace, 1996). Therefore, the amygdala appears to gate more on an event-related basis. In other words, amygdala activation appears to only gate single events, rather than providing a more general type of context-dependent gating of multiple stimuli. What is the nature of this gating influence? Numerous studies have provided evidence that the amygdala is involved in affective or emotional responses (LeDoux & Muller, 1997). Therefore, the premise is that the amygdala will gate events based on their affective valence, or the amount of emotional drive evoked by a given stimulus.

To put this together, the subiculum provides a context-dependent gating to keep an organism on-task and avoid distractions, whereas the amygdala selectively gates stimuli that evoke a strong emotional response (Grace & Moore, 1998). The prefrontal cortex, being the most anterior of the precentral (or motor-related) association cortices, is most likely involved in functions such as advanced motor planning and goal-directed activity in selecting the appropriate response (Grace, 2000b). Thus, the subiculum and the amygdala interact to enable the organism to focus their goals and keep on task, but also enable the organism to respond to an immediate threat (Fig. 5–2). However, the smooth functioning of this system can be disrupted pharmacologically. In particular, drugs such as amphetamine can, in a dose-dependent manner, interfere with this system on several levels. Thus, amphetamine can decrease the effectiveness of the cortical input by presynaptic inhibition in the accumbens, as well as diminish the ability of the hippocampal subiculum to facilitate prefrontal throughput. As a consequence, amphetamine can cause the amygdala drive to predominate (Grace, 2000a; Moore & Grace, 1996). One consequence of such behaviorally activating doses of amphetamine would thus be to distract the organism from staying on-task toward reaching predefined short-term and long-term goals, and instead engage in immediately reinforcing, pleasure-seeking behavior.

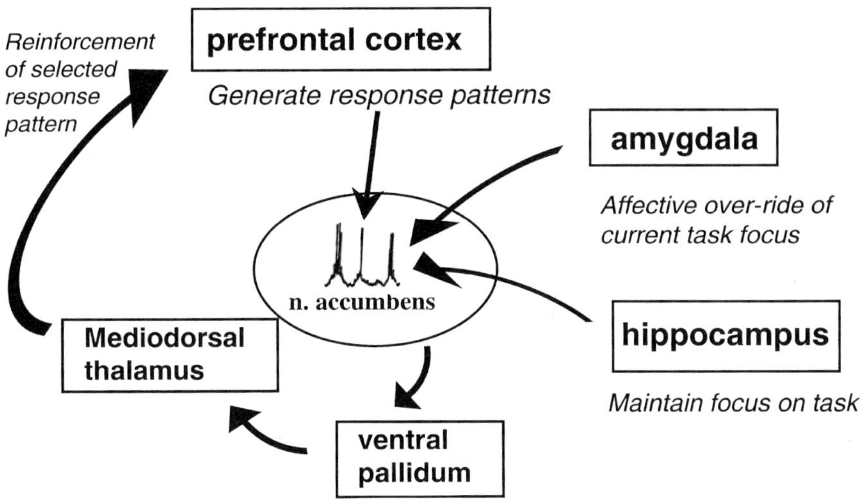

Figure 5–2. The nucleus accumbens is the striatal region in which the limbic system has overlapping inputs. In this model, the accumbens neurons are driven to generate spikes via the interaction of several afferent systems. Among the many functions of the prefrontal cortex is the generation of a set of potential responses that the subject can utilize to react to a stimulus. A specific response plan is then selected within the nucleus accumbens based on several conditions. The hippocampus subiculum, by generating long-duration plateau depolarizations in the accumbens neurons (slow potentials illustrated in electrophysiological tracing in accumbens) gates prefrontal throughput in the accumbens according to the behavioral context of the organism. Therefore, the hippocampal-accumbens projection will keep the subject focused on the current task. In contrast, the amygdala input provides a more brief, event-related gating of prefrontal throughput in the accumbens. Since the amygdala is involved in emotional or affective properties of stimuli, the amygdalo-accumbens projection will provide a type of emotional over-ride to the system, enabling the subject to respond to events that are emotionally charged, and therefore of immediate survival value. The net result of the interaction of these systems is that one specific motor response that best meets the needs of the organism is selected, and by modulating the mediodorsal thalamic nucleus projection to the prefrontal cortex, this specific circuit is reinforced. In this way, the subject can maintain focus on a specific task, but also can break this focus when presented with stimuli that activate the subject emotionally, enabling the subject to divert attention away from the current task. However, in pathological states such as that proposed to exist in ADHD, the amygdala input may be overdriven to the extent that maintenance of focus is severely disrupted by comparatively minor emotional events.

Dual Modulation of DA System Function by Psychostimulants—DA Neuron Autoreceptors

Based on the preceding considerations, evidence seems to suggest that increases in DA transmission caused by amphetamine would lead to a condition in which the amygdalar input would predominate, causing the individual to be easily distracted and driven to behave in an emotionally guided, disruptive manner. However, in the case of ADHD, psychostimulants actually are used to achieve the *opposite* result. How could this come to be? A potential explanation for this circumstance may be derived from a careful consideration of the dose and time course of action of these drugs, and the part of the DA system on which they exert their actions.

As reviewed previously, the behavioral activation produced by psychostimulants most likely is due to their potentiation of spike-dependent DA release secondary to interference with the DA uptake system. However, DA released in this manner also is subject to escape from the synaptic cleft, and can thereby stimulate DA receptors other than those located within the synapse on the postsynaptic target neuron. Thus, in addition to acting on DA receptors located across the synaptic cleft from the DA terminal, this transmitter can potentially affect DA receptors on other structures as well. DA can act on several extrasynaptic sites to alter the function of this system. The most prominent action would be on the DA autoreceptors. DA autoreceptors are DA receptors that are located on the DA neuron itself, and serve to provide feedback inhibition. Three types of DA autoreceptors have been characterized to date: (*1*) firing rate-modulating autoreceptors, (*2*) synthesis-modulating autoreceptors, and (*3*) release-modulating autoreceptors.

Somatodendritic Autoreceptors

Firing rate-modulating autoreceptors are located at the site of spike generation—that is, on the DA cell body in the midbrain. These autoreceptors are tonically stimulated by DA that is released from the dendrites of DA neurons, and thereby provide a continuous, down-modulation of DA neuron firing (Pucak & Grace, 1994, 1996; Mercuri et al., 1990, 1996; Wuerthele et al., 1981). This calcium- and depolarization-dependent release (Chéramy et al., 1981) is thought to be mediated by action potentials within the DA neuron, and provides a feedback inhibition of DA neuron spike activity. The feedback appears to act via a tonic increase in extracellular DA in the cell body region rather than via a rapid feedback onto the single neuron releasing the DA. This model is supported by data showing that 6-hydroxydopamine-induced lesions of a portion of the DA neurons in the substantia nigra lead to an increase in the sensitivity of the remaining neurons to DA (Pucak & Grace, 1991; Harden & Grace, 1995). If the neuron was only subject to inhibition by the DA that the neuron releases onto its own dendrites, this would not be expected to occur. Thus, the DA neuron population must be responding to the pool of DA contributed by the active population of neurons.

DA levels in the midbrain appear to tonically inhibit DA neuron firing. Thus, acute blockade of DA receptors activates DA neuron firing both by local actions within the midbrain as well as by feedback from blockade of postsynaptic receptors (Pucak & Grace, 1994). Moreover, chronic blockade of DA receptors (Bunney & Grace, 1978) as well as lesions of a portion of the DA neurons in the midbrain (Pucak & Grace, 1991; Harden & Grace, 1995) lead to supersensitivity of these firing rate-modulating autoreceptors. Psychostimulants are known to potentiate the actions of dendritic DA, as shown by in vitro studies (Lacey et al., 1990). Because the levels of DA in the midbrain are exceedingly low (Santiago & Westerink, 1992), the autoreceptor action is much more sensitive to DA stimulation than are the postsynaptic receptors (Skirboll et al., 1979). This is not believed to be due to differences in affinity of the receptors, since the autoreceptors are of the standard D2 type. Instead, this is believed to be due to the presence of "spare receptors." In this case, a maximal response can be obtained, even though not all of the receptors are activated by the transmitter (Cox & Waszczak, 1989; Meller et al., 1987; Yokoo et al., 1988). As a consequence, a given concentration of DA produces a proportionately greater response by acting at the D2 autoreceptors compared to the postsynaptic D2 receptors. For example, concentrations of DA that are sufficient to activate only half of the D2 receptors and cause half-maximal response postsynaptically may nonetheless be sufficient to cause maximal inhibition of DA cell firing (Skirboll et al., 1979). It is for this reason that the behavioral response to DA agonists is biphasic in nature, causing an attenuation of activity at low doses, but activating locomotion at higher doses of the agonist (Strombom, 1976; DiChiara et al., 1976).

Terminal Autoreceptors

In addition to DA receptors located at the DA neuron soma, there are also DA autoreceptors located on the synaptic terminal of the DA neuron. There appear to be two types of terminal autoreceptors: synthesis-modulating autoreceptors and release-modulating autoreceptors. As the names imply, the first type regulates synthesis of catecholamines via modulation of the activity of the rate-limiting enzyme in catecholamine synthesis, tyrosine hydroxylase. This autoreceptor appears to be tonically stimulated by DA, since it also shows activation with DA antagonists and becomes supersensitive upon repeated blockade (Nowycky & Roth, 1997). However, one exception to this type of compensatory response appears to be present in the midbrain DA neurons projecting to the prefrontal cortex. In the prefrontal cortex, repeated administration of DA antagonists does not alter DA synthesis, which is believed to be due to the absence of synthesis-modulating autoreceptors from this population of DA terminals (Bannon & Roth, 1983). Therefore, unlike subcortical DA systems, the DA innervation of the prefrontal cortex does not appear to be capable of long-term modulation in terms of alterations in its DA synthetic capacity. This receptor appears to co-vary with the firing rate–modulating autoreceptors, in that both are G protein–linked receptors (Cooper et al., 1996), and are absent only in the mesoprefrontal DA system (Chiodo et al., 1984; Bannon & Roth, 1983).

In contrast to the slow-acting G-protein–linked synthesis-modulating autoreceptors, evidence shows that a more rapidly acting system is present for modulating DA system function—that is, the release-modulating autoreceptors. Upon stimulation, these presynaptic autoreceptors rapidly down-regulate spike-dependent DA release. In contrast to the synthesis-modulating autoreceptors, however, there is little evidence that this system undergoes alterations in sensitivity over extended periods of time. Together, presynaptic autoreceptors exert a powerful control over spike-dependent DA release (Farnebo & Hamberger, 1971; Roth, 1987). Indeed, studies have shown that the tonic level of extracellular DA produces a powerful tonic down-regulation of spike-dependent DA release via the DA autoreceptor (Reimann et al., 1979).

The Tonic/Phasic Model of Psychostimulant Action

As in the case of the DA receptors at the soma, because the presynaptic DA autoreceptors on the DA terminals in the striatum are located outside of the synaptic cleft, they are exposed to much lower concentrations of DA than are the intrasynaptic receptors. Thus, Wightman and colleagues have provided evidence to suggest that the concentration of DA reached within the synaptic cleft by DA cell spike–dependent release is on the order of hundreds of micromolar to low millimolar concentrations (Garris et al., 1994). However, outside of the synaptic cleft, microdialysis studies have shown that the concentration of DA is 5 to 6 orders of magnitude lower, in the range of approximately 4 nM in the nucleus accumbens (Parsons & Justice, 1992). Moreover, evidence indicates that the extracellular concentration of DA is highly regulated, in that it is maintained at baseline concentrations even in the face of up to an 80% loss of the DA innervation to the striatum (Abercrombie et al., 1990; Robinson & Whishaw, 1988). Obviously, at this low concentration, the DA in the extracellular space cannot substitute for the high concentrations of DA that occur in the synapse during spike-dependent release. Nonetheless, these levels of DA are apparently sufficient to stimulate the sensitive presynaptic autoreceptors on DA terminals in the striatum (Richfield et al., 1989) and firing rate-modulating autoreceptors on the DA cell body, as demonstrated by their activation by antagonist administration (Pucak & Grace, 1994). Therefore, these autoreceptor systems exert a powerful tonic down-regulation of spike-dependent DA release.

In contrast to this tonic extracellular DA level, spike-dependent DA release is much more transient in nature. DA neurons are known to fire in two modes: a single-spike firing mode and burst firing. The single-spiking mode is the basal level of firing, and appears to be required for glutamatergic afferents to modulate tonic DA release (Moghaddam & Gruen, 1991; Keefe et al., 1992). In contrast, when the DA system is activated, the neuron emits spikes in bursts. Unlike the self-driven tonic firing of the DA neuron (Grace & Bunney, 1984a), DA neuron burst firing is dependent on afferent drive from rostral brain regions (Pucak & Grace, 1994; Hollerman & Grace, 1992; Smith & Grace, 1992). This burst firing is known to occur under conditions in which there is a DA-dependent behavioral response, such as that associated with reward processes (Grace & Bunney, 1984b; Schultz, 1997, 1998). Thus, these high intrasynaptic concentrations are reached very rapidly following the arrival of bursts of spikes into the DA terminal (Grace & Bunney, 1984a,b; Gonon, 1988). Moreover, because of the DA concentration-dependent activation of the DA transporter, the DA that is released into the synaptic cleft is removed before substantial amounts can escape the synaptic cleft. Thus, studies have shown that stimulation of DA cell axons at stimulation rates that are within the normal range of DA neuron firing (Grace & Bunney, 1984a) are insufficient to modify extracellular DA levels measured by microdialysis, but are nonetheless sufficient to activate DA-dependent behaviors (Fibiger et al., 1987; Gratton et al., 1988; Nakahara et al., 1989). With more intense levels of stimulation, the extracellular concentration of DA near the release site may rise, but only very briefly (Garris et al., 1994). Because the spike-dependent DA release, which appears to be responsible for DA-dependent behavioral processes, is so brief in nature, this process has been termed the *phasic* component of DA release (Fig. 5–3; Grace, 1991, 1995). In contrast, because the extracellular DA level is maintained at a stable steady-state low concentration, this system has been termed *tonic* DA stimulation (Fig. 5–3; Grace, 1991). Because the tonic DA system is sufficient to stimulate presynaptic autoreceptors on DA terminals, the tonic DA system exerts a

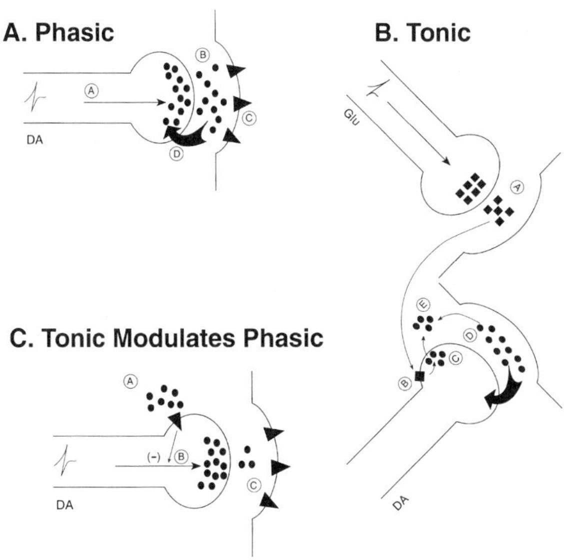

Figure 5–3. An illustration of the tonic/phasic model of DA system function. A: Phasic DA release: This high-amplitude DA release process is triggered by DA neuron firing (*A*) leading to calcium-dependent DA release into the synaptic cleft (*B*) where it reaches sufficiently high concentrations (i.e., up to mM) to stimulate postsynaptic DA receptors (*C*). The DA is then rapidly removed from the synaptic cleft by the DA transporter (D) before it can escape into the extrasynaptic space. B: Tonic DA: Tonic DA is the low-concentration but tightly regulated DA present in the extrasynaptic space. This DA level is derived from two sources: 1) it is modulated by glutamatergic afferents in close proximity to the DA terminal. Glutamate released from these terminals (A) stimulates presynaptic glutamate heteroreceptors on the DA terminal (B) to release DA from an intraterminal pool distal from the synaptic cleft (C); 2) low concentrations of DA also can escape from the synaptic cleft (D) to contribute to this pool, particularly during periods of slow spike discharge. This extrasynaptic DA (E) is maintained at low (nM) concentrations. C: Tonic modulates Phasic: The tonic extracellular DA pool (A) is too low in concentration to stimulate postsynaptic DA receptors. Nonetheless, it is of sufficient amplitude to provide a tonic down-modulation of spike-dependent DA release (B) by stimulating release- and synthesis-modulating DA autoreceptors on the DA terminal. This causes a decrease in spike-dependent DA release into the synaptic cleft (C).

continuous down-modulation of phasic spike-dependent DA release (Fig. 5–3; Grace, 1991, 1995; Moore et al., 1999).

What modulates the tonic DA system? Evidence indicates that this system can be regulated by several processes, most notably the glutamatergic afferents arising from the prefrontal cortex and the hippocampus. Thus, stimulation of glutamate receptors in the striatum can lead to increases in tonic DA levels (Glowinski et al., 1988; Grace, 1991), probably via activation of metabotropic glutamate receptors (Antonelli et al., 1997; Ohno & Watanabe, 1995; Verma & Moghaddam, 1998), which are known to be located presynaptically in other systems (Shigemoto et al., 1996). Moreover, stimulation of the prefrontal cortical afferents (Taber et al., 1995), hippocampal afferents (Blaha et al., 1997), or amygdala afferents (Floresco et al., 1998), all of which utilize glutamate as a transmitter, leads to increases in extracellular DA levels. These data are consistent with findings that lesions of the cortical input to the striatum will decrease extracellular DA and glutamate in the striatum (Smolders et al., 1996) which, via a suppression of tonic stimulation of DA autoreceptors, could lead to hyperlocomotion and an increase in the behavioral response to amphetamine (Adler, 1961; Iversen, 1971; Wolf et al., 1995). Thus, evidence indicates that alterations in tonic DA levels produced by cortical afferents can potently alter spike-dependent DA release, and thereby modulate DA-dependent behaviors (Grace, 1991, 1995; Moore et al., 1999). Tonic down-modulation would be particularly effective when the reuptake system is blocked by a psychostimulant. Thus, although the DA transporter is normally very effective at clearing DA from the synapse before it can escape the synaptic cleft, once the transporter is blocked substantially greater levels of DA can escape the cleft and augment the tonic DA pool.

Given these variables, I would predict that drugs that interfere with DA uptake, such as methylphenidate, when administered i.v. should have two stages in their action (Fig. 5–4): (*1*) a short-latency effect due to blockade of the DAT and accumulation of spike-dependent DA release within the synapse, followed by (*2*) diffusion of DA from the synaptic cleft, an increase in extracellular DA, and subsequent feedback inhibition of DA release and DA cell firing (Grace, 1995; Seeman & Madras, 1998; Solanto, 1986). The first effect would result in reward-related behaviors as well as a disruption of focus and increase in emotionally-driven responses. This would occur before there was a substantial rise in extracellular DA. Indeed, studies examining microdialysis measures of DA overflow show that the locomotor activation (which parallels reward-related behaviors) tends to precede the maximal increase in extracellular DA (Kuczenski & Segal, 1989; Kalivas & Duffy, 1990). During the second stage, the increase in extracellular DA would serve to cause an inhibition in the activating action of DA secondary to an attenuation of spike-dependent DA release. This would correspond to the falling-off in locomotor activation, which microdialysis experiments shows tends to occur as the extracellular DA reaches its peak (Kuczenski & Segal, 1989; Kalivas & Duffy, 1990). This is also consistent with the neuroimaging data comparing subjective experiences of euphoria with psychostimulant transporter occupancy as described in Chapter 11 by Swanson and Volkow. This two-stage process would also provide an explanation for the slight but consistent worsening observed immediately

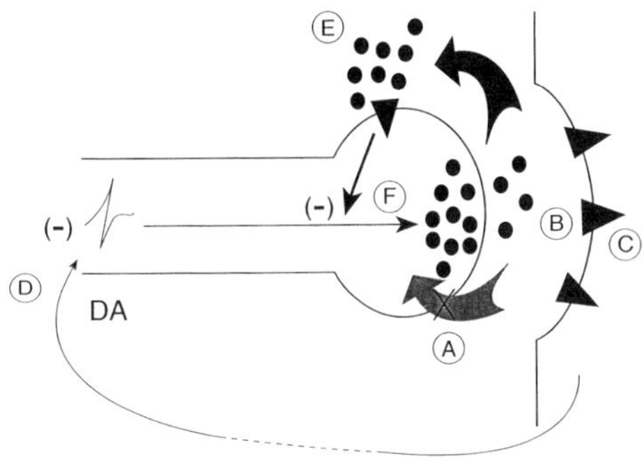

Figure 5–4. Model of methylphenidate actions. By blocking the DA transporter (A), methylphenidate will cause an accumulation of DA in the synaptic cleft (B). Although this may initially cause increased stimulation of postsynaptic DA receptors (C), in the long term the consequence is a down-regulation of DA release. This occurs via several processes. First, the increased postsynaptic DA stimulation will cause a feedback inhibition of DA neuron firing (D) to decrease spike-dependent DA release. More significantly, the blockade of DA uptake will enable much larger quantities of DA to escape from the synaptic cleft and accumulate in the extrasynaptic space (E). This will substantially increase presynaptic DA autoreceptor stimulation, resulting in a potent attenuation of spike-dependent DA release (F). The result is a dramatic decrease in the concentration of DA that is released into the synaptic cleft (B), thereby decreasing the amount of phasic DA that can be released upon activation of DA neuron firing.

after psychostimulants are administered to ADHD patients in rigorously timed pharmacodynamic studies (Fig. 5–5; Pelham et al., 1995; Swanson et al., 1998).

Much of the early phase of psychostimulant action may be circumvented, however, with oral administration of the drug. Thus, with oral administration, brain levels of drugs such as methylphenidate rise more slowly (Volkow et al., 1998). As a consequence, the incremental blockade of the DA transporter prevents a rapid increase in phasic DA release, because the blockade of the transporter would be continually offset by increasing stimulation of autoreceptors on DA cell bodies and terminals. As a consequence, the augmentation of phasic DA release would be blunted, instead being replaced by a developing attenuation of phasic DA release. As such, this form of administration would be predicted to achieve primarily the opposite of what occurs with rapid, intravenous administration of fast-acting psychostimulants; that is, a blunting of the phasic DA response (Fig. 5–5).

Model of ADHD

Considering the characteristics of ADHD in relation to the above model of psychostimulant drug action, a hypothesis regarding the pathophysiology of ADHD may be derived. The hypothesis is that the hyperactivity and impulsivity of ADHD result

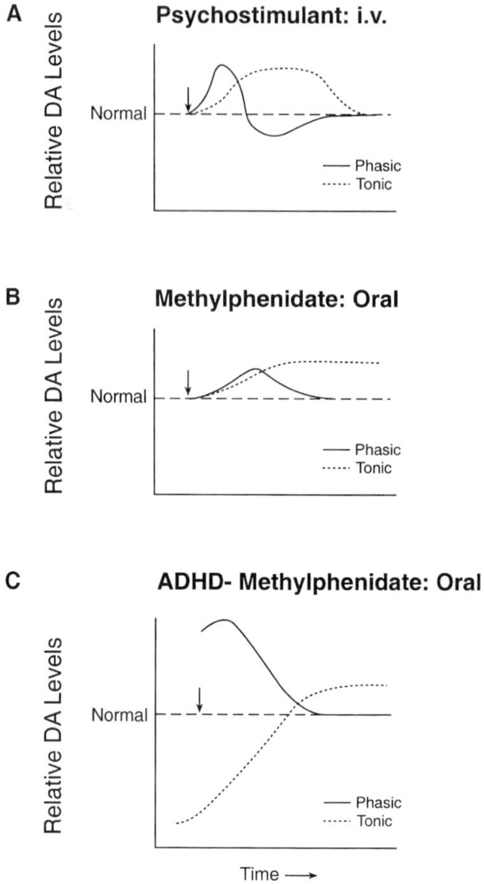

Figure 5–5. A model of psychostimulant drug action in the treatment of ADHD, as derived from the tonic/phasic model of DA system function. The "normal" level of tonic and of phasic DA transmission in control individuals is indicated as a horizontal dashed line in each case. (*A*) The response of a control individual to a psychostimulant administered intravenously. In this case, the psychostimulant causes a rapid rise in phasic DA transmission (solid line). This is due to the blockade of the reuptake process and consequent potentiation of spike-dependent DA release. The increase in phasic DA produces the rapid behavioral activation (or feelings of euphoria) associated with intravenous psychostimulant administration. This is followed by efflux of the released DA from the synaptic cleft, and accumulation in the extrasynaptic space as tonic DA (dotted line). The increased tonic DA causes an inhibition of spike-dependent phasic DA release, causing the phasic DA response to dip below baseline before returning to control levels. (*B*) The response to oral doses of methylphenidate. In contrast to the intravenous route of administration, the oral dose has a much slower onset of action. As a consequence, the phasic response is more blunted, since the slow onset preferentially augments tonic DA by allowing diffusion of DA from the synaptic cleft (the phasic component) into the extracellular space (the tonic component). As a consequence, the euphoric phasic component is blunted due to the parallel increase in tonic

from abnormally low tonic DA activity within the ventral striatum/nucleus accumbens, leading to abnormally high phasic DA responses. This model is supported by several factors. First, tonic DA is proposed to be regulated in part by frontal cortical afferents to the accumbens (Grace, 1991, 1995; Moore et al., 1999). The dorsolateral prefrontal cortex, which projects to the striatum, is proposed to be the origin of this regulation. Moreover, this region is also the cortical area believed to be responsible for the proper function of working memory (Goldman-Rakic 1996). A decrease in prefrontal cortical function in AD/HD would be consistent with the working memory deficits posited in ADHD patients (see Chapter 13 by Mehta et al.). In addition, the smaller striatal volume reported to be present in ADHD (Casey et al., 1997; see Chapter 10 by Castellanos) is also consistent with diminished corticostriatal activity. Moreover, the reports that ADHD symptoms may arise secondary to frontal cortical damage (Faraone & Biederman 1998; see Chapter 10 by Castellanos) is consistent with a deficit in corticostriatal transmission as an etiological factor in this disorder. In summary, a deficit in prefrontal cortical activation that appears to be present in ADHD could underlie the proposed decrease in tonic DA stimulation in the accumbens.

A number of studies have also argued that ADHD may not be a distinct entity in itself. Thus, rather than a clear pathological grouping of affected children apart from normals, the level of ADHD appears to be more of a continuous distribution extending from slightly above normal to severely afflicted (Robbins & Sahakian, 1979; see Chapter 1 by Solanto). This would suggest that the level of attenuation of tonic DA would be highly variable, with the attenuation correlating with the severity of the ADHD (Fig. 5–6). Under such conditions, the response to the psychostimulant would also be predicted to depend on the magnitude of the attenuation of tonic DA in the baseline state, which itself is proposed to correlate with the severity of the ADHD symptoms. Thus, the greater the decrease in tonic DA, the higher the phasic DA response and the more severe the ADHD symptoms. Moreover, psychostimulants would have a proportionately greater effect on tonic DA when the tonic DA level is lowest (Fig. 5–6). This model is thus consistent with the finding that the response to psychostimulants is greatest in patients with the most severe ADHD (Sahakian & Robbins, 1977; Robbins & Sahakian, 1979; Buitelaar et al., 1995) and with the

Figure 5–5. (continued) inhibition of the phasic response. The phasic component then returns to baseline, while the tonic component remains elevated and continues to suppress the phasic response. (*C*) The ADHD patient is proposed to have abnormally low baseline values of tonic DA, leading to an elevation in phasic DA transmission. Upon oral administration of methylphenidate, there will be a brief exacerbation of the elevated phasic response, before the suppressed tonic component can be sufficiently augmented. However, given the very low baseline tonic DA level and the already-elevated phasic DA component, the methylphenidate would be expected to have its most potent action on the tonic DA system. Therefore, with time the tonic component undergoes a dramatic increase in amplitude over its suppressed baseline, thereby causing potent down-regulation of the phasic DA response. The result is that, in the steady-state, the tonic and phasic DA system have returned closer to the normal range. Thus, the response to methylphenidate will be dependent on the baseline ratio of tonic/phasic DA transmission.

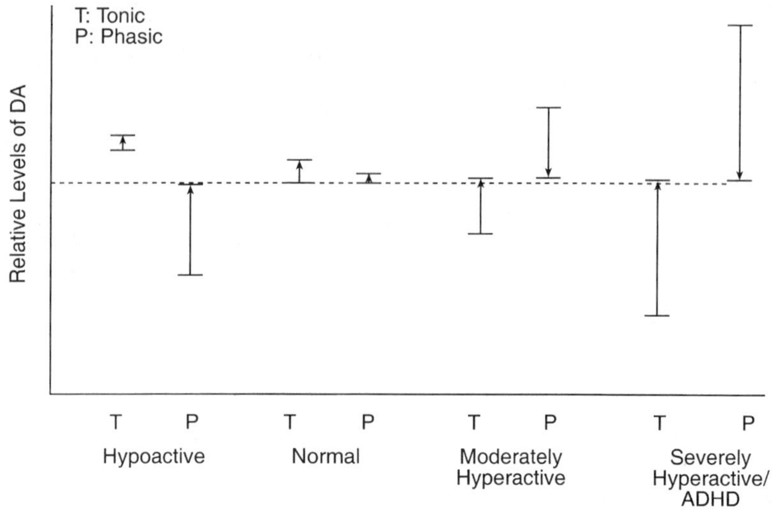

Figure 5–6. The response to methylphenidate is dependent on the basal activity level of the individual, which I propose to be dependent on the tonic/phasic DA ratio present in the subject. The arrows represent the response of each component (tonic and phasic) to methylphenidate for each of the classes of subjects tested, with the horizontal dashed line representing the baseline tonic and phasic levels present in control individuals. In a subject showing comparative hypoactivity levels (far left), the tonic DA is proposed to be elevated, thereby suppressing the phasic DA response. Methylphenidate will produce little action on the already-elevated tonic DA level due to a ceiling effect, but will preferentially potentiate the suppressed phasic DA component, thereby restoring activity levels to normal. In a normal subject, methylphenidate will have small but nearly equal actions on both the tonic and phasic component, thereby keeping the system in equilibrium despite elevated DA levels. An individual that is moderately hyperactive is proposed to exist in this state due to diminished tonic DA levels, and therefore an elevated phasic DA system. By restoring the tonic DA levels to normal, the phasic DA response is also brought back into equilibrium. In a severely hyperactive/ADHD patient (far right), the tonic DA is at greatly diminished levels, causing an elevation in the phasic DA response. Methylphenidate, by interfering with reuptake, will cause the DA released by spikes to escape from the synaptic cleft, thereby producing a rapid augmentation of the tonic DA present in the extracellular space. This will cause a potent down-regulation of the phasic DA response. Therefore, the relative effect of methylphenidate on the DA system is dependent on the ratio of tonic to phasic DA transmission, enabling the restoration of equilibrium in each case.

highest cerebrospinal fluid levels of homovanillic acid (Castellanos et al., 1994, 1996), given that metabolite levels reflect primarily changes in the high-amplitude phasic DA response.

This may also provide an explanation for what is known as "rate-dependency" in the response to psychostimulants (Dews, 1977; Glick & Milloy, 1973; Robbins & Sahakian, 1979; Robbins, 1981). Thus, studies have shown that the response of an animal to a stimulant is proportional to the baseline activity of the organism, in that low rates of responding tend to show increases and high rates of responding are decreased by psychostimulant administration. If the baseline activity of an animal is also dependent on the tonic/phasic balance (i.e., lower tonic/higher phasic corresponding to greater levels of activity), then one would predict that amphetamine would produce the greatest diminution of activity in animals with the lowest baseline tonic/highest baseline phasic DA levels. On the other hand, if phasic DA is depressed and tonic is at or above normal levels, then the stimulant would have a preferential action in augmenting the phasic response, since it is furthest below the normal levels. Thus, a correspondence between tonic/phasic DA balance and activity levels could account for the differential actions of stimulants dependent on the baseline condition of the organism being tested (Fig. 5–6).

It is clear from the previous arguments that hyperactivity can result from abnormally elevated exploratory behavior secondary to the disruption of hippocampal/DA interactions within the accumbens (Wu & Brudzynski, 1995; Brenner & Bardgett, 1998). Furthermore, studies have drawn a close association between heightened exploratory behavior, which may be an equivalent of hyperactivity in general, and the DA-dependent reward system in the accumbens (Wise & Rompre, 1989). Indeed, there are several predictions that could be derived from a condition of abnormally high phasic DA in the reward-related limbic system of ADHD patients. First, because the phasic DA release is already above normal in this disorder, there would be little room for further augmentation. On the other hand, an inhibition of phasic DA release would produce a profound change in the system. Such a condition could be extrapolated from what has been observed with ADHD. One consistent finding is that children with ADHD will not wait additional periods of time; apparently preferring an immediate albeit smaller reward (Sonuga-Barke et al., 1992). One interpretation of this would be that, to an ADHD child, the magnitude of reward expected is not substantially different in both cases, primarily because the reward amplitude saturates at comparatively low levels of stimulation. As a quantitative example, the amount of reward a person experiences from receiving $1000 is much greater than the reward one would experience from receiving $100. However, would a person with an average income experience the same discrepancy in reinforcement if the award were $5 million versus $50 million? In a similar manner, smaller rewards to an ADHD child may already be within the realm of a maximal stimulus, such that a larger reward does not evoke a greater affective response. On the other hand, withholding of an expected reward, which is known to inhibit the firing of DA neurons (Schultz, 1997, 1998) and thereby dramatically decreasing phasic DA release, evokes very strong reactions from ADHD children (Rapport et al., 1982). In the condition of ADHD, in which the basal phasic DA response is elevated, a shut-off of DA cell firing would be expected to produce a much greater dysphoric state than would occur in a normal individual (Figure 5–7).

Similarities to and Differences from Other Pathophysiological States

One condition that is reported to lead to a psychopathology that in many ways resembles ADHD is chronic stress syndrome. Indeed, conditions of a stressful home environment or life circumstances are often used as an exclusion criterion

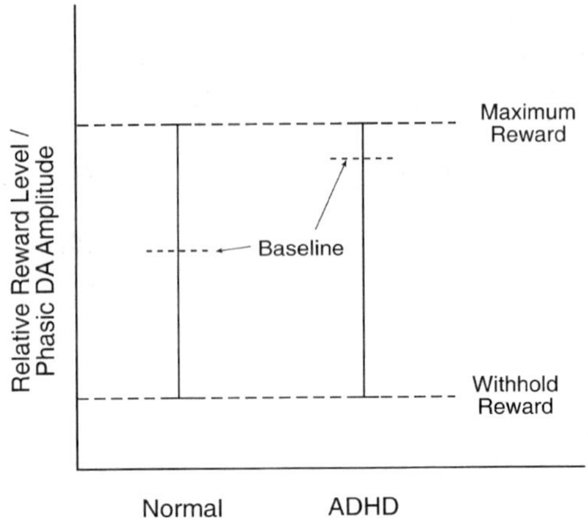

Figure 5–7. A model to account for the relative difference in response of ADHD patients to reward-related stimuli. The Y-axis illustrates the relative reward level, which is also a function of the phasic DA amplitude. In a normal individual, the DA system has the widest dynamic range of response, in that a positively reinforcing stimulus has the capacity to drive the phasic DA response from baseline to maximal reward levels, whereas the withholding of a reward (which is known to inhibit DA neuron discharge; Schults, 1997,1998) will drive the reward system/phasic DA response to minimal levels. However, the patient with ADHD is proposed to already have an elevated phasic DA response system, and therefore should be responding to stimuli as if they had a powerfully rewarding component even at the baseline state. As a result of the diminished dynamic range in the positive direction, minor rewards as well as major rewarding stimuli will both cause the same maximal euphorogenic responses due to a ceiling effect. Therefore, the ADHD patient will not work harder nor will they wait for larger rewards, since the difference in reward amplitude is not detected as significant. In contrast, withholding of a reward will cause a dramatic change from the elevated baseline to the "withhold reward" state of minimal DA cell firing. Therefore, withholding of a reward and its consequent shut down of the phasic DA response would produce a much greater effect on the ADHD patient, whereas positively reinforcing stimuli quickly achieve maximal reward value independent of differences in their inherent rewarding potential.

for ADHD diagnosis, given the similarity in presentation of the two conditions (Glod & Teicher, 1996). In our studies, we found that exposure of rats to a chronically stressful environment will lead to a decrease in the baseline tonic firing of DA neurons (Rose et al., 1998). Thus, with chronic stress, there is a diminution in the tonic DA system leading to excessive phasic response, albeit arising from a different source; that is, decreased basal DA cell activity rather than a decrease in presynaptic regulation of DA at the terminal site.

A model of schizophrenia had been advanced based on a deficit in tonic DA as well (Grace, 1991; Moore et al., 1999). However, the model of ADHD advanced here is distinct from that related to schizophrenia on several grounds. In particular, the positive symptoms of schizophrenia are proposed to arise from abnormal phasic DA regulation primarily in the most ventral portions of the accumbens, which receive projections primarily from orbitofrontal cortex and other regions. Moreover, the dysfunction is believed to involve disturbances in the hippocampal drive (O'Donnell & Grace, 1995, 1998b), which lead to the pathology in corticoaccumbens systems. In contrast, the model advanced here is predicated on the idea that it is primarily the reward systems in the basal ganglia that are disturbed; in particular, the dorsolateral prefrontal cortex (i.e., the site of origin of the working memory deficit; Goldman-Rakic, 1996), the aspects of the accumbens that have a more direct involvement in reward-related behaviors, and the motor aspects of the striatum involved in hyperlocomotor responses. Nonetheless, there are areas in which these disorders show overlap. For example, in both disorders patients are distracted by irrelevant stimuli, and exhibit an inability to focus on the salient aspects of their environment (Gray et al., 1991). Moreover, amphetamine is reported to alleviate a subset of the symptoms of schizophrenia (Angrist et al., 1982; Goldberg et al., 1991) as well. Therefore, these disorders may be manifest by a common disruption in cortico-accumbens modulation, with the specifics of the pathophysiologies leading to their unique psychiatric phenotypes.

I thank Dr. Holly Moore for help in preparing one of the figures, and Dr. T. Celeste Napier for her helpful comments on the manuscript. I also thank Dr. Mary Solanto for including me as a participant in this project. This work was supported by USPHS grants MH01055, MH29670, MH57440, and MH45156.

References

Abercrombie, ED, Bonatz, AE & Zigmond, MJ (1990). Effects of L-DOPA on extracellular dopamine in striatum of normal and 6-hydroxydopamine-treated rats. *Brain Res, 525*, 36–44.

Adler, M (1961). Changes in sensitivity to amphetamine in rats with chronic brain lesions. *J Pharmacol Exp Ther, 134*, 214–224.

Angrist, B, Peselow, E, Rubinstein, M, Corwin, J & Rotrosen, J (1982). Partial improvement in negative schizophrenia symptoms after amphetamine. *Psychopharmacology, 78*, 128–130.

Antonelli, T, Govoni, B, Bianchi, CL, & Beani, L (1997). Glutamate regulation of dopamine release in guinea pig striatal slices. *Neurochem Intl, 30*, 203–209.

Bannon, MJ & Roth, RH (1983). Pharmacology of mesocortical dopamine neurons. *Pharmacol Rev, 35*, 53–68.

Blaha, CD, Yang, CR, Floresco, SB, Barr, AM & Phillips, AG (1997). Stimulation of the ventral subiculum of the hippocampus evokes glutamate receptor-mediated changes in dopamine efflux in the rat nucleus accumbens. *Eur J Neurosci, 9*, 902–911.

Brenner, DM & Bardgett, ME (1998). Haloperidol blocks increased locomotor activity elicited by carbachol infusion into the ventral hippocampal formation. *Pharmacol Biochem Behav, 60*, 759–764.

Brudzynski, SM & Gibson, CJ (1997). Release of dopamine in the nucleus accumbens caused by stimulation of the subiculum in freely moving rats. *Brain Res Bull, 42*, 303–308.

Buitelaar, JK, Van der Gaag, RJ, Swaab-Barneveld, H & Kuiper, M (1995). Prediction of clinical response to methylphenidate in children with attention-deficit/hyperactivity disorder. *J Am Acad Child Adolesc Psychiatry, 34*, 1025–32.

Bunney, BS & Grace, AA (1978). Acute and chronic haloperidol treatment: Comparison of effects on nigral dopaminergic cell activity. *Life Sci, 23*, 1715–1728.

Casey, BJ, Castellanos, FX, Giedd, JN, Marsh, WL, Hamburger, SD, & Schubert, AB, et al. (1997). Implication of right frontostriatal circuitry in response inhibition and attention-deficit/hyperactivity disorder. *J Am Acad Child Adolesc Psychiatry, 36*, 374–383.

Castellanos, FX, Elia, J, Kruesi, MJ, Gulotta, CS, Mefford, IN, Potter, WZ, et al. (1994). Cerebrospinal fluid monoamine metabolites in boys with attention-deficit hyperactivity disorder. *Psychiatry Res, 52*, 305–16.

Castellanos, FX, Elia, J, Kruesim MJ, Marsh, WL, Gulotta, CS, Potter, WZ, et al., (1996). Cerebrospinal fluid homovanillic acid predicts behavioral response to stimulants in 45 boys with attention deficit/hyperactivity disorder. *Neuropsychopharmacology, 14*, 125–37.

Chéramy, A, Leviel, V & Glowinski, J (1981). Dendritic release of dopamine in the substantia nigra. *Nature, 289*, 537–542.

Chiodo, LA, Bannon, MJ, Grace, AA, Roth, RH & Bunney, BS (1984). Evidence for the absence of impulse-regulating somatodendritic and synthesis-modulating nerve terminal autoreceptors on subpopulations of mesocortical dopamine neurons. *Neuroscience, 12*(1), 1–16.

Christie, MJ, James, LB, & Beart, PM (1987). An excitant amino acid projection from the medial prefrontal cortex to the anterior part of nucleus accumbens in the rat. *J Neurochem, 45*, 477–82.

Chronister, RB, Sikes, RW, Trow, TW, & DeFrance, JF (1981). The organization of nucleus accumbens. In RB Chronister & JF DeFrance (Eds.), The neurobiology of the nucleus accumbens. (pp. 97–146). Brunswick, ME: Hauer Institute.

Colle, LM & Wise, RA (1988). Effects of nucleus accumbens amphetamine on lateral hypothalamic brain stimulation reward. *Brain Res, 459*(2), 361–368.

Cooper, DC & White, FJ (1998). Cocaine alters the bistable membrane potential in the nucleus accumbens: An in vivo intracellular study in mice. *Soc Neurosci Abstr, 24*, 1736.

Cooper, JR, Bloom, FE, & Roth, RH (1996). *The biochemical basis of neuropharmacology*, 7th ed. New York: Oxford University Press.

Cox, RF & Waszczak, BL (1989). Differences in dopamine receptor reserve for N-n-propylnorapomorphine enantiomers: Single unit recording studies after partial inactivation of receptors by EEDQ. *Mol Pharmacol, 34*, 125–131.

Dews, PB (1977). Rate-dependency hypothesis. *Science, 198*, 1182–183.

DiChiara, G, Porceddu, ML, Vargiu, L, Argiolas, A, & Gessa, GL (1976) Evidence for dopamine receptors mediating sedation in the mouse brain. *Nature, 264*, 564–567.

Doucet, G, Descarries, L & Garcia, S (1986). Quantification of the dopamine innervation in adult rat neostriatum. *Neuroscience, 19*, 427–445.

Faraone, SV & Biederman, J (1998). Neurobiology of attention-deficit hyperactivity disorder. [Review] [84 refs] *Biol Psychiatry, 44(10)*, 951–958,

Farnebo, L-O & Hamberger, B (1971). Drug-induced changes in the release of 3H-monoamines from field-stimulated rat brain slices. *Acta Physiol Scand Suppl, 371*, 35–44.

Fibiger, HC, Lepiane, FG, Jakubovic, A, & Phillips, AG (1987). The role of dopamine in intracranial self-stimulation of the ventral tegmental area. *J Neurosci*, 7, 3888–3896.

Floresco, SB, Yang, CR, Phillips, AG, & Blaha, CD (1998). Basolateral amygdala stimulation evokes glutamate receptor-dependent dopamine efflux in the nucleus accumbens of the anaesthetized rat. *Eur J Neurosci, 10*, 1241–1251.

Garris, PA, Ciolkowski, EL, Pastore, P, & Wightman, RM (1994). Efflux of dopamine from the synaptic cleft in the nucleus accumbens of the rat brain. *J Neurosci, 14,* 6084–6093.

Gatley, SJ, Pan, D, Chen, R, Chaturvedi, G, & Ding, YS (1996). Affinities of methylphenidate derivatives for dopamine, norepinephrine and serotonin transporters. *Life Sci, 58,* 231–239.

Glick, SD, & Milloy, S (1973). Rate-dependent effects of d-amphetamine on locomotor activity in mice: Possible relation to paradoxical sedation in minimal brain dysfunction. *Eur J Pharmacol, 24,* 266–268.

Glod, CA, & Teicher, MH (1996). Relationship between early abuse, posttraumatic stress disorder, and activity levels in prepubertal children. *J Am Acad Child Adolesc Psychiatry, 35(10),* 1384–1393.

Glowinski, J, Chéramy, A, Romo, R, & Barbeito, L (1988). Presynaptic regulation of dopaminergic transmission in the striatum. *Cell Mol Neurobiol, 8,* 7–17.

Goldberg, TE, Bigelow, LB, Weinberger, DR, Daniel, DG, & Kleinman, JE (1991). Cognitive and behavioral effects of the coadministration of dextroamphetamine and haloperidol in schizophrenia. *Am J Psychiatry, 148,* 78–84.

Goldman-Rakic, PS (1996). Regional and cellular fractionation of working memory. *Proc Natl Acad Sci USA, 93,* 13473–13480.

Gonon, FG (1998). Nonlinear relationship between impulse flow and dopamine released by rat midbrain dopaminergic neurons as studied by in vivo electrochemistry. *Neuroscience, 24,* 19–28.

Grace, AA (1991). Phasic versus tonic dopamine release and the modulation of dopamine system responsivity: A hypothesis for the etiology of schizophrenia. *Neuroscience, 41,* 1–24.

Grace, AA (1993). Cortical regulation of subcortical dopamine systems and its possible relevance to schizophrenia. *J Neural Transm, 91,* 111–134.

Grace, AA (1995). The tonic/phasic model of dopamine system regulation: Its relevance for understanding how stimulant abuse can alter basal ganglia function. *Drug Alcohol Depend, 37,* 111–129.

Grace, AA (2000a). The tonic/phasic model of dopamine system regulation and its implications for understanding drug and alcohol craving. *Addiction.* (In press).

Grace, AA (2000b). Gating of information flow within the limbic system and the pathophysiology of schizophrenia. *Brain Research Reviews, 31*; 330–341.

Grace, AA & Bunney, BS (1984a). The control of firing pattern in nigral dopamine neurons: Single spike firing. *J Neurosci, 4,* 2866–2876.

Grace, AA & Bunney, BS (1984b). The control of firing pattern in nigral dopamine neurons: Burst firing. *J Neurosci, 4,* 2877–2890.

Grace, AA, Kuhar, MJ, Lambert, PD, Couceyro, PR & Onn, S-P (1998). Both amphetamine and cocaine-amphetamine-regulated-transcript (CART) suppress prefrontal cortico-accumbens synaptic transmission. *Soc Neurosci Abstr, 24,* 1644.

Grace, AA & Moore, H (1998). Regulation of information flow in the nucleus accumbens: A model for the pathophysiology of schizophrenia. In MF Lenzenweger & RH Dworkin (Eds.) *Origins and development of schizophrenia: Advances in experimental psychopathology.* (pp. 123–157). Washington, DC: American Psychological Association Press.

Gratton, A, Hoffer, BJ, & Gerhardt, GA (1988). Effects of electrical stimulation of brain reward sites on release of dopamine in rat: An in vivo electrochemical study. *Brain Res Bull, 21,* 319–324.

Gray, JA, Feldon, J, Rawlins, JNP, Hemsely, DR, & Smith, AD (1991). The neuropsychology of schizophrenia. *Behav Brain Sci, 14,* 1–84.

Harden, DG & Grace, AA (1995). Activation of dopamine cell firing by repeated L-DOPA administration to dopamine-depleted rats: Its potential role in mediating the therapeutic response to L-DOPA treatment. *J Neurosci, 15*, 6157–6166.

Heikkila, RE, Orlansky, H, & Cohen, G (1975). Studies on the distinction between uptake inhibition and release of (3H) dopamine in rat brain tissue slices. *Biochem Pharmacol, 24*, 847–852.

Hollerman, JR & Grace, AA (1992). Subthalamic nucleus cell firing in the 6-OHDA-treated rat: Basal activity and response to haloperidol. *Brain Res, 590*, 291–299.

Iversen, SD (1971). The effect of surgical lesions to frontal cortex and substantia nigra on amphetamine responses in rats. *Brain Res, 31*, 295–311.

Kalivas, PW & Duffy, P (1990). Effect of acute and daily cocaine treatment on extracellular dopamine in the nucleus accumbens. *Synapse, 5*, 48–58.

Keefe, KA, Zigmond, MJ, & Abercrombie, ED (1992). Extracellular dopamine in striatum: Influence of nerve impulse activity in medial forebrain bundle and local glutamatergic input. *Neuroscience, 47*, 325–332.

Kincaid, AE, Zheng, T, & Wilson, CJ (1998). Connectivity and convergence of single corticostriatal axons. *J Neurosci, 18*, 4722–4731.

Kuczenski, R (1983). Biochemical actions of amphetamine and other stimulants. In I Creese (Ed.), *Stimulants: Neurochemical, behavioral, and clinical perspectives*, (pp. 31–61). New York: Raven Press.

Kuczenski, R & Segal, DS (1989). Concomitant characterization of behavioral and striatal neurotransmitter response to amphetamine using in vivo microdialysis. *J Neurosci, 9*, 2051–2065.

Lacey, MG, Mercuri, NB, & North, RA (1990). Actions of cocaine on rat dopaminergic neurones in vitro. *Br J Pharmacol, 99*(4), 731–735.

Lavin, A & Grace, AA (1994). The modulation of dorsal thalamic cell activity by the ventral pallidum: Its role in the regulation of thalamocortical activity by the basal ganglia. *Synapse, 18*, 104–127.

Ledoux, JE & Muller, J (1997). Emotional memory and psychopathology. *Philosophical Transactions of the Royal Society of London – Series B: Biological Sciences, 352*, 1719–26.

McDonald, AJ (1991). Organization of amygdaloid projections to the prefrontal cortex and associated striatum in the rat. *Neuroscience, 44*, 1–14.

Meller, E, Bohmaker, K, Namba, Y, Friedhoff, AJ, & Goldstein, M (1987). Relationship between receptor occupancy and response at striatal dopamine autoreceptors. *Mol Pharmacol, 31*, 592–598.

Mercuri, NB, Bonci, A, Siniscalchi, A, Stefani, A, Calabresi, P, & Bernardi, G (1996). Electrophysiological effects of monoamine oxidase inhibition on rat midbrain dopaminergic neurones: An in vitro study. *Br J Pharmacol, 117*(3), 528–532.

Mercuri, NB, Calabresi, P & Bernardi, G (1990). Responses of rat substantia nigra compacta neurones to L-DOPA. *Br J Pharmacol, 100*, 257–260.

Moghaddam, B & Gruen, RJ (1991). Do endogenous excitatory amino acids influence striatal dopamine release? *Brain Res, 544*, 329–330.

Moore, H, & Grace, AA (1996) Interactions between amygdala and prefrontal cortical afferents to the nucleus accumbens and their modulation by dopamine receptor activation. *Soc Neurosci Abstr, 22*, 1088.

Moore, H, West, AR, & Grace, AA (1999) The regulation of forebrain dopamine neurotransmission: Relevance to the pathophysiology of schizophrenia. *Biol Psychiatry, 46*, 40–55.

Nakahara, D, Ozaki, N, Kapoor, V, & Nagatsu, T (1989). The effect of uptake inhibition on dopamine release from the nucleus accumbens of rats during self- or forced stimulation of the medial forebrain bundle. A microdialysis study. *Neurosci Lett, 104*, 136–140.

Nowycky, MC & Roth, RH (1977). Presynaptic dopamine receptors. Development of supersensitivity following treatment with fluphenazine decanoate. *Naunyn Schmiedebergs Arch Pharmacol, 300*, 247–254.

O'Donnell, P & Grace, AA (1993). Physiological and morphological properties of accumbens core and shell neurons recorded *in vitro. Synapse, 13*, 135–160.

O'Donnell, P & Grace, AA (1994). Tonic D_2-mediated attenuation of cortical excitation in nucleus accumbens neurons recorded *in vitro. Brain Res, 634*, 105–112.

O'Donnell, P & Grace, AA (1995). Synaptic interactions among excitatory afferents to nucleus accumbens neurons: Hippocampal gating of prefrontal cortical input. *J Neurosci, 15*, 3622–3639.

O'Donnell, P & Grace, AA (1996a). Hippocampal gating of cortical throughput in the nucleus accumbens: Modulation by dopamine. *Biol Psychiatry, 39*, 632.

O'Donnell, P & Grace, AA (1996b). Dopaminergic reduction of excitability in nucleus accumbens neurons recorded in vitro. *Neuropsychopharmacology, 15*, 87–97.

O'Donnell, P & Grace, AA (1998a). Phencyclidine interferes with the hippocampal gating of nucleus accumbens neuronal activity in vivo. *Neuroscience, 87*, 823–830.

O'Donnell, P & Grace, AA (1998b). Dysfunctions in multiple interrelated systems as the neurobiological bases of schizophrenic symptom clusters. *Schizophr Bull, 24*, 267–283.

O'Donnell, P, Lavin, A, Enquist, LW, Grace, AA, & Card, JP (1997). Interconnected parallel circuits between rat nucleus accumbens and thalamus revealed by retrograde transynaptic transport of pseudorabies virus. *J Neurosci, 17*, 2143–2167.

Ohno, M & Watanabe, S (1995). Persistent increase in dopamine release following activation of metabotropic glutamate receptors in the rat nucleus accumbens. *Neurosci Lett, 200*, 113–6.

Olds, ME (1990). Enhanced dopamine receptor activation in accumbens and frontal cortex has opposite effects on medial forebrain bundle self-stimulation. *Neuroscience, 35*, 313–325.

Parsons, LH & Justice, JB Jr. (1992). Extracellular concentration and in vivo recovery of dopamine in the nucleus accumbens using microdialysis. *J Neurochem, 58*, 212–8.

Pelham, WE Jr, Swanson, JM, Furman, MB, & Schwindt, H (1995). Pemoline effects on children with ADHD: A time-response by dose-response analysis on classroom measures. *J Am Acad Child Adolesc Psychiatry, 34*(11), 1504–13.

Phillips, RG & LeDoux, JE (1992). Differential contribution of amygdala and hippocampus to cued and contextual fear conditioning. *Behav Neurosci, 106*, 274–285.

Pucak, ML & Grace, AA (1991). Partial dopamine depletions result in an enhanced sensitivity of residual dopamine neurons to apomorphine. *Synapse, 9*, 144–155.

Pucak, ML & Grace, AA (1994). Evidence that systemically administered dopamine antagonists activate dopamine neuron firing primarily by blockade of somatodendritic autoreceptors. *J Pharmacol Exp Ther, 371*, 1181–1192.

Pucak, ML & Grace, AA (1996). Effects of haloperidol on the activity and membrane physiology of substantia nigra dopamine neurons recorded *in vitro. Brain Res, 713*, 44–52.

Rapport, MD, Murphy, HA, & Bailey, JS (1982). Ritalin vs response cost in the control of hyperactive children: A within-subject comparison. *J Appl Behav Anal, 15*, 205–216.

Reimann, W, Zumstein, A, Jackisch, R, Starke, K, & Hertting, G (1979). Effect of extracellular dopamine on the release of dopamine in the rabbit caudate nucleus: Evidence for a dopaminergic feed-back inhibition. *Naunyn Schmiedebergs Arch Pharmacol, 306*, 53–60.

Reith, MEA, Meisler, BE, Sershen, J, & Lastha, A (1986) Structural requirements for cocaine congeners to interact with dopamine and serotonin uptake sites in mouse brain and to induce stereotyped behavior. *Biochem Pharmacol, 35*, 1123–1129.

Richfield, EK, Penney, JB, & Young, AB (1989). Anatomical and affinity state comparisons between dopamine D1 and D2 receptors in the rat central nervous system. *Neuroscience, 30*, 767–777.

Robbins, TW (1981). Behavioural determinants of drug action: rate-dependency revisited. In SJ Cooper (Ed.), *Theory in psychopharmacology,* Vol. 1 (pp 1–63). London: Academic Press.

Robbins, TW & Sahakian, BJ (1979). "Paradoxical" effects of psychomotor stimulant drugs in hyperactive children from the standpoint of behavioural pharmacology. *Neuropharmacology, 18*, 931–950.

Robinson, TE & Whishaw, IQ (1988). Normalization of extracellular dopamine in striatum following recovery from a partial unilateral 6-OHDA lesion of the substantia nigra: A microdialysis study in freely moving rats. *Brain Res, 450*, 209–224.

Rose, H, Moore, H, Hollerman, JR, & Grace, AA (1998). Chronic cold stress reduces the spontaneous activity of ventral tegmental dopamine neurons. *Soc Neurosci Abstr, 24*, 1920.

Roth, RH (1987). Biochemical correlates of the electrophysiological activity of dopaminergic neurons: Reflections on two decades of collaboration with electrophysiologists. In *LA Chiodo & AS* Freeman *(Eds.), Neurophysiology of dopaminergic systems—current status and clinical perspectives* (pp. 187–203). Detroit: Lakeshore Publishing Co.

Sahakian, BJ, Robbins, TW (1977). Are the effects of psychomotor stimulant drugs on hyperactive children really paradoxical? *Med Hypotheses, 3*(4), 154–8,

Salzmann, E, Vidyasagar, TR, & Creutzfeldt, OD (1993). Functional comparison of neuronal properties in the primate posterior hippocampus and parahippocampus (area TF/TH) during different behavioural paradigms involving memory and selective attention. *Behav Brain Res, 53*, 133–149.

Santiago, M & Westerink, BH (1992). Simultaneous recording of the release of nigral and striatal dopamine in the awake rat. *Neurochem Intl Suppl, 20*, 107S–110S,

Schultz, W (1997). Dopamine neurons and their role in reward mechanisms. *Curr Opin Neurobiol, 7*, 191–197.

Schultz, W (1998). The phasic reward signal of dopamine neurons. *Adv Pharmacol, 42*, 686–690.

Seeman, P & Madras, BK (1998). Anti-hyperactivity medication: methylphenidate and amphetamine. *Mol Psychiatry, 3*(5), 386–396.

Sesack, SR, Deutch, AY, Roth, RH & Bunney, BS (1989). Topographical organization of the efferent projections of the medial prefrontal cortex in the rat: An anterograde tract-tracing study with Phaseolus vulgaris leucoagglutinin. *J Comp Neurol, 290*, 213–242.

Shigemoto, R, Kulik, A, Roberts, JD, Ohishi, H, Nusser, Z, Kaneko, T, & Somogyi, P (1996). Target-cell-specific concentration of a metabotropic glutamate receptor in the presynaptic active zone. *Nature, 381*, 523–525.

Skirboll, LR, Grace, AA, & Bunney, BS (1979) Dopamine auto- and postsynaptic receptors: Electrophysiological evidence for differential sensitivity to dopamine agonists. *Science, 206*, 80–82.

Smith, ID & Grace, AA (1992). Role of the subthalamic nucleus in the regulation of nigral dopamine neuron activity. *Synapse, 12*, 287–303.

Smolders, I, Sarre, C, Vanhaesendonck, C, Ebinger, G & Michotte, Y (1996). Extracellular striatal dopamine and glutamate after decortication and kainate receptor stimulation, as measured by microdialysis. J *Neuroschem, 66*, 2373–2380.

Solanto, MV (1986). Behavioral effects of low-dose methylphenidate in childhood attention deficit disorder: Implications for a mechanism of stimulant drug action, *J Am Acad Child Adolesc Psychiatry, 25*, 96–101.

Sonuga-Barke, EJ, Taylor, E, Sembi, S, & Smith, J (1992). Hyperactivity and delay aversion—I. The effect of delay on choice. *J Child Psychol Psychiatry, 33*, 387–398.

Strombom, U (1976). Catecholamine receptor agonists: Effects on motor activity and rate of tyrosine hydroxylation in mouse brain. *Naunyn Schmiedebergs Arch Pharmacol, 292*, 167–176.

Swanson, JM, Wigal, S, Greenhill, LL, Browne, R, Waslik, B, & Lerner, M, et al. (1998). Analog classroom assessment of Adderall in children with ADHD [see comments]. *J Am Acad Child Adolesc Psychiatry, 37*(5), 519–26.

Taber, MT, Das, S, & Fibiger, HC (1995). Cortical regulation of subcortical dopamine release: mediation via the ventral tegmental area. *J Neurochem, 65*, 1407–10.

Verma, A & Moghaddam, B (1998). Regulation of striatal dopamine release by metabotropic glutamate receptors. *Synapse, 28*, 220–226.

Volkow, ND, Wang, GJ, Fowler, JS, Hitzemann, R, Gatley, & J, Ding, et al. (1998). Differences in regional brain metabolic responses between single and repeated doses of methylphenidate. *Psychiatry Res, 83*, 29–36.

Weiner, I (1990). Neural substrates of latent inhibition: the switching model. *Psychol Bull, 108*, 42–461.

Wise, RA & Rompre, P-P (1989). Brain dopamine and reward. *Ann Rev Psychol, 40*, 191–225.

Wolf, ME, Dahlin, SL, Hu, X-T, Xue, C-J, & White, FJ (1995). Effects of lesions of prefrontal cortex, amygdala, or fornix on behavioral sensitization to amphetamine: comparison with N-methyl-d-aspartate antagonists. *Neuroscience, 69*, 417–439.

Wu, M & Brudzynski, SM (1995). Mesolimbic dopamine terminals and locomotor activity induced from the subiculum. *Neuroreport, 6*, 1601–1604.

Wuerthele, SM, Freed, WJ, Olson, L, Morihisa, J, Spoor, L, Wyatt, RJ, & Hoffer, BJ (1981). Effect of dopamine agonists and antagonists on the electrical activity of substantia nigra neurons transplanted into the lateral ventricle of the rat. *Exp Brain Res, 44*, 1–10.

Yang, CR & Mogenson, GJ (1987). Hippocampal signal transmission to the pedunculopontine nucleus and its regulation by dopamine D2 receptors in the nucleus accumbens: An electrophysiological and behavioral study. *Neuroscience, 23*, 1041–1055.

Yokoo, J, Goldstein, M, & Meller, E (1988). Receptor reserve at striatal dopamine receptors modulating the release of [3H] dopamine. *Eur J Pharmacol, 155*, 323–327.

6

Arousal- and Attention-Related Actions of the Locus Coeruleus–Noradrenergic System: Potential Target in the Therapeutic Actions of Amphetamine-Like Stimulants

CRAIG W. BERRIDGE

The early demonstration of a neurotransmitter role of norepinephrine (NE) within the brain stimulated intense study of its functions over the subsequent decades. A majority of brain noradrenergic neurons are concentrated in the brainstem nucleus, locus coeruleus (LC). This nucleus gives rise to an extensive and regionally specialized noradrenergic innervation of the CNS. Importantly, the LC provides the sole source of NE to hippocampus and neocortex. Despite intense examination of the LC-noradrenergic (LC-NA) system, and substantial progress in our understanding of the neurobiology of this system, the behavioral functions supported by this neurotransmitter system remained largely enigmatic.

This situation appears to be changing, as recent work provides relatively unambiguous evidence that the LC-NA system modulates behavioral state and state-dependent cognitive processes. Combined, these observations suggest the LC-NA system generates dynamic and widespread patterns of neuronal electrophysiological activity in neocortex and other forebrain areas that are essential substrates of the waking state and for the operation of attention and other state-dependent cognitive processes. Together with previous anatomical, electrophysiological, and behavioral evidence, these observations indicate that central NE serves at least two general behavioral functions. First, noradrenergic systems induce a behavioral state appropriate for the acquisition of sensory information (e.g., waking). Second, within the awake behavioral state, NE enhances/modulates information-processing abilities. Both of these functions likely involve actions of NE distributed across multiple anatomical regions at multiple receptor subtypes.

Based on these observations it is posited that dysregulation of the LC-NA system may result in deficits in the regulation of attentional and/or arousal processes associated with Attention Deficit/Hyperactivity Disorder (ADHD). However, regardless of whether noradrenergic systems contribute to the development of attention/arousal deficits, these systems may nonetheless be an appropriate target for pharmacological-based treatments of attention/arousal disorders, including ADHD. These hypotheses are supported by the fact that both indirect and direct noradrenergic agonists are effective in the treatment of ADHD (for review, see Arnsten et al., 1996; Solanto, 1998; Pliszka, Chapter 14). Recent observations indicate that the actions of amphetamine-like stimulants on monoaminergic neurotransmission are dependent on both the dose and the identity of the amphetamine derivative. The current lack of information on the actions of low-dose amphetamine-like stimulants used in the treatment of ADHD on noradrenergic neurotransmission and noradrenergic-dependent cognitive processes represents a major lacuna in our understanding of the neurobiology of ADHD.

The LC-NA System

The LC is a well-delineated cluster of noradrenergic neurons, located adjacent to the fourth ventricle in the pontine brainstem. It is composed of a small number of neurons: approximately 1,500 per nucleus in rat, several thousand in monkey, and 10,000 to 15,000 in human. However, these cells possess immensely ramified axons such that the nucleus projects widely throughout the neuraxis, from spinal cord to neocortex (reviewed in Foote et al., 1983; Lewis, Chapter 13).

Although these neurons project widely within the brain, there may be substantial specificity and heterogeneity in the innervation pattern for a given subset of LC neurons. For example, cortically projecting LC neurons are found in the caudal portion of LC and these neurons project in a predominantly ipsilateral manner (Waterhouse et al., 1983), whereas subcortical structures receive a more pronounced bilateral LC innervation. Additionally, the cortically projecting subset of LC neurons appears to be further organized in both dorsoventral and anterior-caudal dimensions. Finally, recent evidence suggests that individual LC neurons innervate functionally related, yet discrete, elements within an ascending sensory pathway. For example, LC neurons that project to somatosensory cortex are more likely to innervate somatosensory thalamus than non-somatosensory thalamus (Simpson et al., 1997). Combined, these observations suggest a function-specific organization of LC efferents may be a general feature of the LC-NA system. Within a given terminal field, LC efferents display non-uniform innervation patterns. For example, within neocortex LC axons preferentially innervate particular laminae, especially in primates (reviewed in Foote and Morrison, 1987; Lewis, this volume). The dense innervation of layer I is particularly interesting given that this is considered an important site for cortical integration (Vogt, 1991).

LC efferents are the major source of NE throughout the forebrain. Rates of NE release are linearly related to LC discharge rates, across the range of LC discharge rates typically observed across the sleep–wake cycle (Berridge & Abercrombie, 1999). These observations indicate that relatively small alterations in absolute LC discharge rates (i.e., from 2 to 4 Hz) result in pronounced alterations in NE neurotransmission within the activity range typically observed in normal sleep and waking.

Similar to other neurotransmitter systems, NE acts at multiple receptors. Traditionally, three noradrenergic receptor subtypes have been recognized: alpha-1, alpha-2, and beta. alpha-1- and beta-receptors are thought to exist primarily at postsynaptic sites whereas, alpha-2-receptors exist both pre- and postsynaptically. Presynaptic alpha-2-receptors exert an inhibitory influence on LC neuronal discharge and NE release (Svensson et al., 1975). The distribution across and within cortical regions varies between the specific receptor subtypes, as does second messenger coupling. For example, within neocortex, beta-receptors appear to be more broadly distributed across laminae and are positively coupled to the Gs/cAMP second messenger systems, whereas alpha-1- and alpha-2-receptors are concentrated in the superficial layers and are coupled to the phosphoinositol and Gi/cAMP systems, respectively (Dohlman et al., 1991). Recently, molecular biological studies have revealed an even greater diversity of adrenergic receptors, with multiple subtypes each of beta-, alpha-1-, and alpha-2-receptors identified (Boyajian & Leslie, 1987; Jones & Palacios, 1991; Bylund et al., 1992). The distinct cellular and behavioral functions mediated by the recently identified receptor subtypes, across multiple terminal fields, remain largely unknown.

To summarize, through the actions of NE, the LC-NA system is uniquely situated to influence virtually the entire CNS. The only exception to this is the caudate/putamen, which is virtually devoid of NA fibers (Swanson & Hartman, 1975). The anatomical features combined with the above-described physiological observations, indicate that alterations in activity of a small number of LC neurons can be broadcast globally, affecting neuronal populations of immense number. However, within this system there exists a degree of specificity conferred by the pattern of fiber termination, receptor subtype distribution, and second messenger coupling that provides a potential substrate for the support of multiple cortical and subcortical functions. These observations suggest prominent and global neuromodulatory actions of the LC-NA system. The nature of these actions has been the focus of numerous studies over the past quarter century, a limited selection of which are reviewed below.

LC Neuronal Discharge Activity Displays Both State-Dependent and Sensory-Driven Activity

The compact and homogeneous nature of both the rat and primate LC facilitates recording neuronal discharge activity of these neurons in the unanesthetized animal. This information has been essential in the formulation of hypotheses concerning potential behavioral functions of LC neurons. Past studies demonstrate that LC neurons fire in two distinct activity modes: tonic and phasic. Tonic activity is characterized by relatively low-frequency, sustained, and highly regular discharge patterns. The pioneering work of Hobson and McCarley and colleagues demonstrated that tonic discharge activity is state-dependent: LC neurons exhibit their highest discharge rates during waking (quiet waking <2 Hz; active waking > 2 Hz), evince very slow rates during slow-wave sleep (<1 Hz), and are virtually silent during REM or paradoxical sleep (Hobson et al., 1975; Foote et al., 1980). Furthermore, changes in their discharge rates anticipate changes in behavioral state (Hobson et al., 1975; Foote et al., 1980; Aston-Jones & Bloom, 1981a). Within the waking state, LC neurons discharge most rapidly in anticipation of EEG and behavioral changes that signal enhanced arousal or attentiveness (Foote et al., 1980; Aston-Jones & Bloom, 1981a).

Within waking, LC neurons also display phasic alterations in discharge rates in response to salient sensory stimuli (Foote et al., 1980; Aston-Jones & Bloom, 1981b). Phasic responses are observed with a relatively short latency (15–70 msec in rat) and are comprised of a brief burst of two or three action potentials followed by a more prolonged period of suppression of discharge activity (approximately 300–700 msec). Phasic responses are observed in association with overt attending to a stimulus within a particular environmental location (e.g., an orienting response). These responses habituate with repeated presentation of a simple stimulus. This habituation of LC responses occurs in parallel with habituation of the behavioral response to the stimulus. Further, phasic responses are less robust during lower levels of vigilance, including those associated with sleep, grooming, and eating (Aston-Jones & Bloom, 1981b). Phasic alterations in LC neuronal discharge rates have been posited to participate in processing sensory information and modulation of attention to sensory stimuli (see text that follows).

Noradrenergic Modulation of Cortical and Thalamic Neuronal Activity

To better understand the functions of the LC-NA system, the actions of NE on basal and afferent-driven neuronal activity within the forebrain have been examined extensively. These studies demonstrate that NE facilitates information processing by individual cortical, hippocampal, and thalamic neurons. To briefly summarize a vast literature, these studies show that the application of NE to these neurons can reduce spontaneous or weakly driven activity while at the same time preserve or enhance neuronal responses to potent and specific synaptic input (Foote et al., 1975; Woodward et al., 1979; Moises et al., 1981). This is observed for both inhibitory and excitatory responses to both sensory and pharmacologically driven (excitatory and inhibitory amino acid) alterations in neuronal activity (Mouradian et al., 1991; Segal et al., 1991; Waterhouse et al., 1998a). The enhancement of excitatory responses appears dependent on alpha-1-receptors, whereas the enhancement of inhibitory responses involves actions of beta-receptors (Mouradian et al., 1991; Marwaha & Aghajanian, 1982; Segal & Bloom, 1974). Similar facilitatory actions are observed following enhancement of LC discharge rates, demonstrating that enhanced rates of NE released from LC terminals facilitates information processing similar to that observed with exogenous NE (Waterhouse et al., 1998b). At least some of these actions of NE are dose-dependent, observed only within a specific concentration range: above or below this range neuronal responsivity to specific sensory stimuli is decreased. An example of this is shown in Figure 6–1, which summarizes the dose-dependent nature of the modulatory action of NE on the excitatory effects of the excitatory amino acid, glutamate, on cortical neuronal activity

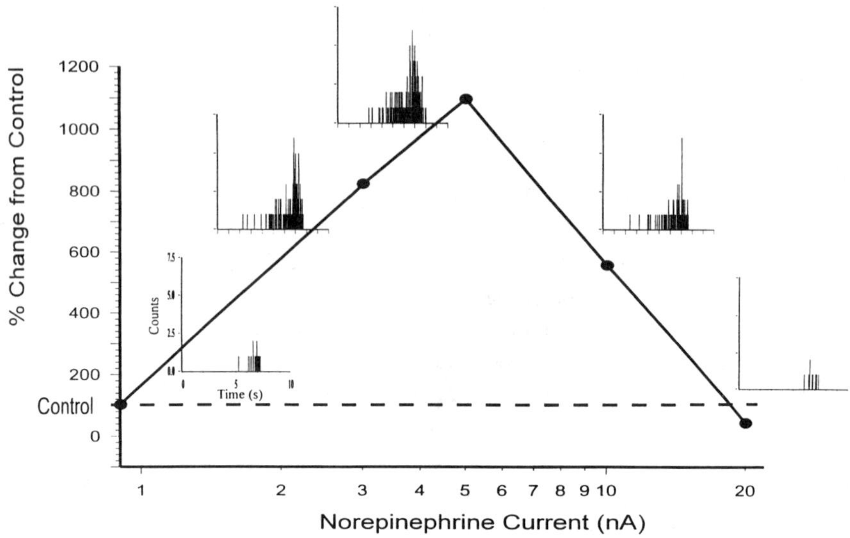

Figure 6–1. Biphasic dose-response relationship for NE modulation of glutamate-evoked responses in rat sensory cortical neurons. Each point on the graph and its associated perievent histogram represent the extracellularly recorded response of a single layer V somatosensory cortical neuron (in vitro tissue slice preparation) to uniform iontophoretic pulses (20 nA, 8-sec duration, 40-sec cycle) of glutamate before (control, at left) and during administration of different tonic levels of iontophoretic NE (1–20 nA). NE increases the glutamate-evoked neuronal response maximally when administered at an intermediate dose. Above and below this concentration, NE does not facilitate glutamate-evoked neuronal responses to a comparable extent. Each histogram sums unit activity during six consecutive glutamate applications. (From Waterhouse et al., 1998a.)

(Waterhouse et al., 1998a). This is particularly intriguing given similar non-monotonic effects of alterations in noradrenergic neurotransmission on certain cognitive processes (see the following).

In addition to the modulatory actions of NE on afferent driven neuronal responses, McCormick and colleagues demonstrated that, in vitro, NE induces a shift in the firing pattern of cortical and thalamic neurons from a burst mode to a single-spike mode (see McCormick et al., 1991). The burst activity mode is associated with slow-wave sleep and a relative insensitivity of these neurons to incoming sensory information, whereas the single-spike mode is associated with the awake behavioral state and the associated enhanced sensitivity of cortical and thalamic neurons to synaptic input (McCormick & Bal, 1997; 1991).

Based on these observations, it is clear that the actions of NE cannot be categorized simply as either excitatory or inhibitory. Rather, this neurotransmitter exerts multiple actions on cortical and thalamic neurons that impact the ability of these neurons to process sensory information. First, via actions at both alpha-1- and beta-receptors, NE induces an activity state appropriate for processing of ascending sensory information (eg., single-spike mode). Second, NE acts within a constricted concentration range to facilitate cortical and thalamic neuronal responses to sensory information. These effects are observed for both inhibitory and excitatory sensory-driven responses as well as for excitatory and inhibitory amino acid-induced alterations in neuronal activity. This facilitation of neuronal responsiveness involves actions of NE at both alpha-1- and beta-receptors. Combined, these observations indicate that through the actions of NE, LC neurons enhance the responsiveness of individual cortical and thalamic neurons to information arising from the environment.

NE-Elicits Long-Term Alterations in Synaptic Efficacy Within the Hippocampal Formation and Neocortex

LC-dependent alterations in responsiveness to afferent information are also observed at the level of large-population neuronal ensembles. For example, potent modulatory actions of NE have been observed in an extensively studied model of the cellular mechanisms of memory, long-term potentiation (LTP). LTP refers to a use-dependent, long-lasting increase in synaptic strength or efficacy. Thus, when excitatory synapses are rapidly and repetitively stimulated for brief periods (tetanic stimulation), the postsynaptic neurons generate action potentials more readily upon subsequent stimulation, which is reflected in an enhancement of the population spike. In the hippocampal formation, three forms of LTP have been described; one involving mossy-fiber input to CA3 of the hippocampus (from the dentate gyrus), one involving the Schaffer collateral input to region CA1 (from the entorhinal cortex), and one involving perforant path input to the dentate gyrus. That LTP is readily observed in a structure critical for memory function has further stimulated interest in LTP as a possible mechanism underlying memory.

NE has been demonstrated repeatedly to influence LTP within CA3 and the dentate gyrus. For example, depletion of NE substantially decreases the population spike observed in dentate gyrus (Stanton & Sarvey, 1985), whereas NE application elicits a frequency-dependent enhancement of LTP in the CA3 subfield (Hopkins & Johnston, 1984). These effects appear to be dependent on actions of NE at beta-receptors (Hopkins & Johnston, 1988).

NE also elicits a long-lasting enhancement of synaptic efficacy in both the dentate gyrus and CA1 region of the hippocampus in vitro in the absence of tetanic stimulation. Thus, NE increases the population spike evoked by stimulation of major afferent pathways to the hippocampal formation; perforant path stimulation in the dentate gyrus (Stanton & Sarvey, 1987) and Schaffer collateral stimulation in the hippocampal CA1 region (Heginbotham & Dunwiddie, 1991). The latter effects were attributable to actions of NE at beta-receptors. In vivo, a similar facilitation of the population spike was observed in the dentate gyrus following enhanced NE neurotransmission induced by either LC activation (Harley & Sara, 1992), alpha-2-antagonist administration (Segal et al., 1991) or direct application of NE (Stanton & Sarvey, 1987). The potentiating effect of NE in the dentate gyrus involves actions of both beta- and alpha-1-receptors (Chaulk & Harley, 1998).

Importantly, enhancement of synaptic strength within the dentate gyrus is observed following behaviorally relevant, sensory-driven increases in LC discharge rate. For example, LC neurons display a short duration (approximately 1 to 2 sec) increase in discharge activity when a rat encounters a novel stimulus within the environment (Vankov et al., 1995). This novelty-induced activation of LC neuronal discharge is accompanied by a short duration (approximately 20 sec) enhancement in the population spike within the dentate gyrus in response to electrical stimulation of the perforant path. This increase in population spike is attenuated by pretreatment with the beta-antagonist, propranolol (Kitchigina et al., 1997).

Thus, these studies demonstrate substantial and coordinated enhancement of synaptic strength within the hippocampal formation following activation of the LC-NA system. An additional form of NE-dependent plasticity has also been described in neocortex (Kirkwood et al., 1999). In these studies, it was observed that NE elicits a long-term synaptic depression (LDP) of the population response recorded from layer III of visual cortex. These actions of NE appear to derive from actions at alpha-1-receptors and are dependent on neurotransmission at NMDA receptors.

Together, these observations indicate a potentially critical role of the LC-NA system in mediating long-lasting modifications in neurotransmission within large populations of forebrain neurons. It is particularly intriguing that these actions are observed in structures that support higher-level cognitive processes, which are dependent on neuronal plasticity. The actions of the LC-NA system within these structures may serve an essential role in the ability of these structures to maintain optimal plasticity, and thus support optimal operation of cognitive processes.

Behavioral Functions of the LC-NA System

The above-described properties and electrophysiological actions of the LC-NA system suggest global actions of the LC-NA system related to the modulation of behavioral state and state-dependent processing of sensory information. Results from recent studies support these hypotheses. In particular, as reviewed below, experimental evidence now suggests at least two general functions of the LC-NA system. First, the LC-NA system is a critical participant in the modulation of behavioral state, with enhanced noradrenergic neurotransmission resulting in an awake and alert behavioral state. Second, within waking, fluctuations in LC neuronal activity appear to contribute to fluctuations in attention to environmental stimuli.

Noradrenergic Modulation of Behavioral State

The observation that changes in LC neuronal discharge rates precede changes in behavioral and EEG indices of behavioral state suggests a causal role for the LC in behavioral and/or EEG state modulation. This hypothesis is supported by work in anesthetized rat that demonstrates a causal relationship between alterations in forebrain EEG and alterations in LC neuronal discharge rates (Berridge & Foote, 1991; Berridge et al., 1993; Page et al., 1993). In these studies, electrophysiological recordings were used to guide placement of an infusion needle in close proximity of the LC. Drugs were then infused through this needle to elicit alterations in LC neuronal discharge rates. Under these conditions, the selective unilateral enhancement of LC neuronal activity results in a robust bilateral activation of forebrain EEG (Fig. 6–2; Berridge & Foote, 1991). The activation of noradrenergic beta-receptors appears to be essential for LC-dependent EEG activation because EEG activation is blocked by pretreatment with the beta-noradrenergic antagonist propranolol (Berridge & Foote, 1991). Conversely, bilateral inhibition, but not unilateral inhibition, of LC neuronal discharge activity decreases forebrain EEG activation (Berridge et al., 1993).

These studies demonstrate that minimal LC neuronal discharge activity within one hemisphere is sufficient to maintain an activated forebrain, bilaterally. This observation is consistent with previous observations demonstrating diminished, yet not absent, LC discharge activity in animals that are clearly awake yet engaged in consummatory or grooming behavior (Aston-Jones & Bloom, 1981a). In both the LC activation and inactivation studies, the onset and recovery of the EEG responses closely followed changes in LC neuronal activity levels, suggesting that the EEG responses were indeed due to alterations in activity of the LC-NA system. Further, localization studies indicated that infusion-induced changes in EEG state are not observed when infusions are placed outside the immediate vicinity of the LC. Combined, these observations indicate that under these experimental conditions LC is a potent modulator of forebrain EEG state, with unilateral LC neuronal discharge activity causally related to the bilateral maintenance of EEG activity patterns associated with arousal.

Additional studies examined which LC terminal fields are involved in the LC-dependent modulation of forebrain activity state. Small unilateral infusions of the beta-agonist, isoproterenol, into the region of the medial septal area (Berridge et al., 1996) or the medial preoptic area (unpublished observations) but not the substantia innominata (Berridge et al., 1996) elicit bilateral activation of forebrain EEG in the anesthetized animal. In the unanesthetized animal, infusions of a beta-agonist (Berridge & Foote, 1996; Alam & Mallick, 1990) or amphetamine (Figure 6–3; Berridge et al., 1999) into the medial septal area or the medial preoptic area increases behavioral and EEG indices of waking. Arousal-enhancing actions of either beta-agonist or amphetamine were not observed when these drugs were infused immediately outside the medial septal area and medial preoptic area, including within the substantia innominata (Berridge & Foote, 1996; Berridge et al., 1999).

Thus, actions of noradrenergic systems contained within an extended region of the medial basal forebrain, which contains the medial septal area anteriorally and the medial preoptic area posteriorally, are sufficient to induce and maintain the awake behavioral state.

Noradrenergic Modulation of Vigilance

The above studies indicate that one function of the LC-NA system is the induction of a behavioral state appropriate for the collection of sensory information (e.g., waking). Additional evidence suggests that within the awake state, LC neurons may

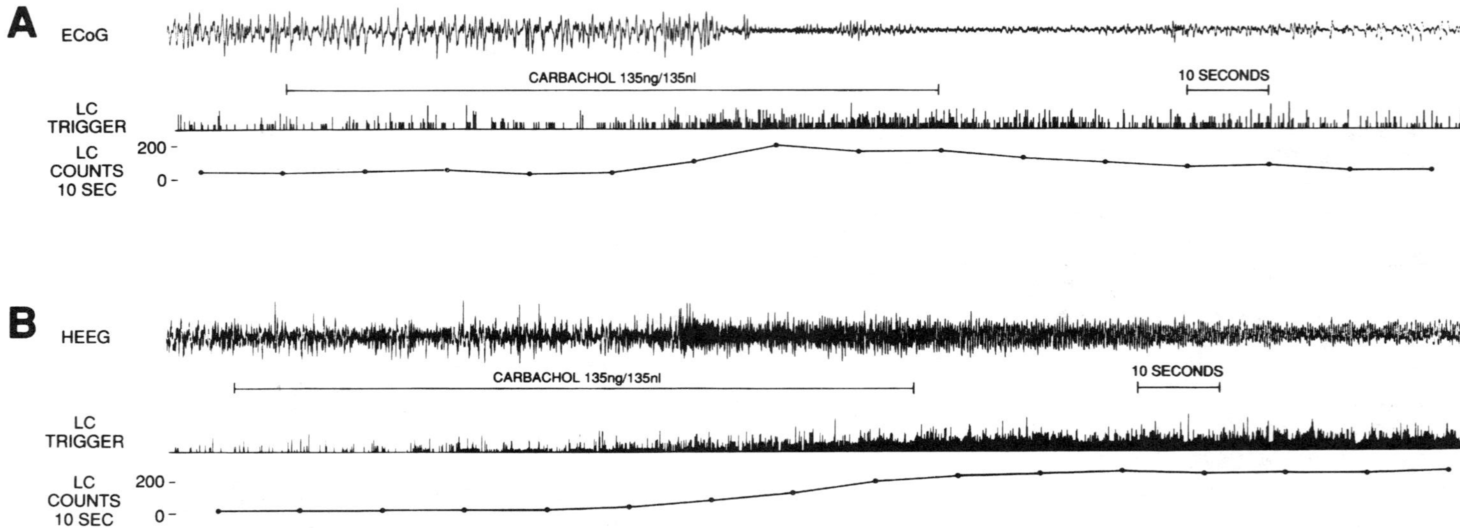

Figure 6–2. Relationship of LC activity to cortical (ECoG; Panel A) and hippocampal (HEEG; Panel B) before, during, and after peri-LC infusions of the cholinergic agonist bethanechol. Cholinergic agonists exert potent excitatory effects on LC neuronal discharge rates. *A* and *B* represent data from separate experiments. In each experiment, bethanechol-induced changes in EEG activity were observed simultaneously in both the ECoG and HEEG recordings. Bethanechol was infused at a constant rate throughout the interval indicated. EEG activity is shown in the *top trace* of each panel, the raw trigger output from LC activity in the *middle trace*, and the integrated trigger output (10-sec intervals) in the *bottom trace*. In *A*, LC activity is seen to increase during the latter part of the infusion. Several seconds later, reduced amplitude and increased frequency become evident in the ECoG trace. As LC activity begins to decrease following the infusion, ECoG amplitude large-amplitude, slow-wave activity increases. In *B*, enhanced LC activity becomes evident in the latter part of the infusion period, and several seconds later thetarhythm begins to dominate the HEEG trace. For the remainder of the trace, LC activity remains elevated and theta rhythm predominates. (From Berridge & Foote, 1991.)

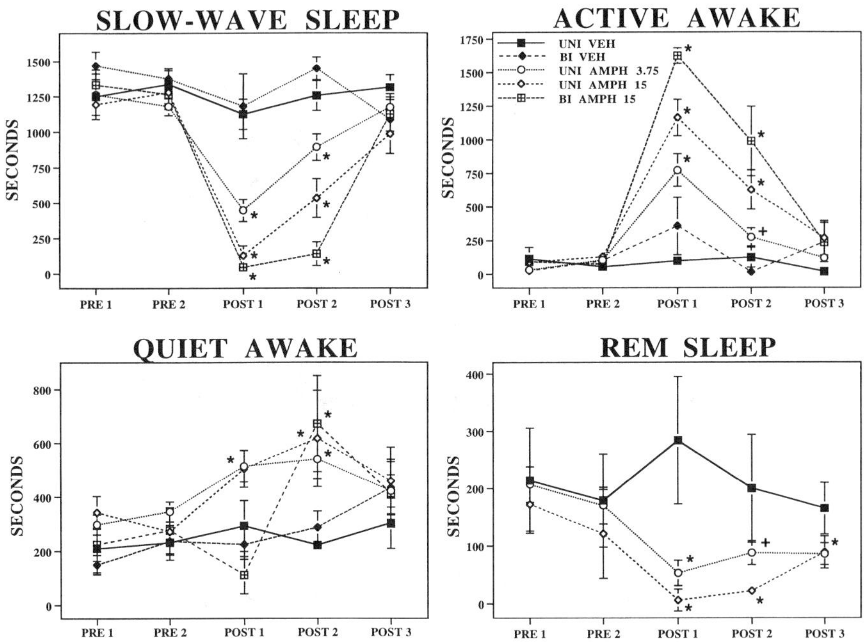

Figure 6–3. Effects of MS unilateral (UNI VEH) and bilateral (BI VEH) vehicle infusions, unilateral 3.75 μg AMPH (UNI AMPH 3.75) and 15.0 μg AMPH (UNI AMPH 15), and bilateral 15 μg AMPH (BI AMPH 15) infusions on REM sleep, slow-wave sleep, quiet waking, and active waking, as defined by ECoG and EMG measures. Symbols represent means (± SEM) of time (secs) spent in the five different behavioral state categories per 30-min epoch. PRE1 and PRE2 represent pre-infusion portions of the experiment. POST1-POST3 represent post-infusions epochs. The lack of visible error-bars indicates that the magnitude of the SEM fell within the range corresponding to the dimensions of the symbol. Due to problems in EEG/EMG recordings in some of the bilateral vehicle animals, REM was not scored in this group. Thus, statistical analyses on the effects of AMPH infusions on REM were conducted only on animals receiving unilateral infusions. The apparent increase in quiet waking within the unilateral 15 μg AMPH group was not statistically significant. $^{+}P < 0.05$, $^{*}P < 0.01$ compared to PRE1. (From Berridge et al., 1999.)

modulate sensory information processing (Rajkowski et al., 1994; Aston-Jones et al., 1997,1998). For example, in monkeys phasic fluctuations in LC neuronal discharge rates are correlated with performance on a vigilance task (Figure 6–4). In this task, animals are required to press a lever when they detect a visual target stimulus embedded in a series of nontarget stimuli. Correct behavioral responses are rewarded with delivery of a food reward (juice). Phasic LC responses are elicited preferentially by target stimuli when an animal responds correctly. These LC responses precede the behavioral response. When the animal fails to respond behaviorally to the target stimulus, phasic LC responses are absent. Reversal training demonstrated that LC neuronal responses to target stimuli are independent of stimulus attributes and

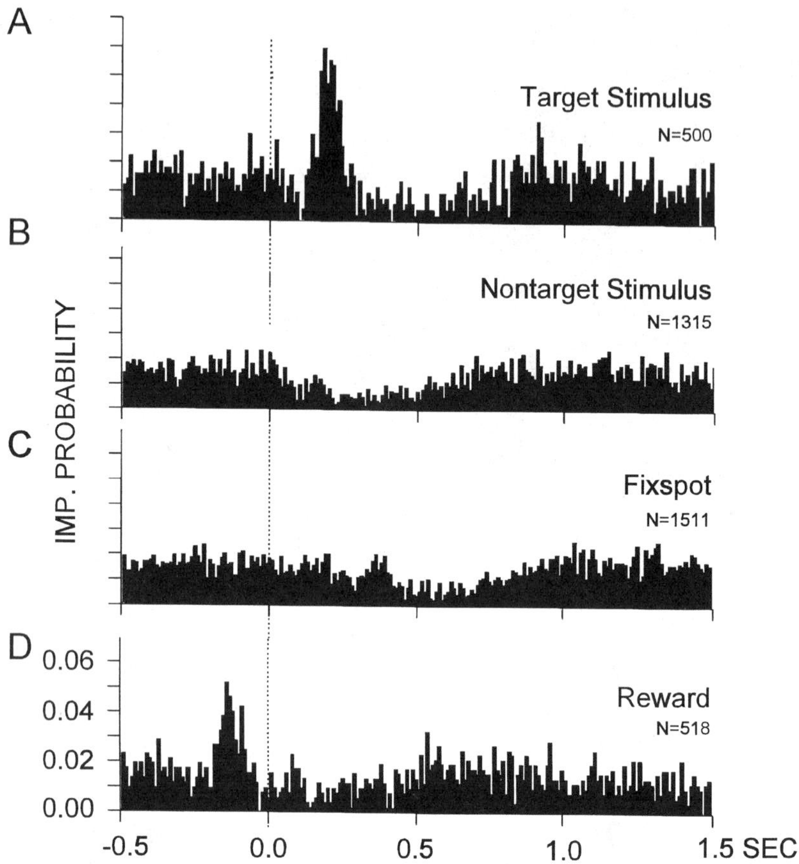

Figure 6–4. Peristimulus time histograms (PSTHs) for a typical individual LC neuron neuron in response to various events during performance of a vigilance task. A monkey was trained to respond to a target visual stimulus embedded in nontarget visual stimuli. The task required the subject to stably foveate a central fixation spot on the video monitor. Correct responses were rewarded with juice delivery. Normalized PSTHs are synchronized with (A) target stimuli, (B) nontarget stimuli, (C) fix spot presentation, and (D) juice solenoid activation. Note the selective activation by target stimuli (A). The activation seen before reward presentation (D) is due to the activation by target cues. LC neurons are activated consistently by target stimuli. Nontarget stimuli either do not elicit alterations in LC discharge activity or, in a subset of neurons, elicits inhibition of LC discharge (B). Inhibition in the Fixspot PSTH is due to frequent nontargets, which follow fix spot presentation. (From Aston-Jones et al., 1998.)

are dependent only on whether a stimulus was a target signal. Thus, in this task, LC neurons display phasic activity in response to salient sensory stimuli that contain information relevant to goal attainment.

In addition to phasic fluctuations, vigilance appears also to be dependent on fluctuations in LC tonic discharge rates. For example, low levels of tonic discharge activity are associated with sedation (Foote et al., 1980; Aston-Jones & Bloom, 1981a) whereas, elevated tonic discharge rates are associated with increased eye movements and decreased ability to foveate on a pretest eye fixation stimulus (e.g., enhanced scanning) and more frequent false alarm errors (Rajkowski et al., 1994). This increase in false alarms results from a decreased ability to discriminate target from distractor (decreased d') and an increase in nontarget responding (decreased beta factor; Aston-Jones et al., 1994; Usher et al., 1999).

These correlational observations suggest that LC discharge rates are causally related to alterations in attention. More direct evidence for a causal relationship between LC discharge rates and vigilance stems from preliminary observations concerning the effects of manipulations of LC discharge rates in monkeys performing in this vigilance task. Thus, in one hyperactive monkey that performed poorly in the vigilance task, suppression of LC discharge activity via local infusion of the alpha-2-agonist, clonidine, improved task performance. Conversely, when LC neurons were activated by local infusion of a cholinergic agonist in animals performing well prior to the infusion, performance was impaired (Ivanova, 1997). Combined, these observations suggest strongly that phasic activation of LC is associated with attending to salient environmental stimuli that guide goal-directed behavior.

Importantly, the relationship between phasic activity and target detection is influenced by tonic discharge rates, being observed only when tonic discharge activity falls within a relatively narrow range. Below this range, an animal appears drowsy and less attentive to the environment, whereas above this range an animal is less able to maintain focused attention on specific environmental stimuli. Arnsten and colleagues have described a similar inverted-U relationship between activity of the LC-NA system and prefrontal cortex-dependent working memory (see the following; Arnsten, Chapter 7). The LC rate-dependent nature of these behaviors may result, at least in part, from dose-dependent effects of NE on neocortical neuronal activity, as described previously (Fig. 6–1).

Additional Behavioral Evidence for NE-Dependent Alterations in Attention

The ability of NE to enhance cortical function by reducing "noise" and/or facilitating processing of relevant signals suggests that the LC-NA system might enhance cognitive function under "noisy" conditions where irrelevant stimuli impair performance. Experimental data from rodents, monkeys, and humans largely supports this hypothesis. For example, in rodents, depletion of forebrain NE generally does not impair performance of simple learning and memory tasks. An exception to this rule is the aged (Leslie et al., 1985; Collier et al., 1988) or combined LC/nucleus basalis lesioned animal (Haroutunian et al., 1990), suggesting that NE actions may become significant to simple learning in the compromised brain. However, NE depletion does produce deficits in the performance of non-aged, otherwise intact animals on a variety of tasks when irrelevant stimuli are presented during testing. Thus, the addition of distracting visual stimuli at the choice point in T-maze produces a much greater disruption of performance in NE-depleted rats than in sham-treated animals (Roberts et al., 1975; Oke & Adams, 1978). Similarly, the presentation of irrelevant, auditory stimuli interrupts the visual discrimination performance of rats with forebrain NE depletion, although these animals perform normally under nondistracting conditions (Carli et al., 1983). Further, NE depletion increases conditioned responses to irrelevant stimuli while decreasing responses to relevant stimuli (Lorden et al., 1980; Selden et al., 1990,1991).

Interpretation of lesion studies is problematic due to the well-documented lesion-induced compensatory responses that occur within central noradrenergic systems. These compensatory responses are observed at the levels of release, postsynaptic number, and second-messenger systems (Sporn et al., 1977; Skolnick et al., 1978; U'Prichard & Kvetnansky, 1980; Harik et al., 1981; Hallman et al., 1984; Logue et al., 1985). Compensation can occur to such a degree as to produce a system that appears hyperactive relative to the pre-lesion state (Diaz et al., 1978; Mogilnicka, 1986; Berridge & Dunn, 1990). Thus, it remains to be determined unambiguously whether lesion-induced disruption of behavior in these paradigms result from hypoactive noradrenergic systems. Nonetheless, the fact that noradrenergic lesions alter attention to relevant stimuli indicates that noradrenergic neurotransmission influences attentional processes.

The LC-NA system may be particularly sensitive to novel environmental stimuli within the environment. For example, enhanced LC discharge rates are observed when rats encounter novel stimuli (Vankov et al., 1995). Further, pharmacological manipulations that enhance NE release increase physical contact/interaction with novel stimuli, whereas pharmacological treatments that decrease NE release have the opposite effects (Devauges & Sara, 1990). This suggests that the LC-NA system influences neural circuits that support exploration of novel events within the environment. Alterations in the extent to which an animal attends to novel stimuli in the environment might also underlie the effects of noradrenergic manipulations on exploration of a complex novel environment (Arnsten et al., 1981; Berridge & Dunn, 1989). In this paradigm, pharmacologically or stress-induced enhancement of NE release decreased the time rats and mice spend in sustained, or focused, contact with a complex, novel object when tested in a novel environment (Arnsten et al., 1985; Berridge & Dunn, 1989, 1990). This NE-dependent alteration in exploration could reflect increased scanning of the environment and inversely, decreased time spent attending to a single stimulus in an environment containing multiple novel stimuli.

Finally, as reviewed elsewhere in this text, Arnsten and colleagues have demonstrated that noradrenergic neurotransmission modulates prefrontal cortex (PFC)-dependent working memory in nonhuman primates. This work suggests an important, facilitatory role of NE in the proper regulation of PFC-dependent behavior through actions at postsynaptic alpha-2-noradrenergic receptors. For example, noradrenergic alpha-2-agonists improve performance in dorsolateral PFC-dependent tasks in monkeys with catecholamine depletion produced by MPTP (Schneider & Kovelowski, 1990), local infusion of 6-OHDA into the PFC (Arnsten & Goldman-Rakic, 1985), or aging (Arnsten & Goldman-Rakic, 1985; Arnsten et al., 1988). As mentioned, it is interesting to note that in this task of higher-order cognitive function there exists an inverted-U relationship for the actions of NE, similar to that described for LC neuronal activity in vigilance.

ERPs/P300

The above-described observations suggest the involvement of the LC-NA system in attention-related cognitive processes. Studies on event-related potentials (ERPs) provide additional support for this hypothesis. ERPs are widely distributed voltage fluctuations time-locked to sensory, motor, or cognitive events that are generated from large populations of neurons and are extracted from EEG recordings using signal-averaging techniques. The P300 component of human ERPs is elicited in response to novel and/or task-related stimuli and is thus thought to have particular relevance to attentional or mnemonic processes. P300-like components are also observed and have been studied in monkeys as an animal model of attention.

Although the origin of the widely distributed P300-like potentials has not been determined, it has been proposed that the LC-NA system, with its anatomical and physiological properties, represents a potential candidate system for the generation of these potentials (see Foote et al., 1983; Pineda et al., 1987; Pineda et al., 1989; Foote et al., 1991). Experimental evidence supports this hypothesis. For example, bilateral LC lesions in squirrel monkeys selectively decreased P300 components of the ERP measured in response to infrequent tone pips (Pineda et al., 1989). Similarly, inhibition of LC firing and NE release with systemic administration of the alpha-2-agonist, clonidine, also decreases P300 components (Swick et al., 1988). These results further suggest an important role of the LC-NA system in the modulation of cortical responsiveness to sensory information and the modulation of attentional processes.

Effects of Acute Amphetamine-like Stimulants on Monoamine Release

Although it has long been known that amphetamine-like stimulants are effective in the treatment of ADHD, the neural mechanisms that underlie these therapeutic actions remain unidentified. These drugs enhance the synaptic concentrations of dopamine, NE, and depending on the drug, serotonin, through inhibition of the respective transporter and/or stimulation of monoamine efflux from the terminal (for review, see Kuczenski & Segal, 1994). However, the extent to which these different neurochemical actions occur is drug-, dose-, and monoamine-dependent. For example, amphetamine stimulates dopamine efflux from the terminal and inhibits dopamine reuptake. Both actions are observed at both low and high doses (for review, see Kuczenski & Segal, 1994). In contrast, at low-to-moderate doses

(0.5–2.0 mg/kg, s.c.) amphetamine acts primarily as an inhibitor of NE reuptake, whereas at higher doses (5.0 mg/kg, s.c.), amphetamine both stimulates NE efflux and inhibits NE reuptake (Kuczenski et al., 1995; Florin et al., 1994).

Recent evidence indicates that the extent to which a given amphetamine derivative influences a given monoamine is highly dependent on the identity of the drug. For example, and of particular relevance for a discussion of ADHD, serotonergic systems are relatively insensitive to moderate-to-high doses (10–30 mg/kg, s.c.) of methylphenidate (see Figure 6–5; Kuczenski et al., 1995). This is in contrast to the particularly potent NE-releasing actions of methylphenidate at these same doses (see Fig. 6–5, area under the curve [AUC] bar graph). This observation suggests that the therapeutic action of methylphenidate in the treatment of ADHD most likely does not involve alterations in extracellular levels of serotonin.

In addition to the issue of drug identity, the dose of amphetamine-like stimulant is also a critical factor determining the extent to which a specific neurotransmitter system is affected by a given drug. This is a particularly critical issue for a discussion of ADHD, given that much of what is known about the neurochemical and behavioral actions of amphetamine derivatives derives from studies using doses of this drug that are relevant to drug abuse and amphetamine psychosis. In contrast, therapeutic doses of amphetamine-like stimulants are substantially below those typically associated with either drug abuse or amphetamine psychosis. A good example of the dose-dependent nature of the qualitative actions of these drugs is the dose-dependent effects of amphetamine on extracellular levels of serotonin. Thus, amphetamine does not increase extracellular serotonin levels except at higher doses (2.5 μg/kg; Kuczenski et al., 1995). Of potentially greater significance to the current discussion, recent observations by Kuczenski and Segal (personal communication) indicate that at low doses (0.5 mg/kg s.c.), methylphenidate elicits significant increases in NE efflux in the *absence* of changes in extracellular dopamine levels, as measured by in vivo microdialysis. These observations indicate that it may well be inappropriate to postulate neuronal mechanisms of therapeutic action of methylphenidate in ADHD on the basis of (*1*) observations made with other amphetamine derivatives and (*2*) observations made with doses of methylphenidate that exceed the range used clinically.

Effects of Acute and Chronic Amphetamine-like Stimulants on the LC-NA System

As discussed above, considerable evidence suggests that the LC-NA system modulates both attention and arousal-related processes. This evidence suggests that dysregulation of the LC-NA system might contribute to the clinical features of arousal/attention disorders. Alternatively, in the absence of a role on etiology, actions of NE might underlie the therapeutic actions of drugs used to treat these disorders. The ability of amphetamine-like stimulants, those drugs most effective in the treatment of ADHD, to alter extracellular concentrations of NE supports these hypotheses. Additional support for these hypotheses is that noradrenergic alpha2-agonist administration is a useful adjunct in the pharmacological treatment of ADHD

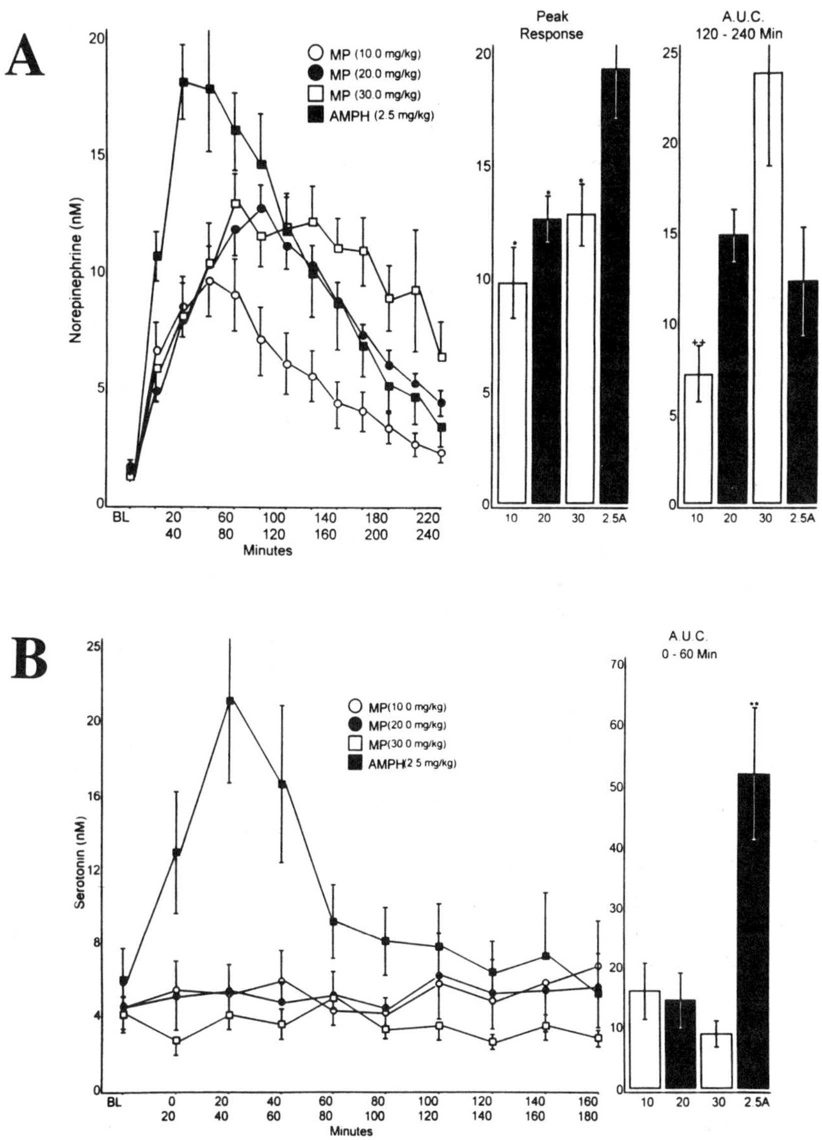

Figure 6–5. Effects of AMPH (2.5 mg/kg) and varying doses of methylphenidate (MP; 10, 20 and 30 mg/kg) administration (intraperitoneal) on NE (*Panel A*) and serotonin (*Panel B*) concentrations in in vivo microdialysis samples. NE samples were collected from hippocampus (*HP*), whereas serotonin samples were collected from caudate-putamen (*CP*). Shown are data from 20-min samples collected immediately preceding (*BL*) and following drug administration. In both A and B, the temporal profile of NE and serotonin responses are shown along with histograms depicting the peak response or the area under the curve (AUC). It is of interest, in the present context, that methylphenidate is particularly potent at enhancing NE release (see *Panel A*, AUC). In contrast, methylphenidate has no detectable effects on serotonin release despite a substantial stimulatory effect of AMPH on serotonin release. (From Kuczenski & Segal, 1997.)

(Hunt et al., 1985; Chappel et al., 1995; Horrigan & Barnhill, 1995; Hunt et al., 1995; Arnsten et al., 1996). Finally, the preliminary observations described above of a selective increase in NE efflux and not dopamine at lower doses of methylphenidate further strengthens the hypothesis that therapeutic actions of amphetamine-like stimulants in the treatment of ADHD involves alterations in NE neurotransmission.

Aside from the potent actions of amphetamine-like stimulants on NE efflux, these drugs also alter spontaneous and sensory-evoked LC neuronal activity. Thus, acute administration of methylphenidate, amphetamine, and cocaine decrease spontaneous and non-noxious sensory-evoked LC discharge activity. These effects are observed at relatively low doses (e.g., 0.25–1.0 mg/kg of amphetamine; Graham & Aghajanian, 1971; Engberg & Svensson, 1979; Lacroix & Ferron, 1988; Surprenant & Williams, 1987; Pitts & Marwah, 1987; Curtis et al., 1993). The suppressive actions of cocaine on LC discharge activity appear to result from NE stimulation of alpha-2-receptors (Surprenant & Williams, 1987). Interestingly, acute cocaine administration does not alter the LC activational effects of either hypotensive stress (nitroprusside-induced) or corticotropin-releasing factor, a stress-related neuropeptide (Curtis et al., 1993).

The clinical benefits of amphetamine-like stimulant administration in ADHD occurs with acute and/or short-term administration. However, given the long-term nature of amphetamine-like stimulant administration in the treatment of ADHD, it is important to understand the effects of long-term administration of these drugs on catecholaminergic neurotransmission. For dopamine, an extensive literature exists that describes an enhanced sensitivity of dopaminergic systems to prolonged administration of amphetamine like-stimulants (e.g., sensitization; for review, see Robinson & Becker, 1986; Robinson & Berridge, 1993; Taylor, this volume). However, it should be noted that virtually all of these studies utilized relatively high doses of these drugs, which do not mimic conditions associated with pharmacological treatment of ADHD. In contrast to that of DA, little is known concerning the effects of repeated amphetamine-like stimulants on NE efflux. Limited information indicates that repeated administration of amphetamine, in an escalating-dose regimen, results in alterations in both basal and amphetamine-induced NE efflux (Camp et al., 1997). However, given the differences between this administration regimen and that utilized in the treatment of ADHD, the relevance of these observations to ADHD is unclear. Long-term methylphenidate administration results in tolerance to the suppressive effects of this drug on spontaneous LC discharge activity in the anesthetized animal (Lacroix & Ferron, 1988). This effect coincides with the methylphenidate-induced tolerance to the suppressive effects of the alpha-2-agonist, clonidine, on LC discharge activity (Lacroix & Ferron, 1988).

At the level of the terminal field, LC stimulation suppresses spontaneous discharge activity of neocortical neurons (Lacroix & Ferron, 1988). Acute intravenous methylphenidate administration (0.65 mg/kg) substantially reduced this inhibitory action of LC activation on spontaneously active cells (Lacroix & Ferron, 1988). The authors do not address whether methylphenidate administration, in the absence of LC stimulation, decreases the percentage of spontaneously active cortical neurons. Similar effects on cortical neuronal activity were observed with repeated (7-day) methylphenidate administration (2.0 mg/kg, i.p.). Finally, although repeated methylphenidate administration substantially decreased the responsiveness of cortical neurons to iontophoretically administered NE, this effect of methylphenidate was smaller than that observed with acute methylphenidate administration (Lacroix & Ferron, 1988).

In addition to alterations in neuronal discharge rates and rates of NE release, repeated amphetamine administration may elicit alterations in postsynaptic receptor number or second messenger responsivity of noradrenergic receptors. For example, treatments that increase NE release, such as stress or NE reuptake blockers, decrease beta- and alpha-2- receptor number and/or sensitivity (for review, see Baudry et al., 1976; Stone, 1983). In contrast, chronic antidepressant administration, which increases NE release, increases the sensitivity of alpha-1-receptors (Menkes & Aghajanian, 1981). Although these observations suggest that longterm increases in noradrenergic neurotransmission may alter noradrenergic receptor number and/or sensitivity, the extent to which chronic, low-dose amphetamine/methylphenidate elicits similar alterations remains to be elucidated.

To summarize, noradrenergic systems are a primary target of amphetamine-like stimulants. Acutely, these drugs increase rates of NE efflux within terminal fields and suppress spontaneous and non-noxious sensory-evoked LC discharge rates while preserving LC responsiveness to certain stress-related events. A substantial limitation in our understanding of the neurobiological mechanisms underlying the therapeutic actions of amphetamine-like stimulants in ADHD is a lack of information concerning the effects of low-dose drug administration across all levels of noradrenergic neurotransmission. The critical nature of dose is highlighted by recent observations that indicate relatively selective actions of methylphenidate on noradrenergic, and not dopaminergic or serotonergic, neurotransmission. These observations call into question the relevance to ADHD of previous studies on the effects of amphetamine-like stimulants that utilize doses that exceed substantially the dose range used in the treatment of ADHD.

Noradrenergic Involvement in Behavioral and Electrophysiological Actions of Amphetamine-like Stimulants

As reviewed above, substantial evidence indicates that noradrenergic systems modulate a variety of arousal- and attention-related electrophysiological activity and behavior. In light of evidence that links attentional and/or arousal deficits associated with ADHD and the actions of amphetamine-like stimulants on NE release, it is surprising that little is known concerning the actions of amphetamine on NE-dependent attention- and arousal-related processes. Based on the above-reviewed evidence, it can be argued that amphetamine-induced alterations in NE release will influence the activity state and information processing ability of neural circuits that are the essential substrates for distinct behavioral states and state-dependent attentional and memory processes. However, for the most part, the extent to which amphetamine-induced alterations in cognitive function are, in fact, dependent on actions of NE remains to be determined.

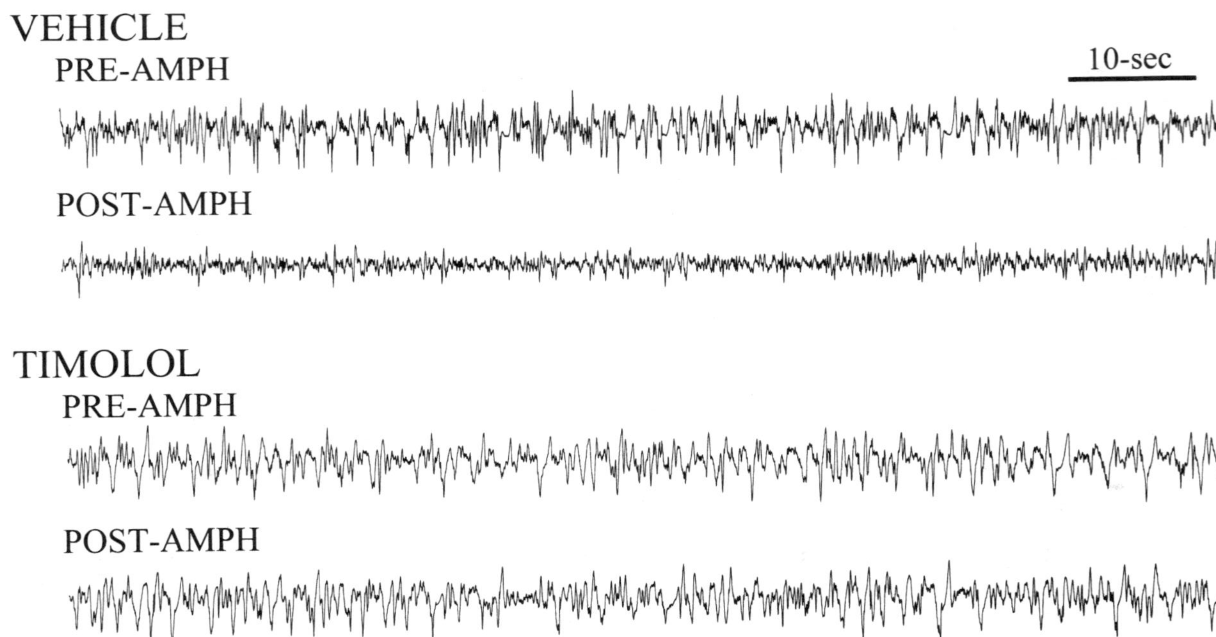

Figure 6–6. Effects of beta-receptor blockade on amphetamine-induced ECoG activation in halothane-anesthetized rat, pretreatment with vehicle or the beta-antagonist, timolol (100 μg), on 0.15 mg/kg IV AMPH-induced ECoG activation. Shown are 60-sec traces of ECoG recordings taken immediately prior to (PRE-AMPH) or 3-min following (POST-AMPH) AMPH infusion. *Top pair* of traces show ECoG activity from an animal pretreated intracerebroventricularly (ICV) with vehicle. *Bottom pair* of traces show ECoG from an animal pretreated ICV with the beta-antagonist, timolol. IV amphetamine was administered 30 min following ICV pretreatment. In vehicle-pretreated animals, AMPH induces a robust activation of ECoG, characterized primarily by a decrease in large-amplitude, slow-wave activity. In contrast, ICV timolol pretreatment prevented this AMPH-induced ECoG activation. (From Berridge et al., 1999.)

Nonetheless, limited evidence supports the hypothesis that amphetamine-like stimulants modulate behavioral state and higher cognitive processes, at least in part, through actions of NE. For example, at the level of behavioral state, amphetamines have potent arousal-enhancing actions. When infused into a region of the medial basal forebrain in rats, amphetamine increases EEG and behavioral indices of arousal (Fig. 6–3; Berridge et al., 1999). The pattern of sites where amphetamine infusions increase and do not increase waking is identical to that observed with infusions of the NE beta-agonist isoproterenol (Berridge et al., 1996; Berridge & Foote, 1996). Evidence of a causal involvement of NE in the arousal-enhancing actions of amphetamine is the observation that central pretreatment with a beta-antagonist blocks intravenous amphetamine-induced EEG activation (Figure 6–6; Berridge & Morris, 2000). Further, work by Waterhouse and colleagues demonstrates that cocaine administration alters thalamic somatosensory neuronal responses to perithreshold sensory stimulation and the time course of these effects parallels cocaine-induced alterations in NE efflux (Bekavac & Waterhouse, 1995). Additionally, within the hippocampal formation, blockade of either alpha-1- or beta-receptors blocked the amphetamine-induced (3.0 mg/kg, i.p.) enhancement of the dentate gyrus population spike in response to perforant path stimulation (Segal et al., 1991). Finally, amphetamine-induced (2.5 mg/kg, i.p.) changes in auditory gating are reversed by pretreatment with either noradrenergic alpha-1- or beta-antagonists (Stevens et al., 1991).

These observations indicate that actions of amphetamine-like stimulants on arousal and attentional processes involve, at least in part, alterations in noradrenergic neurotransmission. These arousal- and/or attention-modulatory actions of amphetamine may involve actions at the level of the noradrenergic cell body and/or within the terminal field.

Summary

To summarize, results from a variety of investigations of LC and NE function reveal a surprising degree of cohesion: whether at the level of the single cell, populations of cells, or whole animal (behavior), NE increases responses to relevant stimuli (signal) and/or suppresses responses to irrelevant stimuli (noise). Some of these effects display a non-monotonic, dose-dependent nature, with both low and high levels of NE release associated with diminished information processing capacity (single-cell level) or diminished selective attention capacity. The LC-dependent effects on attention may derive, in part, from interactions between attentional and arousal states. For example, at lower rates of NE release, impaired vigilance appears to result from sedation, whereas at high levels of NE release, highly alert animals attend to multiple stimuli in the environment and are less able to focus attention on a specific stimulus (e.g., scanning). In contrast to that observed with vigilance, the relationship between arousal levels and LC discharge rates appears to be relatively linear, at least over a broader range of LC discharge rates. Thus, an emerging view is one in which the LC-NA system exerts widespread effects at multiple levels within the brain that modulate various cognitive processes involved in the collection and processing of sensory information that guide behavior.

Combined, these observations suggest that the LC-NA system is a critical component of the neural architecture supporting attentional and arousal processes. As such, it appears reasonable to propose that dysregulation of this system might contribute to dysregulation of attention and/or arousal, including that observed in ADHD. Independent of whether noradrenergic systems contribute to the etiology of ADHD, noradrenergic systems may well be an appropriate target for pharmacological intervention in the treatment of attention and/or arousal dysfunction. Consistent with both of these hypotheses, limited evidence suggests that amphetamine-like stimulants, those drugs currently used in the treatment of ADHD, impact attention and arousal-related processes, at least in part, through alterations in noradrenergic neurotransmission. The contention that amphetamine-like stimulants exert therapeutic actions through targeting of noradrenergic neurotransmission is furthered strengthened by the fact that noradrenergic alpha-2-agonists are a useful adjunct in the treatment of ADHD.

Given the above, it is surprising that little is known concerning the actions of low-dose amphetamine and amphetamine derivatives on noradrenergic neurotransmission, those doses that have therapeutic benefit in the treatment of ADHD. The lack of information concerning the actions of low-dose amphetamine-like stimulants on noradrenergic-sensitive neural circuits and the contribution of specific noradrenergic receptor subtypes to these actions is particularly notable. This lack of information represents a major limitation in our understanding of the neurobiology of ADHD and the neuronal mechanisms underlying the therapeutic actions of pharmacological treatments of this disorder.

References

Alam, MN & Mallick, BN (1990). Differential acute influence of medial and lateral preoptic areas on sleep-wakefulness in freely moving rats. *Brain Res, 525,* 242–248.

Arnsten, AFT, Segal, DS, Loughlin, SE, & Roberts, DCS (1981). Evidence for an interaction of opioid and noradrenergic locus coeruleus systems in the regulation of environmental stimulus-directed behavior. *Brain Res, 222*, 351–363.

Arnsten, AFT, Berridge, CW, & Segal, DS (1985). Stress produces opioid-like effects on investigatory behavior. *Pharm Biochem Behav, 22*, 803–809.

Arnsten, AFT & Goldman-Rakic, PS (1985). Alpha-2 adrenergic mechanisms in prefrontal cortex associated with cognitive decline in aged nonhuman primates. *Science, 230*, 1273–1276.

Arnsten, AFT, Cai, JX, & Goldman-Rakic, PS (1988). The alpha-2 adrenergic agonist guanfacine improves memory in aged monkeys without sedative or hypotensive side effects. *J Neurosci, 8*, 4287–4298.

Arnsten, AFT, Steere, JC, & Hunt, RD (1996). The contribution of alpha-2 noradrenergic mechanisms to prefrontal cortical cognitive function: Potential significance to Attention Deficit Hyperactivity Disorder. *Arch Gen Psychiatry, 53*, 448–455.

Aston-Jones, G & Bloom, FE (1981a). Activity of norepinephrine-containing locus coeruleus neurons in behaving rats anticipates fluctuations in the sleep-waking cycle. *J Neurosci, 1*, 876–886.

Aston-Jones, G & Bloom, FE (1981b). Norepinephrine-containing locus coeruleus neurons in behaving rats exhibit pronounced responses to non-noxious environmental stimuli. *J Neurosci, 1*, 887–900.

Aston-Jones, G, Rajkowski, J, Kubiak, P, & Alexinsky, T (1994). Locus coeruleus neurons in the monkey are selectively activated by attended stimuli in a vigilance task. *J Neurosci, 14*, 4467–4480.

Aston-Jones, G, Rajkowski, J, & Kubiak, P (1997). Conditioned responses of monkey locus coeruleus neurons anticipate acquisition of discriminative behavior in a vigilance task. *Neuroscience, 80*, 697–715.

Aston-Jones, G, Rajkowski, J, Ivanova, S, Usher, M, & Cohen J (1998). Neuromodulation and cognitive performance: Recent studies of noradrenergic locus ceruleus neurons in behaving monkeys. In DS Goldstein, G Eisenhofer, & R McCarty (Eds.), *Catecholamines: Bridging basic science with clinical medicine* (pp. 755–759). San Diego Academic Press.

Baudry, M, Martres, MP, & Schwartz, JC (1976). Modulation in the sensitivity of noradrenergic receptors in the CNS studied by the responsiveness of the cyclic AMP system. *Brain Res, 116*, 111–24.

Bekavac, I & Waterhouse, BD (1995). Systemically administered cocaine selectively enhances long-latency responses of rat barrel field cortical neurons to vibrissae stimulation. *J Pharm Exp Ther, 272*, 333–342.

Berridge, CW & Dunn, AJ (1989). Restraint-stress-induced changes in exploratory behavior appear to be mediated by norepinephrine-stimulated release of CRF. *J Neurosci, 9*, 3513–3521.

Berridge, CW & Dunn, AJ (1990). DSP-4-induced depletion of brain norepinephrine produces opposite effects on exploratory behavior 3 and 14 days after treatment. *Psychopharmacology, 100*, 504–508.

Berridge, CW & Foote, SL (1991). Effects of locus coeruleus activation on electroencephalographic activity in neocortex and hippocampus. *J Neurosci, 11*, 3135–3145.

Berridge, CW, Page, M, Valentino, RJ, & Foote, SL (1993). Effects of locus coeruleus inactivation on forebrain electroencephalographic activity. *Neuroscience, 55*, 381–393.

Berridge, CW, Bolen, SJ, Manley, MS, & Foote, SL (1996). Modulation of forebrain electroencephalographic (EEG) activity in the halothane-anesthetized rat via actions of noradrenergic beta-receptors located within the medial septal region of the basal forebrain. *J Neurosci, 16*, 7010–7020.

Berridge, CW & Foote, SL (1996). Enhancement of behavioral electroencephalographic (EEG), and electromyographic (EMG) indices of waking following stimulation of noradrenergic beta-receptors located within the medial septal region of the basal forebrain in the unanesthetized rat. *J Neurosci, 16*, 6999–7009.

Berridge, CW & Abercrombie, ED (1999). Relationship between locus coeruleus neuronal discharge rate and rates of norepinephrine efflux in cortex. *Neuroscience, 93*, 1263–1270.

Berridge, CW, O'Neil, J & Wifler, K (1999). Amphetamine acts within the basal forebrain to initiate and maintain alert waking. *Neuroscience, 93*, 1885–896.

Berridge, CW & Morris, MF (2000). Amphetamine-induced activation of forebrain EEG is prevented by noradrenergic beta-receptor blockade in the halothane-anesthetized rat. *Psychopharmacology, 148*, 307–313.

Boyajian, CL & Leslie, FM (1987). Pharmacological evidence for alpha-2 adrenoceptor heterogeneity: differential binding properties of [3H]rauwolscine and [3H]idazoxan in rat brain. *J Pharm Exp Ther, 241*, 1092–1098.

Bylund, DB, Blaxall, HS, Iversen, LJ, Caron, MG, Lefkowitz, RJ, & Lomasney, JW (1992). Pharmacological characteristics of alpha-2 adrenergic receptors: comparison of pharmacologically defined subtypes with subtypes identified by molecular cloning. *Mol Pharm, 42*, 1–5.

Camp, DM, DeJonghe, DK, & Robinson, TE (1997). Time-dependent effects of repeated amphetamine treatment on norepinephrine in the hypothalamus and hippocampus assessed with in vivo microdialysis. *Neuropsychopharmacology, 17*, 130–140.

Carli, M, Robbins, TW, Evenden, JL, & Everitt, BJ (1983). Effects of lesions to ascending noradrenergic neurones on performance of a 5-choice serial reaction task in rats; implications for theories of dorsal noradrenergic bundle function based on selective attention and arousal. *Behav Brain Res, 9*, 361–380.

Chappell, PB, Riddle, MA, Scahill, L, Lynch, K, Schultz, R, Arnsten, AFT, Leckman, JF, & Cohen DL (1995). Guanfacine treatment of comorbid Attention Deficit Hyperactivity Disorder and Tourette's Syndrome: Preliminary clinical experience. *J Am Acad Child Adolesc Psychiatry, 34*, 1140–1146.

Chaulk, PC & Harley, CW (1998). Intracerebroventricular norepinephrine potentiation of the perforant path-evoked potential in dentate gyrus of anesthetized and awake rats: A role for both alpha- and beta-adrenoceptor activation. *Brain Res, 787*, 59–70.

Collier, TJ, Gash, DM, & Sladek, JR Jr. (1988). Transplantation of norepinephrine neurones into aged rats improves performance of a learned task. *Brain Res, 448*, 77–87.

Curtis, AL, Conti, E, & Valentino, RJ (1993). Cocaine effects of brain noradrenergic neurons of anesthetized and unanesthetized rats. *Neuropharmacology, 32*, 419–428.

Devauges, V & Sara, SJ (1990). Activation of the noradrenergic system facilitates an attentional shift in the rat. *Behav Brain Res, 39*, 19–28.

Diaz, J, Elison, G, & Masouka, D (1978). Stages of recovery from central norepinephrine lesions in enriched and impoverished environments: a behavioral and biochemical study. *Exp Brain Res, 31*, 117–130.

Dohlman, HG, Thorner, J, Caron, MG, & Lefkowitz, RJ (1991). Model systems for the study of seven-transmembrane segment receptors. *Ann Rev Biochem, 60*, 653–688.

Engberg, G & Svensson, TH (1979). Amphetamine-induced inhibition of central noradrenergic neurons: a pharmacological analysis. *Life Sci, 24*, 2245–53.

Florin, SM, Kuczenski, R, & Segal, DS (1994). Regional extracellular norepinephrine responses to amphetamine and cocaine and effects of clonidine pretreatment. *Brain Res, 654*, 53–62.

Foote, SL, Freedman, FE, & Oliver AP (1975). Effects of putative neurotransmitters on neuronal activity in monkey auditory cortex. *Brain Res, 86*, 229–242.

Foote, SL, G Aston-Jones, & FE Bloom (1980). Impulse activity of locus coeruleus neurons in awake rats and monkeys is a function of sensory stimulation and arousal. *Proc Natl Acad Sci USA, 77*, 3033–3037.

Foote, SL, FE Bloom, & G Aston-Jones (1983). The nucleus locus coeruleus: New evidence of anatomical and physiological specificity. *Physiol Rev, 63*, 844–914.

Foote, SL & JH Morrison (1987). Extrathalamic modulation of neocortical function. *Ann Rev Neurosci, 10*, 67–95.

Foote, SL, Berridge, CW, Adams, LM & Pineda, JA (1991). Electrophysiological evidence for the involvement of the locus coeruleus in alerting, orienting, and attending. In CD Barnes & O Pompeiano (Eds.), *Progress in Brain Research, Vol. 88* pp. 521–532. Amsterdam: Elsevier Science Publishers BV.

Graham, AW & Aghajanian, GK (1971). Effects of amphetamine on single cell activity in a catecholamine nucleus, the locus coeruleus. *Nature, 234*, 100–2.

Hallman, H, Sundstrom, E, & Jonsson G (1984). Effects of the noradrenaline neurotoxin DSP 4 on monoamine neurons and their transmitter turnover in rat CNS. *J Neural Transm, 60*, 89–102.

Harik, SI, Duckrow, RB, LaManna, JC, Rosenthal, M, Sharma, VK, & Banerjee, SP (1981). Cerebral compensation for chronic noradrenergic denervation induced by locus ceruleus lesion: recovery of receptor binding, isoproterenol-induced adenylate cyclase activity, and oxidative metabolism. *J Neurosci, 1*, 641–9.

Harley, CW & Sara, SJ (1992). Locus coeruleus burst induced by glutamate trigger delayed perforant path spike amplitude potentiation in the dentate gyrus. *Exp Brain Res, 89*, 581–587.

Haroutunian, V, Kanof, PD, Tsuboyama, G, & Davis, KL (1990). Restoration of cholinomimetic activity by clonidine in cholinergic plus noradrenergic lesioned rats. *Brain Res, 507*, 261–266.

Heginbotham, LR & Dunwiddie, TV (1991). Long-term increase in the evoked population spike in the CA1 region of rat hippocampus induced by B-adrenergic receptor activation. *J Neurosci, 11*, 2519–2527.

Hobson, JA, RW McCarley, & PW Wyzinski (1975). Sleep cycle oscillation: reciprocal discharge by two brainstem neuronal groups. *Science, 189*, 55–58.

Hopkins, WF & Johnston, D (1984). Frequency-dependent noradrenergic modulation of long-term potentiation in the hippocampus. *Science, 226*, 350–351.

Hopkins, WF & Johnston, D (1988). Noradrenergic enhancement of long-term potentiation at mossy fiber synapses in the hippocampus. *J Neurophys, 59*, 667–687.

Horrigan, JP & Barnhill, LJ (1995). Guanfacine for treatment of Attention-Deficit Hyperactivity Disorder in boys. *J Child Adolesc Psychopharm, 5*, 215–223.

Hunt, RD, Minderaa, RB, & Cohen, DJ (1985). Clonidine benefits children with Attention Deficit Disorder and Hyperactivity: Report of a double-blind placebo-crossover therapeutic trial. *J Am Acad Child Psychiatry, 5*, 617180–629.

Hunt, RD, Arnsten, AF, & Asbell, MD (1995). An open trial of guanfacine in the treatment of attention-deficit hyperactivity disorder. *J Am Acad Child Adolesc Psychiatry, 34*, 50–54.

Ivanova, S, Rajkowski, J, Silakov, V, Watanabe, T, & Aston-Jones, G (1997). Local chemomanipulations of locus coeruleus (LC) activity in monkeys alter cortical event-related potentials (ERPs) and task performance. *Soc Neurosci Abst, 23*, 1587.

Jackson, WJ & Buccafusco, JJ (1991). Clonidine enhances delayed matching-to-sample performance by young and aged monkeys. *Pharm Biochem Behav, 39*, 79–84.

Jones, CR & Palacios, JM (1991). Autoradiography of adrenoceptors in rat and human brain: alpha-adrenoceptor and idazoxan binding sites. *Prog Brain Res, 88*, 271–291.

Kirkwook, A, Rozas C, Kirkwood, J, Perez, F, & Bear, MF (1999). Modulation of long-term synaptic depression in visual cortex by acetylcholine and norepinephrine. *J Neurosci, 19*, 1599–1609.

Kitchigina, V, Vankov, A, Harley, C, & Sara, SJ (1997). Novelty-elicited, noradrenaline-dependent enhancement of excitability in the dentate gyrus. *Eur J Neurosci, 9*, 41–47.

Kuczenski, R & Segal, DS (1994). Neurochemistry of amphetamine. In AK Cho & DS Segal (Eds.), *Amphetamine and its analogues: psychopharmacology, toxicology and abuse* (pp 81–113). San Diego: Academic Press.

Kuczenski R, Segal, DS, Cho, AK, & Melega, W (1995). Hippocampus norepinephrine, caudate dopamine and serotonin, and behavioral responses to the stereoisomers of amphetamine and methamphetamine. *J Neurosci, 15*, 1308–1317.

Kuczenski, R, & Segal, DS (1997). Effects of methylphenidate on extracellular dopamine, serotonin, and norepinephrine: Comparison with amphetamine. *J Neurochem, 68*, 2032–2037.

Lacroix, D & Ferron, A (1988). Electrophysiological effects of methylphenidate on the coeruleo-cortical noradrenergic system in the rat. *Eur J Pharmacol, 149*, 277–285.

Leslie, FM, Loughlin, SE, Sternberg, DB McGaugh, JL, Young, LE, & Zornetzer, SF (1985). Noradrenergic changes in senescent memory loss. *Brain Res, 359*, 292–299.

Logue, MP, Growdon, JH, Coviella, ILG, & Wurtman, RJ (1985). Differential effects of DSP-4 administration on regional brain norepinephrine turnover in rats. *Life Sci, 37*, 403–409.

Lorden, JF, Rickett, EJ, Dawson, R, & Pelleymounter, MA (1980). Forebrain norepinephrine and the selective processing of information. *Brain Res, 190*, 569–573.

Mair, RG & McEntee, WJ (1986). Cognitive enhancement in Korsakoff's psychosis by clonidine: A comparison with 1-dopa and ephedrine. *Psychopharmacology, 88*, 374–380.

Mallick, BN & Alam, MN (1992). Different types of norepinephrinergic receptors are involved in preoptic area mediated independent modulation of sleep-wakefulness and body temperature. *Brain Res, 591*, 8–19.

Marwaha, J & Aghajanian, GK (1982). Relative potencies of alpha-1 and alpha-2 antagonists in the locus ceruleus, dorsal raphe and dorsal lateral geniculate nuclei: an electrophysiological study. *J Pharm Exp Ther, 222*, 287–93.

McCormick, DA, Pape, HC, & Williamson, A (1991). Actions of norepinephrine in the cerebral cortex and thalamus: implications for function of the central noradrenergic system. *Prog Brain Res, 88*, 293–305.

McCormick, DA, & Bal, T (1997). Sleep and arousal: Thalamocortical mechanisms. *Ann Rev Neurosci, 20*, 185–215.

Menkes, DB & Aghajanian, GK (1981). alpha-1-Adrenoceptor-mediated responses in the lateral geniculate nucleus are enhanced by chronic antidepressant treatment. *Eur J Pharmacol, 74*, 27–35.

Mogilnicka, E (1986). Increase in beta- and alpha-1-adrenoceptor binding sites in the rat brain and in the alpha-1-adrenoceptor functional sensitivity after the DSP-4-induced noradrenergic denervation. *Pharm Biochem Behav 25*, 743–746.

Moises, HC, Waterhouse, BD, & Woodward, DJ (1981). Locus coeruleus stimulation potentiates purkinje cell responses to afferent input: the climbing fiber system. *Brain Res, 222*, 43–64.

Mouradian, RD, Seller, FM, & Waterhouse, BD (1991). Noradrenergic potentiation of excitatory transmitter action in cerebrocortical slices: evidence of mediation by an alpha-1-receptor-linked second messenger pathway. *Brain Res, 546*, 83–95.

Oke, AF & Adams, RN (1978). Selective attention dysfunctions in adult rats neonatally treated with 6-hydroxydopamine. *Pharm Biochem Behav, 9*, 429–432.

Page, ME, Berridge, CW, Foote, SL, & Valentino, RJ (1993). Corticotropin-releasing factor in the locus coeruleus mediates EEG activation associated with hypotensive stress. *Neurosci Lett, 164*, 81–84.

Pineda, JA, Foote, SL, & Neville, HJ (1987). The effects of locus coeruleus lesions on a squirrel monkey late positive component: a preliminary study. *Electroencephalogr. Clin Neurophysiol Suppl, 40*, 481–486.

Pineda, JA, Foote, SL, & Neville, HJ (1989). Effects of locus coeruleus lesions on auditory, long-latency, event-related potentials in monkey. *J Neurosci, 9*, 81–93.

Pitts, DK, & Marwah, J (1987). Electrophysiological actions of cocaine on noradrenergic neurons in rat locus ceruleus. *J Pharm Exp Ther, 240*, 345–351.

Rajkowski, J, Kubiak, P & Aston-Jones, G (1994). Locus coeruleus activity in monkey: Phasic and tonic changes are associated with altered vigilance. *Brain Res Bull, 35*, 607–616.

Roberts, DSC, Price, MTC, & Fibiger, HC (1975). The dorsal tegmental noradrenergic projection: An analysis of its role in maze learning. *J Comp Physiol Psychol, 90*, 363–372.

Robinson, TE & Becker, JB (1986). Enduring changes in brain and behavior produced by chronic amphetamine administration: A review and evaluation of animal models of amphetamine psychosis. *Brain Res Rev, 11*, 157–198.

Robinson, TE & Camp, DM (1987). Long-lasting effects of escalating doses of *d*-amphetamine on brain monoamines, amphetamine-induced stereotyped behavior and spontaneous nocturnal locomotion. *Pharm Biochem Behav, 26*, 821–827.

Robinson, TE & Berridge, KC (1993). The neural basis of drug craving: an incentive-sensitization theory of addiction. *Brain Res Rev, 18*, 247–291.

Schneider, JS & Kovelowski, CJ (1990). Chronic exposure to low doses of MPTP. I. Cognitive deficits in motor asymptomatic monkeys. *Brain Res, 519*, 122–128.

Segal, M, Markram, H, & Richter-Levin, G (1991). Actions of norepinephrine in the rat hippocampus. *Prog Brain Res, 88*, 323–30.

Selden, NRW, Everitt, BJ, & Robbins, TW (1991). Telencephalic but not diencephalic noradrenaline depletion enhances behavioural but not endocrine measures of fear conditioning to contextual stimuli. *Behav Brain Res, 43*, 139–154.

Selden, NRW, Robbins, TW & Everitt, BJ (1990). Enhanced behavioral conditioning to context and impaired behavioral and neuroendocrine responses to conditioned stimuli following ceruleocortical noradrenergic lesions: Support for an attentional hypothesis of central noradrenergic function. *J Neurosci, 10*, 531–539.

Simpson, KL, Altman, DW, Wang, L, Kirifides, ML, Lin, RC-S, & Waterhouse, BD (1997). Lateralization and functional organization of the locus coeruleus projection to the trigeminal somatosensory pathway in rat. *J Comp Neurology, 385*, 135–147.

Skolnick, P, Stalvey, LP, Daly, JW, Hoyler, E, & Davis, JN (1978). Binding of alpha and beta adrenergic ligands to cerebral cortical membranes. Effect of 6-hydroxydopamine treatment and relationship to the responsiveness of cyclic AMP generating systems in two rat strains. *Eur J Pharmacol, 47*, 201–210.

Solanto, MV (1998). Neuropsychopharmacological mechanisms of stimulant drug action in attention-deficit hyperactivity disorder: a review and integration. *Behav Brain Res, 94*, 127–52.

Sporn, JR, Wolfe, BB, Harden, TK, Kendall, T, & Molinoff, PB (1977). Supersensitivity in rat cerebral cortex: pre- and post-synaptic effects of 6-hydroxydopamine at noradrenergic synapses. *Mol Pharm, 13*, 1170–1180.

Stanton, PK & Sarvey, JM (1985). Depletion of norepinephrine, but not serotonin, reduces long-term potentiation in the dentate gyrus of rat hippocampal slices. *J Neurosci, 5*, 2169–2176.

Stanton, PK & Sarvey, JM (1987). Norepinephrine regulates long-term potentiation of both the population spike and dendritic EPSP in hippocampal dentate gyrus. *Brain Res Bull, 18*, 115–119.

Stevens, KE, Fuller, LL, & Rose, GM (1991). Dopaminergic and noradrenergic modulation of amphetamine-induced changes in auditory gating. *Brain Res, 555*, 91–98.

Stone EA (1983). Adaptation to stress and brain noradrenergic receptors. *Neurosci Biobehav Rev, 7* 503–9.

Surprenant, A & Williams, JT (1987). Inhibitory synaptic potentials recorded from mammalian neurones prolonged by blockade of noradrenaline uptake. *J. Phys, 382*, 87–103.

Svensson, TH, BS Bunney, & GK Aghajanian (1975). Inhibition of both noradrenergic and serotonergic neurons in brain by the alpha-adrenergic agonist clonidine. *Brain Res, 92*, 291–306.

Swanson, LW & Hartman, BK (1975). The central adrenergic system. An immunofluorescence study of the location of cell bodies and their efferent connections in the rat utilizing dopamine-beta-hydroxylase as a marker. *J Comp Neuroanat, 163*, 467–506.

Swick, D, Pineda, JA, Holmes, TC, & Foote, SL (1988). Effects of clonidine on P300-like potentials in squirrel monkeys. *Soc Neurosci Abstr, 14*, 1014.

U'Prichard, DC & Kvetnansky, R (1980). Central and peripheral adrenergic receptors in acute and repeated immobilization stress. In E Usdin, R Kvetnansky, & I Kopin, (Eds.) *Catecholamines and stress: recent advances* pp. 299–308. New York: Elsevier/North Holland.

Usher, M, Cohen, JD, Rajkowki, P, & Aston-Jones, G (1999). The role of the locus coeruleus in the regulation of cognitive performance. *Science, 283*, 549–554.

Vankov, A, Herve-Minvielle, A, & Sara, SJ (1995). Response to novelty and its rapid habituation in locus coeruleus neurons of the freely exploring rat. *Eur J Neurosci, 7*, 1180–1187.

Vogt, BA (1991). The role of layer I in cortical function. In EG Jones & A Peters (Eds.) *Cerebral cortex, Vol. 9* (pp. 49–80). New York: Plenum Press.

Waterhouse, BD, Lin, C-S, Burne, RA, & Woodward (1983). The distribution of neocortical projection neurons in the locus coeruleus. *J Comp Neurol, 217*, 418–431.

Waterhouse, BD, Devilbiss, D, Fleischer, D, Sessler, FM, & Simpson, KL (1998a). New perspectives on the functional organization and postsynaptic influences of the locus ceruleus efferent projection system. In DS Goldstein, G Eisenhofer, & R McCarty (Eds.) *Catecholamines: Bridging basic science with clinical medicine* (pp. 749–754) San Diego: Academic Press.

Waterhouse, BD, Moises, HC, & Woodward, DJ (1998b). Phasic activation of the locus coeruleus enhances responses of primary sensory cortical neurons to peripheral receptive field stimulation. *Brain Res, 790*, 33–44.

Woodward, DJ, Moises, HC, Waterhouse, BD, Hoffer, BJ, & Freedman, R (1979). Modulatory actions of norepinephrine in the central nervous system. *Fed Proc, 38*, 2109–2116.

7

Dopaminergic and Noradrenergic Influences on Cognitive Functions Mediated by Prefrontal Cortex

AMY F. T. ARNSTEN

Many researchers consider dysfunction of the prefrontal cortex (PFC) a fundamental problem in Attention-Deficit/Hyperactivity Disorder (ADHD). The PFC is critical for guiding behavior using working memory (Goldman-Rakic, 1987; see Rapport, Chapter 12). Working memory is constantly updated; the PFC uses these representations to guide behavior effectively, freeing the organism from its dependence on the environment, inhibiting inappropriate responses or distractions and allowing us to plan and organize our behavior (Robbins, 1996). The PFC, particularly on the right side, is activated during both sustained and selective attention tasks (Coull et al., 1998). Animals or humans with lesions to the PFC can exhibit poor attention regulation, disorganized behavior, hyperactivity and impulsivity (Stuss et al., 1994). For example, humans with PFC lesions exhibit poor sustained attention, especially when tasks are boring and there are long intervals between stimuli (Wilkins et al., 1987). They also have difficulty gating out irrelevant stimuli and are more distractible (Knight et al., 1989; Woods & Knight, 1986). Lesions to the right medial/orbital PFC often produce deficits in the affective realm such as immature behavior, impulsivity, and increased motor activity (e.g., Blumer & Benson, 1975; Stuss et al., 1992). Patients with PFC lesions are particularly impaired when they must inhibit prepotent but inappropriate responses (e.g., Milner, 1982; Perret, 1974). Thus, one could view the PFC as inhibiting inappropriate responses for both overt behaviors (reducing impulsivity) and covert responses (reducing distractibility).

Although ADHD is a complex and varied disorder, many investigators have noted the resemblance between ADHD patients and subjects with lesions to the PFC (reviewed in Arnsten et al., 1996). Data from neuropsychological studies of ADHD patients were the first to indicate that PFC dysfunction is fundamental to this disorder. ADHD patients are consistently impaired on tests of frontal lobe function (e.g., Continuous Performance, go/no-go, Stroop Interference tasks; see Rapport, Chapter 12, or Barkley, 1997, for review), but show only subtle changes on a test of covert attentional shifting that depends on the functional integrity of the parietal cortex (Swanson et al., 1991; and see Mehta et al., Chapter 13). Recently, Barkley has described in depth the importance of PFC dysfunction to ADHD symptoms, with particular focus on the relationship between PFC and impulse control (Barkley, 1997).

Structural imaging studies have also shown evidence of PFC dysfunction. The right PFC has been shown to be consistently smaller in ADHD patients than in age-matched controls (see Castellanos, Chapter 10). There are also abnormalities in projection zones of the PFC including the caudate nucleus, the cerebellar vermis, and the rostral corpus callosum where PFC fibers cross between hemispheres (ibid). Interestingly, the inability to suppress responses to salient but irrelevant stimuli correlated with reduced volume of the right PFC (Casey et al., 1997). Event-related potential studies have shown that ADHD patients (Satterfield et al., 1988), like PFC-lesioned subjects (Woods & Knight, 1986), show reduced attention-related negativity (Nd) during a selective attention task. Similarly, ADHD subjects show deficits in visual evoked potentials during a continuous performance task associated with deficits in prefrontal processing (Silberstein et al., 1998). Functional imaging studies in adults and adolescents with ADHD have often shown a pattern of *decreased* activity in the PFC and striatum and *increased* activity in the posterior sensory/motor cortices that can be normalized by methylphenidate treatment (see Castellanos, Chapter 10). SPECT imaging studies of ADHD children and adolescents have also shown reduced perfusion of the PFC compared with control subjects both during a task requiring concentration and at rest (Amen & Carmichael, 1997). A functional MRI study of unmedicated ADHD adolescents found reduced activity in the right PFC as subjects carried out motor inhibition or timing tasks (Rubia et al., 1999). In contrast, another functional MRI study showed increased activity in the PFC of children with ADHD as they performed a response inhibition task (Vaidya et al., 1998). The differences between these two studies may have been due to task difficulty. Stimulant treatment enhanced performance of the response inhibition task and increased PFC activity (Vaidya et al., 1998). Thus, multiple imaging techniques have shown altered PFC activity in ADHD subjects. Intriguingly, a recent PET study has shown reduced fluorodopa binding in the PFC of adult ADHD patients, indicative of altered NE and/or DA transmission in the PFC in this disorder (Ernst et al., 1998). Thus, both imaging and neuropsychological data demonstrate that the PFC and its subcortical targets contribute significantly to this disorder.

Research in animals has shown that the PFC is tremendously sensitive to its neurochemical environment, and that small changes in neuromodulators can produce large changes in PFC function. In particular, the catecholamines dopamine and norepinephrine can have marked effects on the working memory functions of the PFC. It is likely that at least some of the beneficial effects of stimulant medications in ADHD patients result from altered catecholamine receptor stimulation in the PFC. This chapter will provide a review of how the catecholamines dopamine and norepinephrine influence PFC function.

Catecholamines Are Essential to the Working Memory Functions of the PFC

Catecholamine Depletion

The work of Brozoski et al. (1979) was the first to demonstrate that catecholamines play a critical role in the modulation of the spatial working memory functions of the PFC. DA and NE depletion in the PFC was produced by infusion of the catecholamine neurotoxin, 6-hydroxy-dopamine (6-OHDA), into the dorsolateral PFC of monkeys. Animals with large catecholamine depletion of the PFC were as impaired as those with PFC ablations, highlighting the importance of catecholamine modulatory influences. Animals were not impaired on performance of a non-PFC task, visual pattern discrimination, implicating an effect of catecholamine depletion on PFC cognitive function rather than nonspecific performance deficits. The Brozoski study focused on the importance of DA to PFC function, as monkeys with large NE depletion and small DA depletion did not show deficits. However, it is now appreciated that both catecholamines are important to PFC function, and it is likely that both must be substantially depleted to produce marked impairment. The landmark Brosozki study has been replicated in rats with 6-OHDA lesions of the medial PFC (Simon, 1981), and in marmosets with 6-OHDA lesions to the dorsolateral PFC (Collins et al., 1998; Roberts et al., 1994).

Interestingly, although catecholamine depletion in the dorsolateral PFC of marmosets impaired spatial working memory, it improved performance of a set-shifting task (Roberts et al., 1994), and had no effect on a self-ordered sequencing task (Collins et al., 1998), even though both tasks are sensitive to excitotoxic lesions of the dorsolateral PFC. These data raise the fascinating possibility that different operations within the same region of PFC may have differing neurochemical needs. If this is true, different PFC symptoms in ADHD may require varying doses of stimulant for optimal beneficial effect. This is an important area for future research.

Spatial working memory deficits with preserved visual discrimination function have also been observed with global catecholamine depletion. For example, systemic, chronic reserpine treatment impairs delayed response without altering discrimination performance in young adult monkeys (Cai et al., 1993). Furthermore, aged monkeys and rats with naturally occurring catecholamine depletion exhibit prominent spatial working memory deficits and generally spared performance on discrimination tasks (Bartus et al., 1978; Luine et al., 1990. In both young [Sahakian, et al., 1985) and aged (Luine et al., 1990) rats, cognitive performance correlated with levels of PFC catecholamines. Pharmacological studies in aged animals and young, depleted animals indicate that both DA and NE facilitate PFC working memory function through actions at D1 and alpha-2 adrenergic receptors, respectively.

DA Enhances PFC Working Memory Function via D1 Receptor Stimulation

Brososki and colleagues illustrated the importance of DA mechanisms to PFC working memory function by showing that the mixed D1/D2 agonist, apomorphine, could ameliorate delayed alternation deficits in their catecholamine-depleted monkeys (Brozoski et al., 1979). This interpretation was reinforced by electrophysiological findings in monkeys performing a working memory task, where iontophoresis of DA onto PFC neurons enhanced memory-related cell firing during the delay period (Sawaguchi et al., 1988). DA can also excite PFC neurons that respond to the cue signal and/or to the go signal informing the monkey it is time to respond for reward (Sawaguchi et al., 1990). These data are quite consistent with recordings from DA cell bodies in monkeys performing a working memory task, where DA cells fired to cues associated with reward (Schultz, 1998; Schultz et al., 1993). In highly trained monkeys, the DA neurons fired to the visuospatial cue signal. One thus could speculate that DA release is enhanced in the PFC at the initiation of each trial by the cue signal, in time to modulate cue-, delay-, and go-related activity. Note that this speculation assumes long-acting effects of DA on PFC cells, that is, over many seconds, presumably via second-messenger actions.

Research in monkeys and rodents has now established the importance of the D1 receptor family in the regulation of PFC function (Fig. 7–1). Recordings from PFC neurons in monkeys performing a working memory task first suggested that D1 family mechanisms may be more influential than D2 family mechanisms for working memory function, as mixed D1/D2 family antagonists suppressed delay-related activity of PFC neurons, while a more selective D2 family antagonist was ineffective (Sawaguchi et al., 1988). This hypothesis was supported by behavioral studies that showed that infusions of selective D1, but not D2 /D3, antagonists into the PFC markedly impaired spatial working memory (Sawaguchi & Goldman-Rakic, 1991). Impaired working memory with D1 receptor antagonist infusion has now been replicated in rat PFC (Seamans et al., 1998). Similarly, performance of an attention task is impaired in rats given intra-PFC infusions of D1 but not D2 antagonists (Robbins et al., 1998). Systemic administration of a D1 receptor antagonist can also impair working memory in monkeys (Arnsten et al., 1994) and rats (Murphy et al., 1996a; Figure 7–1A), as can low doses of a D2/D3/D4 agonist that inhibit DA release (Arnsten et al., 1995). Conversely, working memory can be improved by low doses of D1 agonists. For example, systemic administration of low—but not high—doses of D1 agonist improves spatial working memory performance in young monkeys or in aged monkeys with naturally occurring DA depletion (Arnsten et al., 1994; Cai & Arnsten, 1997; Figure 7–1B). Similarly, intra-PFC infusions of low, but not high, doses of a partial D1 agonist improve attentional performance in rats (Robbins et al., 1998). Selective, full D1 agonists have only recently become available for research in animals, and are not yet available for human use; thus, less is known about D1 influences on cognitive function in humans. However, the mixed D1/D2 family agonist, pergolide, appears to produce more consistent improvement in spatial working memory in humans than

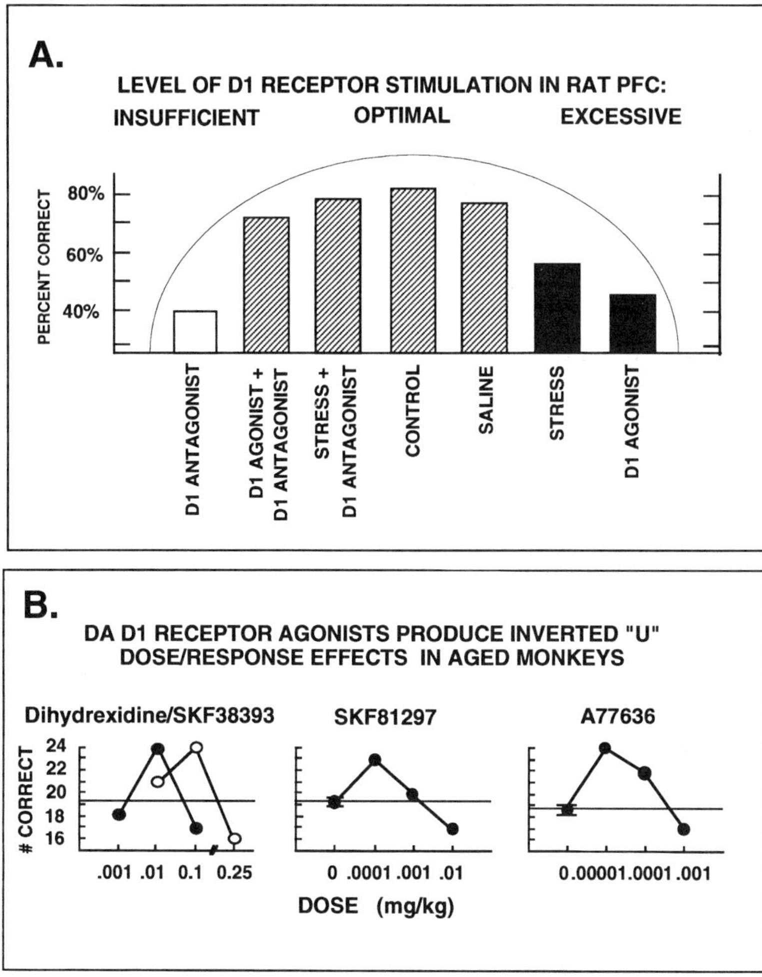

Figure 7–1. DA D1 receptor mechanisms produce an inverted U-shaped dose response curve, where either too little or too much D1 receptor stimulation impairs working memory function. *A.* The effects of D1 receptor agents and stress on delayed alternation performance in rats. Either pharmacological stress (FG7142, 20 mg/kg, i.p.) or infusion of a D1 agonist (0.1 μg/0.5 μl SKF81297) into the PFC impairs performance. These impairments were blocked by pretreatment with a D1 receptor antagonist, SCH23390, which by itself impaired performance at higher doses. Adapted from Murphy et al., 1996a, and Zahrt et al., 1997. *B.* The effects of D1 agonists on delayed response performance in aged monkeys with naturally occurring DA depletion. Low doses improve performance and higher doses impair below baseline levels of responding. Both the improvement and impairment are reversed by D1 receptor antagonist pretreatment (not shown). (Adapted from Arnsten et al., 1994 and Cai & Arnsten, 1997.)

does the D2 family agonist, bromocriptine, suggesting that mechanisms observed with D1 agonists in animals may extend to humans as well (Kimberg et al., 1997; Muller et al., 1998).

The role of the D2 family of receptors in PFC working memory function is less clear. D2/D3 compounds have opposing actions at pre- vs. postsynaptic receptors, and often produce marked side effects (e.g., dyskinesias, nausea) when administered systemically, which can obscure cognitive changes. Further discussion of D2 drug effects can be found in Chapter 13 by Mehta et al.

NE Enchances PFC Function via Post-Synaptic Alpha-2 Receptor Stimulation

Studies in rodents, monkeys, and humans have all shown that NE has an important beneficial influence on PFC function through its actions at postsynaptic alpha-2 receptors (reviewed in Arnsten et al., 1996; Coull, 1994; Fig. 7–2). Young monkeys with working memory impairment induced by local PFC (Arnsten & Goldman-Rakic, 1985) or global (Cai et al., 1993) catecholamine depletion are greatly improved by treatment with alpha-2 agonists such as clonidine and guanfacine. The beneficial effects of alpha-2 NE agonists also have been observed in aged monkeys (Arnsten et al., 1988; Arnsten & Goldman-Rakic, 1985; Rama et al., 1996) and aged rats (Carlson et al., 1992) with naturally occurring catecholamine loss. Alpha-2 agonists also improve working memory performance in intact, young monkeys (Fig. 7–3), but at higher doses than those needed to improve aged or depleted animals (Franowicz & Arnsten, 1998). Indeed, the greater the loss of NE, the lower the dose of alpha-2 agonist needed to improve PFC performance (Franowicz & Arnsten, 1999), a pattern consistent with drug actions at supersensitive, post-synaptic receptors. Improvements with alpha-2 agonists can be reversed with alpha-2, but not alpha-1, antagonists, and alpha-2 antagonists by themselves impair PFC function, consistent with an alpha-2 receptor mechanism (Arnsten & Goldman-Rakic, 1985).

Pharmacological profiles further indicate that the alpha-2A receptor subtype may be the most critical for cognitive enhancement (see Arnsten et al., 1996, for review). For example, the alpha-2A selective agonist, guanfacine, is the most effective compound in enhancing working memory without side effects (Arnsten et al., 1988; Rama et al., 1996). Guanfacine is about 10 times weaker than clonidine in inhibiting firing of the NE cell bodies in the locus coeruleus (LC) or in decreasing NE release (Engberg & Eriksson, 1991), but is 10 to 100 times more potent in improving working memory in aged monkeys (Arnsten et al., 1988). Recent studies in genetically altered mice emphasize the importance of the alpha-2A receptor subtype, as mice with a mutation of the alpha-2A subtype no longer show beneficial effects on working memory (Franowicz et al., 1998), while knockout of the alpha-2C subtype has no effect on drug response (Tanila et al., 1999).

The second-messenger mechanisms underlying the beneficial effects of alpha-2 agonists on PFC function are currently unknown. However, alpha-2 receptors are commonly coupled to G_i proteins, which inhibit adenylyl cyclase/cAMP/pathways (schematically depicted in Fig. 7–2; Duman & Nestler, 1995). As activation of the

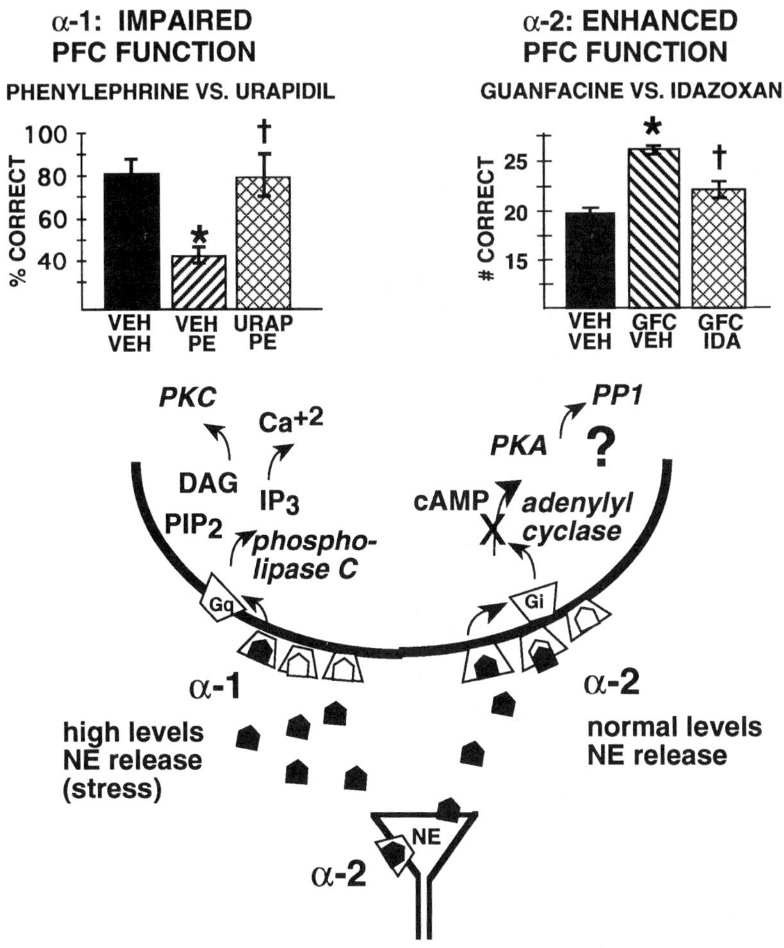

Figure 7–2. Basal levels of NE may enhance PFC functions through actions at postsynaptic alpha-2 receptors; higher levels of NE may engage alpha-1 receptors and impair working memory performance. For example, infusion of the alpha-1 agonist phenylephrine into the PFC of rats impairs delayed alternation performance and this deficit is reversed by co-infusion of the alpha-1 antagonist, urapidil (upper left graph; adapted from Arnsten et al, 1999). Evidence to date suggests that this impairment arises from activation of the protein kinase C pathway (Birnbaum & Arnsten, unpublished data). In contrast, alpha-2 agonists such as guanfacine improve working memory performance. Systemic administration of high doses of guanfacine (0.1-1.0 mg/kg) can even improve the performance of "normal" young adult monkeys, and this response is reversed by the alpha-2 antagonist idazoxan (upper right graph; adapted from Franowicz & Arnsten, 1998). We are currently examining whether this improvement involves inhibition of the cAMP/PKA pathway.

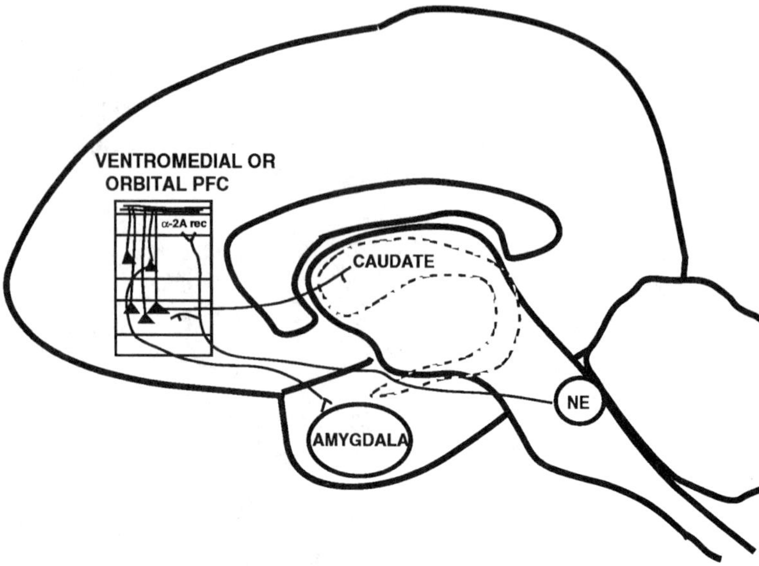

Figure 7–3. Hypothetical circuits whereby alpha-2A agonists such as guanfacine may act to inhibit impulsivity and to calm behavior. The medial and orbital PFC in monkeys are especially important for response inhibition (although it is likely that the dorsolateral PFC also contributes). Alpha-2A agonists are able to improve response inhibition in monkeys (Steere & Arnsten, 1995), and may do so by enhancing the functions of the medial and orbital PFC. The PFC projections to caudate and amygdala may be particularly important in this regard. We have speculated that more dorsal areas of PFC may inhibit distractibility (i.e., covert responses) through inhibitory interactions with posterior sensory cortices (Arnsten et al., 1996).

adenylyl cyclase/cAMP/pathway seems to impair PFC function, inhibition of this intracellular signaling pathway may contribute to the beneficial effects of alpha-2 agonists on PFC cognitive function.

Alpha-2 agonists improve the performance of tasks that challenge the PFC, but often have little beneficial effect under conditions that do not challenge the PFC. For example, alpha-2 agonists are particularly effective in enhancing working memory during distracting conditions (Arnsten & Contant, 1992; Jackson & Buccafusco, 1991), a finding consistent with earlier studies showing that forebrain NE depletion increases distractibility (Carli et al., 1983). Alpha-2 agonists have been shown to improve working memory for both visuospatial (Arnsten et al., 1988) and visuo-feature (Jackson & Buccafusco, 1991) cues, suggesting enhancement of both dorsolateral and ventrolateral PFC function. Higher doses of guanfacine have also been shown to improve performance of an object reversal task associated with orbital PFC function (Steere & Arnsten, 1997). In contrast, alpha-2 agonists do not improve, and sometimes impair, cognitive functions associated with parietal and temporal lobe functions. For example, these drugs have no effect or impair the spatial memory functions of the hippocampus (Sirviö et al., 1991), the visual feature discrimination memory functions of the inferior temporal cortex (Arnsten & Goldman-Rakic, 1985; Steere & Arnsten, 1997), the visual feature recognition memory functions of the perirhinal cortex (Arnsten & Goldman-Rakic, 1990a), and the covert visual-spatial attention-shifting functions of the parietal cortex (Witte & Marrocco, 1997).

Similar profiles with PFC vs. non-PFC tasks have been observed in humans administered alpha-2 agonists (see Arnsten et al., 1996; Coull, 1994 for earlier reviews). Most of this research has been performed with clonidine, a suboptimal drug due to its potent inhibition of LC firing and prominent side effect profile. Clonidine has mixed effects in healthy, young adults, presumably due to competing pre- vs. postsynaptic effects and dose limitations from side effects (Coull, 1994; Jakala et al., 1999a). However, the more selective alpha-2A agonist guanfacine has recently been shown to improve working memory, planning and paired associates learning tasks in healthy young adults (Jakala et al., 1999a; Jakala et al., 1999b). Alpha-2 agonists have also been shown to improve performance of PFC tasks, but not nonPFC tasks, in patients with PFC dysfunction such as Korsakoff's amnesia (Mair & McEntree, 1986), schizophrenia (Fields et al., 1988), and Alzheimer's disease (Riekkinen & Riekkinen, 1999). Interestingly, the alpha-2 antagonist idazoxan has also been shown to enhance performance of PFC tasks in patients with frontal lobe dementia (Coull et al., 1996). Monkey studies suggest that low doses of alpha-2 antagonists can improve PFC function by preferentially increasing NE release and enhancing endogenous stimulation of postsynaptic alpha-2 receptors (Arnsten & Cai, 1993).

A number of studies in rodents and monkeys demonstrate that alpha-2 compounds act directly in the PFC to alter working memory function. As with systemic injection, direct infusion of alpha-2 antagonists, but not alpha-1 or beta antagonists, into the dorsolateral PFC produces a delay-related impairment in spatial working memory (Li & Mei, 1994), demonstrating that endogenous NE stimulation of alpha-2 receptors in the PFC is critical to working memory performance. Conversely, intra-PFC infusion of alpha-2 agonists improves working memory performance in either young (Mao et al., 1999) or aged monkeys (Arnsten, 1997) or in aged rats (Tanila et al., 1996). Electrophysiological studies have observed similar findings at the cellular level. Iontophoresis of the alpha-2 antagonist yohimbine reduces delay-related activity in PFC neurons of monkeys performing spatial working memory tasks (Li et al., 1999; Sawaguchi, 1998). Conversely, systemic clonidine administration enhances delay-related firing in PFC cells, and this effect is reversed by iontophoretic application of yohimbine (Li et al., 1999). Thus, NE actions at alpha-2 receptors in PFC have powerful effects on delay-related firing, the presumed neuronal substrate of working memory function. It is possible that the beneficial effects of alpha-2 receptor stimulation occur on dendritic spines (Fig. 7–4), as alpha-2A receptors have been localized on the postsynaptic membranes of spines in monkey PFC (Aoki et al., 1994; Arnsten et al., 1996). However, there are no electrophysiological data to support or refute this hypothesis.

Imaging studies are also consistent with alpha-2 adrenergic mechanisms enhancing PFC function. Administration of high doses of clonidine to Korsakoff's patients increased frontal regional cerebral blood flow that correlated with improved verbal

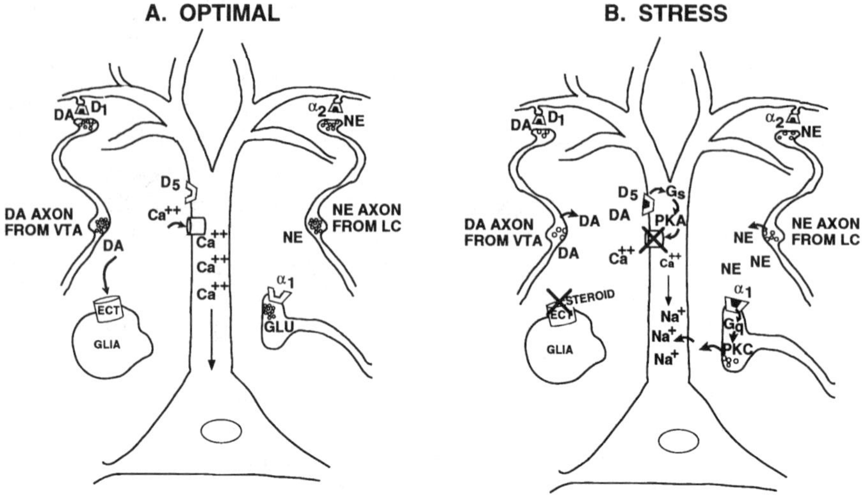

Figure 7–4. Highly schematized representation of possible catecholamine actions on a PFC pyramidal cell. See text for more complete description. (*A*) Under optimal conditions, moderate levels of NE and DA may act on alpha-2 and D1 receptors, respectively, localized on dendritic spines to enhance signal reception. Signals are then conveyed from the dendritic tree to the soma via high threshold calcium currents along the dendritic stem. Extraneuronal catecholamine transporters (ECT) on glia remove excess catecholamines in the extrasynaptic space and thus minimize DA or NE actions on the dendritic stem. (*B*) During stress exposure, catecholamine release is increased in the PFC. We speculate that steroid blockade of ECT on glia may contribute to the rise in extrasynaptic catecholamine levels during stress. DA may engage the D5 receptors concentrated on the dendritic stem and decrease calcium currents through n or p calcium channels, thus reducing signal transfer as observed by Yang and Seamans (1996). High levels of NE may engage alpha-1-adrenoceptors, inducing glutamate release and sodium entry into the dendritic stem (Marek & Aghajanian, 1999). This increase in "noise" would further obscure signal transfer to the cell body. Note that this model suggests that the effects of catecholamines on PFC cognitive function may not depend simply on the overall levels of catecholamine release in PFC, but on the cellular location of stimulated receptors. Conditions that reduce dendritic spines (aging, schizophrenia) may leave the individual particularly vulnerable to the detrimental effects of catecholamines.

fluency performance (Moffoot et al., 1994). However, studies of normal individuals administered lower clonidine doses generally showed changes in thalamus likely related to clonidine's sedating actions (Coull et al., 1997). In contrast, guanfacine enhances PFC metabolic activity in healthy adults as measured by PET imaging (B. Swartz, personal communication). We have also observed increased regional cerebral blood flow in the dorsolateral PFC of monkeys treated with guanfacine prior to performing a working memory task (Avery et al., 2000). Thus, mechanisms observed in animals appear to extend to humans.

In summary, an increasing body of evidence supports the view that NE has powerful beneficial effects on PFC function through actions at postsynaptic alpha-2A receptors in the PFC. Studies to date suggest these effects may apply to many aspects of PFC function, including working memory, behavioral inhibition, verbal fluency, and planning. These effects may explain why guanfacine has had therapeutic effects in ADHD patients in both open-label trials (Chappell et al., 1995; Horrigan & Barnhill, 1995; Hunt et al., 1995), and now in controlled trials as well (trial in ADHD adults, guanfacine favorable with dexedrine, Taylor & Russo, 2000; trial in ADHD children with tics, guanfacine improved tics and ADHD symptoms, L. Scahill, personal communication). The importance of alpha-2 mechanisms to PFC function may also be relevant to the recent finding that polymorphisms in alpha-2 receptors are linked with ADHD symptoms (Comings et al., 1999).

Guanfacine's therapeutic effects may also be viewed in terms of PFC-striatal-thalamic-PFC circuits. Studies in monkeys have accentuated differing contributions of the dorsal PFC and ventral PFC circuits, the dorsal circuits regulating behavior in the sensory/attentional realms and the ventral circuits regulating behavior in the affective realm (Dias et al., 1997; Goldman-Rakic, 1987). We have speculated that guanfacine may enhance concentration and reduce distractibility (covert responding to inappropriate stimuli) by strengthening the inhibitory influences of the dorsolateral PFC on the posterior sensory cortices (Arnsten et al., 1996). Similarly, guanfacine may reduce impulsivity (overt responding to inappropriate stimuli) by strengthening the inhibitory influences of the ventromedial/orbital PFC (schematically depicted in Fig. 7–3). Lesion studies in monkeys have shown that the ventral PFC is particularly important for response inhibition (Dias et al., 1997; Mishkin, 1964), and guanfacine can improve performance of a response inhibition task (Steere & Arnsten, 1997).

High Levels of Catecholamines Impair the Working Memory Functions of the PFC

Small increases in catecholamine receptor stimulation can enhance performance in young adult animals. For example, very low doses of D1 agonists can significantly improve delayed response performance in young adult monkeys (Arnsten et al., 1994), and a low dose of a partial D1 agonist (0.03 μg) infused into the PFC can improve attentional performance in young rats (Robbins et al., 1998). Systemic administration of low, but not high, doses of the alpha-2 antagonist, yohimbine, improve performance in young monkeys by increasing endogenous NE release and stimulating postsynaptic alpha-2 receptors (Arnsten & Cai, 1993). Similarly, low, but not high doses of methylphenidate can improve PFC function in monkeys (Prendergast et al., 1998; see discussion below). However, large increases in catecholamine receptor stimulation in PFC are associated with working memory deficits.

Stress Impairs PFC Working Memory Function

It has been appreciated for more than 20 years that exposure to even relatively mild, uncontrollable stress greatly increases catecholamine release in the PFC (Thierry et al., 1976). Although most studies have focused on the increase in DA release (e.g., Deutch & Roth, 1990), stress exposure also increases the release of several other neuromodulators in the PFC, including NE and serotonin (Finlay et al., 1995; Goldstein et al., 1996).

Studies in monkeys, rats and humans have shown that acute exposure to mild uncontrollable stress impairs cognitive functions associated with the PFC. Thus, exposure to either an environmental stress (loud noise, Arnsten & Goldman-Rakic, 1998) or a pharmacological stressor (Birnbaum et al., 1999; Murphy et al., 1996a), impairs performance of spatial working memory tasks in monkeys and rats (Fig. 7–1A). In contrast, stress exposure has little effect on, or actually improves performance of, tasks dependent on the inferior temporal cortex (Arnsten & Goldman-Rakic, 1998), parietal cortex (Murphy et al., 1996a), or cerebellum (Shors et al., 1992). A similar profile has been observed in human studies. The original work of Broadbent, Hockey and others showed that exposing human subjects to loud noise stress improved reaction time on well-rehearsed or simple tasks, but impaired performance of more complex tasks (Broadbent, 1971), especially when the subjects experienced themselves as having no control over the stressor (Glass et al., 1971). Noise stress impaired the ability to sustain attention (Hockey, 1970), or inhibit prepotent, inappropriate responses on the Stroop interference test (Hartley & Adams, 1974).

Accumulating data from animal studies indicates that the cognitive deficits observed during stress result from excessive stimulation of DA D1 and NE alpha-1 receptors in the PFC.

DA Impairs PFC Working Memory Function via D1 and Possibly D4 Receptor Stimulation

The working memory deficits induced by stress exposure in rats or monkeys can be blocked by pretreatment with DA receptor antagonists, including the D1 receptor antagonist, SCH23390 (Arnsten & Goldman-Rakic, 1990b; Arnsten & Goldman-Rakic, 1998; Murphy et al., 1996a; Fig. 7–1A). Working memory deficits can also be averted by pretreatment with agents such as clonidine that prevent stress-induced increases in DA turnover (Murphy et al., 1996b). Stress-induced working memory deficits correlate with the rise in DA turnover in PFC (NE was not measured, Murphy et al., 1996a). Conversely, stress-induced working memory deficits can be mimicked by administering D1 receptor agonists. Although the systemic administration of low doses of D1 agonists improves working memory performance in young and aged monkeys (Arnsten et al., 1994; Cai & Arnsten, 1997), higher doses impair performance without any evidence of side effects. Furthermore, infusion of a full D1 agonist (0.1 μg) directly into the PFC in rats impairs working memory performance and produces a mildly perseverative pattern of response, much as is seen in stressed rats (Zahrt et al., 1997; Fig. 7–1A). These deficits can be reversed by treatment with a D1 receptor antagonist (Arnsten et al., 1994; Cai & Arnsten, 1997; Zahrt et al., 1997; Fig. 7–1A), consistent with actions at D1 receptors.

Electrophysiological studies of PFC neurons have also shown that high levels of D1 receptor stimulation can erode neuronal function. For example, the iontophoresis of low concentrations of D1 receptor antagonists can *enhance* memory-related neuronal responses in monkeys performing a challenging working memory task (Williams & Goldman-Rakic, 1995). Conversely, intracellular recordings of pyramidal cells from rodent PFC slices have shown that D1 receptor stimulation decreases the n and p calcium currents that convey signals from dendrite to soma (Yang & Seamans, 1996). Optimal levels of D1 receptor stimulation appear to focus signal transmission, conveying only large or temporally coincident signals to the cell body (ibid). However, higher concentrations of dopamine or a D1 agonist abolish calcium currents, effectively "strangling" information transfer from dendrite to soma (Zahrt et al., 1997; Fig. 7–4). This interruption of information transfer may underlie the working memory impairment seen at high levels of D1 receptor stimulation. Animals may become perseverative, as no new spatial information can access the soma to change response patterns.

We are currently testing the hypothesis that high levels of D1 receptor stimulation impair PFC cognitive function by activating a cAMP-dependent protein kinase A (PKA) intracellular pathway. Surmeier and colleagues have shown that D1 receptor stimulation decreases n and p calcium currents in dendrites of striatal neurons through a cAMP/PKA/protein phosphatase 1 intracellular signaling cascade (Surmeier et al., 1995). For example, either a D1 agonist or the PKA activator, Sp-cAMPS, decrease n and p calcium currents (ibid). We examined whether intra-PFC infusion of Sp-cAMPS, like a D1 agonist, would impair working memory performance (Taylor et al., 1999). As predicted, Sp-cAMPS produced a dose-related impairment in delayed alternation performance that could be reversed by the PKA inhibitor, Rp-cAMPS (ibid). The Sp-cAMPS–treated animals exhibited a slightly perseverative pattern of response, just as had been observed with D1 agonist infusion or mild stress. These data suggest the exciting possibility that activation of cAMP/PKA intracellular signaling cascades in the PFC contribute to PFC dysfunction (Fig. 7–4). Interestingly, guanfacine is even more potent than clonidine in preventing stress-induced cognitive deficits (Birnbaum & Arnsten, 1996), even though guanfacine is weaker than clonidine in reducing catecholamine release and reducing LC cell firing (Engberg & Eriksson, 1991). These data suggest the intriguing possibility that guanfacine may protect cognitive performance by engaging G_i proteins and reducing cAMP/PKA activity in PFC.

The role of the D2 family of receptors in the modulation of working memory is less well understood. Agents with potent D2 or D4 receptor-blocking activity such as haloperidol and clozapine are effective in preventing stress-induced cognitive deficits (Arnsten & Goldman-Rakic, 1998; Murphy et al., 1997; Murphy et al., 1996a), but these agents have many other receptor actions, including D1 and alpha-1 receptor blockade. Recently, we have shown that a selective D4 receptor antagonist can reverse stress-induced cognitive deficits in monkeys (Arnsten et al., 2000), and rat studies have shown that infusion of a D2 receptor family agonist into the PFC impairs working memory in rats (Druzin et al., 1999). These findings suggest that high levels of D2/D3/D4 receptor stimulation may also contribute to cognitive deficits.

NE Impairs PFC Working Memory Function via Alpha-1 Receptor Stimulation

New evidence indicates that high concentrations of NE similarly impair PFC function and do so through activation of alpha-1 receptors (Figure 7–2). NE has higher affinity for alpha-2A than alpha-1 receptors: alpha-2A receptors—56 nM (O'Rourke et al., 1994); alpha-1 receptors—330 nM (Mohell et al., 1983). Thus it is likely that low levels of NE (e.g., under basal or nonstress conditions) preferentially engage alpha-2 receptors and improve PFC function, while during conditions of high NE release, alpha-1 receptors would become engaged and override the effects of alpha-2 receptor stimulation. It is well established that high levels of NE are released in the PFC during stress exposure (Finlay et al., 1995), and recent evidence suggests that these high NE levels stimulate alpha-1 receptors and impair PFC function (Birnbaum et al., 1999). Thus, stress-induced cognitive deficits were blocked by infusion of the alpha-1 receptor antagonist, urapidil, into the PFC prior to cognitive testing (Birnbaum et al., 1999). Infusions of urapidil had no effect under nonstress conditions presumably due to little endogenous NE alpha-1 receptor stimulation during nonstressful conditions.

The effects of stress on working memory performance can be mimicked by infusion of an alpha-1 receptor agonist into the PFC. Infusions of the alpha-1 agonist phenylephrine into the PFC in rats markedly impaired working memory performance (Fig. 7–2; Arnsten et al., 1999). This impairment was reversed by co-infusion of the alpha-1 receptor antagonist, urapidil, consistent with actions at alpha-1 receptors. Similar effects have been observed in monkeys, where infusions of phenylephrine into the dorsolateral PFC produced a delay-related impairment in working memory performance (Mao et al., 1999). Infusions were most effective in the caudal two-thirds of the principal sulcal cortex (ibid), the cortical region most tightly associated with spatial working memory performance in monkeys (Goldman & Rosvold, 1970).

Alpha-1 receptors are commonly coupled to the phosphatidyl inositol/protein kinase C (PKC) intracellular pathway via Gq proteins (schematically depicted in Fig. 7–2; Duman & Nestler, 1995). Evidence to date suggests that alpha-1 receptor stimulation impairs PFC function through activation of this second messenger pathway. For example, the cognitive impairment induced by phenylephrine infusions into the rat PFC can be completely reversed by pretreatment with a dose of lithium known to suppress phosphatidyl inositol turnover (Arnsten et al., 1999). However, lithium can alter other second messenger pathways; thus, current studies in animals are focusing on agents that selectively target molecules in the phosphatidyl inositol/PKC cascade. For example, intra-PFC infusion of the PKC inhibitor chelerythrine appears to block the detrimental effects of alpha-1 agonists (S Birnbaum & A Arnsten, unpublished results). These results are consistent with activation of the phosphatidyl inositol/PKC pathway underlying alpha-1 receptor-mediated impairment of PFC cognitive function.

The intracellular recordings of Marek and Aghajanian (1999) suggest an interesting mechanism by which alpha-1 receptor stimulation may impair PFC cognitive functioning. Their studies have shown that NE acts at alpha-1-adrenoceptors to increase glutamate release, which in turn increases excitatory postsynaptic sodium currents in the proximal dendritic stem (schematically represented in Figure 7–4). In other words, alpha-1 receptor stimulation may erode signal transfer by increasing noise in the dendritic stem. This increase in background noise would obscure signal transmission by decreasing the signal/noise ratio of cell response, and would prevent the normal increase in delay-related activity of the soma needed to guide behavior during working memory tasks. In summary, DA D1 receptor stimulation would decrease "signal," whereas NE alpha-1 stimulation may increase "noise." It is likely that these NE and DA actions may be additive or synergistic, and that the signal may be able to survive either modest reduction by DA or modest noise by NE, but could not survive both. This hypothesis is supported by the finding that either a D1 (Arnsten & Goldman-Rakic, 1998; Murphy et al., 1996a) or an alpha-1-adrenoceptor antagonist (Birnbaum et al., 1999) can protect performance from the detrimental effects of stress.

Stress-induced PFC Dysfunction: Relevance to ADHD

The finding that exposure to stress impairs PFC function may explain why highly structured, low stress environments can be especially helpful for subjects with ADHD. These environments may produce the conditions that optimize the neurochemical environment in the PFC, thus helping to reduce impulsivity and distractibility.

The literature demonstrates an association between exposure to stress and symptoms of ADHD. Exposure to adverse home environments increases the risk of ADHD in children (Biederman et al., 1995). Children with multiple stressors are particularly vulnerable. Our new neurobiological perspective suggests that these behavioral problems in children could arise from stress-induced PFC dysfunction.

It is possible that ADHD-like symptoms can arise either from external factors, such as living in an adverse home environment, or from inherent changes in PFC circuits, such as genetic changes in dopamine or norepinephrine transporters or receptors, or smaller size of the PFC itself. Given the requirements of DSM-IV, we may diagnose children with inherent changes as having ADHD, while those with visible causes for their behavioral problems would be less likely to be given this diagnosis. It is hoped that one would try to help the children in the latter group by identifying and lessening their stressors rather than by medication. However, current research suggests that neurochemical changes in the PFC could underlie these problems irrespective of the cause. The similarity between ADHD symptoms and stress-induced PFC deficits may help to explain why ADHD is often not taken seriously as a true biological disorder—ADHD is often dismissed as simply the result of "bad parenting." Understanding the biological underpinnings of these symptoms may facilitate respect for the genuine nature of this disorder.

The Effects of Stimulants on PFC Function

Although there have been numerous studies on the effects of stimulants on long-term memory consolidation, there have been surprisingly few studies evaluating the effects of stimulants on working memory and/or PFC function in monkeys. Indeed, many studies of stimulants in monkeys have focused on very high dose, chronic regimens as potential models of schizophrenia and are not relevant to ADHD. A few early studies found that relatively high doses of d-amphetamine (0.2–0.4 mg/kg i.m.; Bauer & Fuster, 1978; Glick & Jarvik, 1970) or methylphenidate (0.4–0.8 mg/kg, i.m.; Bartus, 1979) impaired working memory performance. A lower dose of methylphenidate (0.1 mg/kg) had no effect on performance (Bartus, 1979). However, these monkeys were tested under conditions with minimal distractions (in the dark) and with very long delays, that is, conditions that challenge medial temporal cortical function. More recently, investigators have tested monkeys in distracting conditions using lower stimulant doses and found enhanced performance following 0.125 mg/kg methylphenidate (Prendergast et al., 1998). Both lower and higher doses were ineffective, demonstrating the familiar inverted U-shaped dose-response curve (ibid). An inverted U-shaped curve was also observed by Schneider et al. (1993), who administered methylphenidate to low dose MPTP–treated monkeys. These findings fit well with the human literature (see Mehta et al., Chapter 13).

High doses of stimulants that induce "attentional sticking" and perseverative responding likely induce excessive catecholamine stimulation in the PFC, resulting in loss of PFC regulation of attention and motor responses. Furthermore, high levels of catecholamines in sensory cortices might exacerbate "attentional sticking" by increasing signal/noise processing of salient stimuli in the environment (e.g., see Chapter 6 by Berridge).

PFC Regulation of Other Brain Structures Relevant to ADHD

The PFC has extensive interconnections with numerous brain structures, many of which are directly relevant to ADHD symptoms (e.g., caudate, nucleus accumbens, cerebellum via the pons, catecholamine cell bodies). Thus, PFC dysfunction may deregulate the activity of many subcortical and posterior cortical regions and worsen behavior (reviewed in Arnsten, 1998; Arnsten et al., 1996). Indeed, PFC dysfunction may actually facilitate processing in some subcortical regions, thus transferring control of behavior from the PFC to subcortical structures (see Chapter 5 by Grace). Conversely, optimization of PFC function with appropriate medication may provide regulation of posterior cortical and subcortical brain regions.

Relevance of PFC Catecholamines to ADHD

Changes in PFC catecholamines may be relevant to both the treatment of ADHD and possibly to the etiology of ADHD symptoms. However, it is still unclear as to how stimulants produce their therapeutic actions, and what mechanisms may produce ADHD symptomology. For example, the prevailing view is that ADHD may arise from insufficient catecholamines in the PFC as suggested by Ernst (1998), and that stimulants ameliorate symptoms by increasing catecholamine release and normalizing function. Alternatively, ADHD symptoms could arise from excessive catecholamine activity in the PFC, and low doses of stimulants might be therapeutic by decreasing catecholamine release through presynaptic actions. Finally, changes in PFC catecholamines could be therapeutic even if catecholamine activity in PFC is completely normal, for example, by optimizing the functioning of a smaller or otherwise immature PFC. These ideas are discussed in more detail in the closing Chapter 15 by Solanto, Arnsten, and Castellanos.

In summary, our increasing awareness of the neurochemical mechanisms influencing the PFC may be particularly helpful in understanding and treating the symptoms of ADHD. However, there is a great need for further research, particularly on low-dose stimulant actions. We need to learn how low doses of stimulants alter the neurochemical environment in the PFC, and how these changes in turn influence the working memory, attentional, and inhibitory functions of this important cortical area.

References

Amen, DG, & Carmichael, BD (1997). High-resolution brain SPECT imaging in ADHD. *Ann Clin Psychiatry, 9*, 81–86.

Aoki, C, Go, C-G, Venkatesan, C, & Kurose, H (1994). Perikaryal and synaptic localization of alpha-2A-adrenergic receptor-like immunoreactivity. *Brain Res, 650*, 181–204.

Arnsten, AFT (1997). Catecholamine regulation of the prefrontal cortex. *J Psychopharmacology, 11*, 151–162.

Arnsten, AFT (1998). Catecholamine modulation of prefrontal cortical cognitive function. *Trends in Cognitive Science, 2*, 436–447.

Arnsten, AFT & Cai, JX (1993). Post-synaptic alpha-2 receptor stimulation improves working memory in aged monkeys: Indirect effects of yohimbine vs. direct effects of clonidine. *Neurobiol Aging, 14*, 597–603.

Arnsten, AFT, Cai, JX, & Goldman-Rakic, PS (1988). The alpha-2 adrenergic agonist guanfacine improves memory in aged monkeys without sedative or hypotensive side effects. *J Neurosci, 8*, 4287–4298.

Arnsten, AFT, Cai, JX, Murphy, BL, and Goldman-Rakic, PS (1994). Dopamine D1 receptor mechanisms in the cognitive performance of young adult and aged monkeys. *Psychopharmacology, 116*, 143–151.

Arnsten, AFT, Cai, JX, Steere, JC, & Goldman-Rakic, PS (1995). Dopamine D2 receptor mechanisms contribute to age-related cognitive decline: The effects of quinpirole on memory and motor performance in monkeys. *J Neuroscience, 15*, 3429–3439.

Arnsten, AFT & Contant, TA (1992). Alpha-2 adrenergic agonists decrease distractibility in aged monkeys performing a delayed response task. *Psychopharmocology, 108*, 159–169.

Arnsten, AFT & Goldman-Rakic, PS (1985). Alpha-2 adrenergic mechanisms in prefrontal cortex associated with cognitive decline in aged nonhuman primates. *Science, 230*, 1273–1276.

Arnsten, AFT & Goldman-Rakic, PS (1990a). Analysis of alpha-2 adrenergic agonist effects on the delayed nonmatch-to-sample performance of aged rhesus monkeys. *Neurobiol Aging, 11*, 583–590.

Arnsten, AFT & Goldman-Rakic, PS (1990b). Stress impairs prefrontal cortex cognitive function in monkeys: Role of dopamine. *Soc. Neurosci Abstr, 16*, 164.

Arnsten, AFT, & Goldman-Rakic, PS (1998). Noise stress impairs prefrontal cortical cognitive function in monkeys: Evidence for a hyperdopaminergic mechanism. *Arch Gen Psychiatry, 55*, 362–369.

Arnsten, AFT, Mathew, R, Ubriani, R, Taylor, JR, & Li, B-M (1999). Alpha-1 noradrenergic receptor stimulation impairs prefrontal cortical cognitive function. *Biol Psychiatry, 45*, 26–31.

Arnsten, AFT, Murphy, BL & Merchant, K (2000). The selective dopamine D4 receptor antagonist, PNU-101387G, prevents stress-induced cognitive deficits in monkeys. *Neuropsychopharmacology,* (In press).

Arnsten, AFT, Steere, JC, and Hunt, RD (1996). The contribution of a-2 noradrenergic mechanisms to prefrontal cortical cognitive function: potential significance to Attention Deficit Hyperactivity Disorder. *Arch Gen Psychiatry, 53*, 448–455.

Avery, RA, Franowicz, JS, Studholme, C, van Dyck, CH & Arnsten, AFT (2000). The alpha-2A-adenoceptor agonist, guanfacine, increases regional cerebral blood flow in dorsolateral prefrontal cortex of monkeys performing a spatial working memory task. *Neuropsychopharmacology,* (In press).

Barkley, RA (1997). *ADHD and the nature of self-control.* New York: Guilford Press.

Bartus, RT (1979). Four stimulants of the central nervous system: Effects on short-term memory in young vs. aged monkeys. *J American Geriatrics Society, 27*, 289–297.

Bartus, RT, Fleming, D, & Johnson, HR (1978). Aging in the rhesus monkey: Debilitating effects on short-term memory. *J Gerontol, 33*, 858–871.

Bauer, RH & Fuster, JM (1978). Effects of d-amphetamine and prefrontal cortical cooling on delayed matching-to-sample behavior. *Pharmacol Biochem Behav, 8,* 243–249.

Biederman, J, Milberger, S, Faraone, SV, Kiely, K, Guite, J, & Mick, et al. (1995). Family-environment risk factors for Attention-Deficit Hyperactivity Disorder. *Arch Gen Psychiatry, 52,* 464–470.

Birnbaum, SG & Arnsten, AFT (1996). The alpha-2A noradrenergic agonist, guanfacine, reverses the working memory deficits induced by pharmacological stress (FG7142). *Soc Neurosci Abstracts, 22,* 1126.

Birnbaum, SG, Gobeske, KT, Auerbach, J, Taylor, JR, & Arnsten, AFT (1999). A role for norepinephrine in stress-induced cognitive deficits: Alpha-1-adrenoceptor mediation in prefrontal cortex. *Biol Psychiatry, 46*, 1266–1274.

Blumer, D & Benson, DF (1975). Personality changes with frontal and temporal lobe lesions. In DF Benson & D. Blumer (eds). *Psychiatric aspects of neurologic disease.* New York: Grune and Stratton.

Broadbent, D (1971). *Decision and stress.* London: Academic Press.

Brozoski, T, Brown, RM, Rosvold, HE, and Goldman, PS (1979). Cognitive deficit caused by regional depletion of dopamine in prefrontal cortex of rhesus monkey. *Science, 205*, 929–931.

Cai, JX & Arnsten, AFT (1997). Dose-dependent effects of the dopamine D1 receptor agonists A77636 or SKF81297 on spatial working memory in aged monkeys. *J Pharmacol Exp Ther, 282*, 1–7.

Cai, JX, Ma, Y, Xu, L, & Hu, X (1993). Reserpine impairs spatial working memory performance in monkeys: Reversal by the alpha-2 adrenergic agonist clonidine. *Brain Res, 614*, 191–196.

Carli, M, Robbins, TW, Evenden, JL, & Everitt, BJ (1983). Effects of lesions to ascending noradrenergic neurons on performance of a 5-choice serial reaction task in rats; implications for theories of dorsal noradrenergic bundle function based on selective attention and arousal. *Behav Brain Res, 9*, 361–380.

Carlson, S, Ranila, H, R, P, Mecke, E, & Pertovaara, A (1992). Effects of medetomidine, an alpha-2 adrenoceptor agonist, and atipamezole, an alpha-2 antagonist, on spatial memory performance in adult and aged rats. *Behav Neural Biol, 58*, 113–119.

Casey, BJ, Castellanos, FX, Giedd, JN, Marsh, WL, Hamburger, SD, Schubert, AB, Vauss, YC, Vaituzis, AC, Dickstein, DP, Sarfatti, SE, & Rapoport, JL (1997). Implication of right frontostriatal circuitry in response inhibition and attention-deficit/hyperactivity disorder. *J Am. Acad Child Adolesc Psychiatry, 36*, 374–383.

Chappell, PB, Riddle, MA, Scahill, L, Lynch, KA, Schultz, R, & Arnsten, A, et al. (1995). Guanfacine treatment of comorbid attention deficit hyperactivity disorder and Tourette's Syndrome: Preliminary clinical experience. *J Am. Acad Child Adolesc Psychiatry, 34*, 1140–1146.

Collins, P, Roberts, AC, Dias, R, Everitt, BJ, & Robbins, TW (1998). Perseveration and strategy in a novel spatial self-ordered sequencing task for nonhuman primates: effects of excitotoxic lesions and dopamine depletions of the prefrontal cortex. *J Cognitive Neuroscience, 10*, 332–354.

Comings, DE, Gade-Andavolu, R, Gonzalez, N, & MacMurray, JP (1999). Additive effect of three noradrenergic genes (ADRA2A, ADRA2C, DBH) on attention-deficit hyperactivity disorder and learning disabilities in Tourette syndrome subjects. *Clin Genet, 55*, 160–172.

Coull, JT (1994). Pharmacological manipulations of the a-2 noradrenergic system: Effects on cognition. *Drugs Aging, 5*, 116–126.

Coull, JT, Frackowiak, RS, & Frith, CD (1998). Monitoring for target objects: activation of right frontal and parietal cortices with increasing time on task. *Neuropsychologia, 36*, 1325–1334.

Coull, JT, Frith, CD, Dolan, RJ, Frackowiak, RS, & Grasby, PM (1997). The neural correlates of the noradrenergic modulation of human attention, arousal and learning. *Eur J Neurosci, 9*, 589–598.

Coull, JT, Sahakian, BJ, & Hodges, JR (1996). The alpha-2 antagonist idazoxan remediates certain attentional and executive dysfunction in patients with dementia of frontal type. *Psychopharmacology, 123*, 239–249.

Deutch, AY, & Roth, RH (1990). The determinants of stress-induced activation of the prefrontal cortical dopamine system. *Prog Brain Res, 85*, 367–403.

Dias, R, Robbins, TW, & Roberts, AC (1997). Dissociable forms of inhibitory control within prefrontal cortex with an analog of the Wisconsin Card Sort Test: restriction to novel situations and independence from "on-line" processing. *J Neurosci, 17*, 9285–9297.

Druzin, MY, Kurzina, NP, Malinina, EP, & Kozlov, AP (2000). The effects of local application of D2 selective dopaminergic drugs into the medial prefrontal cortex of rats in a delayed spatial choice task. *Behavioural Brain Res, 109*, 99–111.

Duman, RS & Nestler, EJ (1995). Signal transduction pathways for catecholamine receptors. In FE Bloom & DJ Kupfer (Eds.), *Psychopharmacology: The fourth generation of progress* (pp. 303–320). New York: Raven Press.

Engberg, G & Eriksson, E (1991). Effects of alpha-2-adrenoceptor agonists on locus coeruleus firing rate and brain noradrenaline turnover in EEDQ-treated rats. *Naunyn-Schmiedebergs Arch Pharmaco., 343*, 472–477.

Ernst, M, Zametkin, AJ, Matochik, JA, Jons, PH, & Cohen, RM (1998). DOPA decarboxylase activity in attention deficit disorder adults. A [fluorine-18] fluorodopa positron emission tomographic study. *J Neurosci, 18*, 5901–5907.

Fields, RB, Van Kammen, DP, Peters, JL, Rosen, J, Van Kammen, WB, & Nugent, et al. (1988). Clonidine improves memory function in schizophrenia independently from change in psychosis. *Schiz Res, 1*, 417–423.

Finlay, JM, Zigmond, MJ & Abercrombie, ED (1995). Increased dopamine and norepinephrine release in medial prefrontal cortex induced by acute and chronic stress: Effects of diazepam. *Neuroscience, 64*, 619–628.

Franowicz, JCS & Arnsten, AFT (1998). The alpha-2A noradrenergic agonist, guanfacine, improves delayed response performance in young adult rhesus monkeys. *Psychopharmacology, 136*, 8–14.

Franowicz, JS & Arnsten, AFT (1999). Treatment with the noradrenergic alpha-2 agonist clonidine, but not diazepam, improves spatial working memory in normal young rhesus monkeys. *Neuropsychopharmacology, 21*, 611–621.

Franowicz, JS, Kessler, L, Morgan, C, & Arnsten, AFT (1998). The alpha-2 noradrenergic agonist, guanfacine, improves spatial working memory in wild type mice but not mice with point mutations of the gene for the alpha-2A receptor. *Soc Neurosci Abstracts, 24*, 711.

Glass, DC, Reim, B, & Singer, JE (1971). Behavioral consequences of adaptation to controllable and uncontrollable noise. *J Exp Social Psychol, 7*, 244–257.

Glick, SD & Jarvik, ME. (1970). Differential effects of amphetamine and scopolamine on matching performance of monkeys with lateral frontal lesions. *J Comp Physiol Psychol, 73*, 307–313.

Goldman, PS & Rosvold, HE (1970). Localization of function within the dorsolateral prefrontal cortex of the rhesus monkey. *Exp Neurol, 27*, 291–304.

Goldman-Rakic, PS (1987). Circuitry of the primate prefrontal cortex and the regulation of behavior by representational memory. In F Plum (Ed.), *Handbook of physiology, the nervous system, higher functions of the brain, Vol. V* (pp. 373–417). Bethesda: American Physiological Society.

Goldstein, LE, Rasmusson, AM, Bunney, SB, & Roth, RH (1996). Role of the amygdala in the coordination of behavioral, neuroendocrine and prefrontal cortical monoamine responses to psychological stress in the rat. *J Neurosci, 16*, 4787–4798.

Hartley, LR & Adams, RG (1974). Effect of noise on the Stroop test. *J Exp Psychol, 102*, 62–66.

Hockey, GRJ (1970). Effect of loud noise on attentional selectivity. *Q J Exp Psychol, 22*, 28–36.

Horrigan, JP & Barnhill, LJ (1995). Guanfacine for treatment of Attention-Deficit-Hyperactivity Disorder in boys. *J Child Adolesc Psychopharmacol, 5*, 215–223.

Hunt, RD, Arnsten, AFT, & Asbell, MD (1995). An open trial of guanfacine in the treatment of attention deficit hyperactivity disorder. *J Am. Acad Child Adolesc Psychiatry, 34*, 50–54.

Jackson, WJ & Buccafusco, JJ (1991). Clonidine enhances delayed matching-to-sample performance by young and aged monkeys. *Pharmacol Biochem Behav, 39*, 79–84.

Jakala, P, Riekkinen, M, Sirvio, J, Koivisto, E, Kejonen, K, Vanbanen, M, & Riekkinen, PJ (1999a). Guanfacine, but not clonidine, improves planning and working memory performance in humans. *Neuropsychopharmacology, 20*, 460–470.

Jakala, P, Sirvio, J, Riekkinen, M, Koivisto, E, Kejonen, K, Vanhanen, M, & Riekkinen, PJ (1999b). Guanfacine and clonidine, alpha-2 agonists, improve paired associates learning, but not delayed matching to sample, in humans. *Neuropsychopharmacology, 20*, 119–130.

Kimberg, DY, D'Esposito, M, & Farah, MJ (1997). Effects of bromocriptine on human subjects depend on working memory capacity. *Neuroreport, 8*, 3581–3585.

Knight, RT, Scabini, D, & Woods, DL (1989). Prefrontal cortex gating of auditory transmission in humans. *Brain Res., 504*, 338–342.

Li, B-M, Mao, Z-M, Wang, M, & Mei, Z-T (1999). Alpha-2 adrenergic modulation of prefrontal cortical neuronal activity related to spatial working memory in monkeys. *Neuropsychopharmacology, 21*, 601–610.

Li, B-M & Mei, Z-T (1994). Delayed response deficit induced by local injection of the alpha-2 adrenergic antagonist yohimbine into the dorsolateral prefrontal cortex in young adult monkeys. *Behav Neural Biol, 62*, 134–139.

Luine, V, Bowling, D, & Hearns, M (1990). Spatial memory deficits in aged rats: contributions of monoaminergic systems. *Brain Res, 537*, 271–278.

Mair, RG & McEntree, WJ (1986). Cognitive enhancement in Korsakoff's psychosis by clonidine: A comparison with l-dopa and ephedrine. *Psychopharmacology, 88*, 374–380.

Mao, Z-M, Arnsten, AFT, & Li, B-M (1999). Local infusion of alpha-1 adrenergic agonist into the prefrontal cortex impairs spatial working memory performance in monkeys. *Biol. Psychiatry, 46*, 1259–1265.

Marek, GJ & Aghajanian, GK (1999). 5-HT2A receptor or alpha1-adrenoceptor activation induces EPSCs in layer V pyramidal cells of the medial prefrontal cortex. *Eur J Pharmacol, 367*, 197–206.

Milner, B (1982). Some cognitive effects of frontal-lobe lesions in man. *Phil Trans R Soc Lond B, 298*, 211–226.

Mishkin, M (1964). Perseveration of central sets after frontal lesions in monkeys. In: JM Warren, & K Akert (Eds.), *The frontal granular cortex and behavior* (pp. 219–241). New York: McGraw-Hill.

Moffoot, A, O'Carroll, RE, Murray, C, Dougall, N, Ebmeier, K, & Goodwin, GM (1994). Clonidine infusion increases uptake of Tc-exametazime in anterior cingulate cortex in Korsakoff's psychosis. *Psychol. Med, 24*, 53–61.

Mohell, N, Svartengren, J & Cannon, B (1983). Identification of [3H]prazosin binding sites in crude membranes and isolated cells of brown adipose tissue as alpha-1 adrenergic receptors. *Eur J Pharmacology, 92*, 15–25.

Muller, U, von Cramon, DY, & Pollmann, S (1998). D1- versus D2-receptor modulation of visuospatial working memory in humans. *J Neuroscience, 18*, 2720–2728.

Murphy, B, Roth, R, & Arnsten, AFT (1997). Clozapine reverses the spatial working memory deficits induced by FG7142 in monkeys. *Neuropsychopharmacology, 16*, 433–437.

Murphy, BL, Arnsten, AFT, Goldman-Rakic, PS, & Roth, RH (1996a). Increased dopamine turnover in the prefrontal cortex impairs spatial working memory performance in rats and monkeys. *Proc Nat Acad Sci USA, 93*, 1325–1329.

Murphy, BL, Arnsten, AFT, Jentsch, JD, & Roth, RH (1996b). Dopamine and spatial working memory in rats and monkeys: Pharmacological reversal of stress-induced impairment. *J Neurosci, 16*, 7768–7775.

O'Rourke, MF, Blaxall, HS, Iversen, LJ, & Bylund, DB (1994). Characterization of [3H]RX821002 binding to alpha-2 adrenergic receptor subtypes. *J Pharmacol Exp Ther, 268*, 1362–1367.

Perret, E (1974). The left frontal lobe of man and the suppression of habitual responses in verbal categorical behavior. *Neuropsychologia, 12*, 323–330.

Prendergast, MA, Jackson, WJ, Terry, AVJ, Kille, NJ, Arneric, SP, Decker, MW, & Buccafusco, JJ (1998). Age-related differences in distractibility and response to methylphenidate in monkeys. *Cerebral Cortex, 8*, 164–172.

Rama, P, Linnankoski, I, Tanila, H, Pertovaara, A, & Carlson, S (1996). Medetomidine, atipamezole, and guanfacine in delayed response performance of aged monkeys. *Pharmacol Biochem Behav, 54*, 1–7.

Riekkinen, P & Riekkinen, M (1999). THA improves word priming and clonidine enhances fluency and working memory in Alzheimer's disease. *Neuropsychopharmacology, 20*, 357–364.

Robbins, TW (1996). Dissociating executive functions of the prefrontal cortex. *Phil Trans R Soc Lond, 351*, 1463–1471.

Robbins, TW, Granon, S, Muir, JL, Durantou, F, Harrison, A, & Everitt, BJ (1998). Neural systems underlying arousal and attention: Implications for drug abuse. *Ann N Y Acad Sci, 846*, 222–237.

Roberts, AC, Salvia, MA, Wilkinson, LS, Collins, P, Muir, JL, Everitt, BJ, & Robbins, TW (1994). 6-Hydroxydopamine lesions of the prefrontal cortex in monkeys enhance performance on an analog of the Wisconsin Card Sort Test: Possible interactions with subcortical dopamine. *Neuroscience, 14*, 2531–2544.

Rubia, K, Overmeyer, S, Taylor, E, Brammer, M, Williams, SCR, Simmons, A, & Bullmore, ET (1999). Hypofrontality in Attention Deficit Hyperactivity Disorder during higher-order motor control: A study with functional MRI. *Am J Psychiatry, 156*, 891–896.

Sahakian, BJ, Sarna, GS, Kantamaneni, BD, Jackson, A, Hutson, PH & Curzon, G (1985). Association between learning and cortical catecholamines in non-drug-treated rats. *Psychopharmacology, 86*, 339–343.

Satterfield, JH, Schell, AM, Nicholas, T, & Backs, RW (1988). Topographic study of auditory event-related potentials in normal boys and boys with attention deficit disorder with hyperactivity. *Psychophysiology, 25*, 591–606.

Sawaguchi, T (1998). Attenuation of delay-period activity of monkey prefrontal cortical neurons by an alpha-2 adrenergic antagonist during an oculomotor delayed-response task. *J Neurophysiology, 80*, 2200–2205.

Sawaguchi, T. & Goldman-Rakic, PS (1991). D1 dopamine receptors in prefrontal cortex: Involvement in working memory. *Science, 251*, 947–950.

Sawaguchi, T, Matsumura, M, & Kubota, K (1988). Dopamine enhances the neuronal activity of spatial short-term memory task in the primate prefrontal cortex. *Neurosci Res, 5*, 465–473.

Sawaguchi, T, Matsumura, M, & Kubota, K (1990). Catecholaminergic effects on neuronal activity related to delayed response task in monkey prefrontal cortex. *J Neurophys, 63*, 1385–1400.

Schneider, JS, Sun, Z-Q, & Roeltgen, DP (1993). Effects of dopamine agonists on delayed response performance in chronic low dose MPTP-treated monkeys. *Pharmacol Biochem Behav, 48*, 235–240.

Schultz, W (1998). The phasic reward signal of primate dopamine neurons. *Advances in Pharmacology, 42*, 686–690.

Schultz, W, Apicella, P, & Ljungberg, T (1993). Responses of monkey dopamine neurons to reward and conditioned stimuli during successive steps of learning a delayed response task. *J Neuroscience, 13*, 900–913.

Seamans, JK, Floresco, SB, & Phillips, AG (1998). D1 receptor modulation of hippocampal-prefrontal cortical circuits integrating spatial memory with executive functions in the rat. *J Neurosci, 18*, 1613–1621.

Shors, TJ, Weiss, C, & Thompson, RF (1992). Stress-induced facilitation of classical conditioning. *Science, 257*, 537–539.

Silberstein, RB, Farrow, M, Levy, F, Pipingas, A, Hay, DA, & Jarman, FC (1998). Functional brain electrical activity mapping in boys with attention-deficit/hyperactivity disorder. *Arch Gen Psychiatry, 55*, 1105–1112.

Simon, H (1981). Dopaminergic A10 neurons and the frontal system. *J de Physiol, 77*, 81–95.

Sirviö, J, Riekkinen, P, Vajanto, I, Koivisto, E, & Riekkinnen, PJ (1991). The effects of guanfacine, alpha-2 agonist, on the performance of young and aged rats in spatial navigation task. *Behav Neural Biol, 56*, 101–107.

Steere, JC & Arnsten, AFT (1995). The alpha-2A noradrenergic agonist, guanfacine, improves visual object discrimination reversal in rhesus monkeys. *Soc Neurosci Abstracts, 21*, 195.

Steere, JC & Arnsten, AFT (1997). The alpha-2A noradrenergic agonist, guanfacine, improves visual object discrimination reversal performance in rhesus monkeys. *Behav. Neuroscience, 111*, 1–9.

Stuss, DT, Eskes, GA, & Foster, JK (1994). Experimental neuropsychological studies of frontal lobe functions. *Handbook of Neuropsychology, 9*, 149–185.

Stuss, DT, Gow, CA, & Hetherington, CR (1992). "No longer gage": Frontal lobe dysfunction and emotional changes. *J Consult Clin Psychol, 60*, 349–359.

Surmeier, DJ, Bargas, J, Hemmings, HCJ, Nairn, AC, & Greenguard, P (1995). Modulation of calcium currents by a D1 dopaminergic protein kinase/phosphatase cascade in rat neostriatal neurons. *Neuron, 14*, 385–397.

Swanson, JM, Posner, M, Potkin, S, Bonforte, S, Youpa, D, Fiore, C, Cantwell, D, & Crinella, F (1991). Activating tasks for the study of visual-spatial attention in ADHD children: a cognitive anatomic approach. *J Child Neurol, 6*, S119–S127.

Tanila, H, Mustonen, K, Sallinen, J, Scheinin, M, & Riekkinen, P (1999). Role of alpha-2C-adrenoceptor subtype in spatial working memory as revealed by mice with targeted disruption of the alpha-2C-adrenoceptor gene. *Eur J Neurosci, 11*, 599–603.

Tanila, H, Rama, P, & Carlson, S (1996). The effects of prefrontal intracortical microinjections of an alpha-2 agonist, alpha-2 antagonist and lidocaine on the delayed alternation performance of aged rats. *Brain Res Bull, 40*, 117–119.

Taylor, JR, Birnbaum, SG, Ubriani, R, & Arnsten, AFT (1999). Activation of protein kinase A in prefrontal cortex impairs working memory performance. *J Neuroscience (Online), 19*, RC 23.

Taylor, FB & Russo, J (2000). Comparing guanfacine and dextroamphetamine for the treatment of adult attention deficit hyperactivity disorder. *J Clinical Psychopharm*, (In press).

Thierry, AM, Tassin, JP, Blanc, G, & Glowinski, J (1976). Selective activation of the mesocortical DA system by stress. *Nature, 263*, 242–244.

Vaidya, CJ, Austin, G, Kirkorian, G, Ridlehuber, HW, Desmond, JE, Glover, GH, & Gabrieli, JDE (1998). Selective effects of methylphenidate in attention deficit hyperactivity disorder: A functional magnetic resonance study. *Proc Natl Acad Sci USA, 95*, 14494–14499.

Wilkins, AJ, Shallice, T, & McCarthy, R (1987). Frontal lesions and sustained attention. *Neuropsychologia, 25*, 359–365.

Williams, GV & Goldman-Rakic, PS (1995). Blockade of dopamine D1 receptors enhances memory fields of prefrontal neurons in primate cerebral cortex. *Nature, 376*, 572–575.

Witte, EA & Marrocco, RT (1997). Alterations of brain noradrenergic activity in rhesus monkeys affects the alerting component of covert orienting. *Psychopharmacology, 132*, 315–323.

Woods, DL & Knight, RT (1986). Electrophysiological evidence of increased distractibility after dorsolateral prefrontal lesions. *Neurology, 36*, 212–216.

Yang, CR & Seamans, JK (1996). Dopamine D1 receptor actions in layers V-VI rat prefrontal cortex neurons in vitro: modulation if dendritic-somatic signal integration. *J Neurosci, 16*, 1922–1935.

Zahrt, J, Taylor, JR, Mathew, RG, & Arnsten, AFT (1997). Supranormal stimulation of dopamine D1 receptors in the rodent prefrontal cortex impairs spatial working memory performance. *J Neurosci, 17*, 8528–8535.

8

A Review of Rodent Models of ADHD

SHERRY A. FERGUSON

Studies using animal models of ADHD provide some necessary inherent advantages over human studies. For example, socioeconomic status, parental behavior, and school environment are potential confounds in human studies of ADHD. As noted by Smith (1989), over- or undercontrolling for these confounding variables can diminish or magnify effects. Development of subjects in longitudinal studies cannot be hastened and such studies proceed slowly. There may be interactions of ADHD with aging during the geriatric period, which would be extremely difficult for a longitudinal study to assess. Finally, human studies investigating the etiology of ADHD are often correlational in nature and thus cannot establish causal relationships.

Animal models can causally establish effects resulting from documented exposure or treatment. In addition, they permit more invasive manipulations in order to investigate underlying *neurochemical or neuropathological* alterations. More specific to ADHD, animal models allow assessment of potential new therapeutics. Models described in this chapter include the *coloboma* strain of mouse, use of chemical lesions (e.g., 6-hydroxydopamine treatment) or irradiation, treatments that induce developmental cerebellar stunting as well as the most thoroughly studied animal model, the Spontaneously Hypertensive Rat (SHR).

Neonatal 6-Hydroxydopamine (6-OHDA) Treatment

When combined with desmethylimipramine treatment (to prevent norepinephrine depletion), intracerebroventricular or intracisternal injection of the catecholamine neurotoxin 6-hydroxydopamine (6-OHDA) in neonatal rats causes selective and permanent dopamine depletion. Depending on the dose of 6-OHDA, adult dopamine levels can be reduced to 5% to 40% of control in the frontal cortex, nucleus accumbens, septum, striatum, and amygdala (Archer et al., 1988; Luthman et al., 1997).

This severe dopamine depletion results in functional effects. Rats treated with 6-OHDA exhibit increased locomotor activity concurrent with decreased rearing and severe learning deficits (Archer et al., 1988; Luthman et al., 1997, 1989; Shaywitz et al. 1978, 1976). The learning deficits are apparent as early as 48 hours after treatment and are unrelated to sensory or motor deficits (Wool et al., 1987).

While relatively high doses of methylphenidate (i.e., 2 mg/kg) decrease the locomotor activity of 6-OHDA–treated rats, this effect appears to occur only at a specific age (postnatal day 26) (Shaywitz et al., 1978), although later ages have not been studied. The same dose of methylphenidate in postnatal day 26 control rats had little effect on activity (see Table 3 in Shaywitz et al., 1978). Amphetamine doses in the clinically relevant range (0.25 mg/kg) transiently decreased the locomotor activity of 6-OHDA–treated rats tested as adults (Luthman et al., 1989).

Learning deficits exhibited by 6-OHDA–treated rats appear to respond to amphetamine treatment as well. When treated with 0.5 mg/kg during conditioning and testing, the olfactory conditioning learning deficit in 6-OHDA treated rats was somewhat alleviated (Wool et al., 1987). A relatively high dose of methylphenidate (2 mg/kg) also alleviated a T-maze learning deficit in 6-OHDA treated rats (Shaywitz et al., 1978).

The results of investigations of neonatal 6-OHDA treatment are encouraging but there remain many areas where additional research is needed. Thus far, most studies have assessed only males and there are few data available on the effects of 6-OHDA treatment in females. A longitudinal assessment of locomotor activity as well as the effects of clinically relevant stimulant doses across the life span would prove beneficial in the validation of this model. The learning deficits are troubling because they are severe in nature. However, because they seem to respond to stimulant treatment, these may prove to be more attentional in nature than true learning deficits. This hypothesis awaits the results of a stringent measure of attention and impulsivity in 6-OHDA-treated rats.

Spontaneously Hypertensive Rat

The Spontaneously Hypertensive Rat (SHR) was developed in Kyoto, Japan, for use in studies of hypertension by selectively inbreeding rats of the Wistar-Kyoto strain that exhibited high systolic blood pressure. During the course of such inbreeding, the SHR was noted to be significantly more active in open field tests than its normotensive strain, the Wistar-Kyoto (Hendley et al., 1985; Van Den Buuse & De Jong, 1989; Wultz et al., 1990).

Much of the work examining the use of the SHR as a model of ADHD has been done by Sagvolden and colleagues at the University of Oslo and several of the findings do seem to parallel ADHD-related behaviors. For example, when reinforced for remaining immobile in a specific area of an operant chamber, SHR perform as well as Wistar-Kyoto rats when the required duration of immobility is relatively short (<8 seconds) (Wultz & Sagvolden, 1992). However, when the required duration is increased (e.g., 8–10 seconds), SHR make more movements, seemingly unable to remain immobile for the duration. Such findings closely resemble the "fidgety" movements of ADHD children. SHR are more active in operant tests, pressing the levers more frequently even though reinforcement is not contingent on frequent pressing (Sagvolden, et al., 1992). Detailed analyses of operant performance indicate that SHR appear to be more sensitive to *immediate* reinforcement and less sensitive to *delayed* reinforcement (Sagvolden, Metzger, Schiorbeck et al., 1992). This finding parallels that in ADHD children in which a smaller, but immediate, reward is chosen over a larger, but delayed, reward (Sonuga-Barke et al., 1992). Finally, differences in response rates are noted only during times of infrequent reinforcement, indicating the SHR are more sensitive to reward processes (Sagvolden, Metzger, & Sagvolden, 1993).

Results from learning assessments of the SHR appear to parallel those of ADHD children as well. For example, when the learning task requires *varying* sequences of responses, SHR are as accurate as Wistar-Kyoto rats. However, when the task requires *repetitive* responses, SHR perform poorly (Hunziker et al., 1996; Low et al., 1984; Mook et al. 1993; Mook & Neuringer, 1994). Similar findings have also been noted in ADHD children (Hynd et al., 1989).

Such findings suggest that the behavioral alterations exhibited by the SHR are similar to those of the ADHD child. However, it is unclear whether those alterations result from the *hypertension* or the *hyperactivity*. For example, SHR make more errors in acquisition and performance of an 8-arm radial maze, a task specifically designed to measure learning and memory (Nakamura-Palacios et al., 1996). Similarly, hypertensive humans have learning and memory problems (Elias et al., 1987; Goldstein et al., 1990). However, no study has consistently demonstrated such deficits in ADHD children (although some ADHD children do have learning disabilities). SHR exhibit a greater magnitude of the normal decrements in learning and memory that occur with aging than either Wistar-Kyoto or Sprague-Dawley rats (Meneses et al., 1996; Wyss et al., 1992). The hypertension appears to be responsible for those exaggerated learning and memory deficits since if SHR are treated with antihypertensive drugs, those deficits are attenuated (Wyss et al., 1992).

Several researchers have questioned the use of the Wistar-Kyoto strain as the "control" for SHR rats. For example, evidence of sufficient genetic polymorphism (Johnson et al., 1992) has caused some to suggest that continued use of the Wistar-Kyoto as a control strain for the SHR is no longer useful in studies of hypertension (St. Lezin et al., 1992). Others, however, note that such genetic polymorphism is typical for inbred disease strains relative to their control strains (Lindpaintner et al., 1992). Thus, this potential confound is not unique to studies of the SHR.

The hyperactivity commonly described in SHR is environment-specific. In activity tests, SHR were more active than Wistar-Kyoto only when the open field is of the "free-exploration" type (Sagvolden, Pettersen, & Larsen, 1993). For example, when the home cage is accessible (i.e., the rat can move from its home cage to the open field and back), SHR were more active than Wistar-Kyoto. However, in that same paradigm, Sprague-Dawley rats were *more* active than either SHR or Wistar-Kyoto (Sagvolden, Pettersen, & Larsen, 1993). When the test is of the "forced-exploration" type which is more common (i.e., the rat is placed into the open field and cannot leave or escape the apparatus), then SHR and Sprague-

Dawley rats are equivalent in activity. Thus, when compared with other common laboratory rat strains in typical laboratory tests, SHR are not necessarily a particularly hyperactive strain.

SHR offspring exhibit a unique response to their maternal environment. Specifically, if SHR rats are cross-fostered to normotensive dams (Sprague-Dawley or Wistar-Kyoto dams), they exhibit lower blood pressures as adults than if reared by their natural SHR dams (Cierpial & McCarty, 1987; DiNicolantonio et al., 1986; McCarty et al., 1992). More detailed investigations determined that SHR dams behave differently with their natural offspring, licking/grooming the pups and assuming a nursing position more often than Wistar-Kyoto dams (Myers, Brunelli, Squire et al., 1989; Sanders & Gray, 1997).

Some of the behavioral alterations exhibited by SHR resemble those of ADHD, but others do not (e.g., learning deficits), and it is unclear which alterations are associated with the hypertension. Thus, some have attacked these potential confounds from another route. Hendley and colleagues at the University of Vermont have selectively bred hypertensive only and hyperactive only rats by crossbreeding SHR with Wistar-Kyoto rats (Hendley & Ohlsson, 1991; Hendley, Wessel & Van Houten, 1986). Behavioral profiles of the resulting WKHA (hyperactive without hypertension) and WKHT (hypertensive without hyperactivity) show that the original SHR is different from either of the two new strains (Sagvolden, Hendley, & Knardahl, 1992), but additional work is needed to validate the WKHA as a potential animal model of ADHD.

Irradiation

Focal x-irradiation of the hippocampus in young rats (PND 2–15) also appears to cause ADHD-like behaviors in the adult (Diaz-Granados, Greene & Amsel, 1994); however, such treatment also induces severe learning deficits (Czurko, Czeh, Seress et al., 1997). Again, those learning deficits limit the usefulness of this particular model.

DAT Knockout Mice

Caron and colleagues recently described a knockout strain of mouse that lacks the gene that encodes for the dopamine transporter (DAT) (Giros, Jaber, Jones et al., 1996). The DAT controls dopaminergic activity via reuptake of dopamine from the synapse into the presynaptic neuron. In mice lacking functional DATs, extracellular striatal levels of dopamine are approximately 5 times that of wild-type mice (Jones et al., 1998).

Behavioral alterations in DAT knockout mice include profound hyperactivity in novel and stressful environments, but not in the home cage, and this hyperactivity is exhibited throughout the circadian period (Gainetdinov, Wetsel, Jones et al., 1999; Giros et al., 1996; Jaber, Jones, Giros et al., 1997). Originally described as virtually unresponsive to the locomotor-increasing effects of cocaine and amphetamine (Giros et al., 1996), a more recent report described attenuation of the hyperactivity after treatment with methylphenidate, amphetamine and cocaine (Gainetdinov et al., 1999). These recent results were described as "paradoxical" in that the same doses of methylphenidate, amphetamine, and cocaine induced hyperactivity in wild-type mice. However, these results argue against the validity of this model since methylphenidate and amphetamine do not produce "paradoxical" effects in ADHD children. Both ADHD and non-ADHD children exhibit decreased activity and improved attention after clinically relevant stimulant doses (Rapoport, Buchsbaum, Zahn et al., 1978). Additionally, the clinical dose of methylphenidate ranges from 0.3 to 0.7 mg/kg (Smith, Pelham, Gnagy et al., 1998) while the dose of methylphenidate used (30 mg/kg) in the DAT knockout mice was approximately 40 to 100 times greater.

The hyperactivity exhibited by DAT knockout mice has not been described with regard to sex; in fact, from studies to date, it is unclear which sex has been tested or whether the results described apply to both sexes. Further investigation into sex differences in activity of DAT knockout mice will be helpful in determining its usefulness as a model of ADHD.

DAT knockout mice have severe learning impairments as assessed in the 8-arm radial maze (Gainetdinov et al., 1999). Even after 21 sessions, most DAT knockout mice were unable to retrieve all reinforcers within the allotted time. This learning impairment was directly related to an increase in perseverative errors as well as a decrease in locomotor activity in the maze. Such learning impairments prevent the usefulness of this model. As described above, ADHD children do not consistently exhibit learning deficits. Additionally, perseveration is not typically associated with ADHD children; in fact, non-perseverative or alternating-type behaviors are one trait that distinguishes ADHD children (Hynd et al., 1989). Finally, the hypoactivity exhibited by DAT knockout mice began to be exhibited by the fifteenth session; DAT knockout mice exhibited normal levels of activity during the first 12 sessions with no indication of hyperactivity during the first few sessions of a presumably novel environment (Gainetdinov et al., 1999)

DAT knockout mice show profound growth retardation and impaired postnatal survival (Giros et al., 1996; Gainetdinov, Jones, Fumagalli et al, 1998). By 10 weeks of age, there is only 68% survival. Female DAT knockout mice do not lactate and thus, pups must be fostered to wild-type mice (Gainetdinov et al., 1998). The lactational problems appear to be related to the anterior pituitary hypoplasia and thus, there may be other endocrinological problems as well (Bosse, Fumagalli, Jaber et al., 1997). Such physiological alterations must be further investigated with regard to potential behavioral alterations.

Coloboma Strain of Mouse

In 1966, a new mutant strain of mouse resulting from neutron irradiation was described (Searle, 1966). The *coloboma* mouse is a pigmented mouse with severe ophthalmic dysmorphology, making it essentially blind (E. Hess, personal communication). The mutation results from deletion of the SNAP-25 gene, a neuron-specific plasma membrane protein thought to facilitate synaptic vesicle fusion at the presynaptic membrane (reviewed by Hess, Collins, Copeland et al., 1994). The *coloboma* mouse is extremely hyperactive in open field tests (Hess, Jinnah, Kozak et al., 1992; Hess, Collins & Wilson, 1996) and delayed in achieving certain developmental milestones such as the righting reflex and bar holding (Heyser, Wilson, & Gold, 1995). In addition, it exhibits a particularly prominent stereotyped behavior, near-constant head bobbing (Hess et al., 1992; Heyser et al., 1995). Because the exact genetic mutation is known, the human analogue of the SNAP-25 gene was investigated in five families in which a genetic inheritance of ADHD was suspected (Hess, Rogan, Domoto et al., 1995). However, there was no significant linkage detected between the human SNAP-25 and *coloboma* mouse loci.

Developmental Cerebellar Stunting

The cerebellum is typically described as having a regulatory function in movement activities, being critically involved in eye movements, body equilibrium during stance and gait, control of the ongoing execution of limb movement, and planning and initiation of movement. More recent evidence suggests its involvement in higher cognition, such as IQ (Andreasen et al., 1993; Bracke-Tolkmitt et al., 1989; Fiez, 1996; Leiner et al. 1986), emotion (Lalonde & Botez, 1990), and language (Leiner, et al., 1993). Disorders such as autism and schizophrenia have also been correlated with cerebellar alterations (Bauman, 1991; Courchesne, 1989; Courchesne, Saitoh, Yeung-Courchesne et al., 1994; and reviewed by Martin & Albers, 1995, and Katsetos, Hyde & Herman, 1997).

A recent review of developmental cerebellar insults and the resulting neuropathology and behavioral alterations described an apparent age-dependent principle of disruption: there is developmental stage specificity in the severity and type of cerebellar structural and functional effects (Ferguson, 1996). Specifically, lesions or toxicant exposure in the developing rodent produce neuroanatomical and functional consequences that are strikingly different depending on when the insult is produced. In general, early neonatal insults (birth to postnatal day 4) produce more severe cerebellar neuropathology in that Purkinje cells are misoriented and cerebellar foliation malformations are apparent. Such severe neuropathology is accompanied by the classic cerebellar syndrome: ataxia, motor incoordination, hypoactivity, tremors, and learning deficits. However, lesions or toxicant exposure that occur later in the neonate (postnatal day 4 through postnatal day 10 or 12) result in much less severe neuropathology but the granule cell population is drastically depleted. The behavioral alteration most often associated with these later insults was hyperactivity with few or no learning deficits. More recent work from our laboratory and others investigating the hypothesis that later cerebellar insults produce ADHD-like behaviors is described below.

The potent antimitotic compound methylazoxymethanol acetate is commonly used either prenatally or postnatally to induce regional neural hypoplasia. Postnatal treatment that resulted in a permanent 8% reduction in cerebellar weight left other brain regions unaffected (as assessed by gross wet weight) (Ferguson, Paule, & Holson, 1996). Concurrent with this, a mild hyperactivity was apparent in several assessments (e.g., open field and running wheel tests) and this hyperactivity was more prominent in males than females. Similarly, the synthetic glucocorticoid dexamethasone when administered during the developmental period caused a 9% to 14% reduction in cerebellar weight concurrent with mild hyperactivity, again particularly prominent in males (Ferguson & Holson, 1999). Finally, oral administration of all-*trans* retinoic acid on gestational days 11 through 13 decreased cerebellar weight by approximately 5% to 8% and produced hyperactivity in running wheel assessments (Holson et al., 1997). In none of these studies was learning ability impaired. These results parallel those of many years of study in the laboratory of Joseph Altman indicating focal irradiation of the cerebellum during the second postnatal week caused hyperactivity in open field and running wheel tests but spared learning ability (reviewed in Ferguson, 1996).

The hyperactivity resulting from developmental cerebellar stunting was reflective of the type seen in ADHD in that it was more prominent in male rats. However, the studies described did not include measures of attention or impulsivity and such assessments are imperative in order to validate this particular ADHD rodent model. In addition, the behavioral and neurochemical response to the typical stimulants used to treat ADHD should be studied. Finally, a broad spectrum of assessments designed to examine learning, motor coordination, anxiety, social behaviors, etc. would further contribute to the understanding of this animal model and its similarity to the clinical syndrome.

The cerebellum is rarely the focus of human or laboratory animal studies investigating ADHD; however, a careful review of the literature indicates that there is sufficient evidence to suggest its involvement in the disorder (see Castellanos, 2000). Scanning studies indicated mild cerebellar atrophy in adults diagnosed with ADHD as children (Nasrallah et al., 1986). In ADHD boys, the cerebellum was approximately 7% smaller even after correcting for differences in total cerebral volume (Castellanos et al., 1996) and this size decrease was particularly apparent in the vermal area (Berquin, Giedd, Jacobsen et al., 1998). Additionally, some studies have suggested deficits in fine motor control, suggestive of cerebellar dysfunction (Hartsough & Lambert, 1985; Stewart, Pitts, Craig et al., 1966; Ullman, Barkley & Brown, 1978; Whitmont & Clark, 1996). Finally, a recent report of dose-dependent decreases in blood flow in the cerebellar vermis of ADHD boys after methylphenidate treatment further implicates this area (Anderson, Polcari, McGreenery et al., 1999). Further work of cerebellar alterations in humans with ADHD will certainly guide and assist the laboratory animal research.

Summary

ADHD is a troublesome childhood and sometimes lifelong disorder: the etiology is unknown and the only treatment with substantial evidence of efficacy is chronic medication. Animal models are necessary in order to evaluate potential causes as well neurological alterations which cannot be examined in humans. Several of the described rodent models appear very promising and deserve further research.

Solanto (2000) has outlined six characteristics that animal models of ADHD should meet in order to adequately reflect not only the untreated clinical picture of ADHD, but also the response to pharmacological treatments. While no one rodent model is likely to meet all six criteria, these serve to guide animal research in this area.

References

Anderson, CM, Polcari, AM, McGreenery, CE, Maas, LC, Renshaw, PF, & Teicher, MH (1999). Methylphenidate dose-dependently decreases blood flow in the cerebellar vermis of children with ADHD. Presented at the annual meeting of the New Drug Clinical Evaluation Unit of NIMH, Boca Raton, Florida, May 1999.

Andreasen, NC, Flaum, M, Swayze, V, 2nd, O'Leary, DS, Alliger, R, & Cohen, G et al., (1993). Intelligence and brain structure in normal individuals. *Am J Psychiatry, 150*, 130–134.

Archer, T, Danysz, W, Fredriksson, A, Jonsson, G, Luthman, J, Sundstrom, E, & Teiling, A (1988). Neonatal 6-hydroxydopamine-induced dopamine depletions: Motor activity and performance in maze learning. *Pharmacol Biochem Behav, 31*, 1988.

Bauman, ML (1991). Microscopic neuroanatomic abnormalities in autism. *Pediatrics, 87* (Suppl.), 791–796.

Berquin, PC, Giedd, JN, Jacobsen, LK, Hamburger, SD, Krain, AL, Rapoport, JL, & Castellanos, FX (1998). Cerebellum in attention-deficit hyperactivity disorder: A morphometric MRI study. *Neurology, 50*, 1087–1093.

Bosse, R, Fumagalli, F, Jaber, M, Giros, B, Gainetdinov, RR, Wetsel WC, Missale C, & Caron, MG (1997). Anterior pituitary hypoplasia and dwarfism in mice lacking the dopamine transporter. *Neuron, 19,* 127–138.

Bracke-Tolkmitt, R, Linden, A, Canavan, AGM, Rockstroh, B, Scholz, E, Wessel, K, & Diener, H-C (1989). The cerebellum contributes to mental skills. *Behav Neurosci, 103*, 442–446.

Castellanos, FX. (2000). Neuroimaging studies of ADHD. In MV Solanto, AFT Arnsten, & FX Castellanos (Eds.), Stimulant drugs and ADHD: Basic and clinical neuroscience (pp. X-X). New York: Oxford University Press.

Castellanos, FX, Giedd, JN, Marsh, WL, Hamburger, SD, Vaituzis, AC, & Dickstein, DP et al. (1996). Quantitative brain magnetic resonance imaging in Attention-Deficit Hyperactivity Disorder. *Arch Gen Psychiatry, 53*, 607–616.

Cierpial, MA, & McCarty, R (1987). Hypertension in SHR rats: Contribution of maternal environment. *Am J Physiol, 253*, H980–H984.

Courchesne, E (1989). Neuroanatomical systems involved in infantile autism: The implications of cerebellar abnormalities. In G Dawson (Ed.), *Autism: New perspectives on diagnosis, nature and treatment* (pp. 119–143). New York: Guilford Press.

Courchesne, E, Saitoh, O, Yeung-Courchesne, R, Press, GA, Lincoln, AJ, & Haas, RH, et al. (1994). Abnormality of cerebellar vermian lobules VI and VII in patients with infantile autism: Identification of hypoplastic and hyperplastic subgroups by MR imaging. *Am J Roentgenol,162*, 123–130.

Czurko, A, Czeh, B, Seress, L, Nadel, L, & Bures, J (1997). Severe spatial navigation deficit in the Morris water maze after single high dose of neonatal x-ray irradiation in the rat. *Proc Natl Acad Sci USA, 94*, 2766–2771.

Diaz-Granados, JL, Greene, PL, & Amsel, A (1994). Selective activity enhancement and persistence in weanling rats after hippocampal X-irradiation in infancy: Possible relevance for ADHD. *Behav Neural Biol, 61*, 251–259.

DiNicolantonio, R, Marshall, SJ, Nicolaci, JA, & Doyle, AE (1986). Blood pressure and saline preference of cross-suckled genetically hypertensive and normotensive rats: Role of milk electrolytes. *J Hypertens, 4* (Suppl 3), S253–S254.

Elias, MF, Robbins, MA, Schultz, NR, Jr., Streeten, DH, & Elias, PK (1987). Clinical significance of cognitive performance by hypertensive patients. *Hypertension, 9*, 192–197.

Ferguson, SA (1996). Neuroanatomical and functional alterations resulting from early postnatal cerebellar insults in rodents. *Pharmacol Biochem Behav, 55*, 663–671.

Ferguson, SA, & Holson, RR (1999). Neonatal dexamethasone on day 7 but not 3, causes mild hyperactivity and cerebellar stunting. *Neurotoxicol Teratol, 21*, 71–76.

Ferguson, SA, Paule, MG, & Holson, RR (1996). Functional effects of methylazoxymethanol-induced cerebellar hypoplasia in rats. *Neurotoxicol Teratol, 18*, 529–537.

Fiez, JA (1996). Cerebellar contributions to cognition. *Neuron, 16*, 13–15.

Gainetdinov, RR, Jones, SR, Fumagalli, F, Wightman RM, & Caron, MG (1998). Re-evaluation of the role of the dopamine transporter in dopamine system homeostasis. *Brain Res. Rev., 26*, 148–153.

Gainetdinov, RR, Wetsel, WC, Jones, SR, Levin, ED, Jaber, M, & Caron, MG (1999). Role of serotonin in the paradoxical calming effect of psychostimulants on hyperactivity. *Science, 283*, 397–401.

Giros, B, Jaber, M, Jones, SR, Wightman, RM, & Caron, MG (1996). Hyperlocomotion and indifference to cocaine and amphetamine in mice lacking the dopamine transporter. *Nature, 379*, 606–612.

Goldstein, G, Materson, BJ, Cushman, WC, Reda, DJ, Freis, ED, & Ramirez EA, et al. (1990). Treatment of hypertension in the elderly: II. Cognitive and behavioral function: Results of a Department of Veterans Affairs Cooperative Study. *Hypertension, 15*, 361–369.

Hartsough, CS, & Lambert, NM (1985). Medical factors in hyperactive and normal children: Prenatal, developmental, and health history findings. *Am J Orthopsychiat, 55*, 190–201.

Hendley, ED, & Ohlsson, WG (1991). Two new inbred rat strains derived from SHR: WKHA, hyperactivity, and WKHT, hypertensive, rats. *Am J Physiol, 261*, H583–H589.

Hendley, ED, Wessel, DJ, Atwater, DG, Gellis, J, Whitehorn, D, & Low, WC (1985). Age, sex and strain differences in activity and habituation in SHR and WKY rats. *Physiol Behav, 34*, 379–383.

Hendley, ED, Wessel, DJ, & Van Houten, J (1986). Inbreeding of Wistar-Kyoto rat strain with hyperactivity but without hypertension. *Behav. Neural Biol., 45*, 1–16.

Hess, EJ, Collins, KA, Copeland, NG, Jenkins, NA, & Wilson, MC (1994). Deletion map of the Coloboma (Cm) locus on mouse chromosome 2. *Genomics, 21*, 257–261.

Hess, EJ, Collins, KA, & Wilson, MC (1996). Mouse model of hyperkinesis implicates SNAP-25 in behavioral regulation. *J Neurosci, 16*, 3104–3111.

Hess, EJ, Jinnah, HA, Kozak, CA, & Wilson, MC (1992). Spontaneous locomotor hyperactivity in a mouse mutant with a deletion including the Snap gene on chromosome 2. *J Neurosci, 12*, 2865–2874.

Hess, EJ, Rogan, PK, Domoto, M, Tinker, DE, Ladda, RL, & Ramer, JC (1995). Absence of linkage of apparently single gene mediated ADHD with the human syntenic region of the mouse mutant coloboma. *Am J Med Genet, 60*, 573–579.

Heyser, CJ, Wilson, MC, & Gold, LH (1995). Coloboma hyperactive mutant exhibits delayed neurobehavioral developmental milestones. *Dev Brain Res, 89*, 264–269.

Holson, RR, Gazzara, RA, Ferguson, SA, & Adams, J (1997). Behavioral effects of low-dose gestational day 11–13 retinoic acid exposure. *Neurotoxicol Teratol, 19*, 355–362.

Hunziker, MHL, Saldana, RL, & Neuringer, A (1996). Behavioral variability in SHR and WKR rats as a function of rearing environment and reinforcement contingency. *J Exp Anal Behav, 65*, 129–144.

Hynd, GW, Nieves, N, Connor, RT, Stone, P, Town, P, & Becker, MG, et al. (1989). Attention deficit disorder with and without hyperactivity: Reaction time and speed of cognitive processing. *J Learning Disabil, 22*, 573–580.

Jaber, M, Jones, S, Giros, B, & Caron, MG (1997). The dopamine transporter: A crucial component regulating dopamine transmission. *Movement Disorders, 12*, 629–633.

Johnson, ML, Ely, DL, & Turner, ME (1992). Genetic divergence between the Wistar-Kyoto rat and the spontaneously hypertensive rat. *Hypertension, 19*, 425–427.

Jones, SR, Gainetdinov, RR, Wightman, RM, & Caron, MG. (1998). Mechanisms of amphetamine action revealed in mice lacking the dopamine transporter. *J. Neurosci., 18*, 1979–1986.

Katsetos, CD, Hyde, TM, & Herman, MM (1997). Neuropathology of the cerebellum in schizophrenia — An update: 1996 and future directions. *Biol Psychiatry, 42*, 213–224.

Kostrzewa, RM, Brus, R, Kalbfleisch, JH, Perry, KW, & Fuller, RW (1994). Proposed animal model of attention deficit hyperactivity disorder. *Brain Res Bull, 34*, 161–167.

Lalonde, R, & Botez, MI (1990). The cerebellum and learning processes in animals. *Brain Res Rev, 15*, 325–332.

Law, SF, & Schachar, RJ (1999). Do typical clinical doses of methylphenidate cause tics in children treated for attention-deficit hyperactivity disorder? *J Am Acad Child Adolesc Psychiatry, 38*, 944–951.

Leiner, HC, Leiner, AL, & Dow, RS (1986). Does the cerebellum contribute to mental skills? *Behav Neurosci, 100*, 443–454.

Leiner, HC, Leiner, AL, & Dow, RS (1993). Cognitive and language functions of the human cerebellum. *Trends Neurosci, 16*, 444–447.

Lindpainter, K, Kreutz, R, & Ganten, D (1992). Genetic variation in hypertensive and "control" strains: What are we controlling for anyway? *Hypertension, 19*, 428–430.

Low, WC, Whitehorn, D, & Hendley, ED (1984). Genetically related rats with differences in hippocampal uptake of norepinephrine and maze performance. *Brain Res Bull, 12*, 703–709.

Luthman, J, Bassen, M, Fredriksson, A, & Archer, T (1997). Functional changes induced by neonatal cerebral 6-hydroxydopamine treatment: Effects of dose levels on behavioral parameters. *Behav Brain Res, 82*, 213–221.

Luthman, J, Fredriksson, A, Sundstrom, E, Jonsson, G, & Archer, T (1989). Selective lesion of central dopamine or noradrenaline neuron systems in the neonatal rat: Motor behavior and monoamine alterations at adult stage. *Behav Brain Res, 33*, 267–277.

Martin, P, & Albers, M (1995). Cerebellum and schizophrenia: A selective review. *Schizophr Bull, 21*, 241–250.

McCarty, R, Cierpial, MA, Murphy, CA, Lee, JH, & Fields-Okotcha, C (1992). Maternal involvement in the development of cardiovascular phenotype. *Experientia, 48*, 315–322.

Meneses, A, Castillo, C, Ibarra, M, & Hong, E (1996). Effects of aging and hypertension on learning, memory, and activity in rats. *Physiol Behav, 60*, 341–345.

Mook, DM, Jeffrey, J, & Neuringer, A (1993). Spontaneously Hypertensive Rats (SHR) readily learn to vary but not repeat instrumental responses. *Behav Neural Biol, 59*, 126–135.

Mook, DM, & Neuringer, A (1994). Different effects of amphetamine n reinforced variations versus repetitions in spontaneously hypertensive rats (SHR). *Physiol Behav, 56*, 939–944.

Myers, MM, Brunelli, SA, Squire, JM, Shindeldecker, RD, & Hofer, MA (1989). Maternal behavior of SHR rats and its relationship to offspring blood pressures. *Dev Psychobiol, 22*, 29–53.

Nakamura-Palacios, EM, Caldas, CK, Fiorini, A, Chagas, KD, Chagas, KN, & Vasquez, EC (1996). Deficits of spatial learning and working memory in spontaneously hypertensive rats. *Behav Brain Res, 74*, 217–221.

Nasrallah, HA, Loney, J, Olson, SC, McCalley-Whitters, M, Kramer, J, & Jacoby, CG (1986). Cortical atrophy in young adults with a history of hyperactivity in childhood. *Psychiatry Res, 17*, 241–246.

Rapoport, JL, Buchsbaum, MS, Zahn, TP, Weingartner, H, Ludlow, C, & Mikkelsen, EJ (1978). Dextroamphetamine: Cognitive and behavioral effects in normal prepubertal boys. *Science, 199*, 560–563.

Sagvolden, T, Hendley, ED, & Knardahl, S (1992). Behavior of hypertensive and hyperactive rat strains: Hyperactivity is not unitarily determined. *Physiol Behav, 52*, 49–57.

Sagvolden, T, Metzger, MA, & Sagvolden, G (1993). Frequent reward eliminates differences in activity between hyperkinetic rats and controls. *Behav Neural Biol, 59*, 225–229.

Sagvolden, T, Metzger, MA, Schiorbeck, HK, Rugland, A-L, Spinnangr, I, & Sagvolden, G (1992). The Spontaneously Hypertensive Rat (SHR) as an animal model of childhood hyperactivity (ADHD): Changed reactivity to reinforcers and to psychomotor stimulants. *Behav Neural Biol, 58*, 103–112.

Sagvolden, T, Pettersen, MB, & Larsen, MC (1993). Spontaneously hyertensive rats (SHR) as a putative animal model of childhood hyperkinesis: SHR behavior compared to four other rat strains. *Physiol Behav, 54*, 1047–1055.

Sanders, BJ, & Gray, MJ (1997). Early environmental influences can attenuate the blood pressure response to acute stress in borderline hypertensive rats. *Physiol Behav, 61*, 749–7654.

Searle, AJ (1966). New mutants, Vol. 2: Coloboma. *Mouse News Letter, 35*, 27.

Shaywitz, BA, Klopper, JH, & Gordon, JW (1978). Methylphenidate in 6-hydroxydopamine-treated developing rat pups: Effects on activity and maze performance. *Arch Neurol, 35*, 463–469.

Shaywitz, BA, Yager, RD, & Klopper, JH (1976). Selective brain dopamine depletion in developing rats: An experimental model of minimal brain dysfunction. *Science, 191*, 305–307.

Smith, BH, Pelham, WE, Gnagy, E, & Yudell, RS (1998). Equivalent effects of stimulant treatment for attention-deficit hyperactivity disorder during childhood and adolescence. *J Am Acad Child Adolesc Psychiatry, 37*, 314–321.

Smith, M (1989). The effects of low-level lead exposure on children. In MA Smith, LD Grant, AI Sors (Eds.), *Lead exposure and child development: An international assessment* (pp. 3–47). Dordrecht: Kluwer Academic Publishers.

Solanto, MV (2000). Clinical psychopharmacology of AD/HD: Implications for animal models. *Neurosci Biobehav Rev, 24*, 27–30.

Sonuga-Barke, EJ, Taylor, E, Sembi, S, & Smith, J (1992). Hyperactivity and delay aversion. I. The effect of delay on choice. *J Child Psychol. Psychiatry, 33*, 387–398.

St. Lezin, E, Simonet, L, Pravenec, M, & Kurtz, TW (1992). Hypertensive strains and normotensive "control" strains: How closely are they related? *Hypertension, 19*, 419–424.

Stewart, MA, Pitts, FN, Jr., Craig, AG, & Dieruf, W (1966). The hyeractive child syndrome. *Amer J Orthopsych, 36*, 861–867.

Ullman, DG, Barkley, RA, & Brown, HW (1978). The behavioral symptoms of hyperkinetic children who successfully responded to stimulant drug treatment. *Amer J Orthopsych, 48*, 425–437.

Van Den Buuse, M, & De Jong, W (1989). Differential effects of dopaminergic drugs on open-field behavior of spontaneously hypertensive rats and normotensive Wistar-Kyoto rats. *J Pharmacol Exp Ther, 248*, 1189–1196.

Whitmont, S, & Clark, C. (1996). Kinaesthetic acuity and fine motor-skills in children with attention deficit hyperactivity disorder: A preliminary report. *Dev Med Child Neurol, 38*, 1091–1098.

Wool, RS, Weldon, DA, Shaywitz, BA, Anderson, GM, Cohen, DJ, & Teicher, MH (1987). Amphetamine reverses learning deficits in 6-hydroxydopamine-treated rat pups. *Dev Psychobiol, 20*, 219–232.

Wultz, B, & Sagvolden, T (1992). The hyperactive spontaneously hypertensive rat learns to sit still, but not to stop bursts of responses with short interresponse times. *Behav Genet, 22*, 415–433.

Wultz, B, Sagvolden, T, Moser, EI, & Moser, M-B (1990). The spontaneously hypertensive rat as an animal model of Attention-Deficit Hyperactivity Disorder: Effects of methylphenidate on exploratory behavior. *Behav Neural Biol, 53*, 88–102.

Wyss, JM, Fisk, G, & Van Groen, T (1992). Impaired learning and memory in mature spontaneously hypertensive rats. *Brain Res, 592*, 135–140.

9

The Spontaneously Hypertensive Rat as a Model of ADHD

TERJE SAGVOLDEN

In all areas of research, one is trying to find or build models that can be used to summarize and simplify the understanding of a disorder, research problem, or phenomenon under investigation. In clinical research there are several advantages associated with animal models of disorders (Sagvolden et al., 1992a; Sagvolden, 1996; Sagvolden & Sergeant, 1998a; Solanto, 1998; Taylor, 1998). Ethical, methodological and economic factors restrict research on human subjects. For example, ethical factors limit studies of drug effects on children with ADHD and comparison children. The heterogeneity of subjects within groups of ADHD children creates methodological problems. Economic factors limit the possibility of examining behavioral processes in highly controlled settings. An animal model of ADHD is therefore needed. It should be quite similar to the disorder in terms of etiology, biochemistry, symptomatology, and treatment (McKinney & Bunney, 1969). Still, the researcher undoubtedly will deal with a simpler nervous system and behavior that may be easier to interpret than the full-blown clinical case; the groups may be genetically more homogeneous; environmental control is simpler; and interventions of various sorts are possible. The Spontaneously Hypertensive Rat (SHR) is the most frequently used ADHD model. This is a strain that has been bred from progenitor Wistar Kyoto rats (WKY) (Okamoto & Aoki, 1963).

Validation Criteria

In the literature on animal models of ADHD, there has not been any systematic discussion on how to assess the validity of models. Thus, the most adequate model is the one that (*a*) best mimics, although in a simpler form than the full-blown clinical case, the fundamentals of the behavioral characteristics of ADHD—*face validity*, (*b*) conforms with a theoretical rationale for ADHD—*construct validity*, and (*c*) is able to predict aspects of ADHD behavior, genetics, and neurobiology previously uncharted in the clinics—*predictive validity* (Sarter et al., 1992; Willner, 1984). Of these, construct validity is the most problematic as there is only limited agreement on how to explain ADHD (Barkley, 1997; Castellanos, 1997; Sagvolden & Sergeant, 1998a; Swanson et al., 1998; Taylor, 1998). In order to achieve such goals, there have to be (*a*) close interactions between clinical researchers defining the symptoms of the disorder and basic researchers trying to validate animal models and (*b*) systematic developments of comparative behavioral tasks that can be used both with humans and animals in order to evaluate behavioral homology.

Face Validity: Mimicking the Fundamental Aspects of the ADHD Symptomatology

In order to discuss Willner's validation criteria (Willner, 1984) as well as Taylor's (Taylor, 1998) requirement of homology of symptoms, ADHD symptoms will be compared with SHR behavioral deviations applying a task where ADHD children, SHR, and appropriate controls have been tested with homologous behavioral paradigms. The design of behavioral tests that can be used comparatively is based on operant analyses of behavior originated by B. F. Skinner (Ferster & Skinner, 1957). Operant analyses are used because the methods and results are generally easy to replicate. These analyses enable study of only one or a few aspects of the behavior, apply a common nomenclature of behavior, and make possible comparative research (Berger & Sagvolden, 1998; Matthews et al., 1977; Sagvolden et al., 1998b).

We have systematically investigated behavioral characteristics of children with ADHD and control children, and compared their behavior to that of SHR and a large number of rat strains. Both children with ADHD and control children (Sagvolden et al., 1998b) as well as SHR and other strains (Berger & Sagvolden, 1998; Boix et al., 1998; Holene et al., 1998; Sagvolden et al., 1992b; Sagvolden et al., 1993a) have been tested with the very same multiple fixed-interval/extinction schedule of reinforcement. All experiments were run with the same computer programs, with same type of interface, but of course with different manipulanda and reinforcers. A schedule is termed multiple when two or more schedule components operate in alternation, each in the presence of a different stimulus. The fixed-interval component delivered reinforcers (e.g., trinkets to children or drops of water to rats) at fixed time intervals if the organism performs the required response, e.g., pressing a lever switch. Fixed-interval schedules measure aspects of the time course of reinforcement processes (Catania et al., 1988; Sagvolden et al., 1992a). It is possible to measure both the efficacy of the reinforcer (the reinforcement-delay-gradient factor) as well as the maximum attainable response rate in the given organism and situation (the response-rate-limiting factor) (Sagvolden et al., 1992a). Finally, the fixed-interval component makes it possible to measure motor impulsiveness, that is, responses with short interresponse times (IRTs). The other schedule component was in effect when the signal was turned off. Then no reinforcer was ever delivered. This is called an extinction component. The extinction component measures sensitivity to stimulus change and sustained attention.

Deficient Sustained Attention

In the clinics, the problems with *sustained* attention in ADHD occur only in situations where stimuli are widely spaced in time (van der Meere, 1996) or when the ability to concentrate is stressed by the task being unwelcome or uninteresting (Taylor, 1998). In the extinction component of the multiple fixed-interval/extinction schedule of reinforcement, a peculiar behavior developed in the extinction component. The ADHD children (as well as the SHR) did not differ from comparison children at the start of testing. Furthermore, at the start of every extinction component, both ADHD and comparison children noticed when the light signaled that the schedule changed to the extinction component by ceasing to respond, but the ADHD children consistently resumed responding after a short time (Sagvolden et al., 1998b). Closely similar results were obtained with SHR. This peculiar extinction behavior seems to be in accordance with common definitions of deficient sustained attention. In order to show if other strains have a similar sustained attention deficit, a number of strains were run with the same schedule (Berger & Sagvolden 1998; Boix et al., 1998; Holene et al., 1998; Sagvolden et al., 1992b; Sagvolden et al., 1993a). These studies showed normal sensitivities to stimulus change, but a deficient sustained attention in ADHD children and SHR only (Sagvolden, 1999).

Overactivity

ADHD children show restlessness, fidgeting and generally unnecessary gross body movements (Porrino et al., 1983; Taylor, 1998). This overactivity might not always be present (Taylor, 1998) for it seems to be absent in novel situations (Sagvolden et al., 1998b; Sleator & Ullman, 1981).

The SHR is hyperactive in a variety of behavioral paradigms (Knardahl & Sagvolden, 1979; McCarty & Kopin, 1979; Moser et al., 1988; Myers et al., 1982; Sagvolden et al., 1992b; Sagvolden et al., 1993a; Wultz et al., 1990; Wultz & Sagvolden, 1992). When tested with the multiple fixed-interval/extinction schedule of reinforcement, the ADHD children had the same activity level as the comparison children at the start of testing. The overactivity in the fixed-interval component developed as gradually more reinforcers were given (Sagvolden et al., 1998b). Closely similar results were obtained with the SHR.

A series of studies has been run in order to ensure that it is the SHR that is hyperactive, not the WKY comparison group that is hypoactive (Berger & Sagvolden, 1998; Boix et al., 1998; Holene et al., 1998; Sagvolden et al., 1992b; Sagvolden et al., 1993a). The group treated with PCB 153 (Holene et al., 1998) is the only other group showing an overactivity similar to that of the SHR.

The ADHD and SHR activity is only about twice that of the comparison groups. The reason for the limited increase in response output in these studies compared to open fields might be that the behavior is under rather tight control by scheduled reinforcers. Results from other activity paradigms often yield larger differences in activity between SHR and other strains (Knardahl et al., 1979; McCarty & Kopin, 1979; Moser et al., 1988; Myers et al., 1982; Sagvolden et al., 1992b; Sagvolden et al., 1993a; Wultz et al., 1990; Wultz & Sagvolden, 1992). However, the main problem with open fields is the lack of standardization of equipment and procedures.

Motor Impulsiveness

Impulsiveness is increasingly seen as the symptom of greatest significance (Sagvolden & Sergeant, 1998a; Taylor, 1998). The peculiar ADHD style of brief, short sequences of activity on tasks and rapid change is typical in the SHR. This strain responds with bursts of activity rather than careful, well-controlled moves. In the open field, the SHR typically exhibits short runs entering 5 to 15 sections without stopping, seemingly without exploring its path. SHR motor impulsiveness is also seen in the home cage as quick, jerky movements (Sagvolden et al., 1992b; Wultz & Sagvolden 1992).

With the multiple fixed-interval/extinction schedule, motor impulsiveness is operationalized as bursts of responses with short interresponse times. These responses are maladaptive because they will not result in any increase in the number of reinforcers. The ADHD children do not show any motor impulsiveness at the start of testing. The impulsiveness develops gradually (Sagvolden et al., 1998b). Motor impulsiveness is seen in *both male and female* SHR (Sagvolden, 2000). Neither WK-HA nor any of the other untreated strains show motor impulsiveness (Berger et al., 1998; Boix et al., 1998; Holene et al., 1998; Sagvolden et al., 1992b). Male *Lewis* rats treated with PCB 153 is the only other group showing impulsiveness similar to that of SHR (Holene et al., 1998; Sagvolden, 2000).

Other Behavioral Similarities Between ADHD Children and SHR

ADHD behavior is generally more variable than what is normal (Oosterlaan & Sergeant, 1998; Rubia et al., 1998). Also SHR behavior shows increased intra- as well as intersubject variability (Mook et al., 1993). Finally, the deficient response re-engagement (i.e., inhibit one response and start another) in ADHD children (Oosterlaan & Sergeant, 1998) is similar to the deficient response re-engagement seen in SHR when high response rates are induced by the reinforcement schedule. This aspect of SHR behavior is seen in the reduced response rates of the SHR when a high reinforcement density is inducing quick changes of responses (Sagvolden et al., 1992a) and in the longer minimum IRTs of the SHR compared to the WKY (Metzger & Sagvolden, 1994).

The SHR Hypertension

It has been discussed whether the essential hypertension causes the behavioral changes of the SHR. In order to separate hyperactivity and hypertension, Hendley (Hendley et al., 1986) developed a hyperactive, but not hypertensive, strain (WK-HA) and a hypertensive, but not hyperactive, strain (WK-HT) by crossbreeding SHR with WKY followed by selective inbreeding of the desired traits. However, the behavior of WK-HA differs from that of SHR (below). Part of the reason why it seems to be difficult to separate these traits is the fact that dopamine is a major neuromodulator in the control of cortical microcirculation (Krimer et al., 1998) as well as behavior (below). In addition, hypertension (when there is a family history of hypertension) is associated with poorer attention and short-term memory, suggesting once more a genetic link between hypertension and impaired cognitive abilities (Thyrum et al., 1995).

Conclusions on the Face Validity of the SHR as an Animal Model of ADHD

It is obvious from the results of studies that have been using the same behavioral schedule for testing ADHD children and various rat strains, that male SHR shows all the behavioral characteristics of ADHD as operationalized: A sustained-attention deficit without obvious sensory problems, a normal activity level in a novel situation that develops into overactivity after some time, and motor impulsiveness that is not present in a novel situation but develops. Finally, just like ADHD children, the male SHR shows increased behavioral variability and a deficient response re-engagement. Thus, it seems reasonable to conclude that face validity of the SHR as an animal model of ADHD has been established.

Other Animal Models of ADHD

Researchers have suggested other animal models of ADHD besides the SHR (Ferguson, 2000). By crossbreeding of SHR with WKY, followed by selected inbreeding of the desired traits, Hendley (Hendley et al., 1986) developed SHR and WKY into a hyperactive, but not hypertensive strain (WK-HA) and a hypertensive, but not hyperactive strain (WK-HT). With the addition of these strains it seemed possible to compare the SHR (impulsive, hyperactive, deficient in sustained attention as well as hypertensive) with WK-HA (hyperactive), WK-HT (hypertensive), and WKY controls. Although the SHR, WK-HA, and WK-HT were hyperactive in activity tests, it turned out that the hyperactivity depends on the parameters of the test situation (free- vs. forced open fields, duration of testing, illumination of the field, etc.) (Sagvolden et al., 1993a). All three strains develop hyperactivity in free-exploration open fields (Sagvolden et al., 1992b). When tested with the multiple fixed-interval/extinction schedule of reinforcement, however, only the SHR was overactive (Holene et al., 1998; Sagvolden et al., 1992b; Sagvolden et al., 1993a). Thus, the WK-HA strain is hyperactive in open fields, but evidently not when the behavior is under control of the fixed-interval operant reinforcement schedule.

Other models that have been suggested are: Rats selected from a general population (Puumala et al., 1997); rats reared in social isolation (Robbins et al., 1989); rats exposed to environmental pollutants like lead (Silbergeld & Goldberg, 1974) or polychlorinated biphenyls (PCBs) (Holene et al., 1995, 1998); rats that have undergone neonatal anoxia (Dell'Anna et al., 1993); rats that have undergone hippocampal

X-irradiation in infancy (Diaz-Granados et al., 1994); rats that have undergone neurotoxic brain lesions (Archer, 1989; Shaywitz et al., 1976b; Shaywitz et al., 1976a) other genetic models such as the Naples High/Low Excitability Rats (Sadile, 1993; Sadile et al., 1996), or knockout mice (Giros et al., 1996).

In general, it seems that these other models of ADHD have not yet been advanced beyond face validity. In some cases, not even that—e.g., the fact that ADHD overactivity is not present at all times, but develops, makes the overactivity of some of these animal models differ fundamentally from that of children with ADHD and even violates aspects of face validity.

In conclusion, it seems that the general lack of systematic and parallel investigations of these animal models and ADHD children hamper a thorough discussion of the validity of these models at present.

Construct Validity: Do Changes in SHR Genetics, Neurobiology and Behavior Conform with a Theoretical Rationale for ADHD?

In comprehensive, long-term studies of neural systems underlying arousal and attention, Robbins and coworkers (1998) have tested rats with selective neurochemical manipulations behaviorally, using an analogue of an attentional task developed for human subjects. They suggest that the nigro-striatal and mesolimbic dopaminergic systems are involved in response speed and vigor, the coeruleo-cortical noradrenergic system in divided and selective attention, the basal forebrain cholinergic system in stimulus detection, and the mesencephalic serotoninergic systems in response inhibition. However, all these systems are closely linked and an imbalance in one will inevitably affect the other.

It is increasingly clear that ADHD is associated with altered monoamine function. A role for dopamine and other monoamine dysfunctions in ADHD has been suspected since 1969 when Le Moal showed that dopamine depletion following ventral tegmental-area lesions, and Shaywitz later showed that massive dopamine depletion by the neurotoxin 6-OHDA in juvenile rats, produce hyperactivity (Shaywitz et al., 1976b; Stinus et al., 1992) that is reversible by amphetamine (Shaywitz et al., 1976a).

More recently, dopamine gene polymorphism has been shown in ADHD (Cook, Jr. et al., 1995; LaHoste et al., 1996). It is therefore interesting that dopamine hypofunctioning has been shown in SHR, too (Carey et al., 1998; Linthorst et al., 1991; Russell et al., 1998). The time course of dopamine release into and clearance from the synaptic cleft seems to be critical for the long-term synaptic functions that are the neuronal substrates of reinforcement (cf., Di Chiara & Imperato, 1988; Schultz, 1998). Thus the reduced dopamine function may cause the repeatedly suggested altered reinforcement processes in ADHD symptomatology (Douglas, 1983; Douglas & Parry, 1994; Sagvolden, 1996; Sagvolden et al., 1998b; Sagvolden & Archer, 1989; Sonuga-Barke et al., 1992; Wender, 1971). It is, however, unlikely that any one gene will account for the whole of the ADHD syndrome (Taylor, 1998).

SHR Reactivity to Psychomotor Stimulants

ADHD is usually treated with psychostimulants (methylphenidate [Ritalin], d-amphetamine, or pemoline). Methylphenidate is the drug of choice (see, e.g., Solanto, 1998). Due to the detailed and refined methods available with animal models, it is possible to describe effects of drugs mathematically. SHR shows reduced reactivity to methylphenidate and to d-amphetamine compared to WKY when tested with the multiple fixed-interval/extinction schedule (Sagvolden et al., 1992a). A dose of methylphenidate brings SHR fixed-interval behavior into substantial conformity, not with WKY baseline behavior, but with WKY behavior following a lower dose. There is no "paradoxical" effect because the dose affects the magnitudes, not the directions of the effects. The methylphenidate effects were shown to be complex as they depend not only on the group variable, but on time till the next reinforcer (s) and dose (d) as well as their interactions sd, d^2 and s^2. Methylphenidate effects on motor impulsiveness were not measured in this study. However, the monoamine oxidase inhibitor l-deprenyl abolishes the development of motor impulsiveness of the SHR (Boix et al., 1998). These results are in accordance with the suggestion that psychomotor stimulants may lengthen the delay gradient (Sagvolden et al., 1988; below).

The reduced reactivity to methylphenidate and to d-amphetamine by the SHR may be associated with dopamine hypofunctioning. Russell has presented evidence in favor of an impaired vesicular storage in the SHR. By using an in vitro superfusion technique, reduced dopamine release was shown in the main projection areas of the dopaminergic systems in the *male* SHR (Russell et al., 1995, 1998). The exact mechanisms of action of the enhanced catecholamine neurotransmission by psychomotor stimulants were studied in more detail (Russell et al., 1998). It was shown that methylphenidate releases dopamine from vesicle stores only and is 7 to 17 times less potent than d-amphetamine, thus making it possible to adjust the dose and thereby "normalize" reduced dopamine functions more precisely than is possible with d-amphetamine.

Boix (Boix et al., 1998) has shown that treatment with the MAO inhibitor l-deprenyl increased dopamine levels in all areas of the male SHR brain. Without treatment, the male SHR showed a lower dopamine turnover (homovanillic acid, HVA, to DA) ratio in the neostriatum as well as an asymmetric HVA/DA ratio in the accumbens with a lower ratio in the right accumbens.

The central nervous system of the juvenile, prehypertensive, male SHR may compensate for the reduced release of dopamine by increasing the numbers of dopamine D-1/D-5 receptor binding sites in the rostral portions of the neostriatum, the core and the shell of the accumbens, and the olfactory tubercle in SHR as shown by quantitative receptor autoradiography (Carey et al., 1998). Such increased levels are associated with lower receptor affinities.

An impaired dopamine storage in the male SHR might not only have caused reduced release and lower D-1 receptor affinities of DA, but also reduced Ca^{2+}/calmodulin-dependent protein kinase II (CaMKII), the major protein of the postsynaptic density and of signal transduction, in the anterior portions of the accumbens shell and a lower expression of peptide products of the immediate early genes of the *FOS* family (*c-FOS*, in particular) and *ZIF-268* (Papa et al., 1998a). The lower expressions of both CaMKII and the immediate early genes are most likely due to reduced neuronal activity in these areas (Papa et al., 1998a).

Finally, there might be an impaired cross-talk between the various brain structures of an ADHD child (Oades, 1998). Such an impaired dialogue is seen as uniformly high correlations between the cytochrome oxidase activities of the various brain areas in the SHR, but not in the WKY (Papa et al., 1998b). This lack of modulation in the SHR could be the result of reduced dopaminergic functions.

A 15-day sub-chronic treatment with methylphenidate normalizes the dopamine D-1/D-5 receptors by decreasing the number of binding sites and increasing the affinity to control level (Carey et al., 1998) and normalizes CaMKII levels (Papa et al., 1997) in the male SHR without affecting the WKY. The impaired cross-talk between various brain structures of the SHR is also improved by methylphenidate treatment (Papa et al., 1997).

The Roles of Other Amines in ADHD

If the impaired vesicular storage of dopamine (Russell et al., 1998) is due to a deficient vesicular amine transporter in SHR as well as in ADHD children, not only dopaminergic systems, but also the other aminergic systems, norepinephrine, adrenaline, serotonin, and histamine, should be hypofunctioning (see references in Carey et al., 1998). However, if the genetic alteration in ADHD is coding for the dopamine receptors and/or membrane transporters only (cf. Cook, Jr. et al., 1995; LaHoste et al., 1996), the primary changes may well be quite restricted to dopamine. Boix (Boix et al., 1998) found that both SHR and WKY had higher noradrenergic concentrations in the left neostriatum and that this asymmetry was less pronounced in the SHR. There is also a higher serotoninergic activity in the neostriatum and the frontal cortex of the SHR (Boix et al., 1998), and altered metabolite/serotonin ratios in several brain areas of the SHR (de Villiers et al., 1995).

In conclusion, SHR shows reduced reactivity to methylphenidate (Ritalin) and to d-amphetamine compared to WKY. It is unlikely that this is a result of dopamine hypofunctioning only. By allowing precise and detailed measurements of behavioral as well as neurobiological effects of drugs, the SHR may be a very valuable tool in the future development of drugs for treating ADHD.

A Shorter Delay-of-Reinforcement Gradient

Reinforcers act *retroactively* on responses that already took place in the sense that they increase the probability of future occurrences of responses of the same operant class (Catania et al., 1988). The reinforcing effect is largest when the reinforcer is delivered immediately after the response was emitted and wanes as a function of the delay of the delivery of the reinforcer. This relation between the effect of the reinforcer and the time interval between response and reinforcer is commonly known as the "delay-of-reinforcement gradient" or simply as the "delay gradient" (Catania et al., 1988; Sagvolden et al., 1998b) (Fig. 9–1). Further, a reinforcer acts not only on the response that produced it but also, to a lesser degree, on responses emitted earlier (Catania, 1971).

Effects of reinforcers on SHR and WKY behavior have been quantified. It has been shown that differences between SHR and WKY behavior are solely due to increased reactivity to reinforcers and not to higher *maximal* response rates in the SHR (Sagvolden et al., 1992a; Sagvolden et al., 1993b). Immediate reinforcers are more effective in SHR than in WKY. The effects of delayed reinforcers are reduced in SHR. It will be argued that at least partially, all three symptoms of ADHD as well as the behavioral characteristics of the SHR, deficient sustained attention, overactivity, and motor impulsiveness, may be due to one factor: a steeper and shorter delay gradient. This may explain why these three symptoms *sometimes* emerge as one factor in factor analyses (cf. Taylor, 1998).

Overactivity Explained as Caused by a Shorter and Steeper Delay-of-Reinforcement Gradient

The hypothesized steeper and shorter delay gradient in children with ADHD and the SHR (Sagvolden, 1996; Sagvolden et al., 1989, 1998b; Sagvolden & Sergeant, 1998a) is probably rooted in a genetically based hypofunctional dopamine system

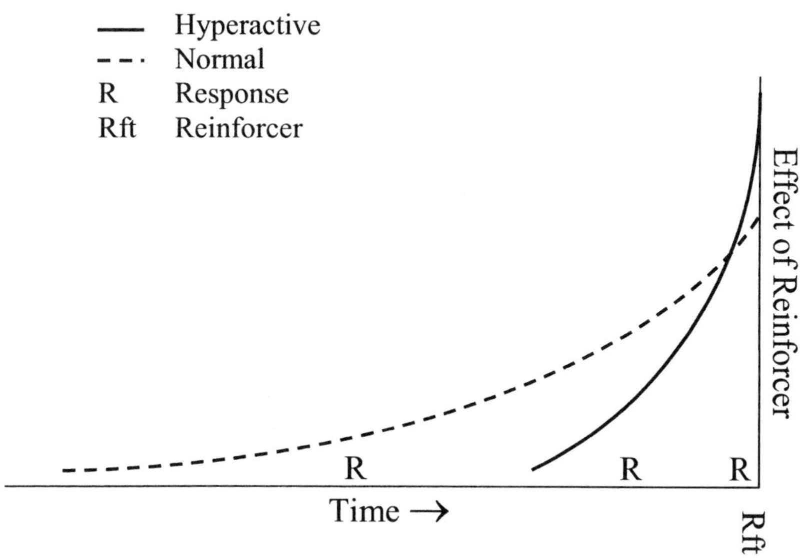

Figure 9–1. Theoretical delay-of-reinforcement gradients. The effect of a reinforcer is more potent when there is a short delay between the response and the reinforcer than when there is a longer delay. The delay gradient may be shorter and steeper in the ADHD children than in normal children. (Modified after Sagvolden et al., 1998b.)

(de Villiers et al., 1995; Russell et al., 1995, 1998). A reinforcer in close proximity to a response may therefore be more effective in ADHD children and in SHR than in controls. This will generate overactivity as gradually more of these reinforcers act on the behavior emitted in a particular situation.

Increased Behavioral Variability

ADHD and SHR behavior are generally more variable than what is normal (Mook et al., 1993; Oosterlaan & Sergeant, 1998; Rubia et al., 1998). Variability is adaptive in some situations and acts as an operant response that may be modified by reinforcers (Mook et al., 1993; Saldana & Neuringer, 1998). Thus, it is not surprising that there are systematic differences in the behavioral problems of children with ADHD depending on the type of situation (Taylor, 1998). The increased inter- and intrasubject variability may be explained as an altered delay gradient: the behavior that frequently is reinforced in children with ADHD (and in SHR) is the behavior that randomly occurs just before a reinforcer is delivered. A more effective immediate reinforcer in children with ADHD (and in SHR) will have a greater impact and therefore increases the likelihood also of the randomly occurring behavior, not only the systematically occurring normal behavior.

Motor Impulsiveness Caused by a Shorter and Steeper Delay-of-Reinforcement Gradient

Not only *single* responses, but also *sequences of* responses are conditioned and maintained by reinforcers (Catania, 1971; Catania et al., 1988; Sagvolden et al., 1998b; Sagvolden & Sergeant, 1998a). In contrast to the normal delay gradient, a short gradient can only reinforce and maintain responses emitted with short interresponse times (IRTs) (Fig. 9–1). This explains why motor impulsiveness, responses emitted with short IRTs, is not present in a novel situation, but develops gradually as more reinforcers modify the behavior of children with ADHD and SHR (Berger & Sagvolden, 1998; Sagvolden et al., 1998b; Sagvolden & Sergeant, 1998a).

Analyses of activity level as functions of reinforcement densities (Sagvolden et al., 1993b) showed no difference between hyperactives and controls when the reinforcers were frequent, only when they were infrequent. The reason for this is that the "tail" of the gradient of each reinforcer is truncated by the previous reinforcer (Catania et al., 1988). Responses emitted between any two reinforcers are only reinforced by one reinforcer, the first one following these responses. A second or third reinforcer following these responses will have no reinforcing effect on these responses. Consequently, frequent reinforcers will curtail the operative part of the normal delay gradient, making it more ADHD-like. This will cause overactive and impulsive behavior in normal organisms. Thus, contrary to previously assumptions (Taylor, 1998), motor impulsiveness may be due to altered reactivity to reinforcers, not vice versa.

Impaired Sustained Attention May Be Caused by a Shorter and Steeper Delay-of-Reinforcement Gradient

Frequently, a reinforcement contingency operates in one situation and not in another. This stimulus-response-consequence relation, the three-term contingency (Catania, 1992), is important for understanding the impaired sustained attention in ADHD. Especially when reinforcers are infrequent, an abnormally short delay gradient may not bridge the time interval between consecutive reinforcers. If, at the same time, the reinforcement contingency is signaled by a discriminative stimulus, a short delay gradient may cause an apparent lack of correlation between the discriminative stimulus and the reinforcement contingency. This results in poorer stimulus control and possibly impaired "sustained attention" in children with ADHD as well as in the SHR (Berger & Sagvolden, 1998; Catania et al., 1997; Sagvolden et al., 1998b).

Motor Control Problems

The neostriatum (caudate and putamen) plays a key role in integrating motor output and receives the most dense dopamine input of all central nervous structures. Striatal dopamine hypofunction has been shown in the SHR (de Villiers et al., 1995; Russell et al., 1995, 1998). Such changes predict subtle motor control problems in the SHR and in children with ADHD. Consistent with this prediction, ADHD children with a pervasive problem are more likely to show retarded development of language and motor functions and to have an onset in the first two years of life (Taylor, 1998). This might implicate neostriatal hypofunctioning also in ADHD. It might well be that results previously ascribed to response 'disinhibition' problems due to frontal-lobe dysfunction might be due to impaired motor control functions associated with dopamine hypofunctioning of the neostriatum. The longer and more variable reaction times (Brandeis et al., 1998; Oosterlaan & Sergeant, 1998; Rubia et al., 1998), the increased variability in speed, and the less accurate response re-engagement (inhibit one response and start another) (Oosterlaan & Sergeant, 1998), impaired orienting responses, and the increased number of responses with very long reaction times (Rubia et al., 1998; Saldana et al., 1998; Sonuga-Barke et al., 1998), language delays (Castellanos, 1997; Taylor, 1998) and so forth may be explained as motor control problems.

Conclusions on the Construct Validity of the SHR as an Animal Model of ADHD

It might be premature to conclude on the construct validity of SHR as an animal model of ADHD because there is only limited agreement on the theoretical rationale for AD/HD, but a promising hypothesis is the one relating dopamine hypofunction to behavior (Sagvolden & Sergeant, 1998a). A hypofunctioning mesolimbic dopamine system may produce a shorter and steeper delay gradient that on a behavioral level may cause deficient sustained attention, hyperactivity, and motor impulsiveness. A hypofunctioning mesocortical dopamine system may produce frontal cortical hypofunctioning resulting in cognitive impulsiveness. A hypofunctioning nigrostriatal dopamine system may cause several "extrapyramidal" symptoms associated with ADHD: poor motor control (clumsiness), longer reaction times, poor response timing, abnormal control of eye saccades, poor handwriting, poor correlation of the activity of different body parts, and so forth.

Predictive Validity: Predicting Aspects of ADHD Genetics, Neurobiology, and Behavior Previously Unknown

Although several authors (Douglas, 1983; Douglas et al., 1994; Sagvolden & Archer, 1989; Sonuga-Barke et al., 1992; Wender, 1971) have argued that altered reinforcement processes are important in ADHD symptomatology, details of these processes and their consequences were not known before the availability of the animal model (Sagvolden & Archer, 1989; Sagvolden et al., 1992a; Sagvolden et al., 1993b; Sagvolden, 1996; Sagvolden et al., 1998b). The most striking predictions are currently found on the behavioral level in terms of more precise descriptions of the altered delay gradient and resulting predictions of motor impulsiveness as the development of an overabundance of responses with short IRTs, development of hyperactivity, and nature of sustained attention in ADHD as discussed above. Future tests of the SHR as an animal model of ADHD may come in genetics, neurobiology, and pharmacology. Genetic studies may help localize candidate genes underlying various disorders. The very sensitive neurobiological methods available in the animal laboratory may help point out exactly which brain structures are malfunctioning in ADHD children (Carey et al., 1998; Papa et al., 1998b). Finally, the model may be used for testing new drugs.

Conclusions on the Validity of the SHR as an Animal Model of ADHD

An animal model should be quite similar to the disorder in terms of etiology, biochemistry, symptomatology, and treatment. Still, compared to the full-blown clinical case, there are several advantages associated with animal models. The researcher undoubtedly will deal with a simpler nervous system and behavior that may be easier to interpret than the full-blown clinical case, the groups may be genetically more homogeneous, environmental control is simpler, interventions of various sorts are possible, and so forth. It is argued that the best model of Attention-Deficit/Hyperactivity Disorder (ADHD) is the one that (*a*) best mimics, although in a simpler form than the full-blown clinical case, the fundamentals of the behavioral characteristics of ADHD (*face validity*), (*b*) conforms with a theoretical rationale for ADHD (*construct validity*), and (*c*) is able to predict aspects of ADHD behavior, genetics, and neurobiology previously uncharted in the clinics (*predictive validity*).

This chapter discusses the validity of the Spontaneously Hypertensive Rat (SHR) as an animal model of ADHD. It is argued that although other strains and species may be hyperactive and show attention deficits following genetic, environmental or pharmacological interventions, the SHR is at present the only strain shown to have the major behavioral symptoms of ADHD. This does not mean that investigating other animals cannot give valuable information.

References

Archer, T (1989). Neurotoxin-induced cognitive and motor activity modifications: A catecholamine connection. In T Sagvolden & T Archer (Eds.), *Attention deficit disorder: Clinical and basic research* (pp 287–322). Hillsdale, NJ: Lawrence Erlbaum Associates.

Barkley, RA (1997). Behavioral inhibition, sustained attention, and executive functions: Constructing a unifying theory of ADHD. *Psychol Bull, 121*, 65–94.

Berger, DF & Sagvolden, T (1998). Sex differences in operant discrimination behaviour in an animal model of Attention-Deficit Hyperactivity Disorder. *Behav Brain Res, 94*, 73–82.

Boix, F, Qiao, S-W, Kolpus, T, & Sagvolden, T (1998). Chronic L-deprenyl treatment alters brain monoamine levels and reduces impulsiveness in an animal model of Attention-Deficit/Hyperactivity Disorder. *Behav Brain Res, 94*, 153–162.

Brandeis, D, van Leeuwen, TH, Rubia, K, Vitacco, D, Steger, J, Pascual-Marqui, RD, & Steinhausen, H-C (1998). Neuroelectric mapping reveals precursor of stop failures in children with attention deficits. *Behav Brain Res, 94*, 111–125.

Carey, MP, Diewald, LM, Esposito, F, Pellicano, MP, Gironi Carnevale, UA, Sergeant, JA, Papa, M, & Sadile, AG (1998). Differential distribution, affinity and plasticity of dopamine D-1 and D-2 receptors in the target sites of the mesolimbic system in an animal model of ADHD. *Behav Brain Res, 94*, 173–185.

Castellanos, FX (1997). Toward a pathophysiology of Attention-Deficit/Hyperactivity Disorder. *Clin Pediatr (Phila.), 36*, 381–393.

Catania, AC (1971). Reinforcement schedules: The role of responses preceding the one that produces the reinforcer. *J Exp Anal Behav, 15*, 271–287.

Catania, AC (1992). *Learning*. Englewood Cliffs, NJ: Prentice-Hall, Inc.

Catania, AC, Sagvolden, T, & Aase H (1997). Delay of reinforcement and the hyperactivity syndrome. *Society for the Quantitative Analysis of Behavior, 8* (abstract).

Catania, AC, Sagvolden, T, & Keller, KJ (1988). Reinforcement schedules: Retroactive and proactive effects of reinforcers inserted into fixed-interval performance. *J Exp Anal Behav, 49*, 49–73.

Cook, EH, Jr., Stein, MA, Krasowski, MD, Cox, NJ, Olkon, DM, Kieffer JE, & Leventhal, BL (1995). Association of attention-deficit disorder and the dopamine transporter gene. *Am J Hum Genet, 56*, 993–998.

de Villiers, AS, Russell, VA, Sagvolden, T, Searson, A, Jaffer, A, & Taljaard, JJF (1995). alpha-2-Adrenoceptor mediated inhibition of [^{3}H]dopamine release from nucleus accumbens slices and monoamine levels in a rat model for Attention Deficit Hyperactivity Disorder. *Neurochem Res, 20*, 357–363.

Dell'Anna, ME, Luthman, J, Lindqvist, E, & Olson, L (1993). Development of monoamine systems after neonatal anoxia in rats. *Brain Res Bull, 32*, 159–170.

Di Chiara, G & Imperato, A (1988). Drugs abused by humans preferentially increase synaptic dopamine concentrations in the mesolimbic system of freely moving rats. *PNAS, 85*, 5274–5278.

Diaz-Granados, JL, Greene, PL, & Amsel, A (1994). Selective activity enhancement and persistence in weanling rats after hippocampal X-irradiation in infancy: Possible relevance for ADHD. *Behav Neural Biol, 61*, 251–259.

Douglas, VI (1983). Attentional and cognitive problems. In Rutter M, (Ed.), *Developmental neuropsychiatry* (pp 280–329). New York: Guilford Press.

Douglas, VI & Parry, PA (1994). Effects of reward and nonreward on frustration and attention in attention deficit disorder. *J Abnorm Child Psychol, 22*, 281–301.

Ferguson, SA. A review of rodent models of ADHD. In MV Solanto, AFT Arnstein & FX Castellanos (Eds.), *Stimulant drugs and ADHD: Basic and clinical neuroscience*. New York, NY: Oxford University Press (pp.).

Ferster, CB & Skinner, BF (1957). *Schedules of reinforcement.* NY: Appleton.

Giros, B, Jaber, M, Jones, SR, Wightman, RM, & Caron, MG (1996). Hyperlocomotion and indifference to cocaine and amphetamine in mice lacking the dopamine transporter. *Nature, 379*, 606–612.

Hendley, ED, Wessel, DJ, & Van Houten, J (1986). Inbreeding of Wistar-Kyoto rat strain with hyperactivity but without hypertension. *Behav Neural Biol, 45*, 1–16.

Holene, E, Nafstad, I, Skaare, JU, Bernhoft, A, Engen P, & Sagvolden, T (1995). Behavioral effects of pre- and postnatal exposure to individual polychlorinated biphenyl congeners in rats. *Environ Toxicol Chem, 14*, 967–976.

Holene, E, Nafstad, I, Skaare, JU, & Sagvolden T (1998). Behavioural hyperactivity in rats following postnatal exposure to sub-toxic doses of polychlorinated biphenyl congeners 153 and 126. *Behav Brain Res, 94*, 213–224.

Knardahl, S & Sagvolden T (1979). Open-field behavior of spontaneously hypertensive rats. *Behav Neural Biol, 27*, 187–200.

Krimer, LS, Muly, C, Williams, GV, & Goldman-Rakic, PS (1998). Dopaminergic regulation of cerebral cortical microcirculation. *Nature Neuroscience, 1*, 286–289.

LaHoste, GJ, Swanson, JM, Wigal, SB, Glabe, C, Wigal, T, King, N, & Kennedy, JL (1996). Dopamine D4 receptor gene polymorphism is associated with attention deficit hyperactivity disorder. *Mol Psychiatry, 1*, 121–124.

Linthorst, ACE, De Lang, H, De Jong, W, & Versteeg, DH (1991). Effect of the dopamine D_2 receptor agonist quinpirole on the in vivo release of dopamine in the caudate nucleus of hypertensive rats. *Eur J Pharmacol, 201*, 125–133.

Matthews, BA, Shimoff, E, Catania, AC, & Sagvolden, T (1977). Uninstructed human responding: Sensitivity to ratio and interval contingencies. *J Exp Anal Behav, 27*, 453–467.

McCarty, R & Kopin, IJ (1979). Patterns of behavioral development in spontaneously hypertensive rats and Wistar-Kyoto normotensive controls. *Dev Psychobiol, 12*, 239–243.

McKinney, WT & Bunney, WE (1969). Animal model of depression. *Arch Gen Psychiatry, 21*, 240–248.

Metzger, MA & Sagvolden, T (1994). Killeen's theory provides an answer—and a question. *Behav Brain Sci, 17*, 144–145.

Mook, DM, Jeffrey, J, & Neuringer, A (1993). Spontaneously hypertensive rats (SHR) readily learn to vary but not repeat instrumental responses. *Behav Neural Biol, 59*, 126–135.

Moser, MB, Moser, EI, Wultz, B, & Sagvolden, T (1988). Component analyses differentiate between exploratory behaviour of spontaneously hypertensive rats and Wistar Kyoto rats in a two-compartment free-exploration open field. *Scand J Psychol, 29*, 200–206.

Myers, MM, Musty, RE, & Hendley, ED (1982). Attenuation of hyperactivity in the spontaneously hypertensive rat by amphetamine. *Behav Neural Biol, 34*, 42–54.

Oades, RD (1998). Frontal, temporal and lateralised brain function in children with attention-deficit hyperactivity-disorder (ADHD): A psychophysiological and neuropsychological viewpoint on development. *Behav Brain Res, 94*, 83–95.

Okamoto, K & Aoki, K (1963). Development of a strain of spontaneously hypertensive rats. *Jpn Circ J, 27*, 282–293.

Oosterlaan, J & Sergeant, JA (1998). Response inhibition and response re-engagement in ADHD, disruptive, anxious and normal children. *Behav Brain Res, 94*, 33–43.

Papa, M, Berger, DF, Sagvolden, T, Sergeant, JA, & Sadile, AG (1998b). A quantitative cytochrome oxidase mapping study, cross-regional and neurobehavioural correlations in the anterior forebrain of an animal model of Attention Deficit Hyperactivity Disorder. *Behav Brain Res, 94*, 197–211.

Papa, M, Sergeant, JA, & Sadile, AG (1997). Subchronic methylphenidate treatment reverses the reduced expression of CaMKII in anterior forebrain sites of an animal model of ADHD. *Society for Neuroscience* (abstract).

Papa, M, Sergeant, JA, & Sadile, AG (1998a). Reduced transduction mechanisms in the anterior accumbal interface of an animal model of Attention-Deficit Hyperactivity Disorder. *Behav Brain Res, 94*, 187–195.

Porrino, LJ, Rapoport, JL, Behar, D, Sceery, W, Ismond, DR, & Bunney, WE (1983). A naturalistic assessment of the motor activity of hyperactive boys. I. Comparison with normal controls. *Arch Gen Psychiat, 40*, 681–687.

Puumala, T, Ruotsalainen, S, Jakala, P, Koivisto, E, Riekkinen, P, Jr., & Sirviö J (1997). Behavioral and pharmacological studies on the validation of a new animal model for attention deficit hyperactivity disorder. *Neurobiol Learn Mem, 66*, 198–211.

Robbins, TW, Granon, S, Muir, JL, Durantou, F, Harrison, A, & Everitt, BJ (1998). Neural systems underlying arousal and attention. Implications for drug abuse. *Ann NY Acad Sci, 846*, 222–237.

Robbins, TW, Jones, GH, & Sahakian, BJ (1989). Central stimulants, transmitters and attentional disorder: A perspective from animal studies. In T Sagvolden & T Archer (Eds.), *Attention deficit disorder: Clinical and basic research* (pp 199–222). Hillsdale, NJ: Lawrence Erlbaum Associates.

Rubia, K, Oosterlaan, J, Sergeant, JA, Brandeis, D, & van Leeuwen, T (1998). Inhibitory dysfunction in hyperactive boys. *Behav Brain Res, 94*, 25–32.

Russell, V, de Villiers, A, Sagvolden, T, Lamm, M, & Taljaard, J (1995). Altered dopaminergic function in the prefrontal cortex, nucleus accumbens and caudate-putamen of an animal model of Attention-Deficit Hyperactivity Disorder—the spontaneously hypertensive rat. *Brain Res, 676*, 343–351.

Russell, V, de Villiers, A, Sagvolden, T, Lamm, M, & Taljaard, J (1998). Differences between electrically-, ritalin- and D-amphetamine-stimulated release of [^{3}H]dopamine from brain slices suggest impaired vesicular storage of dopamine in an animal model of Attention-Deficit Hyperactivity Disorder. *Behav Brain Res, 94*, 163–171.

Sadile, AG (1993). What can genetic models tell us about behavioral plasticity? *Rev Neurosci, 4*, 287–303.

Sadile, AG, Pellicano, MP, Sagvolden, T, & Sergeant, JA (1996). NMDA and non-NMDA sensitive [L-^{3}H] glutamate receptor binding in the brain of the Naples high- and low-excitability rats: An autoradiographic study. *Behav Brain Res, 78*, 163–174.

Sagvolden, T (1996). The attention deficit disorder might be a reinforcement deficit disorder. In J Georgas, M Manthouli, E Besevegis, & A Kokkevi (Eds.), *Contemporary psychology in Europe: Theory, research, and application* (pp 131–143). Göttingen: Hogrefe and Huber.

Sagvolden, T (2000). Behavioral validation of the spontaneously hypertensive rat (SHR) as an animal model of attention-deficit/hyperactivity disorder (AD/HD). *Neurosci Biobehav Rev, 24*, 31–39.

Sagvolden, T, Aase, H, Zeiner, P, & Berger, DF (1998b). Altered reinforcement mechanisms in Attention-Deficit/ Hyperactivity Disorder. *Behav Brain Res, 94*, 61–71.

Sagvolden, T & Archer, T (1989). Future perspectives on ADD research—An irresistible challenge. In T Sagvolden & T Archer (Eds.), *Attention deficit disorder: Clinical and basic research* (pp 369–389). Hillsdale, NJ: Lawrence Erlbaum Associates.

Sagvolden, T, Hendley, ED, & Knardahl, S (1992b). Behavior of hypertensive and hyperactive rat strains: Hyperactivity is not unitarily determined. *Physiol Behav, 52*, 49–57.

Sagvolden, T, Metzger, MA, & Sagvolden, G (1993b). Frequent reward eliminates differences in activity between hyperkinetic rats and controls. *Behav Neural Biol, 59*, 225–229.

Sagvolden, T, Metzger, MA, Schiørbeck, HK, Rugland, AL, Spinnangr, I, & Sagvolden, G (1992a). The spontaneously hypertensive rat (SHR) as an animal model of childhood hyperactivity (ADHD): Changed reactivity to reinforcers and to psychomotor stimulants. *Behav Neural Biol, 58*, 103–112.

Sagvolden, T, Pettersen, MB, & Larsen, MC (1993a). Spontaneously hypertensive rats (SHR) as a putative animal model of childhood hyperkinesis: SHR behavior compared to four other rat strains. *Physiol Behav, 54*, 1047–1055.

Sagvolden, T, & Sergeant, JA (1998a). Attention Deficit/Hyperactivity Disorder—From brain dysfunctions to behaviour. *Behav Brain Res, 94*, 1–10.

Sagvolden, T, Slåtta, K, & Arntzen, E (1988). Low doses of methylphenidate (Ritalin) may alter the delay-of-reinforcement gradient. *Psychopharmacology (Berlin), 95*, 303–312.

Saldana, RL & Neuringer A (1998). Is instrumental variability abnormally high in children exhibiting ADHD and aggressive behavior? *Behav Brain Res, 94*, 51–59.

Sarter, M, Hagan, J, & Dudchenko, P (1992). Behavioral screening for cognition enhancers: From indiscriminate to valid testing: Part I. *Psychopharmacology, 107*, 144–159.

Schultz, W (1998). Predictive reward signal of dopamine neurons. *J Neurophysiol, 80*, 1–27.

Shaywitz, BA, Klopper, JH, Yager, RD, & Gordon, JW (1976a). Paradoxical response to amphetamine in developing rats treated with 6-hydroxydopamine. *Nature, 261*, 153–155.

Shaywitz, BA, Yager, RD, & Klopper, JH (1976b). Selective brain dopamine depletion in developing rats: An experimental model of minimal brain dysfunction. *Science, 191*, 305–308.

Silbergeld, EK & Goldberg, AM (1974). Lead-induced behavioral dysfunction: An animal model of hyperactivity. *Exp Neurol, 42*, 146–157.

Sleator, EK & Ullman, RK (1981). Can a physician diagnose hyperactivity in the office? *Pediatrics, 67*, 13–17.

Solanto, MV (1998). Neuropsychopharmacological mechanisms of stimulant drug action in Attention-Deficit Hyperactivity Disorder: A review and integration. *Behav Brain Res, 94*, 127–152.

Sonuga-Barke, EJS, Saxton, T, & Hall, M (1998). The role of interval underestimation in hyperactive children's failure to suppress responses over time. *Behav Brain Res, 94*, 45–50.

Sonuga-Barke, EJS, Taylor, E, Sembi, S, & Smith, J (1992). Hyperactivity and delay aversion I: The effect of delay on choice. *J Child Psychol Psychiatry, 33*, 387–398.

Stinus, L, Cador, M, & Le Moal, M (1992). Interaction between endogenous opioids and dopamine within the nucleus accumbens. *Ann NY Acad Sci, 654*, 254–273.

Swanson, JM, Sergeant, JA, Taylor, E, Sonuga-Barke, EJS, Jensen, PS, & Cantwell, DP (1998). Attention-deficit hyperactivity disorder and hyperkinetic disorder. *Lancet, 351*, 429–433.

Taylor, E (1998). Clinical foundations of hyperactivity research. *Behav Brain Res, 94*, 11–24.

Thyrum, ET, Blumenthal, JA, Madden, DJ, & Siegel, W (1995). Family history of hypertension influences neurobehavioral function in hypertensive patients. *Psychosomatic Medicine, 57*, 496–500.

van der Meere, JJ (1996). The role of attention. In ST Sandberg (Ed.), *Monographs in child and adolescent psychiatry. Hyperactivity disorders of childhood* (pp 109–146). Cambridge: Cambridge University Press.

Wender, PH (1971). *Minimal brain dysfunction in children.* New York: Wiley.

Willner, P (1984). The validity of animal models of depression. *Psychopharmacology (Berlin), 83*, 1–16.

Wultz, B & Sagvolden, T (1992). The hyperactive spontaneously hypertensive rat learns to sit still, but not to stop bursts of responses with short interresponse times. *Behav Genet, 22*, 415–433.

Wultz, B, Sagvolden, T, Moser, EI, & Moser M-B (1990). The spontaneously hypertensive rat as an animal model of attention-deficit hyperactivity disorder: Effects of methylphenidate on exploratory behavior. *Behav Neural Biol, 53*, 88–102.

III

CLINICAL NEUROSCIENCE

Introduction

MARY V. SOLANTO

Neuroimaging technology has opened a window on the functioning human brain that is already beginning to revolutionize our understanding of psychiatric disorders. Studies using these methods have identified, for the first time, differences in the brains of individuals with ADHD. F. Xavier Castellanos begins the first chapter in Part III with a helpful orientation to the cortico-striatal-thalamo-cortical circuitry that appears to be central to the pathophysiology of ADHD. He proceeds to review, critique, and integrate anatomic MRI studies, including his own major investigations at NIMH, which have reported differences between individuals with ADHD and normal volunteers in corpus callosum, prefrontal brain, caudate nucleus, globus pallidus, and cerebellum. Dr. Castellanos also describes the results of very recent functional MRI studies showing divergent patterns of metabolic activity in ADHD and normal subjects during performance of critical tasks, and he discusses recent PET studies revealing differences in catecholamine neurochemistry. Also reviewed in this chapter are results of EEG studies of event-related potential (ERP), which have allowed for sensitive investigation of attentional processes in humans.

Pharmacokinetics has until late been a relatively neglected area in the study of stimulant drug effects in ADHD. Early research was disappointing in revealing no relationship between blood levels and therapeutic efficacy beyond what could be predicted from dose alone. The recent quest for better long-acting forms of methylphenidate, however, led to an elegant series of pharmacokinetic/pharmacodynamic studies by Jim Swanson and colleagues that have contributed knowledge of basic methylphenidate mechanisms, particularly the surprising and important finding that acute tolerance to methylphenidate can occur and dissipate within hours following closely spaced doses. Recent highly innovative PET studies of normal adult volunteers and primates following oral and/or IV doses of labeled methylphenidate by Nora Volkow and colleagues have shown that methylphenidate binds to dopamine transporter primarily in striatum, with low levels in cortex and cerebellum. These studies have further identified interesting dysjunctions between changes over time in methylphenidate serum levels and certain behavioral effects, which also may reflect acute tolerance. The careful pharmacodynamic and pharmacokinetic studies described by Swanson and Volkow in this chapter have documented convergence of T_{max} for serum, brain, and behavior, and have succeeded in identifying the serum concentrations of methylphenidate and corresponding dopamine transporter occupancy levels that maximize therapeutic behavioral effects.

In their chapter on cognitive pharmacology, Colin Denney and Mark Rapport argue that precise delineation and differentiation of the core and secondary features of the cognitive phenotype of ADHD is prerequisite to understanding the fundamental mechanisms of stimulant drug action on these functions. They illustrate a structural equation modeling approach to determining the relative contributions of inattention, impulsivity, and related behavioral constructs to suboptimal academic performance in children with ADHD. These investigators advocate development of a "cognitive taxonomy" of tasks and task variables that best differentiate the performance of children with and without ADHD, and that are differentially sensitive to stimulant drug treatment. In this context they present an original review of 98 studies of stimulant effects on neurocognitive measures, which suggests an interesting divergence of stimulant dosage effects on lower versus higher-order cognitive functions with potential implications for differences in mechanisms of action. Denney and Rapport further discuss studies that illustrate how polypharmacy may help to delineate effects of drugs that, like the stimulants, affect multiple neurotransmitter systems. The climax of this chapter is the presentation of a provocative new model that postulates, based on a developing cognitive taxonomy of ADHD, that *working memory* is the area of primary deficit in this disorder.

Following development of a cognitive taxonomy of tasks, and identification of the core deficits, the next stage in the elaboration of a comprehensive neuroscience of stimulants in ADHD is the discernment of the neurobiological substrates of these cognitive effects. Mehta, Robbins, and Sahakian do a masterful job of reviewing the research in their chapter on comparative psychopharmacolgy, much of which originated in their laboratory at the University of Cambridge, which is delineating the roles of dopamine and norepinephrine in mediating specific cognitive functions. Many of these studies have been accomplished with primates and normal human volunteers via administration of receptor-selective agonists and antagonists, and, in some cases, coordinated PET neuroimaging, which may not feasible in children. Thus far, studies with D1/D2 and D2 agonists and antagonists have shown that dopamine mediates effects on spatial working memory and attentional set-shifting. Studies with alpha-2 adrenergic agonists and antagonists indicate that NE modulates performance on tests of paired-associate learning and sustained attention. The polypharmacy approach has been utilized by administering methylphenidate along with selective DA antagonists such as droperidol in order to dissociate dopaminergic from noradrenergic mediation.

A scientifically useful as well as ethically appropriate means of ascertaining the roles of specific neurotransmitters in mediating stimulant effects in children with

ADHD is to study the effects of the *non*stimulant monoaminergic agents that have been employed in treatment or clinical trials. In the final chapter in this section, Pliszka provides a thorough and insightful review of the effects of an array of dopaminergic and noradrenergic agents, including MAO inhibitors, tricyclic antidepressants, novel noradrenergic antidepressants, dopamine agonists, neuroleptics, alpha-adrenergic agonists, and serotonergic agents on cognition and behavior, as well as urinary and CSF metabolites in children with ADHD. Comparison of the clinical effects of these drugs with those of the stimulants support the conclusion that both norepinephrine *and* dopamine, but *not* serotonin, are critical to the mediation of stimulant therapeutic effects.

10

Neuroimaging Studies of ADHD

F. XAVIER CASTELLANOS

The attempt to understand the pathophysiology of Attention-Deficit/Hyperactivity Disorder (ADHD) must include a multitude of perspectives and approaches. Perhaps the one that captured the greatest interest during the 1990s was the use of a variety of neuroimaging techniques. Because no particular neuroimaging technique is sufficient to answer all pertinent questions, a number of approaches have been pursued, often in parallel. Thus, a strictly historical review will be avoided. I will first focus on studies of brain anatomy, followed by a summary of recent functional imaging reports, and touch briefly on electrophysiological approaches. Recent exciting work studying methylphenidate in human volunteers by Volkow and colleagues is described in detail elsewhere (see Swanson & Volkow, this volume). Before proceeding to anatomic empirical studies, we will digress to a discussion of the circuitry that is believed to underlie nearly all neuropsychiatric disorders, the cortico-striatal-thalamo-cortical (CSTC) circuits.

Cortico-Striatal-Thalamo-Cortical Circuits

Multiple parallel distributed and segregated circuits functionally link the prefrontal cortex to the appropriate basal ganglia input nuclei (striatum, which includes the caudate nucleus, putamen, and nucleus accumbens), from basal ganglia output nuclei (globus pallidus internal segment [GP_i] or substantia nigra pars reticulata [SNr]) to thalamus, and from the various thalamic nuclei back to cortex (Alexander et al., 1986) (see also chapter by Mehta et al., this volume). CSTC have been implicated in ADHD (Swanson et al., 1998), in obsessive-compulsive disorder (Saxena et al., 1998), and in most movement disorders, including Tourette syndrome (Hallett, 1993). Because our understanding of these intricate circuits remains inadequate, important questions of discriminant validity cannot yet be answered satisfactorily.

Central to the concept of a circuit is the notion that any element of the circuit can provide a starting point. We can thus begin with the "key" cells of the prefrontal cortex (PFC), the cortical pyramidal cells. The neurophysiologic functioning of these cells has been elucidated in greatest detail in the performance of working memory tasks, but these cells have other roles besides providing for working memory (see chapters by Taylor & Jentsch and by Arnsten for more detail). Cortical pyramidal cells must integrate multiple inputs in order to provide top-down regulation of other lower-order neuronal processes. They respond to their complex inputs in a graded fashion, by releasing varying amounts of the excitatory neurotransmitter, glutamate, rather than in a binary all-or-none manner (Houk & Wise, 1995). Though these cells are exclusively excitatory, their net effect is typically inhibitory, through their effects on subcortical regions, particularly including basal ganglia relay stations.

The primary striatal input cells are the medium spiny neurons, which also have a high convergence ratio, receiving an average of 10,000 cortical inputs per neuron (Graybiel, 1998). Striatal cells are "bistable," i.e., they tend to respond in a binary manner in releasing the inhibitory neurotransmitter, GABA (Beiser & Houk, 1998). Though all striatal spiny neurons are classified as inhibitory, the net effect of activity in a particular CSTC circuit can be excitatory or inhibitory, thus allowing for the selection of one specific response pattern, and the simultaneous inhibition of all nonselected responses (Redgrave et al., 1999). The dual nature of these circuits is shown in Figure 10–1.

In Figure 10–1, PFC is schematized as a single rectangular box, with excitatory outputs to other cortical regions (here limited schematically to parietal cortex)

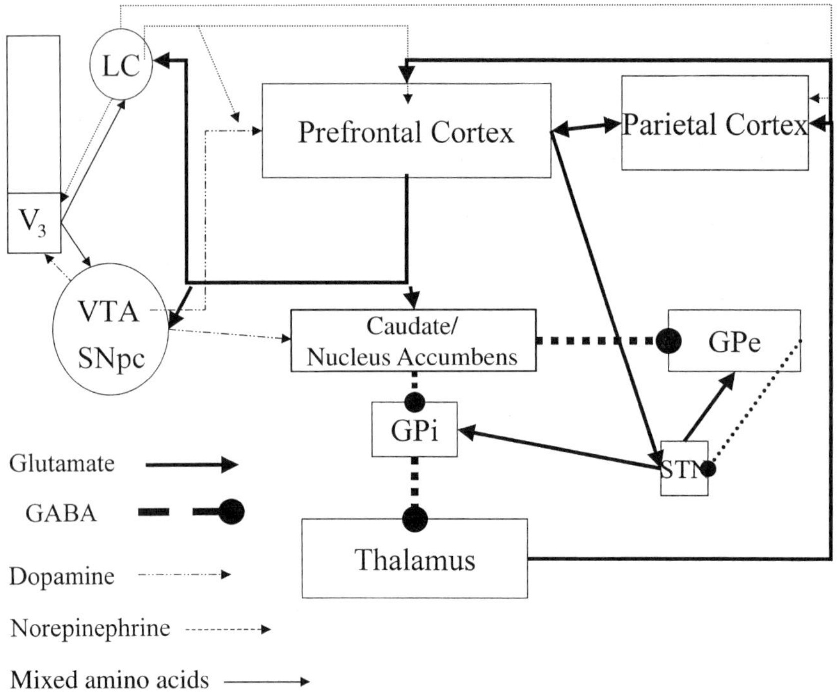

Figure 10–1. Prefrontal-striatal-thalamo-cortical circuit with associated monoaminergic efferents. See text for abbreviations and detailed description.

(Halperin et al., 1997), striatum (Parent, 1990; Mink & Thach, 1993), amygdala (Haber & Fudge, 1997), locus coeruleus (Arnsten, 1997), and subthalamic nucleus (STN) (Smith & Grace, 1992; Wichmann et al., 1994).

Signals that travel from striatum *directly* to GP_i disinhibit thalamic excitatory fibers, resulting in a net positive feedback to cortex. The *indirect pathway* includes the globus pallidus external segment (GP_e) and STN, with the latter receiving an important input directly from PFC. Cortical, thalamic, and STN efferents are marked as heavy unbroken lines with arrowheads depicting glutamatergic excitatory effects. As noted above, all striatal input and pallidal output neurons are GABAergic and inhibitory at each specific synapse. The heavy broken lines with rounded-heads denote these circuits. However, the net effect of these GABAergic cells can be either excitatory or inhibitory, depending on which pathway is activated. Thus, the direct pathway has a net facilitatory or excitatory effect because two GABAergic cells are connected in series, so that the striatal cell inhibits the GP_i neuron, which disinhibits the thalamus. This tendency is counteracted by the indirect circuit, the heart of which is the excitatory subthalamic nucleus. By contrast, the net effect of ventral tegmental area (VTA) and substantia nigra pars compacta (SNpc) DA neurons is excitatory regardless of the pathway as is demonstrated by the functional consequences of profound dopamine depletion in parkinsonism.

The VTA and locus coeruleus (LC) receive multiple inputs, including efferents from the cerebellar vermis, which has been termed "the head afferent ganglion of the catecholamine systems" (Snider et al., 1976a) (V_3 in Figure 10–1 refers to the posterior-inferior vermal lobules). As noted below, abnormalities in the vermis may be central to the pathophysiology of ADHD.

Anatomic Studies of Brain Differences in ADHD

With one seminal, though still qualitative, exception (Yakovlev & Lecours, 1967), the quantitative study of brain development during childhood and adolescence was not feasible until magnetic resonance imaging (MRI) became available in the late 1980s (Jernigan & Tallal, 1990; Jernigan et al., 1991). Subsequent cross-sectional (Reiss et al., 1996; Giedd et al., 1996; Sowell et al., 1999a, b) and mixed longitudinal/cross-sectional studies (Giedd et al., 1999) have demonstrated that although total brain volume changes between ages 5 and 18 years are negligible, there are robust and complex changes in white and gray matter compartments. White matter density increases linearly, reflecting increasing myelination (Paus et al., 1999), and gray matter volume increases until early-to-mid adolescence before decreasing during late adolescence, presumably from continued synaptic pruning (Huttenlocher, 1979; Huttenlocher & Dabholkar, 1997). These studies have also shown that even normal individuals who are closely matched for age, sex, height, and weight can differ in total brain volume by as much as a factor of 2 (Lange et al., 1997). Thus, it is now acknowledged that analyses of regional brain abnormalities should control statistically for differences among individuals in total brain volume, preferably by using analyses of covariance rather than simple ratios or proportions (Arndt et al., 1991), although most of the studies summarized below have not done so.

Corpus Callosum

Most research groups have started by quantifying the mid-sagittal area of the corpus callosum. Although total corpus callosum area has not differed from controls in any study, smaller anterior regions have generally been found (Hynd et al., 1991; Giedd et al., 1994; Baumgardner et al., 1996) in ADHD subjects. The largest MRI study did not confirm these differences when measurements were obtained from untransformed images (Castellanos et al., 1996). However, reanalysis with spatially normalized resampled images confirmed a smaller corpus callosum rostrum in ADHD (mean rostral area 25.5 mm^2 (SD = 10.3) in 50 boys with ADHD, versus 30.0 (12.1) in 50 matched controls ($t = 1.98$, $df = 98$, $p = .05$, uncorrected; Castellanos et al., 1999, unpublished data). The rostrum is the most anterior (and inferior) portion of the corpus callosum, and nearly all findings in ADHD (except Semrud-Clikeman et al., 1994) have been anterior.

Prefrontal Brain

The corpus callosum is the primary connecting structure for the cerebral hemispheres and smaller anterior corpus callosal areas are consistent with involvement of prefrontal cortical regions. Normally, the right anterior brain is slightly but consistently larger than the left (Weinberger et al., 1982). Significant decreases of this asymmetry in ADHD have been reported using computed tomography (Shaywitz et al., 1983) and MRI (Hynd et al., 1990; Reiss et al., 1996; Castellanos et al., 1996; Filipek et al., 1997). Volumetric measures have also detected smaller right-sided prefrontal brain regions measured "en bloc" (Castellanos et al., 1996; Filipek et al., 1997) in boys with ADHD which were correlated with neuropsychological performance on tasks that required response inhibition (Casey et al., 1997). In the only published study to date to report gray-white segmentation, right anterior white matter was also reduced in ADHD boys (Filipek et al., 1997). However, this difference must be interpreted with caution because the patient group was 2 years younger on average than the controls ($p = .12$, $n = 15$ per group) and because white matter volume increases linearly with age within this age range (Reiss et al., 1996; Paus et al., 1999).

Caudate Nucleus

The caudate nucleus and its associated circuits have long been suspected to play a pivotal role in ADHD (Pontius, 1973). Abnormalities of caudate nucleus volume (Castellanos et al., 1996; Filipek et al., 1997) or asymmetry (Hynd et al., 1993; Castellanos et al., 1996; Mataró et al., 1997) have been reported although the studies differ in whether the normal caudate is asymmetric, and whether this asymmetry normally favors the right (Castellanos et al., 1996) or the left caudate (Hynd et al.,

1993; Filipek et al., 1997; Mataró et al., 1997; Castellanos et al., 2000). These inconsistencies may reflect differences in methodology and comorbidity. For example, the mean coefficient of variation in the two studies that used single-slice area measures of the caudate (Hynd et al., 1993; Mataró et al., 1997) was 0.22 (SD 0.04), which is significantly greater than the coefficient of variation in the studies that reported volumes (Castellanos et al., 1996; Filipek et al., 1997) (*mean* 0.14, SD 0.03, $t = 3.20$, $df = 6$, $p = .02$).

Putamen

None of the anatomic MRI studies that reported putamen volumes detected significant diagnostic group differences (Castellanos et al., 1996; Aylward et al., 1996; Castellanos et al., 2000) although statistical power was insufficient in one study to rule out type II error (Aylward et al., 1996).

Globus Pallidus

The output nuclei of the basal ganglia are the internal segment of the globus pallidus (GP_i) and the substantia nigra pars reticulata, but the volume of the latter cannot generally be measured with MRI, and the size of the globus pallidus can only be measured as a unit (lateral and medial segments together), and then only with difficulty. Still, this region has been found to be significantly reduced in size in ADHD (Castellanos et al., 1996; Aylward et al., 1996), although these two studies differed in finding the larger difference on the left and right sides, respectively.

Cerebellum

An early computed tomography study found a trend towards greater cerebellar atrophy in adults with a prior history of hyperkinetic minimal brain dysfunction (Nasrallah et al., 1986). In a quantitative study of 112 subjects, the volumes of the cerebellar hemispheres were found to be significantly smaller in ADHD boys (Castellanos et al., 1996). In a follow-up study within the same sample, the cerebellar vermis as a whole, and particularly the posterior-inferior lobules (lobules VIII-X) were found to be significantly smaller in ADHD (Berquin et al., 1998). Smaller lobules VIII-X were independently replicated in boys with ADHD (Mostofsky et al., 1998), and in a sample of 50 girls with ADHD with the same effect size as in boys (d=0.63) even after controlling for total cerebral volume and a measure of intelligence (Castellanos et al., 2000). The posterior-inferior vermis has also been found to be abnormal in childhood-onset schizophrenia (Jacobsen et al., 1997), and in multiple-episode adult affective disorder patients (DelBello et al., 1999). The potential relevance of the posterior-inferior vermis for dopaminergically–related psychiatric disorders such as ADHD and schizophrenia has been highlighted by the selective finding of dopamine transporter (DAT) immunoreactivity in ventral cerebellar vermis, particularly in lobules VIII-X, in non-human primate (Melchitzky & Lewis, 2000). DAT immunoreactivity was not present in cerebellar hemispheres. The function and origin of these dopaminergic fibers are not known, but they likely form the afferent portion of a cerebellar circuit that powerfully influences the VTA and the LC (Snider et al., 1976b; Dempesy et al., 1983).

Summary of Anatomic Studies

Although most studies have used small samples and quantitative methodology is still evolving, anatomic MRI studies support the notion that a distributed circuit underlies some of the manifestations of ADHD. At least in boys, this circuit appears to include right prefrontal brain regions, the caudate nucleus, globus pallidus, and a sub-region of the cerebellar vermis. With one exception (Mataró et al., 1997), all groups have reported reduced volumes (or areas), which is consistent with the broad notion that the relevant brain regions are hypofunctioning. The CSTC circuits select, initiate, and execute complex motor and cognitive responses (Graybiel, 1998), and lateral cerebellar circuits provide on-line guidance of these functions (Jueptner & Weiller, 1998). The role of the cerebellar vermis remains unclear, although it appears likely that it is one of the key regulators of both catecholaminergic systems.

However, the theoretical implications of current anatomic findings in ADHD are tentative. Besides issues of inadequate statistical power in most studies, which is a considerable problem because of the high degree of variability in anatomic brain measures (Lange et al., 1997), none of the studies published to date has accounted for possible sources of confounding such as prior medication exposure. While future anatomic studies will focus on these issues, functional imaging studies are likely to become increasingly important in understanding the neural substrates of ADHD.

Functional Brain Imaging Studies of ADHD

Positron-emission tomography (PET) with [^{18}F]-fluoro-2-deoxy-D-glucose (FDG) was used to demonstrate decreased frontal cerebral metabolism in adults with ADHD (Zametkin et al., 1990), although inconsistent results in adolescents (Zametkin et al., 1993; Ernst et al., 1994; Ernst et al., 1997) led the authors to explore other techniques in ADHD (Ernst et al., 1998). Other investigators have measured local cerebral blood flow, which is closely linked to neuronal activity and tissue metabolism, with a variety of techniques including ^{133}Xenon inhalation and single photon emission tomography. Decreased blood flow has been found in ADHD subjects in the striatum (Lou et al., 1990) and in prefrontal regions (Amen et al., 1993). However, these results must be interpreted cautiously because ethical constraints make it difficult to obtain truly independent observations from normal control children. A more promising technique is blood oxygenation level dependent (BOLD) functional magnetic resonance imaging (fMRI), which obviates the need to use ionizing radiation.

BOLD fMRI technique was used in a study of 10 boys with ADHD and 6 controls, all of whom were scanned on and off methylphenidate while they performed go/no-go tasks (Vaidya et al., 1998). The authors extended to methylphenidate the observation that stimulants improve performance in normal children as they do in patients with ADHD (Rapoport et al., 1978). In caudate and putamen, Vaidya and colleagues found a striking group difference. In the task with the faster stimulus presentation rate, methylphenidate increased the number of activated pixels in caudate and putamen in ADHD subjects, but it had the opposite effect in the controls. In both caudate and putamen, controls activated significantly fewer pixels when scanned while on methylphenidate compared to drug-free scans. Perhaps equally interesting was the finding that patients as well as controls activated significantly larger numbers of pixels in prefrontal cortex on drug. This regional dissociation between prefrontal cortex and striatum is consistent with the finding that ventral tegmental area dopaminergic neurons, which mostly innervate prefrontal cortex, lack autoreceptors, while nigralstriatal dopamine neurons have abundant numbers of autoreceptors (Meador-Woodruff et al., 1994). Differences in neuronal autoreceptor regulation have been hypothesized to underlie the therapeutic effects of psychostimulants (Castellanos, 1997; Solanto, 1985, 1998) and the pattern of findings in the normal controls fits the prediction that methylphenidate would increase activation in prefrontal neurons by increasing synaptic and extrasynaptic dopamine levels, but that it would have the reverse effect in the striatum by producing a regulatory inhibition of firing. If replicated, these findings suggest that ADHD children differ qualitatively in striatal dopamine regulation and that such a difference may reflect etiological factors. However, before accepting this interpretation, we must note that all the patients had been medicated with methylphenidate until 36 hours prior to their scans. Because the normal controls had by definition never been previously exposed to stimulants, the possibility that these findings reflect medication withdrawal effects must first be excluded in replications and extensions of this work.

Utilizing a sample of seven adolescent boys with ADHD who were unmedicated or medication free for at least 1 week before scanning and nine controls, frontal striatal circuits were targeted in another fMRI study (Rubia et al., 1999). Subjects were scanned while performing the Stop Task (Schachar & Tannock, 1993) and a delay task that required synchronization of a motor response to an intermittently appearing visual stimulus. The hyperactive subjects showed less brain activity, predominantly in the right medial frontal cortex during both tasks, and in the right inferior prefrontal cortex and left caudate nucleus during the stop task. They concluded, "the right inferior frontal lobe—and its projections to the caudate—has been related to response inhibition. . . . It thus seems that the brake system of the brain is localized to the right prefrontal lobe, and its underactivation in ADHD seems to be the neural correlate of a less efficient inhibitory motor control" (Rubia et al., 1999).

The principal limitation of fMRI explorations of ADHD is the exquisite sensitivity of the technique to even minimal movement during scanning. Vaidya and colleagues (1998) found that using a bite-bar was essential in their study of children with ADHD and child controls. Rubia et al. (1999) included only adolescents who were able to remain sufficiently immobile in the scanner. Because physical restlessness decreases with age (Levy, 1980), Bush and colleagues (1999) studied eight adults who had a history of childhood onset and persistence into adulthood of ADHD and eight matched controls using the Counting Stroop during fMRI. The Counting Stroop was used to avoid verbal responses; rather than color words, subjects were shown words that were repeated one to four times per presentation. Subjects were required to press the button corresponding to the number of words, and interference was provided by presenting number words that did not match the number of presented words (Bush et al., 1998). Although both groups of subjects showed the expected slowing of response times in the interference condition, significant activation of bilateral anterior cingulate was only found in the normal controls. In contrast, ADHD subjects significantly activated right and left inferior frontal gyrus, right and left insula, left caudate, right putamen, right thalamus, and left pulvinar. Thus, the absence of cingulate activation could not be ascribed to a simple failure to activate a neural network, although the authors noted that possible anatomic differences in cingulate volume, and stimulant medication history could have confounded their results. Also, the cognitive task did not result in sufficiently robust activation patterns to allow for single-subject analyses, which is typically a strength of fMRI studies. Thus, absence of activation could simply reflect greater anatomic variability in the subjects with ADHD. Nevertheless, further exploration with this task and technique in younger subjects is clearly warranted.

Despite the natural advantages of magnetic resonance scanning for ADHD, positron emission tomography still offers the only practical way of assaying the neurochemistry of the human brain *in vivo*. For example, [fluorine-18] fluorodopa ([^{18}F]F-DOPA) was used to label catecholamine terminals in 17 unmedicated adults with ADHD (Ernst et al., 1998). [^{18}F]F-DOPA uptake was significantly diminished in the left and medial prefrontal cortex of the ADHD adults compared to the 23 controls, with no differences in striatum or midbrain regions. By contrast, in 10 adolescents with ADHD, [^{18}F]F-DOPA uptake in right midbrain was significantly elevated compared to 10 controls ($p = .04$, uncorrected for multiple comparisons) (Ernst et al., 1999). Taken together, these preliminary results further support the notion that catecholamine dysregulation is central to the pathophysiology of ADHD, and not just to its treatment. They also highlight the proposition that dopaminergic and noradrenergic systems cannot be understood without taking developmental effects into account (Castellanos, 1997).

The utility of single photon computed emission tomography (SPECT), which is more available and less expensive than PET, was highlighted in an exciting preliminary report (Dougherty et al., 1999). The highly selective dopamine transporter ligand [123]-I-Altropane was used in six adults with ADHD who were compared to a database of 30 healthy controls. Striatal binding potential (Bi_{max} / K_d) was elevated in all six ADHD patients, with each patient exceeding the mean values of the corresponding age-matched controls by at least 2 standard deviations.

This finding has since been replicated in 10 previously untreated adults with ADHD who were studied before and after 4 weeks of daily treatment with methylphenidate with [Tc-99m]TRODAT-1, another SPECT ligand that binds selectively to the dopamine transporter. Specific binding of Tc-99m-TRODAT-1 was significantly elevated at baseline and significantly reduced by 4 weeks of treatment with

3 daily doses of 5 mg methylphenidate (Krause et al., 2000). This study forecloses the possibility that the elevated levels of dopamine transporter in ADHD adults might be secondary to prior medication exposure, but still leaves open the possibility that these alterations in striatal dopamine transporters may be secondary compensations, rather than primary etiologic factors.

Understanding the role of dopamine in normal behavior and in ADHD will also be facilitated by the application of a technique recently used to demonstrate robust striatal dopamine release in adult normal volunteers as they played a video game (Koepp et al., 1998). The authors used tracer doses of radiolabeled raclopride, a dopamine-2 receptor antagonist, to indirectly measure endogenous dopamine release. While this technique has not yet been used in ADHD in combination with tracers such as Altropane or [^{18}F]F-DOPA, it is simply a question of time.

Neuroimaging at Higher Temporal Resolution

Spatial techniques such as MRI or positron emission tomography provide images that correspond to brain with varying degrees of spatial resolution and quite limited temporal resolution and at high cost. Electrophysiological approaches have the complementary strength of excellent temporal resolution, which extends down into the millisecond range, but with fundamental limits on spatial resolution deriving from the inverse problem. The inverse problem refers to the mathematical impossibility of arriving at a single solution for any given configuration of electrical or magnetic activity recorded at the scalp.

Despite this unavoidable limitation, electrophysiologic techniques are noninvasive, potentially quantitative, and relatively inexpensive. There are two main types of electrophysiological studies. Event-related potentials (ERP) are averaged over multiple trials, with each "tracing" anchored to a time point, such as stimulus presentation. ERP can be performed in any sensory modality, but visual or auditory stimuli are predominantly used. Clinically obtained electroencephalograms (EEG) in ADHD children often report nonspecific abnormal slowing, but the qualitative nature of these observations has not yielded much (Nogueira de Melo & Niedermeyer, 1999). The development of quantitative EEG (QEEG) was an attempt to harness the potential power of this noninvasive technique. In QEEG, EEG is recorded digitally and short artifact-free segments (2 s or so) are selected and then processed by Fourier analyses to obtain spectral information averaged over time. Spectral gradients can then be displayed topographically, giving rise to the term "brain mapping." Prior to the rise of managed care, there was an explosive increase in the use of such "brain mapping" in some communities, despite concerns about excessive false positives and variable training of practitioners (Levy & Ward, 1995; Nuwer, 1997, 1998).

Despite protestations (Hoffman et al., 1999), some papers in this field continue to demonstrate little regard for statistical principles. For example, a recent study of methylphenidate effects on global and complex measures of electroencephalography (EEG) acknowledged that methylphenidate did "not produce any clear-cut overall change in EEG measures" but then went on to argue that changes in correlations between EEG and measures of task performance from $r = -.62$ to $r = -.51$ while off- and on-methylphenidate, respectively, indicated that methylphenidate had induced regional changes in EEG (Lubar et al., 1999). However, this degree of change in correlation magnitude does not approach statistical significance ($z_{diff} = 0.52$, $n = 23$, $p = .60$).

In response to some of these criticisms, more rigorously designed studies have recently been conducted. The best of these studied 482 individuals, including 85 controls, in order to validate an index of cortical slowing, the theta/beta power ratio, recorded at the vertex (Monastra et al., 1999). The authors conclude that the theta/beta power ratio is strongly related to age and to the diagnosis of ADHD. They also report highly accurate rates of classifying subjects with an overall positive predictive power of 99%. The authors describe a number of follow-on studies that are proceeding, particularly to standardize procedures, determine test-retest reliability, cross-validate results with independent samples, and assess the discriminant validity of their measure with comorbid conditions such as oppositional defiant disorder or mood disorders. If subsequent studies pursue these objectives, it would represent an important advance.

Another approach to using electrophysiological methods to probe brain function in ADHD has been taken by an Australian group that has pioneered the use of steady-state visually evoked potentials (SSVEP) (Silberstein et al., 1998). This technique uses a spatially diffuse 13-Hz visual flicker in which are embedded the stimuli for the cognitive task, which in this case was the A-X variation of the Continuous Performance Task. In this task, subjects should respond to an X only if it follows an A. Normal subjects exhibited changes in SSVEP latency and amplitude in right frontal lobe suggestive of enhanced speed of processing. Subjects with ADHD did not demonstrate this rapid temporal shift, and also committed significantly more commission errors. The authors acknowledge that their technique cannot interrogate subcortical structures, particularly basal ganglia. At the same time, their results in cortex are strongly supportive of other neuroimaging findings of abnormal right prefrontal anatomy and function in ADHD.

Conclusions

Accumulating evidence supports the hypothesis that ADHD reflects dysfunction and dysregulation of cerebellar-striatal/adrenergic-prefrontal circuitry, which is partially ameliorated by pharmacological treatments. At the same time, much work remains to be done in sorting out the potential relevance of factors such as age, prior medication exposure, gender, intelligence quotient, socioeconomic status, and comorbidity.

Despite the apparent convergence of current results, fundamental caveats must be kept in mind. Nearly all neuroimaging studies have so far used region-of-interest analyses, which can lead to premature closure. For example, the absence of findings in parietal brain regions or with non-dopamine neurotransmitters mainly reflects the absence of relevant studies. Functional MRI techniques are powerful, but the statistical properties of various modes of analyses are still not well understood, and the number of degrees of freedom available to investigators during statistical analysis should give us pause. Fortunately, one safe prediction is that technological advances will continue to deliver new methods at an accelerating pace. For example, a novel fMRI technique, T_2 relaxometry, has been used to obtain steady-state blood flow measures and to test for medication effects in 11 boys with ADHD and 6 healthy controls (Teicher et al., 2000). The authors found decreased steady state blood flow in ADHD, particularly in putamen. Further, methylphenidate significantly affected putamen blood flow in a manner dependent on the child's unmedicated activity level. Magnetic resonance spectroscopic imaging (Moore et al., 1998), diffusion tensor imaging (Pierpaoli et al., 1996; Baratti et al., 1999), near-infrared optical spectroscopy (Chance et al., 1993; Villringer & Chance, 1997; Franceschini et al., 2000), and techniques as yet unknown will eventually be applied to understanding the pathophysiology of this most common of all childhood psychiatric disorders.

References

Alexander, GE, DeLong, MR, & Strick PL (1986). Parallel organization of functionally segregated circuits linking basal ganglia and cortex. *Annu Rev Neurosci, 9*, 357–381.

Amen, DG, Paldi, JH, & Thisted RA (1993). Brain SPECT imaging. *J Am Acad Child Adolesc Psychiatry, 32*, 1080–1081.

Arndt, S, Cohen, G, Alliger, RJ, Swayze, VW, & Andreasen NC (1991). Problems with ratio and proportion measures of imaged cerebral structures. *Psychiatry Res, 40*, 79–89.

Arnsten, AF (1997). Catecholamine regulation of the prefrontal cortex. *J Psychopharmacol (Oxf), 11*, 151–162.

Aylward, EH, Reiss, AL, Reader, MJ, Singer, HS, Brown, JE, & Denckla, MB (1996). Basal ganglia volumes in children with attention-deficit hyperactivity disorder. *J Child Neurol, 11*, 112–115.

Baratti, C, Barnett, AS, & Pierpaoli C (1999). Comparative MR imaging study of brain maturation in kittens with T1, T2, and the trace of the diffusion tensor. *Radiology, 210*, 133–142.

Baumgardner, TL, Singer, HS, Denckla, MB, Rubin, MA, Abrams, MT, Colli, MJ, & Reiss, AL (1996). Corpus callosum morphology in children with Tourette syndrome and attention deficit hyperactivity disorder. *Neurology, 47*, 477–482.

Beiser, DG, & Houk, JC (1998). Model of cortical-basal ganglionic processing: encoding the serial order of sensory events. *J Neurophysiol, 79*, 3168–3188.

Berquin, PC, Giedd, JN, Jacobsen, LK, Hamburger, SD, Krain, AL, Rapoport, JL, & Castellanos, FX (1998). The cerebellum in attention-deficit/hyperactivity disorder: a morphometric study. *Neurology, 50*, 1087–1093.

Bush, G, Frazier, JA, Rauch, SL, Seidman, LJ, Whalen, PJ, Jenike, MA, Rosen, BR, & Biederman, J (1999). Anterior cingulate cortex dysfunction in attention-deficit/hyperactivity disorder revealed by fMRI and the counting Stroop. *Biol Psychiatry, 45*, 1542–1552.

Bush, G, Whalen, PJ, Rosen, BR, Jenike, MA, McInerney, SC, & Rauch, SL (1998). The counting Stroop: an interference task specialized for functional neuroimaging—validation study with functional MRI. *Hum Brain Mapp, 6*, 270–282.

Casey, BJ, Castellanos, FX, Giedd, JN, Marsh, WL, Hamburger, SD, Schubert, AB, Vauss, YC, Vaituzis, AC, Dickstein, DP, Sarfatti, SE, & Rapoport, JL (1997). Implication of right frontostriatal circuitry in response inhibition and attention-deficit/hyperactivity disorder. *J Am Acad Child Adolesc Psychiatry, 36*, 374–383.

Castellanos, FX (1997). Toward a pathophysiology of attention-deficit/hyperactivity disorder. *Clin Pediatr, 36*, 381–393.

Castellanos, FX, Giedd, JN, Marsh, WL, Hamburger, SD, Vaituzis, AC, Dickstein, DP, Sarfatti, SE, Vauss, YC, Snell, JW, Lange, N, Kaysen, D, Krain, AL, Ritchie, GF, Rajapakse, JC, & Rapoport, JL (1996). Quantitative brain magnetic resonance imaging in attention-deficit/hyperactivity disorder. *Arch Gen Psychiatry, 53*, 607–616.

Castellanos FX, Giedd JN, Berquin PC, Walter JM, Sharp W, Tran T, Vaituzis AC, Blumenthal J, Nelson J, Bastain TM, Zijdenbos A, Evans AC, & Rapoport JL (2000). Quantitative brain magnetic resonance imaging in girls with attention-deficit/hyperactivity disorder. *Arch Gen Psychiatry* (In press).

Chance, B, Zhuang, Z, UnAh, C, Alter, C, & Lipton, L (1993). Cognition-activated low-frequency modulation of light absorption in human brain. *Proc Natl Acad Sci U S A, 90*, 3770–3774.

DelBello, MP, Strakowski, SM, & Zimmerman, ME, Hawkins, JM, Sax, KW (1999). MRI analysis of the cerebellum in bipolar disorder: a pilot study. *Neuropsychopharmacology, 21*, 63–68.

Dempesy CW, Tootle DM, Fontana CJ, Fitzjarrell AT, Garey RE, & Heath RG (1983). Stimulation of the paleocerebellar cortex of the cat: Increased rate of synthesis and release of catecholamines at limbic sites. *Biol Psychiatry, 18*, 127–132.

Dougherty, DD, Bonab, AA, Spencer, TJ, Rauch, SL, Madras, BK, & Fischman, AJ (1999). Dopamine transporter density is elevated in patients with attention deficit hyperactivity disorder. *Lancet, 354*, 2132–2133.

Ernst, M, Cohen, RM, Liebenauer, LL, Jons, PH, & Zametkin, AJ (1997). Cerebral glucose metabolism in adolescent girls with attention-deficit/hyperactivity disorder. *J Am Acad Child Adolesc Psychiatry, 36*, 1399–1406.

Ernst, M, Liebenauer, LL, King, AC, Fitzgerald, GA, Cohen, RM, & Zametkin, AJ (1994). Reduced brain metabolism in hyperactive girls. *J Am Acad Child Adolesc Psychiatry, 33*, 858–868.

Ernst, M, Zametkin, AJ, Matochik, JA, Jons, PH, & Cohen, RM (1998). DOPA decarboxylase activity in attention deficit hyperactivity disorder adults. A [fluorine-18]fluorodopa positron emission tomographic study. *J Neurosci, 18*, 5901–5907.

Ernst, M, Zametkin, AJ, Matochik, JA, Pascualvaca, D, Jons, PH, & Cohen, RM (1999). High midbrain 18F-DOPA accumulation in children with ADHD. *Am J Psychiatry, 156*, 1209–1215.

Filipek, PA, Semrud-Clikeman, M, Steingard, RJ, Renshaw, PF, Kennedy, DN, & Biederman, J (1997). Volumetric MRI analysis comparing attention-deficit hyperactivity disorder and normal controls. *Neurology, 48*, 589–601.

Franceschini, MA, Toronov, V, Filiaci, ME, Gratton, E, & Fantini, S (2000). On-line optical imaging of the human brain with 160-ms temporal resolution. *Optics Express 6*: 49–57.

Giedd, JN, Blumenthal, J, Jeffries, NO, Castellanos, FX, Liu, H, Zijdenbos, A, Paus, T, Evans, AC, & Rapoport, JL (1999). Cerebral cortical gray matter changes during childhood and adolescence: a longitudinal MRI study. *Nat Neurosci, 2*, 861–863.

Giedd, JN, Castellanos, FX, Casey, BJ, Kozuch, P, King, AC, Hamburger, SD, & Rapoport, JL (1994). Quantitative morphology of the corpus callosum in attention deficit hyperactivity disorder. *Am J Psychiatry, 151*, 665–669.

Giedd, JN, Snell, JW, Lange, N, Rajapakse, JC, Casey, BJ, Kozuch, PL, Vaituzis, AC, Vauss, YC, Hamburger, SD, Kaysen, D, & Rapoport, JL (1996). Quantitative magnetic resonance imaging of human brain development: ages 4–18. *Cereb Cortex, 6*, 551–560.

Graybiel, AM (1998). The basal ganglia and chunking of action repertoires. *Neurobiol Learn Mem, 70*, 119–136.

Haber, SN & Fudge, JL (1997). The interface between dopamine neurons and the amygdala: implications for schizophrenia. *Schizophr Bull, 23*, 471–482.

Hallett, M (1993). Physiology of basal ganglia disorders: an overview. *Can J Neurol Sci, 20*, 177–183.

Halperin, JM, Newcorn, JH, Koda, VH, Pick, L, McKay, KE, & Knott, P (1997). Noradrenergic mechanisms in ADHD children with and without reading disabilities: a replication and extension. *J Am Acad Child Adolesc Psychiatry, 36*, 1688–1697.

Hoffman, DA, Lubar, JF, Thatcher, RW, Sterman, MB, Rosenfeld, PJ, Striefel, S, Trudeau, D, & Stockdale, S (1999). Limitations of the American Academy of Neurology and American Clinical Neurophysiology Society paper on QEEG. *J Neuropsychiatry Clin Neurosci, 11*, 401–407.

Houk, JC & Wise, SP (1995). Distributed modular architectures linking basal ganglia, cerebellum, and cerebral cortex: their role in planning and controlling action. *Cereb Cortex, 5*, 95–110.

Huttenlocher, PR (1979). Synaptic density in human frontal cortex—developmental changes and effects of aging. *Brain Res, 163*, 195–205.

Huttenlocher, PR, & Dabholkar, AS (1997). Regional differences in synaptogenesis in human cerebral cortex. *J Comp Neurol, 387*, 167–178.

Hynd, GW, Hern, KL, Novey, ES, Eliopulos, D, Marshall, R, Gonzalez, JJ, & Voeller, KKS (1993). Attention deficit hyperactivity disorder and asymmetry of the caudate nucleus. *J Child Neurol, 8*, 339–347.

Hynd, GW, Semrud-Clikeman, M, Lorys, AR, Novey, ES, & Eliopulos, D (1990). Brain morphology in developmental dyslexia and attention deficit disorder/hyperactivity. *Arch Neurol, 47*, 919–926.

Hynd, GW, Semrud-Clikeman, M, Lorys, AR, Novey, ES, Eliopulos, D, & Lyytinen, H (1991). Corpus callosum morphology in attention-deficit hyperactivity disorder: morphometric analysis of MRI. *J Learn Disabil, 24*, 141–146.

Jacobsen, LK, Giedd, JN, Berquin, PC, Krain AL, Hamburger, SD, Kumra, S, & Rapoport, JL (1997). Quantitative morphology of the cerebellum and fourth ventricle in childhood-onset schizophrenia. *Am J Psychiatry, 154*, 1663–1669.

Jernigan, TL & Tallal, P (1990). Late childhood changes in brain morphology observable with MRI. *Dev Med Child Neurol, 32*, 379–385.

Jernigan, TL, Trauner, DA, Hesselink, JR, & Tallal, PA (1991). Maturation of human cerebrum observed in vivo during adolescence. *Brain, 114*, 2037–2049.

Jueptner, M & Weiller, C (1998). A review of differences between basal ganglia and cerebellar control of movements as revealed by functional imaging studies. *Brain, 121*, 1437–1449.

Koepp, MJ, Gunn, RN, Lawrence, AD, Cunningham, VJ, Dagher, A, Jones, T, Brooks, DJ, Bench, CJ, & Grasby, PM (1998). Evidence for striatal dopamine release during a video game. *Nature, 393*, 266–268.

Krause KH, Dresel SH, Krause J, Kung HF, & Tatsch K (2000). Increased striatal dopamine transporter in adult patients with attention deficit hyperactivity disorder: Effects of methylphenidate as measured by single photon emission computed tomography. *Neurosci Lett, 285*, 107–110.

Lange, N, Giedd, JN, Castellanos, FX, Vaituzis, AC, & Rapoport, JL (1997). Variability of human brain structure size: ages 4 to 20. *Psychiatry Res Neuroimaging, 74*, 1–12.

Levy, F (1980). The development of sustained attention (vigilance) and inhibition in children: some normative data. *J Child Psychol Psychiatry, 21*, 77–84.

Levy, F & Ward, PB (1995). Neurometrics, dynamic brain imaging and attention deficit hyperactivity disorder. *J Paediatr Child Health, 31*, 279–283.

Lou, HC, Henriksen, L, & Bruhn, P (1990). Focal cerebral dysfunction in developmental learning disabilities. *Lancet, 335*, 8–11.

Lubar, JF, White, JN, Swartwood, MO, & Swartwood, JN (1999). Methylphenidate effects on global and complex measures of EEG. *Pediatr Neurol, 21*, 633–637.

Mataró, M, García-Sánchez, C, Junqué, C, Estévez-González, A, & Pujol, J (1997). Magnetic resonance imaging measurement of the caudate nucleus in adolescents with attention-deficit hyperactivity disorder and its relationship with neuropsychological and behavioral measures. *Arch Neurol, 54*, 963–968.

Meador-Woodruff, JH, Damask, SP, & Watson, SJ, Jr (1994). Differential expression of autoreceptors in the ascending dopamine systems of the human brain. *Proc Natl Acad Sci USA, 91*, 8297–8301.

Mink, J & Thach, T (1993). Basal ganglia intrinsic circuits and their role in behavior. *Curr Opin Neurobiol, 3*, 950–957.

Monastra, VJ, Lubar, JF, Linden, M, VanDeusen, P, Green, G, Wing, W, Phillips, A, & Fenger, TN (1999). Assessing attention deficit hyperactivity disorder via quantitative electroencephalography: An initial validation study. *Neuropsychology, 13*, 424–433.

Moore, GJ, MacMaster, FP, Stewart, C, & Rosenberg, DR (1998). Case study: caudate glutamatergic changes with paroxetine therapy for pediatric obsessive-compulsive disorder. *J Am Acad Child Adolesc Psychiatry, 37*, 663–667.

Mostofsky, SH, Reiss, AL, Lockhart, P, & Denckla, MB (1998). Evaluation of cerebellar size in attention-deficit hyperactivity disorder. *J Child Neurol, 13*, 434–439.

Nasrallah, HA, Loney, J, Olson, SC, McCalley-Whitters, M, Kramer, J, & Jacoby, CG (1986). Cortical atrophy in young adults with a history of hyperactivity in childhood. *Psychiatry Res, 17*, 241–246.

Nogueira de Melo, A & Niedermeyer, E (1999). The EEG in infantile brain damage, cerebral palsy, and minor cerebral dysfunctions of childhood. In E Niedermeyer & FH Lopes da Silva (Eds), *Electroencephalography: basic principles, clinical applications, and related fields,* 4th ed. Baltimore: Williams & Wilkins, pp 383–392.

Nuwer, M (1997). Assessment of digital EEG, quantitative EEG, and EEG brain mapping: report of the American Academy of Neurology and the American Clinical Neurophysiology Society. *Neurology, 49*, 277–292.

Nuwer, MR (1998). Assessing digital and quantitative EEG in clinical settings. *J Clin Neurophysiol, 15*, 458–463.

Parent, A (1990). Extrinsic connections of the basal ganglia. *Trends Neurosci, 13*, 254–258.

Paus, T, Zijdenbos, A, Worsley, K, Collins, DL, Blumenthal, J, Giedd, JN, Rapoport, JL, & Evans, AC (1999). Structural maturation of neural pathways in children and adolescents: In vivo study. *Science, 283*, 1908–1911.

Pierpaoli, C, Jezzard, P, Basser, PJ, Barnett, A, & Di Chiro, G (1996). Diffusion tensor MR imaging of the human brain. *Radiology, 201*, 637–648.

Pontius, AA (1973). Dysfunction patterns analogous to frontal lobe system and caudate nucleus syndromes in some groups of minimal brain dysfunction. *J Am Med Wom Assoc, 28*, 285–292.

Rapoport, JL, Buchsbaum, MS, Zahn, TP, Weingartner, H, Ludlow, C, & Mikkelsen, EJ (1978). Dextroamphetamine: cognitive and behavioral effects in normal prepubertal boys. *Science, 199*, 560–563.

Redgrave, P, Prescott, TJ, & Gurney, K (1999). The basal ganglia: a vertebrate solution to the selection problem? *Neuroscience, 89*, 1009–1023.

Reiss, AL, Abrams, MT, Singer, HS, Ross, JL, Denckla, MB (1996). Brain development, gender and IQ in children. A volumetric imaging study. *Brain, 119*, 1763–1774.

Rubia, K, Overmeyer, S, Taylor, E, Brammer, M, Williams, SC, Simmons, A, & Bullmore, ET (1999). Hypofrontality in attention deficit hyperactivity disorder during higher-order motor control: A study with functional MRI. *Am J Psychiatry, 156*, 891–896.

Saxena, S, Brody, AL, Schwartz, JM, & Baxter, LR (1998). Neuroimaging and frontal-subcortical circuitry in obsessive-compulsive disorder. *Br J Psychiatry Suppl 35*, 26–37.

Schachar, RJ & Tannock, R (1993). Inhibitory control, impulsiveness, and attention deficit hyperactivity disorder. *Clin Psychol Rev, 13*, 721–739.

Schmahmann, JD (1996). From movement to thought: anatomic substrates of the cerebellar contribution to cognitive processing. *Human Brain Mapping, 4*, 174–198.

Semrud-Clikeman, M, Filipek, PA, Biederman, J, Steingard, R, Kennedy, D, Renshaw, P, & Bekken, K (1994). Attention-deficit hyperactivity disorder: magnetic resonance imaging morphometric analysis of the corpus callosum. *J Am Acad Child Adolesc Psychiatry, 33*, 875–881.

Shaywitz, BA, Shaywitz, SE, Byrne, T, Cohen, DJ, & Rothman, S (1983). Attention deficit disorder: quantitative analysis of CT. *Neurology, 33*, 1500–1503.

Silberstein, RB, Farrow, M, Levy, F, Pipingas, A, Hay DA, & Jarman, FC (1998). Functional brain electrical activity mapping in boys with attention-deficit/hyperactivity disorder. *Arch Gen Psychiatry, 55*, 1105–1112.

Smith, ID & Grace, AA (1992). Role of the subthalamic nucleus in the regulation of nigral dopamine neuron activity. *Synapse, 12*, 287–303.

Snider RS, Maiti A, & Snider SR (1976a). Cerebellar connections to catecholamine systems: Anatomical and biochemical studies. *Trans Am Neurol Assoc, 101*, 295–297.

Snider RS, Maiti A, & Snider SR (1976b). Cerebellar pathways to ventral midbrain and nigra. *Exp Neurol, 53*, 714–728.

Solanto, MV (1985). Neuropharmacological basis of stimulant drug action in attention deficit disorder with hyperactivity: A review and synthesis. *Psychol Bull, 95*, 387–409.

Solanto, MV (1998). Neuropsychopharmacological mechanisms of stimulant drug action in attention-deficit/hyperactivity disorder: a review and integration. *Behav Brain Res, 94*, 127–152.

Sowell, ER, Thompson, PM, Holmes, CJ, Batth, R, Jernigan TL, & Toga, AW (1999b). Localizing age-related changes in brain structure between childhood and adolescence using statistical parametric mapping. *Neuroimage, 9*, 587–597.

Sowell, ER, Thompson, PM, Holmes, CJ, Jernigan, TL, & Toga, AW (1999a). *In vivo* evidence for post-adolescent brain maturation in frontal and striatal regions. *Nature Neuroscience, 2*, 859–861.

Swanson, J, Castellanos, FX, Murias, M, LaHoste, GJ, & Kennedy, J (1998). Cognitive neuroscience of attention deficit hyperactivity disorder and hyperkinetic disorder. *Curr Opin Neurobiol, 8*, 263–271.

Teicher, MH, Anderson, CM, Polcari, A, Glod, CA, Maas, LC, & Renshaw, PF (2000). Functional deficits in basal ganglia of children with attention-deficit hyperactivity disorder shown with functional magnetic resonance imaging relaxometry. *Nat Med 6*: 470–473.

Vaidya, CJ, Austin, G, Kirkorian, G, Ridlehuber, HW, Desmond, JE, Glover, GH, & Gabrieli, JD (1998). Selective effects of methylphenidate in attention deficit hyperactivity disorder: a functional magnetic resonance imaging study. *Proc Natl Acad Sci USA, 95*, 14494–14499.

Villringer, A & Chance, B (1997). Non-invasive optical spectroscopy and imaging of human brain function. *Trends Neurosci, 20*, 435–442.

Weinberger, DR, Luchins, DJ, Morihisa, J, & Wyatt, RJ (1982). Asymmetrical volumes of the right and left frontal and occipital regions of the human brain. *Neurology, 11*, 97–100.

Wichmann, T, Bergman, H, & DeLong MR (1994). The primate subthalamic nucleus. I. Functional properties in intact animals. *J Neurophysiol, 72*, 494–506.

Wichmann, T & DeLong, MR (1993). Pathophysiology of Parkinsonian motor abnormalities. In Narabayashi H, Nagatsu T, Yanagisawa N, & Mizuno Y (Eds.), *Advances in neurology*, Vol. 60. (pp 53–61.) New York: Raven Press.

Yakovlev, PI & Lecours, AR (1967). The myelogenetic cycles of regional maturation of the brain. In A Minkowski (Ed.), *Regional development of the brain in early life,* (pp 3–70.) Oxford & Edinburgh: Blackwell Scientific.

Zametkin, AJ, Liebenauer, LL, Fitzgerald, GA, King, AC, Minkunas, DV, Herscovitch, P, Yamada, EM, & Cohen, RM (1993). Brain metabolism in teenagers with attention-deficit hyperactivity disorder. *Arch Gen Psychiatry, 50*, 333–340.

Zametkin, AJ, Nordahl, TE, Gross, M, King, AC, Semple, WE, Rumsey, J, Hamburger, SD, & Cohen RM (1990). Cerebral glucose metabolism in adults with hyperactivity of childhood onset. *N Engl J Med, 323*, 1361–1366.

11

Pharmacokinetic and Pharmacodynamic Properties of Methylphenidate in Humans

JAMES SWANSON AND NORA VOLKOW

The stimulant drug methylphenidate (MPH) has been available for use in the United States since the 1960s, but approval for use was granted by the U.S. Food and Drug Administration (FDA) before the establishment of the current guidelines that require pharmacokinetic (PK) information on the intended patient population. Therefore, in the United States there was widespread use of MPH to treat children with Attention Deficit/Hyperactivity Disorder (ADHD) (Committee Report, 1971) before the PK properties of MPH were well defined in humans.

Starting in the 1990s, the recognition and treatment of ADHD increased dramatically in the United States, and so did the production and prescription of MPH (see Swanson, Lerner, & Williams, 1995; Safer et al., 1996). Pharmaceutical companies became interested in the increased market for MPH, and several companies initiated drug development programs (see Greenhill et al., 1998). We will review the information from some recent clinical trials that provide new information about the PK properties as well as the pharmacodynamic (PD) properties of clinical doses of MPH used to treat children with ADHD.

We will also review some studies of MPH based on positron emission tomography (PET) (Volkow et al., 1990, 1995, 1998; Ding et al., 1994), which quantify the effect at a major site of action in the human brain (i.e., occupancy of the dopamine transporter) and show how the time course of this central effect is related to PK and PD properties of MPH (Volkow et al., 1996).

Pharmacokinetic and Pharmacodynamic Studies in Children with ADHD

PK Studies

The initial PK studies of MPH in humans (Faraj et al., 1974; Milberg et al., 1975) were hampered by insensitive assays for the parent drug. In the late 1970s, several research groups worked to increase the sensitivity of assays for measuring the serum concentrations of MPH.

One of these research groups, at the Hospital for Sick Children (HSC) in Toronto, developed sensitive PK assays of MPH and its primary metabolite, ritalinic acid (RA), to go along with some behavioral assays that were in use for measuring the PD properties of MPH in children with ADHD (Swanson & Kinsbourne, 1976; Dalby et al., 1977; Swanson et al., 1978; Humphries et al., 1978). Soldin et al. (1979) developed a high pressure liquid chromatography method that was sensitive enough to quantify the serum concentrations of the metabolite, RA, but this method was not sensitive enough to quantify the much lower serum concentrations of the drug itself, MPH. Chan et al. (1980) developed a gas chromatograph/mass spectrometer method that was sensitive enough for quantifying serum concentrations of MPH. Similar improvements in the techniques for measuring the PK properties of MPH were made by research teams at Yale University (see Shaywitz et al., 1982), University of North Carolina (see Gualiteri et al., 1982; Wargin et al., 1983; Patrick et al., 1987a,b), Columbia University (see Greenhill et al., 1982, 1987), University of Saskatchewan (Srinivas et al., 1987), and National Institute of of Mental Health (see Borcherding et al., 1989).

The early parallel research programs defined the PK properties of MPH in humans, but the sample sizes used in these studies were small. For example, Chan et al. (1980, 1983) in an evaluation of only six children with ADHD, showed the serum half-life was short ($T_{1/2} = 2.1$ hr) and that it, as well as the time of maximum serum concentration ($T_{max} = 1.3$ hr), matched the PD characteristics (i.e., rapid onset and short duration of action) that had been documented in earlier studies in the clinical component of this research program (Swanson et al., 1978). Shaywitz et al. (1982) reported similar values for $T_{1/2}$ in 14 children with ADHD, and showed that half-life was about the same after acute and chronic dosing of MPH. Gualtieri et al. (1982) confirmed the short $T_{1/2}$ in 11 normal adults, as well as in three adults and one adolescent with ADHD.

Did the early studies suggest that serum concentrations of MPH would be helpful in clinical management of stimulant therapy? These studies showed that serum concentrations of MPH were proportional to oral dose, and that individuals requiring high doses for clinical treatment also had high serum concentrations of MPH. Thus, a narrow therapeutic window for serum concentration was not identified to account for the wide variation in clinically effective dose across individuals. The general conclusion was that that PK information was redundant and would not change clinical decisions based on titration of dose.

However, these studies did have significant impact on clinical practices related to the frequency (e.g., QD, BID, or TID) and timing (e.g., before or with meals) of doses. In the 1970s, once-a-day dosing was common, even in research settings (see Sprague & Sleator, 1977). The clear demonstration that the PK half-life of MPH was less than 3 hours supported the shift from once-a-day (QD) to twice-a-day (BID) or thrice-a-day (TID) dosing, which is now standard clinical practice. Also, up until the 1980s, it was generally recommended that MPH be administered on an empty stomach (i.e., 30 minutes to 1 hour before meals), which unfortunately maximized one of its most common side effects (anorexia). This clinical practice was based on an early clinical impression from treatment of patients with narcolepsy (Yoss & Daly, 1968) that was contradicted by the PK and PD evidence. For example, in an evaluation of the PK effects of taking MPH with or before breakfast, Chan et al. (1983) showed that food did not affect the absorption or elimination of MPH; and, in a companion paper, Swanson et al. (1983) showed that the PD effects of clinical doses of MPH were the same whether taken with or before breakfast. The PK information generated in the 1970s and 1980s contributed to important changes in clinical practice to increase efficacy and decrease side effects. The PK information suggested that MPH should be administered more than once a day (at about the half-life of the drug, to increase efficacy) and with or after meals (instead of before meals, to decrease side effects).

Apparently, additional investigations of the PK properties of MPH were stimulated after the patent on Ritalin expired. In the 1980s (Patrick et al., 1989; Meyer et al., 1989), MD Pharmaceuticals gained FDA approval for generic MPH, based on demonstration of bioequivalence. The FDA guideline for bioequivalence is defined as less than 20% difference in PK properties (such as maximum serum concentration [C_{max}] or area under the curve [AUC] for serum concentrations profiles) for the generic and branded forms of a drug; and despite some question in clinical use, this was confirmed by a controlled PK study in adult volunteers (Therapeutic Inequivalency Action Coordinating Committee, 1991). However, these bioequivalence studies did not evaluate the efficacy or effectiveness (i.e., PD properties) of branded and generic MPH.

In the late 1980s (see Greenhill et al., 1987), several research groups evaluated the PK properties of a sustained-release (SR) preparation of MPH. SR-MPH was based on a wax matrix to control absorption. In a study of 18 normal adults in the research program in North Carolina, Patrick et al. (1989) found that absorption was delayed ($T_{max} = 3.3$ hr) but that $T_{1/2}$ for SR-MPH was not statistically different than for immediate-release (IR) MPH in this sample of adults. In a study of 13 children with ADHD in the research program at Columbia, Birmaher et al. (1989) also reported delayed absorption ($T_{max} = 3.4$ hr) but longer values for $T_{1/2}$ (4.1 hr) for SR-MPH than values reported for IR-MPH (2 to 3 hr) in children with ADHD.

In the late 1980s and early 1990s, a research program at the University of Saskatchewan addressed the PK properties of the isomers of MPH (Srinivas et al., 1987). With 2 chiral centers, MPH has 4 isomers: d-threo, l-threo, d-erythro, and l-erythro (Patrick et al., 1987a, b). The erythro racemates were removed in the 1950s because of effects on blood pressure (see Patrick et al, 1987), and the commercially available (branded and generic) forms of MPH contain a 50:50 mixture of d-threo-MPH and l-threo-MPH (see Swanson et al., 1995).

Srinivas et al. (1987) and Hubbard et al. (1989) reported that following oral dosing the concentrations of the *l*-isomer were almost an order of magnitude lower than the concentrations of the *d*-isomer, but that the PK properties of the two isomers were about the same. Hubbard et al. (1989) also reported similar $T_{1/2}$ values for the *d*-isomer (3.69 hr) and the *l*-isomer (3.77 hr) for SR-MPH. Srinivas et al. (1992) administered the pure *l*- isomer, which was contrasted with the *d*-isomer and the d,l racemic mixture. The reported concentrations of the *l*-isomer were low (about 10% of the *d*-isomer levels) when administered alone or in combination with *d*-isomer, and the behavioral (PD) effects were attributed solely to the *d*-isomer. Srinivas et al. (1993) confirmed this PK pattern for IR-MPH and SR-MPH. A separate research program (in the Netherlands) evaluated *d*-isomer and *l*-isomer concentrations in ADHD children. After oral clinical doses, Jonkman and colleagues, (1998) reported higher concentrations of the *l*-isomer than those reported for the investigations at the University of Saskatchewan (Srinivas et al., 1987, 1992, 1993; Hubbard et al., 1989), which were about 50% (rather than 10%) of *d*-isomer concentrations. This research left a basic PK question unanswered: what is the relative serum concentration of *l*-isomer after oral administration of racemic mixture of d,l-threo-MPH?

Recently, Shader et al. (1999) reported the results of a population PK study in a very large group of ADHD children ($n = 273$). This sample is larger than the combined samples of the other studies in the literature on ADHD children. The population PK method (or sparse sampling) is based on the collection of a single sample from many patients at various times after dosing. Then, modeling is used to estimate the population PK values. This investigation yielded a surprising finding: they estimated a greater $T_{1/2}$ (4.5 hr) than the values reported in the literature (2 to 3 hr) from individual PK studies based on multiple samples from each subject. Shader et al. (1999) speculated that this difference might be due to sampling techniques (sparse sampling over the terminal phase of the PK curve) or to an unrecognized difference in PK after chronic dosing compared to PK after a single dose of MPH. Thus, this research identified another critical PK question about MPH: what is $T_{1/2}$ for IR-MPH in clinical settings?

PD Studies

A few early (pre-1990) studies addressed the PD properties of MPH. The basic design of a PD study addresses the time course of effects by administering repeated probes of efficacy across the day. For example, in the absence of a sensitive drug assay for MPH (see Soldin et al., 1979a, b), the HSC research program emphasized the measurement of time-course effects of MPH. Swanson et al. (1978) used a special laboratory learning test (derived from the A-B/C-B paired-associate paradigm) that could be repeated multiple times across a full day of testing. In a study of 36 subjects, they estimated that the maximum effect occurred about 2 hours after administration of an acute dose of MPH, and that the "behavioral half-life" was about 3 hours.

Solanto and Conners (1982) investigated time-course course effects of acute doses of MPH with multiple measures of response (e.g., attention on a RT task, activity by seat movements, physiological status by HR and finger temperature). They found that estimates of "behavioral half-life" differed across these measures. The PD effects on attention paralleled the PK serum concentrations (with a maximum effect at about 1 hour after administration of an acute dose of MPH and a half-life of about 3 hours), but the PD effects on activity showed peak effects at about 1 hour followed by a longer half-life (or even continued increases in efficacy over time after high doses).

Greenhill et al. (1987) and Pelham et al. (1987) initiated investigations of the PD effects of SR-MPH. Greenhill et al. (1987) reported reduced effectiveness of SR-MPH compared to IR-MPH. In the controlled setting of a multimodality Summer Treatment Program (STP), Pelham et al. (1987) evaluated nine children with ADHD and showed that SR-MPH had a delayed onset compared to IR-MPH. Evidence of delayed onset was not replicated in a subsequent study of 22 ADHD children in the same setting (Pelham et al., 1990), but the maximum effects of SR-MPH were not as great as for IR-MPH on some behavioral measures in this setting. Fitzpatrick et al. (1992) evaluated 19 children with ADHD and reported equal effectiveness in the natural environment and on laboratory tests. In clinical practice, the effectiveness of SR-MPH has not been remarkable, and it has not been widely used (see Greenhill & Osman, 1999). This limited research on SR-MPH identified a critical clinical question about MPH: does SR-MPH have limited effectiveness compared to IR-MPH?

At the University of California, Irvine (UCI), Swanson et al. (1998a,b,c and 1999a,b) extended the methods for measuring PD effects of MPH that were initially developed in the HSC research program (see Swanson, 1989 and Pelham et al., 1995). The UCI Laboratory School Protocol (see Swanson et al., 1999a) was developed to monitor subjective impressions (i.e., ratings) of effects in the classroom setting as well as to generate objective measures of academic productivity (i.e., number of math problems worked in a 10 minute written assignment) (see Swanson et al., 1998b,c, and Wigal et al., 1998a,b). In a study of amphetamine (Adderall) in 32 children with ADHD in which MPH was a positive control condition, the UCI Lab School Protocol was used to document the rapid onset (1.5 hr) and short duration of action (4.0 hr) of MPH for both ratings of behavior and academic productivity (see Swanson et al., 1998a).

The PD effects of *d*-isomer and *l*-isomer of MPH have been investigated by Srinivas et al. (1992, 1993). Srinivas et al. (1992) contrasted the pure *l*-isomer, the pure *d*-isomer, and the d,l-racemic mixture, and found that the PD effects of the *d*-isomer and d,l-mixture were similar and the *l*-isomer had no effect. Srinivas et al. (1993) confirmed this pattern that the full effect of MPH was due to the *d*-isomer. However, given the difference in serum concentration (see above), it is not clear whether the lack of efficacy of the *l*-isomer might be a dose effect rather than lack of efficacy of the *l*-isomer.

Correlation of PK and PD Measures

The correlation of PK and PD measures of efficacy can be evaluated in several ways. For example, a correlation can be calculated based on multiple paired PK/PD measures obtained from an individual (i.e., a "within-subject" method) or on multiple paired PK/PD measures obtained from different individuals (i.e., a "between-subject" method). Typically, the within-subject method is used to estimate the time-response characteristics of response to MPH (see Swanson et al., 1978). Paired measures are used to compare the rise and fall of serum concentration (the PK measure) to the onset and offset of efficacy (the PD measure). Based on the number of observations across the day, a PK/PD correlation can be calculated for each individual to evaluate how acute changes in efficacy are related to changes in serum concentration. In contrast, typically a between-subject method is used to estimate the dose-response characteristics of response to MPH. Paired PK and PD measures are obtained at a specific time point (or as summary measures integrating over time) for each subject. Based on the number of subjects evaluated, a correlation can be calculated to estimate the linear relationship between PK and PD measures.

Several studies in the literature suggest that PK and PD measures are highly correlated. In a small ($n = 5$) sample of children with ADHD, Kupietz et al. (1980) used a within-subject method to evaluate time-response: within-day MPH concentrations and paired measures of efficacy (based on errors on a paired-associate learning test) were highly correlated. Also, Shaywitz et al. (1982) reported a significant between-subject correlation based on single point estimates of MPH concentrations and effectiveness (on teacher ratings of symptoms) in a sample of 21 children with ADHD, and suggested that the relationship between effectiveness and MPH concentration may hold for dose-response evaluations.

Gualtieri et al. (1982, 1984) questioned the conclusions that PK and PD measures were correlated. In a dose-response study of 55 children with ADHD, the serum concentration at 1 hour after administration of an acute dose of MPH did not predict between-subject estimates of global effectiveness, side effects, performance on laboratory tests, or classroom ratings. In a subset of subjects from which serial blood samples were obtained, PK parameters (e.g., $T_{1/2}$ and T_{max}) were not significantly correlated with measures of efficacy or effectiveness. Four patients were studied intensively over 3 weeks, and repeated measures of serum concentrations of MPH were obtained at 1 hour after multiple administrations of the same dose across this time period. High variability of MPH concentrations was reported, but these concentrations were not significantly correlated with measures of clinical effectiveness or efficacy. Based on this work, Gualtieri et al. (1982, 1984) concluded that MPH serum concentrations were not related to clinical response.

These early studies have some weaknesses associated with timing of PK and PD measures. For example, Shaywitz et al. (1982) relied on single measures of serum concentration from each subject, but the administration of medication was at home, so the timing was not tightly controlled. In the Gualtieri et al. (1984) studies, the blood sample was taken at 1 hour after administration, which is before the estimated maximum ($T_{max} = 2$ hours after oral dosing). This may have contributed to a high within-subject variability of serum concentration of MPH when measures were taken on the same individual across multiple days.

The UCI Laboratory School Protocol was developed to provide blood samples (for PK measures) and behavioral probes (for PD measures) at regular intervals across the day, with tight control of when the measures were to be taken (i.e., of timing) and of where the child is and what the child is doing when the measures were to be taken (i.e., of the context). In the Lab School (see Swanson et al., 1999a), on each day of testing a cycle of activities is specified that lasts from 1 to 1.5 hours (e.g., for a 1.5 hour cycle, the time would be divided into a 30-minute classroom session, a 30-minute recess session, a 20-minute session for blood sample or lab test, and a 10-minute session for other activities). Cycles are repeated, so classroom probes occur at regular intervals and in a constant context across the day. Different conditions are established on different days, and the same context and timing is used on each test day. Surrogate measures of response were developed and include subjective ratings of Attention and Deportment (on the SKAMP rating scale) and objective measures of productivity (on the 10-minute math test). These measures are obtained during each classroom probe.

The basic methodology of the Lab School Protocol has been described in several publications (see Swanson et al., 1998a,b,c; Wigal et al., 1998a,b; Swanson et al., 1999a,b). A manual has been developed for multisite investigations (Swanson et al., 1999a), so only a brief description will be given here.

A typical schedule is shown in Figure 11–1A, which is from an investigation of the PK and PD properties of generic MPH (Greenhill et al., 1999). In this study, 27 subjects were evaluated on two successive Saturdays in a double-blind crossover study of placebo and 20 mg BID of MPH. Two cohorts were tested on each Saturday, and a cycle time of 1.5 hours was used to track the time course of serum concentrations and classroom behavior from 7:30 am until 5:30 pm. The results for Attention and Deportment subscales of the SKAMP rating scale (Figure 11–1B) show two common patterns that we have found in studies using the Lab School setting: on placebo, the SKAMP rating tend to increase across the day as behavior in the classroom deteriorates, and the effects of MPH on Deportment are greater than the effects on Attention. The expected pattern of response to a BID dose of MPH is also shown in Figure 11–1B: the lowest ratings are at the PK peak times (1.5 hours after dosing), and the effects dissipate over a 4 hour period after each dose.

Recently, the UCI Lab School protocol has been used to perform efficacy studies that offer answers to several of the important questions highlighted in our review of PK and PD properties of MPH (see above). These carefully controlled efficacy studies were not intended to replace the usual effectiveness studies in the natural environment of the home and school. Instead, these studies were designed to provide information on which to base subsequent effectiveness studies. This combination and staging of investigations (i.e., efficacy studies using surrogate measures followed by effectiveness studies using clinical measures) has been used in several development programs for pharmaceutical companies to evaluate new forms of MPH. We will describe the initial efficacy studies in two of these programs that provided surprising new information about the PK and PD relationships after acute oral doses of MPH.

Recent Efficacy Studies of MPH

Development of a New Once-a-Day Dosing Form of MPH

In the prior review, one of the unanswered questions was about the lack of clinical effectiveness and acceptance of SR-MPH compared to IR-MPH. This question was addressed in the first research program that used the Lab School methods at UCI (Swanson et al., 1999b). This program of research was conducted to evaluate the targeted flat or constant serum concentration profile of the existing form of SR-MPH, and to investigate why this drug delivery profile might result in limited efficacy compared to the drug delivery profile of IR-MHP. The long-term goal was to recommend a drug delivery pattern for a new once-a-day dosing form of MPH.

The PK properties of SR-MPH and IR-MPH differ in several ways. One of the most obvious is the lack of an afternoon bolus of drug after morning administra-

6:45 AM	Arrival Older Group	7:30 AM	Arrival Younger Group
7:30 (0)	Medication Dispensed, Class 1	8:15	Medication Dispensed, Class 1
8:15	Recess 1	9:00	Recess 1
9:00 (1.5)	Class 2	9:45	Class 2
9:45	Recess 2	10:30	Recess 2
10:30 (3.0)	Class 3	11:15	Class 3
11:15	Recess 3 (Lunch)	12:00	Recess 3 (Lunch)
12:00 (4.5)	Medication Dispensed, Class 4	12:45 PM	Medication Dispensed, Class 4
12:45	Recess 4	1:30	Recess 4
1:30 (6.0,1.5)	Class 5	2:15	Class 5
2:15	Recess 5 (Snack)	3:00	Recess 5 (Snack)
3:00 (7.5, 3.0)	Class 6	3:45	Class 6
3:45	Recess 6	4:30	Recess 6
4:30 (9.1,4.5)	Class 7	5:15	Class 7
5:15 PM	Dismissal	6:00 PM	Dismissal

A

Figure 11–1. *A:* Schedule showing the pharmacodynamic effects of twice-daily administration of 20 mg methylphenidate.

tion of SR-MPH, which produces a smooth serum concentration for SR-MPH in contrast to the peaks and valleys of multiple IR-MPH doses. In the first study (see Swanson et al., 1999b), we addressed two questions: "Is a bolus of MPH required to elicit full clinical efficacy?" and "Will a constant serum concentration maintain full clinical efficacy across the day?"

To set the conditions to address these questions, a "sipping" study methodology was used. The precise control of Lab School protocol allowed us to administer capsules every 30 minutes. The doses of MPH in these capsules were chosen by PK simulation (see Park et al., 1998) to generate three experimental serum concentration profiles over the day for different ways of delivering the same amount of medication. A standard profile was established by administering each subject's clinical dose of MPH twice a day (BID, at 7:30 a.m. and 12:00 noon), which was expected to generate peaks and troughs in the morning and in the afternoon. One experimental profile (called Flat) was generated by administering a morning bolus (0.8 of a subject's usual morning clinical dose) followed by small but equal doses across the day to remove the peaks and troughs of the BID condition. This Flat profile was

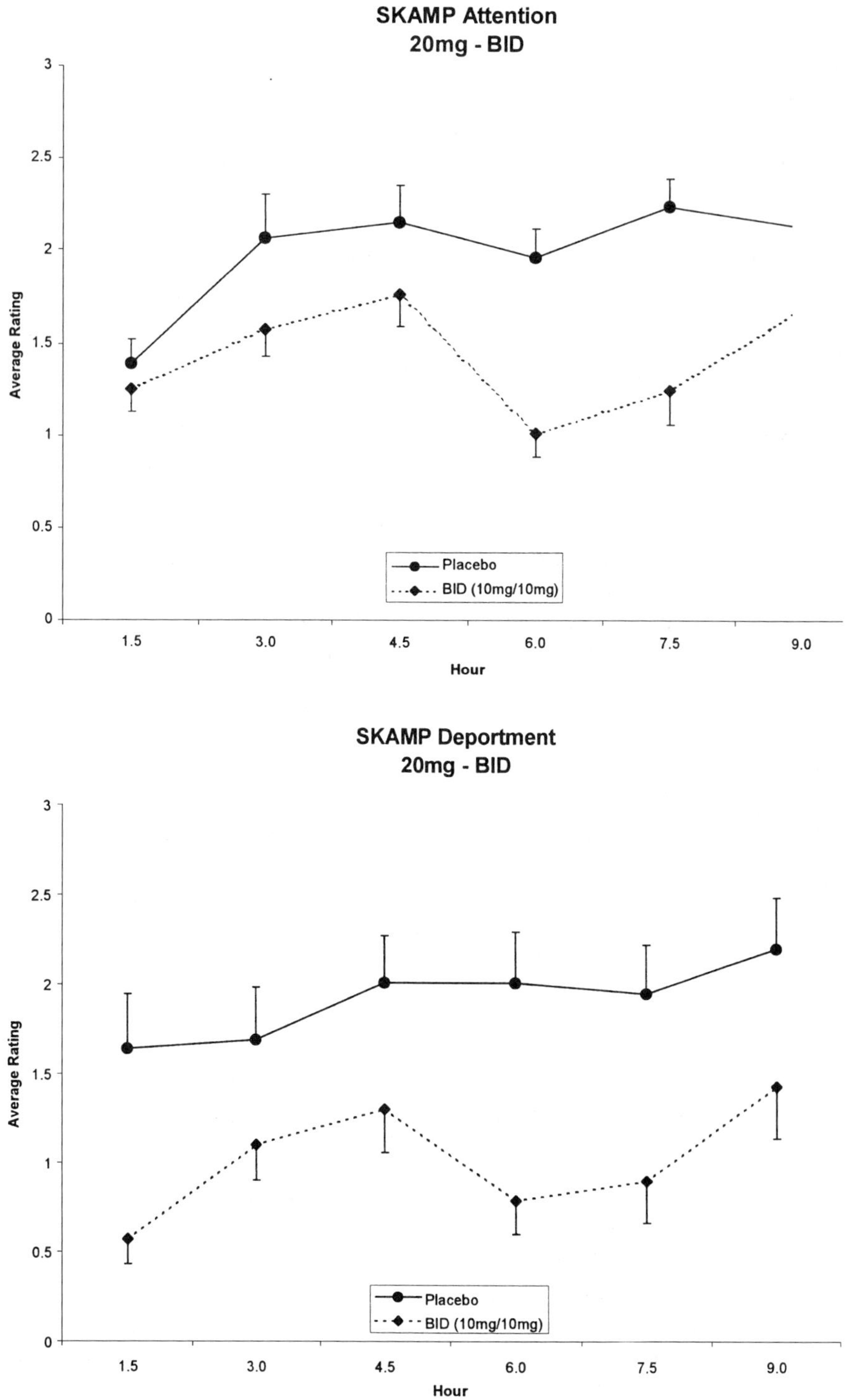

B

Figure 11–1—*continued.* *B:* Comparison with placebo of the pharmacodynamic effects of twice-daily administration of 20 mg methylphenidate on attention and deportment ratings.

intended to mimic the SR-MPH pattern of serum concentrations across the day. Another experimental profile (called Ascending) was generated by administering small ascending doses across the day to achieve a peak serum concentration without a bolus. A placebo condition was included, and the difference between the active control (BID) and the inactive control (Placebo) was used to gauge the effects of the Flat and Ascending experimental profiles at each time point of interest (in this study, at the expected BID peaks and troughs in the afternoon). The results of this "concept" study were clear: in the afternoon, the PD measures showed that the full clinical efficacy could be achieved by the Ascending profile (without a bolus dose) and that the Flat profile (without the peaks and troughs) lost about 40% of full efficacy compared to the standard BID regimen (see Swanson et al., 1999b).

The lack of maintenance of full effects by the Flat profile suggested that acute tolerance might occur after oral administration of clinical doses of MPH. It may be surprising that acute (short-term) tolerance occurs after clinical doses of MPH, because these doses do not produce long-term tolerance (see Safer & Allen, 1989). However, this is consistent with the literature on stimulants and tachyphylaxis (see Avram et al., 1969). The existence of acute tolerance is not inconsistent with the absence of long-term tolerance in the usual treatment of ADHD, since the typical dosing regimens have long periods (10 to 12 hours at night) when serum concentrations levels are negligible. This would allow ample time for the acute tolerance to dissipate. But if very high doses are administered more than 3 times a day and sufficient time is not allowed for drug concentrations to dissipate, the concept of acute tolerance suggests that long-term tolerance may develop (see Swanson et al., 1986).

To provide the data for a rigorous test of the acute tolerance hypothesis, a study was designed to provide a wide variation in simulated serum concentrations. In a between-group design, two experimental TID dosing conditions were used to provide different "recent serum concentration histories" before achieving the standard peaks MPH serum concentrations expected for standard TID dosing. In each experimental condition, calculated doses were administered to generate a middle peak that matched the expected height of the peak of each subject's usual TID regime. In this example, we used the modal clinical dose (10 mg TID) to demonstrate the PK-PD relationships for the group, while in a prior presentation of the PK-PD relationships (Swanson et al., 1999), we used a lower dose (5 mg TID). The actual doses varied across subjects. In standard TID dosing an equal dose is given at 4-hour intervals (e.g., 10 mg at 7:30 am, 11:30 am and 3:30 pm). In the experimental TID conditions, the relative size and timing of the middle dose (e.g., 6.4 mg at 9:30 am or 11.8 mg at 1:30 pm) were varied to generate an earlier (TID–am) or later (TID–pm) middle peak in the simulated serum concentration profile shown as solid lines in the left panels of Figure 11–2.

For the TID-am condition, an early morning peak (see Fig. 11–2a) was established, and for the TID-pm, a late afternoon peak was established (see Fig. 11–2b). The levels of "placebo adjusted" efficacy measures (based on the difference in the SKAMP Attention ratings in the two conditions) at specific time points are presented as dotted lines in Figure 11–2. A placebo-drug difference score was used so a positive effect at the expected peak time would match the expected peak in serum concentration of MPH. These same data are re-plotted in the right panels of Figure 11–2 as serum concentration-efficacy plots, with the arrows showing the time

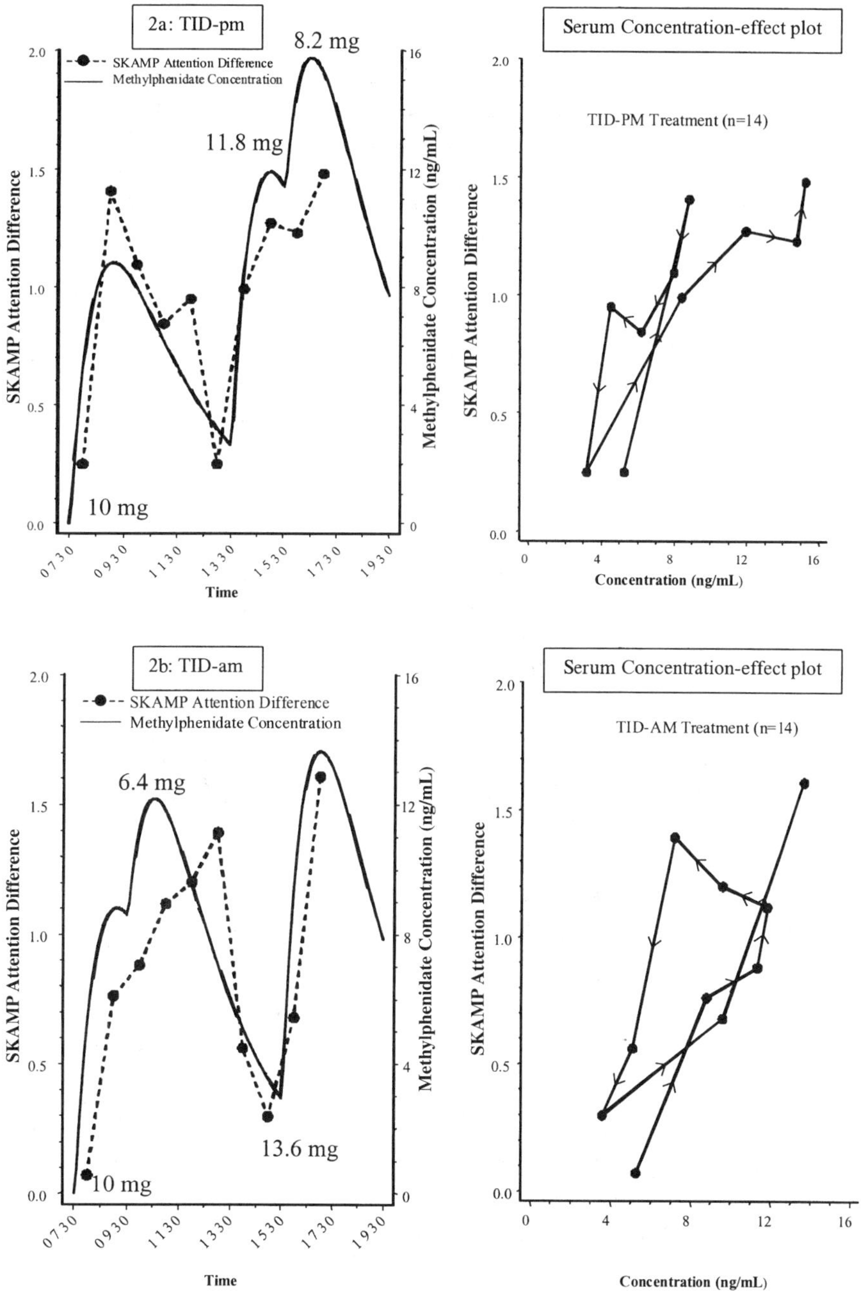

Figure 11–2. Acute tolerance to methylphenidate administration in the morning and the evening.

course of the observations. Inspection and interpretation of these figures must take into account a between-group design (due to different subjects being assigned to the TID-am and TID-pm conditions), a lag in the PD response (due to possible differences in the time of peak concentrations in the serum and at the site of action in the brain), and nonlinear dose-response functions (due to the possibility that the optimal drug concentration might be exceeded at some points in time). The E_{max} model was used to describe the PK-PD relationships shown in Figure 11–2 (see Park et al., 1998 and Swanson et al., 1999b).

In the portrayal of the PK and PD data in Figure 11–2, there are several ways to relate serum concentration to efficacy. We will consider two here. First, consider the maximum response on the surrogate measures in Figure 11–2a (i.e., a difference of about 1.5 on the SKAMP ratings, which is represented by a point on the ordinate in these graphs). This concentration-efficacy plot suggests that the maximum response occurs at about 9 ng/ml after the first exposure to this serum level, but that a higher serum concentration (16 ng/ml) is required to re-establish the maximum response after recent exposure to a 12 ng/ml serum concentration at the end of the day. Second, consider the serum concentration that produces the maximum response after the first peak (e.g., 9 ng/ml, which is represented by a point on the abscissa in these graphs). The first exposure to this level produces maximum response (a difference in the SKAMP rating of about 1.5 on the Attention measure), but a later exposure to the same serum concentration produces a smaller effect (a difference of about 1.0). The same pattern is illustrated in Figure 11–2b when the second exposure to methylphenidate occurs in mid-morning. As discussed in some detail by Swanson et al. (1999b), this suggests that acute tolerance develops rapidly after clinical doses of MPH, and that this tolerance dissipates rapidly across the day. Nonlinear PK/PD modeling was used to estimate the onset and offset of tolerance (see Swanson et al., 1999b, & Park et al., 1998).

The findings of the "concept" studies reported by Swanson et al. (1999) were used to identify a pattern of drug delivery that might produce a smooth, sustained effect across the day: an initial bolus (to elicit the rapid response as to IR-MPH) followed by a smooth but increasing amount of drug (to maintain the effect in the face of acute tolerance). The next step in the program of research was to conduct a "proof of concept" study. The "sipping study" methodology was used again to administer calculated doses of MPH in capsules given every 30 minutes to achieve the target ascending serum concentration profile. In a Lab School study of 32 ADHD children, the efficacy of the "bolus plus ascending" pattern matched the efficacy of standard TID administration of MPH (Swanson et al., 1999c)—thus establishing "proof of concept."

These concept studies provided sufficient information to develop a new once-a-day dosing form of MPH using technology such as the Osmotic Release Oral System (OROS). The OROS system is based on the principle of an osmotic pump: A semi-permeable membrane surrounds the layered combination of a drug and a polymer, which protects the drug from the contents to the stomach and intestine and (most importantly) controls the intake of water after oral administration. Upon contact with water, the polymer layer expands at a programmed rate and acts as a piston to push the drug out of a small laser-drilled hole in the membrane. A combination of characteristics of the OROS membrane, the composition of layers of MPH and polymers, and the type of overcoat of MPH, was developed that was expected to deliver the targeted pattern of drug to the bloodstream (i.e., the "bolus plus ascending" drug delivery profile designed to produce rapid onset and to counteract acute tolerance).

OROS MPH has been evaluated in two "proof of product" studies, which were reported at the 1999 meeting of the American Academy of Child and Adolescent Psychiatry. The first was a PK study in normal adult volunteers, which documented that a morning oral administration of the product (called Concerta, a small, capsule-shaped pill) achieved the intended pattern of ascending serum concentrations across a 12-hour period. The second was a PK study of the intended patient population, which verified an ascending serum concentration pattern in 30 ADHD children evaluated in the UCI Lab School setting.

This research program demonstrates innovative ways that PD and PK studies can be staged in a drug development program (see Levy et al., 1994). The initial "concept" studies were based on PD methods, and the PK information was simulated using the available literature. These PD studies led to a basic discovery (acute tolerance) that directed the next step in designing a drug delivery system. The first PK study in the program was conducted to verify that desired serum concentration pattern was achieved. Then, another PD study was conducted to demonstrate the efficacy of the new drug delivery system that produced the targeted PK pattern. After this initial development phase, the usual studies required for FDA approval were planned and executed.

Evaluation of the Pure *d*-threo Isomer of MPH

Another question identified in our literature review (see above) was about the PK properties of the isomers of MPH. In comparison to the *d*-isomer, some studies of the *l*-isomer reported very low (10%) levels and others reported relatively high (50%) levels. Evaluation of a few patients suggested that the impact of the *l*-isomer was negligible. However, more than a decade ago, the enantiomeric disposition of circulating MPH had been suggested as a possible key to understanding individual differences in response to MPH (see Patrick et al., 1987a,b). Sufficient information had not been generated over the subsequent decade to fully address this question. We used the Lab School setting to evaluate the PK and PD properties of the pure *d*-isomer of MPH (Swanson et al., 1999e).

This research program was conducted for refinement of approved racemic forms of medications by removal of unnecessary or harmful isomers (Greenhill et al., 1999). We designed the first multisite (UC Irvine, U Saskatchewan, and Northwestern U) Lab School study to evaluate the removal of the *l*-isomer from the available racemic mixture of MPH. In a double-blind, crossover study of 32 ADHD children, we contrasted the PK and PD properties of low (5 mg), medium (10 mg), and high (20 mg) doses of standard (racemic) *d,l*-threo MPH to matched doses (2.5 mg, 5 mg, and 10 mg) of chirally pure *d*-threo MPH. On each of the seven test days (the six drug conditions plus placebo), PK and PD measurements were taken at six time points across the day. Included in these cycles of activities was a computer-administered math test of 50 problems with difficulty set for each subject. The results for the low (2.5 mg d and 5 mg d,l), medium doses (5 mg d and 10 mg d,l), and high doses (10 mg d and 20 mg d,l) are shown in Figure 11–3.

This figure presents the PK (serum concentrations of *d*-threo-MPH) and PD (% correct on the math test) curves for the three dose conditions. As shown in Figure 11–3a, the peak serum concentration for both low doses was about 5 ng/ml of d-threo-MPH, for both medium doses was about 10 ng/ml, and for both high doses about 16 ng/ml. The serum levels of *l*-threo-MPH were negligible for all doses. This was the first surprise of this study. This is not consistent with the low but quantifiable concentrations of *l*-threo-MPH reported by Srinivas and colleagues (1987, 1992, 1993), or with the high *l*-threo-MPH concentrations reported by Jonkman et al. (1998), but it is consistent with the notion that the *l*-threo-MPH isomer is rapidly and completely metabolized before reaching systemic circulation (Chan et al., 1983; Coffee et al., 1983; Patrick et al., 1987a).

The second surprise was that the PD response (about 82%) at the peak serum concentration (about 16 ng/ml) of the high dose conditions (see Fig 11–3B) was not greater than in the PD response in the medium-dose conditions (also about 82% correct) at the peak serum concentration (about 10 ng/ml). The high dose condition appeared to increase the duration of response, perhaps because it maintained serum concentrations above the effective level of 8 to 10 ng/ml for an extended time.

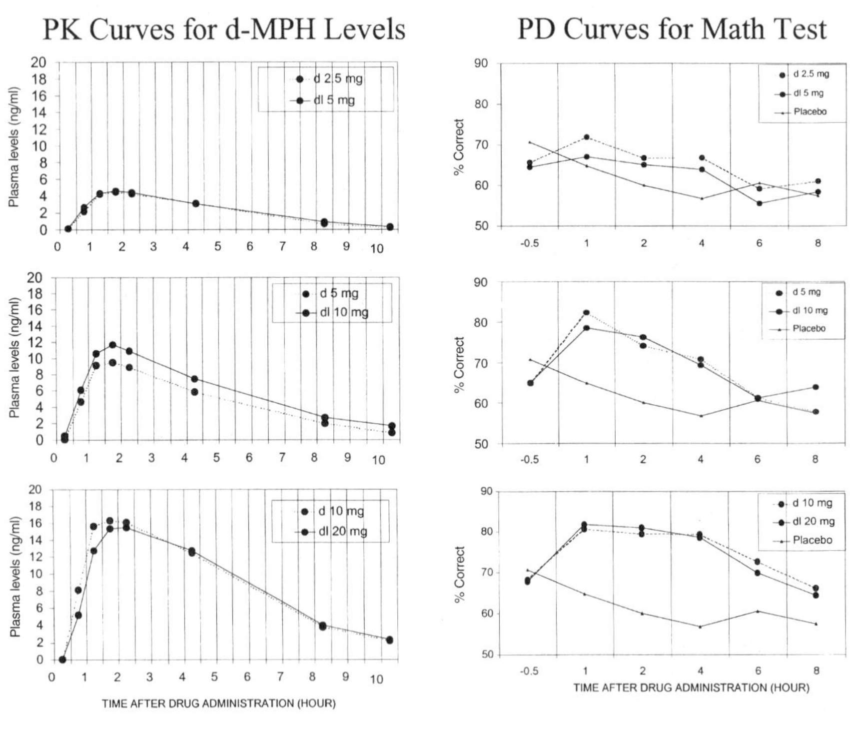

Figure 11–3. Pharmacokinetic and pharmacodynamic effects of d-threo-methylphenidate, the pure isomer, as measured at various time points after administration and during a computer-administered math test.

This was a study of a single dose given on each study day. It is interesting to speculate what might have occurred in a multiple-dosing study. With multiple exposures, the existing PK data suggest that a higher serum concentration would be required to elicit the same PD response of a lower concentration achieved after the initial dose. If the initial peak serum concentration were adjusted for tachyphalaxis (acute tolerance), then a greater optimal serum concentration would be expected in the afternoon than in the morning. Taken together, the two studies described in this section suggests that future studies should investigate the relationship of afternoon serum concentrations to efficacy and side effect to investigate individual differences in response to MPH.

Summary of PK/PD Studies of MPH

The initial PK studies of MPH provided some basic information about MPH (e.g., T_{max} = 1.5 hr and $T_{1/2}$ = 3 hr). Based on a small number of subjects, these studies provided answers to some clinically relevant questions (i.e., should MPH be given on a BID or TID schedule; should MPH be administered before or with meals). However, these small, university-based studies did not address many of the questions that would be asked in modern drug development programs by pharmaceutical companies, so many questions were left unanswered.

In recent investigations of PK and PD properties of MPH funded by pharmaceutical companies, two of these unanswered questions were addressed: "What are the PK/PD properties of sustained-release MPH?" and "What are the PK/PD properties of the pure *d*-isomer of MPH?" These research programs suggest that PK information may provide a good prediction of PD response to MPH, but that the relationship will be more complex than just a narrow therapeutic window. The initial response may be to a narrow range of serum concentrations of d-MPH (around 8 to 10 ng/ml), but the time course of this PD response appears to be affected by acute tolerance of the therapeutic response (requiring higher concentrations later in the day to maintain efficacy) and by insensitivity or rapid tolerance to the side effects of MPH (allowing higher doses to lengthen duration of action).

These two research programs show how the PK and PD properties of MPH are used to develop new commercial products. However, these research programs addressed applied rather than basic questions about the mechanism of action of MPH. Basic questions about the mechanisms of action of MPH have been addressed in another area of investigation (i.e., investigations of drug of abuse), which will be discussed in the next section of this chapter.

PET Imaging Studies of MPH

Positron emission tomography (PET) is a noninvasive imaging method that allows the assessment of concentration of radiotracers in the human brain. PET methods have been used to assess the bioavailability, distribution, pharmacokinetics, and binding properties of MPH, as well as the in vivo potency of MPH to block dopamine transporters (DAT) in the human and the baboon brain. These studies are made possible by labeling MPH with carbon-11, a positron emitter with a 20-minute half-life (Ding et al., 1994), without changing its pharmacological properties.

Distribution of MPH in the Brain

PET and [^{11}C]methylphenidate ([^{11}C]MPH) were used to define the distribution of MPH in the human brain (Volkow et al., 1995). After intravenous doses administered to adult volunteers, the uptake of [^{11}C]MPH was high and corresponded to 7% to 10% of the injected dose. The binding of MPH was heterogeneous and the highest concentration was observed in striatum, the brain area with the highest concentration of DAT (see Fig. 11–4). Relatively low concentrations of MPH were detected in cortex and cerebellum. This regional distribution of [^{11}C]MPH corresponds to that previously described in autoradiographic studies of the human brain (Unis et al., 1985).

Pretreatment of subjects with pharmacological doses of MPH decreased the binding only in the striatum. In the baboon brain, pretreatment with MPH decreased [^{11}C]MPH binding to the same extent as pretreatment with drugs that inhibit DAT (Ding et al., 1994), thus suggesting that the accumulation of MPH in striatum is consistent with its binding to DAT.

Ding and colleagues separately labeled d-MPH and l-MPH with ^{11}C to assess the distribution in brain of the two enantiomers of MPH (Ding et al., 1997). These studies showed that [^{11}C]d-MPH binding was much higher in striatum than in cortex and cerebellum whereas the binding of [^{11}C]l-MPH was homogeneous throughout the human brain. Moreover, although pretreatment with pharmacological doses of MPH or of other DAT blocker drugs inhibited the binding of [^{11}C]d-MPH in striatum they did not affect the binding of [^{11}C]l-MPH, indicating that the binding of [^{11}C]l-MPH in the human brain was mostly nonspecific. This study also showed significant differences in the pharmacokinetics of the *d* and the *l* isomers in the brain. The clearance of [^{11}C]d-MPH in striatum was much slower that that of [^{11}C]l-MPH; for [^{11}C]d-MPH, only 10% to 15% clearance from peak activity had occurred at 74 minutes post injection, whereas for [^{11}C]l-MPH more than 50% clearance had occurred at 74 minutes. The pattern of clearance of l-threo was homogeneous throughout the brain. The pattern of clearance for [^{11}C]d-MPH was significantly slower for striatum than for cerebellum, where the clearance of [^{11}C]d-MPH did not differ from that observed for [^{11}C]l-MPH. The kinetics of [^{11}C]d-MPH in cerebellum (a brain region devoid of DA transporters) mostly reflect the temporal course of nonspecific binding. The fact that the kinetics of [^{11}C]d-MPH in cerebellum were equivalent to those seen for [^{11}C]l-MPH throughout the brain further corroborates the nonspecific binding of [^{11}C]l-MPH. These findings in the human brain are in agreement with the animal microdialysis studies showing that d-threo but not l-threo increased extracellular DA in striatum (Ding et al., 1997). They also provide evidence that the lack of efficacy of [^{11}C]l-MPH may not be the result of its very low serum concentration after oral administration (see above) but of its lack of intrinsic efficacy.

Pharmacokinetics of MPH After IV and Oral Administration

Studies on the pharmacokinetics of intravenous MPH in the human brain were conducted using PET and [^{11}C]MPH, which showed the distribution of the radiotracer in the brain at different times after its administration (Volkow et al., 1996). Figure

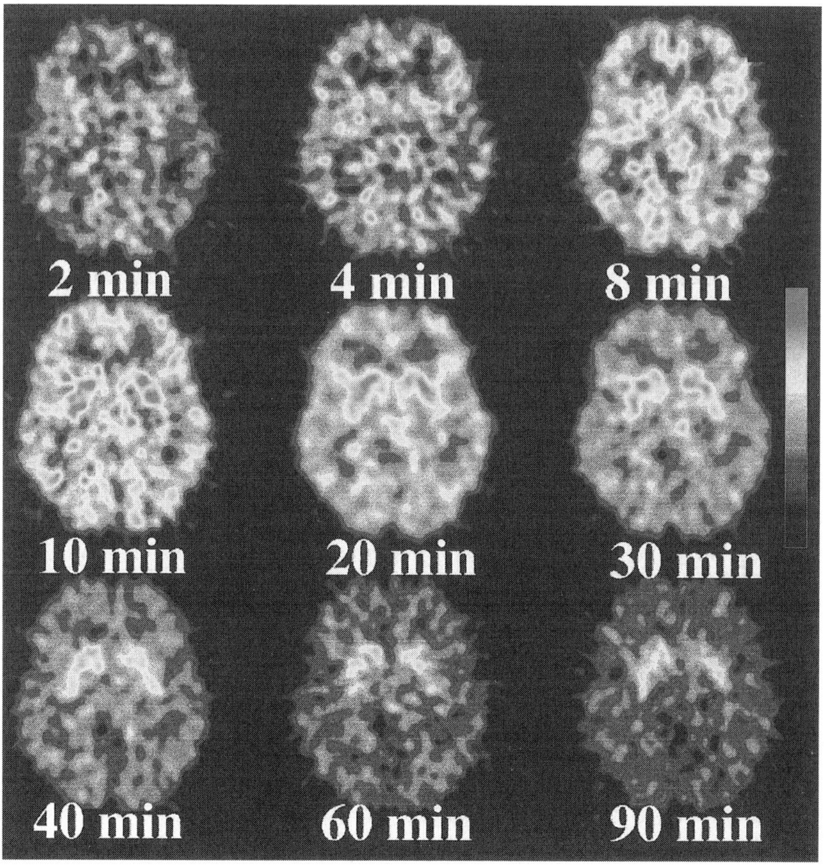

Figure 11–4. Timed results of [^{11}C]methylphenidate administration at the level of the striatum.

11–4 shows the images obtained for [^{11}C]MPH at the level of the striatum at different times after administration of the radiotracer.

The uptake of [^{11}C]MPH in human brain was very fast (peak concentration 6–10 minutes) but the rate of clearance was relatively slow (half peak clearance in striatum = 90 minutes). The temporal course of subjective impressions of "high," "rush," or "restlessness" experienced after IV administration of MPH paralleled the fast uptake of the MPH in striatum, and these effects returned to baseline (although more slowly and less completely for "restlessness") while there was still a significant presence of MPH in brain (see Fig. 11–5). This dissociation suggests rapid tolerance to MPH.

For safety reasons related to radiation exposure to the upper gastrointestinal tract, [^{11}C]MPH PET studies using oral administration of MPH were done in the baboon subjects, not humans. These studies showed that MPH did not reach peak concentration in brain until 60 minutes or more after oral administration (see

Fig. 11–6). This value corresponds well with the T_{max} of serum concentrations of MPH that we derived from the literature. The slow rate of brain uptake of orally administered MPH contrasts markedly with the very fast rate of brain uptake (i.e., peak brain uptake 8–10 minutes) of intravenously administered MPH (Ding et al., 1994).

In Vivo Potency of IV and Oral MPH to Block DAT in the Human Brain

Studies on the potency of MPH to block DAT have been accomplished with PET using [^{11}C]cocaine, which serves as a ligand for the DAT (Fowler et al., 1989). The placebo condition provides an indication of the levels of DAT at baseline, and reductions after different doses of MPH reflect the occupancy (blockade) of the DAT. When MPH was administered intravenously, it significantly blocked the DAT at all the doses tested; the levels of blockade 5 to 7 minutes after its administration corresponded to 35% ± 5% for a 0.05 mg/kg dose, to 62 ± 13% for a 0.1 mg/kg dose, 67 ± 3% for a 0.25 mg/kg dose and 78% ± 11% for a 0.5 mg/kg dose (Volkow et al., 1990). The estimated ED_{50} (dose required to block 50% of the DAT) was 0.075 mg/kg.

When MPH was administered orally it also produced DAT blockade (Volkow et al., 1998). At 2 hours after oral MPH, the concentrations of blockade corresponded to 12 ± 4% for a 5 mg dose, 40 ± 12% for a 10 mg dose, 54 ± 5% for a 20 mg dose, 72 ± 3% for a 40 mg dose, and 74 ± 2% for a 60 mg dose. The levels of DAT occupancy achieved after adjusting the doses for the weight of the subjects are shown in Figure 11–7. The estimated ED_{50} (dose required to block 50% of the DAT) was 0.25 mg/kg.

As shown in Figure 11–8, the serum MPH concentrations of 6 ng/ml of d-MPH blocked an average 50% of the DAT. Serum MPH concentrations in excess of 6 ng/ml occur in ADHD children receiving typical MPH doses of .3 to .6 mg/kg, so this suggests that greater than 50% DAT blockade may be required for therapeutic efficacy. Also, after oral doses of MPH, this study revealed a high correlation between peak plasma concentration of d-MPH and DAT occupancy ($r = 0.80$; df, 11; $p < 0.002$). Thus, plasma concentrations of d-MPH (the active enantiomer) should provide good measures of brain concentrations of MPH.

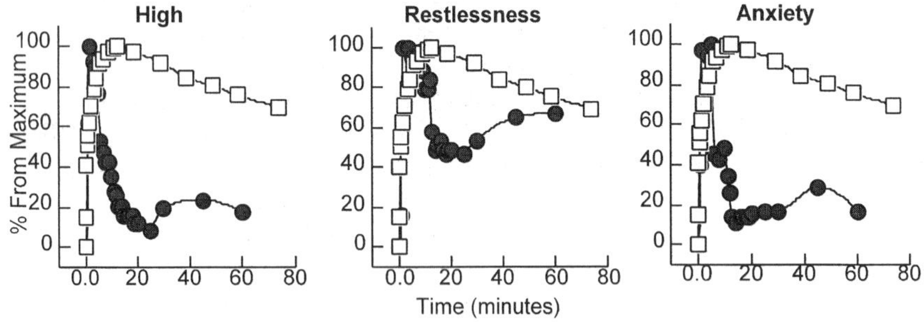

Figure 11–5. Patients' subjective reports of characteristic feelings of "high," "restlessness," and "anxiety" after IV administration of methylphenidate parallel the fast uptake of MPH in the striatum, which is illustrated in Figure 11–4.

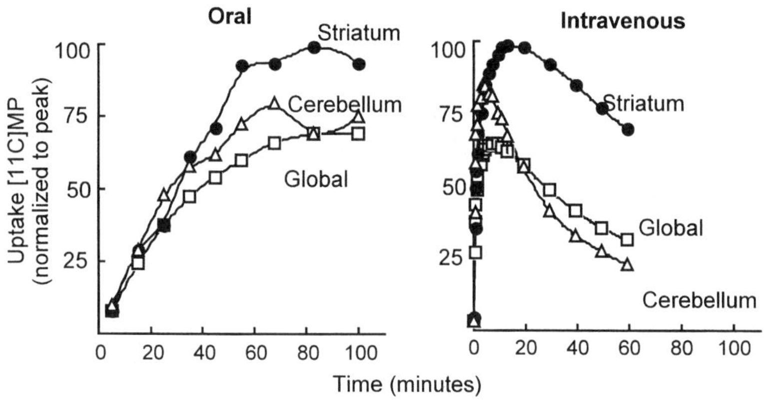

Figure 11–6. Comparison of brain uptake of [^{11}C]methylphenidate administered orally and intravenously.

Summary

Based on the PK/PD and PET studies reviewed in this chapter, several converging points appear to emerge that may be relevant to the clinical responses of ADHD children to MPH:

1. After oral doses of MPH, the PET studies suggest that peak brain concentrations of MPH occur between 1 and 2 hours after dosing (see Fig. 11–6). The PK/PD studies reveal about the same time for peak PK (serum concentration) and peak PD (behavioral) effects of clinical doses of MPH (see Fig. 11–1).
2. After oral administration of MPH, the PET studies suggest a dissociation between the brain concentrations of MPH and some behavioral effects (see

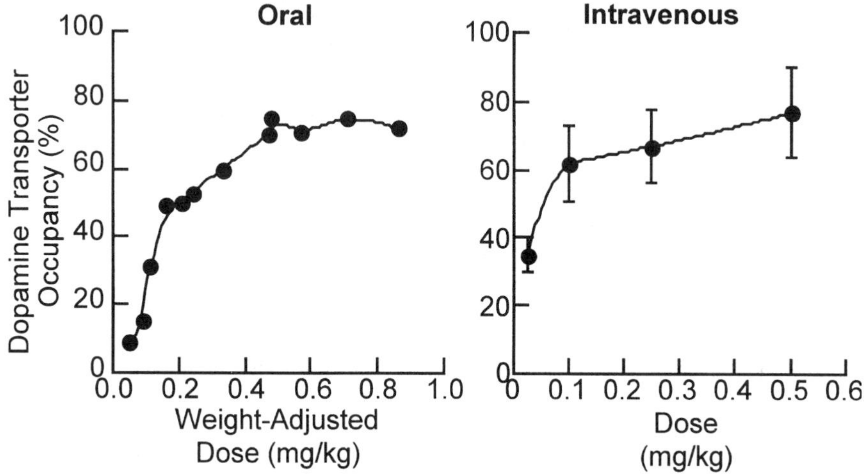

Figure 11–7. Comparison of effects of oral and intravenous doses of methylphenidate on dopamine transporter occupancy in the human brain.

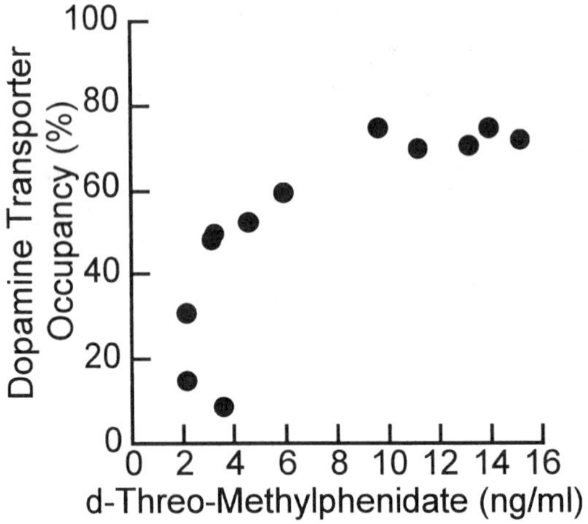

Figure 11–8. Effects of serum concentration of the pure isomer of methylphenidate on dopamine transporter occupancy in the human brain.

Fig. 11–5), which appears as a diminution in effect even when relative high concentrations are maintained in plasma. The PK/PD studies document a similar phenomenon that has been called acute tolerance (see Fig. 11–2). This is poorly understood, but it may reflect adaptation responses at the synaptic level in responses to blockade of the DAT.

3. The PET studies suggest a maximum for DAT-mediated effects of MPH centered around 60% to 80% occupancy, and this corresponds to serum concentrations of about 8 to 10 ng/ml (see Fig. 11–8). One implication is that higher concentrations are not likely to be very effective in further blocking DAT but may result in an increase in side effects. The PK/PD studies also suggest that maximum efficacy also occurs at about 10 ng/ml (see Fig. 11–3). Higher serum concentrations may modulate side effects and duration of action of bolus doses of MPH.

Taken together, the PET and PK/PD studies of MPH offer a clear hypothesis that could be tested in future studies: a therapeutic window for acute response to oral MPH is about 8 ng/ml, and this corresponds to about 60% to 80% occupancy of DAT. After exposure to this concentration of MPH, acute tolerance may emerge, so increasing serum concentration and DAT occupancy may be required to maintain the full effects.

References

Avram, D, Aronow, L, and Kalman, SM (1969). *Principles of drug action*. New York: Harper and Row.

Birmaher, B, Greenhill, LL, Cooper, TB, Fried, J, & Maminski, B (1989). Sustained release methylphenidate: pharmacokinetic studies in ADDH males. *J Am Acad Child Adolesc Psychiatry, 28*, 768–772.

Borcherding, BG, Keysor, CS, Cooper, TB, & Rapoport, JL (1989). Differential effects of methylphenidate and dextroamphetamine on the motor activity level of hyperactive children. *Neuropsychopharmacology, 2*, 255–263.

Chan, YM, Soldin, SJ, Swanson, JM, Deber, CM, Thiessen, JJ, & Macleod, S (1980). Gas chromatographic/mass spectrometric analysis of methylphenidate (Ritalin) in serum. *Clin Biochem, 13*, 266–272.

Chan, YP, Swanson, JM, Soldin, SS, Thiessen, JJ, Macleod, SM, & Logan, W (1983). Methylphenidate hydrochloride given with or before breakfast: II. Effects on plasma concentration of methylphenidate and ritalinic acid. *Pediatrics, 72*, 56–59.

Coffee, B, Shader, RI, & Greenblatt, DJ (1983). Pharmacokinetics of benzodiazepines and psychostimulants in children. *J Clin Psychopharmacol, 3*, 217–225.

Committee report on the psychopharmacology in children in the USA (1971). U.S. Congress, Committee Hearings on the Use of Stimulant Medication for the Treatment of Hyperactive Children. Washington, DC.

Dalby, JT, Kinsbourne, M, Swanson, JM, & Sobol, MP (1977). Hyperactive children's underuse of learning time: correction by stimulant treatment. *Child Dev, 48*, 1448–1453.

Ding, YS, Fowler, JS, Volkow, ND, Dewey, SL, Wang, GJ, & Logan, J, et al. (1997). Chiral drugs: Comparison of the pharmacokinetics of [^{11}C]d-threo and L-threo-methylphenidate in the human and baboon brain. *Psychopharmacology (Berl), 131*, 71–78.

Ding, YS, Fowler, JS, Volkow, ND, Gatley, SJ, Logan, J, Dewey, SL, Alexoff, D, Fazzini, E, & Wolf, AP (1994). Pharmacokinetics and in vivo specificity of [^{11}C]dl-threo-methylphenidate for the presynaptic dopaminergic neuron. *Synapse, 18*, 152–160.

Faraj, BA, Israili, ZH, Perel, JM, Jenkins, ML, Holtzman, SG, Cucinell, SA, & Dayton, PG (1974). Metabolism and disposition of methylphenidate-14C: studies in man and animals. *J Pharmacol Exp Ther, 191*, 535–547.

Fitzpatrick, PA, Klorman, R, Brumaghim, JT, & Borgstedt, AD (1992). Effects of sustained-release and standard preparations of methylphenidate on attention deficit disorder. *J Am Acad Child Adolesc Psychiatry, 31*, 226–234.

Fowler, JS, Volkow, ND, Wolf, AP, Dewey, SL, Schlyer, DJ, Macgregor, RR, Hitzemann, R, Logan, J, Bendriem, B, Gatley, SJ, et al. (1989). Mapping cocaine binding sites in human and baboon brain in vivo. *Synapse, 4*, 371–377.

Greenhill, LL (1991). Methylphenidate in the clinical office practice of child psychiatry. In LL Greenhill & BB Osman, (Eds.), *Ritalin: Theory and patient management*. New York: May Ann Liebert, Inc. (pp. 97–117.)

Greenhill, LL, Cooper, T, Solomon, M, Fried, J, & Cornblatt, B (1987). Methylphenidate salivary levels in children. *Psychopharmacol Bull, 23*, 115–119.

Greenhill LL, Gupta S, Manos M, Pelham WE, Pliszka S, Spencer T, Swanson JM (1999). Do Time-Response Effects vary among Stimulant Medications? 46th Annual Meeting, American Academy of Child and Adolescent Psychiatry, October, Chicago, IL.

Greenhill, LL, & Osman, BB (1999). *Ritalin: Theory and practice, Second Edition*. New York: May Ann Liebert.

Greenhill, LL, Pine, D, March, J, Birmaher, B, & Riddle, M (1998). Assessment issues in treatment research of pediatric anxiety disorders: what is working, what is not working, what is missing, and what needs improvement. *Psychopharmacol Bull, 34*, 155–164.

Greenhill, LL, Puig-Antich, J, Novacenko, H, Solomon, M, Anghern, C, Florea, J, Goetz, R, Fiscina, B, & Sachar, EJ (1984). Prolactin, growth hormone and growth responses in boys with attention deficit disorder and hyperactivity treated with methylphenidate. *J Am Acad Child Adolesc Psychiatry, 23*, 58–67.

Gualtieri, CT, Hicks, RE, Patrick, K, Schroeder, SR, & Breese, GR (1984). Clinical correlates of methylphenidate blood concentrations. *Ther Drug Monit, 6*, 379–392.

Gualtieri, CT, Wargin, W, Kanoy, R, Patrick, K, Shen, CD, Youngblood, W, Mueller, RA, & Breese, GR (1982). Clinical studies of methylphenidate serum levels in children and adults. *J Am Acad Child Adolesc Psychiatry, 21*, 19–26.

Hubbard, JW, Srinivas, NR, Quinn, D, & Midha, KK (1989). Enantioselective aspects of the disposition of dl-threo-methylphenidate after the administration of a sustained-release formulation to children with attention deficit-hyperactivity disorder. *J Pharm Sci, 78*, 944–947.

Humphries, T, Kinsbourne, M, & Swanson, JM (1978). Stimulant effects on cooperation and social interaction between hyperactive children and their mothers. *J Child Psychol Psychiatry, 19*, 13–22.

Jonkman, LM, Verbaten, MN, de Boer, D, Maes, RA, & Buitelaar, JK, et al. (1998). Differences in plasma concentrations of the D- and L-threo methylphenidate enantiomers in responding and non-responding children with attention deficit hyperactivity disorder. *Psychiatry Res, 78*, 115–118.

Kupietz, SS, Richardson, E, Gadow, KD, & Winsberg, BG (1980). Effects of methylphenidate on learning a 'beginning reading vocabulary' by normal adults. *Psychopharmacology (Berl), 69*, 69–72.

Levy, G, Ebling, WF, & Forest, A (1994). Concentration- or effect-controlled clinical trials with sparse data. *Clin Pharmacol Ther, 56*, 1–8.

Meyer, MC, Straughn, AB, & Wood, GC (1989). *Immediate-release methylphenidate bioequivalence study*. FDA Contract No. 223-87-1802, Technical Progress Report No. 7.

Milberg, RM, Rinehart, KL Jr, Sprague, RL, & Sleator, EK (1975). A reproducible gas chromatographic mass spectrometric assay for low levels of methylphenidate and ritalinic acid in blood and urine. *Biomed Mass Spectrom, 2*, 2–8.

Park, K, Verotta, D, Gupta, SK, & Sheiner, LB (1998). Use of a pharmacokinetic/pharmacodynamic model to design an optimal dose input profile. *J Pharmacokinet Biopharm, 26*, 471–492.

Patrick, KS, Caldwell, RW, Ferris, RM, & Breese, GR (1987a). Pharmacology of the enantiomers of threo-methylphenidate. *J Pharmacol Exp Ther, 241*, 152–158.

Patrick, KS, Muellar, RA, Gualtieri, CT, & Breese, GR (1987b). Pharmacokinetics and actions of methylphenidate. In H Y Meltzer (ed.), *Psychopharmacology: the third generation of progress*. New York: Raven, pp. 1387–1395.

Patrick, KS, Straughn, AB, Jarvi, EJ, Breese, GR, & Meyer, MC (1989). The absorption of sustained-release methylphenidate formulations compared to an immediate-release formulation. *Biopharm Drug Dispos, 10*, 165–171.

Pelham, WE, Aronoff, HR, Midlam, JK, Shapiro, CJ, Gnagy, EM, & Chronis, AM, et al. (1999). A comparison of Ritalin and Adderall: efficacy and time-course in children with attention-deficit/hyperactivity disorder. *Pediatrics, 103*, e43.

Pelham, WE Jr, Greenslade, KE, Vodde-Hamilton, M, Murphy, DA, Greenstein, JJ, & Gnagy, EM, et al. (1990). Relative efficacy of long-acting stimulants on children with attention deficit-hyperactivity disorder: A comparison of standard methylphenidate, sustained-release methylphenidate, sustained-release dextroamphetamine, and pemoline. *Pediatrics, 86*, 226–237.

Pelham, WE Jr, Sturges, J, Hoza, J, Schmidt, C, Bijlsma, JJ, & Milich, R, et al. (1987). Sustained release and standard methylphenidate effects on cognitive and social behavior in children with attention deficit disorder. *Pediatrics, 80*, 491–501.

Pelham, WE Jr, Swanson, JM, Furman, MB, & Schwindt, H (1995). Pemoline effects on children with ADHD: A time-response by dose-response analysis on classroom measures. *J Am Acad Child Adolesc Psychiatry, 34*, 1504–1513.

Safer, DJ, & Allen, RP (1989). Absence of tolerance to the behavioral effects of methylphenidate in hyperactive and inattentive children. *J Pediatr, 115*, 1003–1008.

Safer, DJ, Zito, JM, & Fine, EM (1996). Increased methylphenidate usage for attention deficit disorder in the 1990s. *Pediatrics, 98*, 1084–1088.

Shader, RI, Harmatz, JS, Oesterheld, JR, Parmelee, DX, Sallee, FR, & Greenblatt, DJ (1999). Population pharmacokinetics of methylphenidate in children with attention-deficit hyperactivity disorder. *J Clin Pharmacol, 39*, 775–785.

Shaywitz, SE, Hunt, RD, Jatlow, P, Cohen, DJ, Young, JG, Pierce, RN, Anderson, GM, & Shaywitz, BA (1982). Psychopharmacology of attention deficit disorder: Pharmacokinetic, neuroendocrine, and behavioral measures following acute and chronic treatment with methylphenidate. *Pediatrics, 69*, 688–694.

Solanto, MV, & Conners, CK (1982). A dose-response and time-action analysis of autonomic and behavioral effects of methylphenidate in attention deficit disorder with hyperactivity. *Psychophysiology, 19*, 658–667.

Soldin, SJ, Chan, YP, Hill, BM, & Swanson, JM (1979a). Liquid-chromatographic analysis for methylphenidate (Ritalin) in serum. *Clin Chem, 25*, 401–404.

Soldin, SJ, Hill, BM, Cahn, YP, Swanson, JM, & Hill, JG (1979b). A liquid-chromatographic analysis for ritalinic acid [alpha-phenyl-alpha-(2-piperidyl) acetic acid] in serum. *Clin Chem, 25*, 51–54.

Sprague, RL, & Sleator, EK (1977). Methylphenidate in hyperkinetic children: differences in dose effects on learning and social behavior. *Science, 198*, 1274–1276.

Srinivas, NR, Hubbard, JW, Korchinski, ED, & Midha, KK (1993). Enantioselective pharmacokinetics of dl-threo-methylphenidate in humans. *Pharm Res, 10*, 14–21.

Srinivas, NR, Hubbard, JW, Quinn, D, & Midha, KK (1992). Enantioselective pharmacokinetics and pharmacodynamics of dl-threo-methylphenidate in children with attention deficit hyperactivity disorder. *Clin Pharmacol Ther, 52*, 561–568.

Srinivas, NR, Quinn, D, Hubbard, JW, & Midha, KK (1987). Enantioselective aspects of the disposition of methylphenidate in children with attention-deficit disorder. *J Pharmacol Exp Ther, 241*, 300–306.

Swanson, JM (1985). Measures of cognitive functioning appropriate for use in pediatric psychopharmacological research studies. *Psychopharmacol Bull, 21*, 887–890.

Swanson, JM (1989). Paired associate learning in the assessment of ADD-H children. In JM Swanson & LM Bloomingdale (Eds.), *Attention deficit disorder: Volume VI. Current concepts and emerging trends in attentional ad behavioral disorders of childhood*. New York: Pergamon Press.

Swanson, JM, Agler, D, Fineberg, E, Wigal, S, Flynn, D, Fineberg, K, Quintana, Y, & Talebi, H (1999a). UCI laboratory school protocol for PK/PD studies. In LL Greenhill & B Osman (Eds.), *Ritalin: Theory and practice, Second Edition* (pp. 405–430) New York: Mary Ann Liebert.

Swanson, JM, Gupta, S, Guinta, D, Flynn, D, Agler, D, & Lerner, M et al. (1999b). Acute tolerance to methylphenidate in the treatment of Attention Deficit Hyperactivity Disorder in children. *Clin Pharmacol Ther, 66*, 295–305.

Swanson, JM, & Kinsbourne, M (1976). Stimulant-related state-dependent learning in hyperactive children. *Science, 192*, 1354–1357.

Swanson, JM, Kinsbourne, M, Roberts, W, & Zucker, K (1978). Time-response analysis of the effect of stimulant medication on the learning ability of children referred for hyperactivity. *Pediatrics, 61*, 21–29.

Swanson, JM, Lerner, M, & Cantwell, D (1986). Blood levels and tolerance to stimulants in ADDH children. *Clin Neuropharmacol, 9*, 523–525.

Swanson, JM, Lerner, M, March, J, & Gresham, F (1999c). Assessment and intervention for ADHD in the schools: Lessons from the MTA study. In AM Morgan (Ed.), *Pediatric clinics of North America* (pp. 993–1009) Philadelphia, PA: WB Saunders.

Swanson, JM, Lerner, M, & Williams, L (1995). Letter to the Editor. More frequent diagnosis of Attention Deficit Hyperactivity Disorder. *N Engl J Med, 333*, 944.

Swanson JM, McBurnett K, Christian DL, Wigal, T (1995). Stimulant medications and the treatment of children with ADHD. In TH Ollendick & RJ Prinz (Eds.), *Advances in Clinical Child Psychology, 17*, 265–322. New York: Plenum Press.

Swanson, JM, McBurnett, K, Wigal, T, Pfiffner, LJ, Lerner, MA, Williams, L, Christian, D, Tamm, L, Willcutt, E, Crowley, K, Clevenger, W, Khouzam, N, Woo, C, Crinella, FM, & Fisher, TD (1993). The effect of stimulant medication on ADD children: A "Review of Reviews." *Except Child, 60*, 154–162.

Swanson, JM, Sandman, CA, Deutsch, C, & Baren, M (1983). Methylphenidate hydrochloride given with or before breakfast: I. Behavioral, cognitive, and electrophysiologic effects. *Pediatrics, 72*, 49–55.

Swanson, JM, Wigal, SB, Greenhill, LL, Browne, R, Waslik, B, & Lerner, M, et al. (1998a). Analog classroom assessment of Adderall in children with ADHD. *J Am Acad Child Adolesc Psychiatry, 37*, 519–526.

Swanson, JM, Wigal, SB, Greenhill, LL, Browne, R, Waslick, B, & Lerner, M, et al. (1998b). Objective and subjective measures of the pharmacodynamic effects of Adderall in the treatment of children with ADHD in a controlled laboratory classroom setting. *Psychopharmacol Bull, 34*, 55–60.

Swanson, JM, Wigal, SB, Udrea, D, Lerner, M, Agler D, & Flynn, et al. (1998c). Evaluation of individual subjects in the analog classroom setting: I. Examples of graphical and statistical procedures for within-subject ranking of responses to different delivery patterns of methylphenidate. *Psychopharmacol Bull, 34*, 825–832.

Therapeutic Inequivalency Action Coordinating Committee (1991). 87-TI-09 Summary Report: Product: Methylphenidate tablets. Manufacturer: Medeva Pharmaceuticals. Food and Drug Administration, Washington, DC.

Unis, AS, Dawson, TM, Gehlert, DR, and Wamsley, JK (1985). Autoradiographic localization of [^{3}H]methylphenidate binding sites in rat brain. *Eur J Pharmacol, 113*, 155–157.

Volkow, ND, Ding, YS, Fowler, JS, Wang, GJ, Logan, J, & Gatley, JS, et al. (1995). Is methylphenidate like cocaine? Studies on their pharmacokinetics and distribution in the human brain. *Arch Gen Psychiatry, 52*, 456–463.

Volkow, ND, Wang, GJ, Fowler, JS, Gatley, SJ, Logan, J, & Ding et al. (1990). Blockade of striatal dopamine transporters by intravenous methylphenidate is not sufficient to induce self-reports of "High." *J Pharmacol Exp Ther, 288*, 14–20.

Volkow, ND, Wang, GJ, Fowler, JS, Gatley, SJ, Logan, J, Ding, YS, & Hitzemann, R, et al. (1998). Dopamine transporter occupancies in the human brain induced by therapeutic doses of oral methylphenidate. *Am J Psychiatry, 155*, 1325–1331.

Volkow, ND, Wang, GJ, Gatley, SJ, Fowler, JS, Ding, YS, & Logan, J et al. (1996). Temporal relationships between the pharmacokinetics of methylphenidate in the human brain and its behavioral and cardiovascular effects. *Psychopharmacology (Berl), 123*, 26–33.

Wargin, W, Patrick, K, Kilts, C, Gualtieri, CT, Ellington, K, & Mueller et al. (1983). Pharmacokinetics of methylphenidate in man, rat and monkey. *J Pharmacol Exp Ther, 226*, 382–286.

Wigal, SB, Gupta, S, Guinta, D, & Swanson, JM (1998a). Reliability and validity of the SKAMP rating scale in a laboratory school setting. *Psychopharmacol Bull, 34*, 47–53.

Wigal, SB, Swanson, JM, Greenhill, L, Waslick, B, Cantwell, D, & Clevenger, W, et al. (1998b). Evaluation of individual subjects in the analog classroom setting: II. Effects of dose of amphetamine (Adderall). *Psychopharmacol Bull, 34*, 833–838.

Wigal, T, Swanson, JM, Regino, R, Lerner, M, Soliman, I, & Steinhoff, K et al. (1999). Stimulant medications for the treatment of ADHD: Efficacy and limitations. *Ment Retard Dev Disabil Res Rev, 5*, 215–224.

Yoss, RE & Daly, DD (1968). On the treatment of narcolepsy. *Med Clin North Am, 52*, 781–787.

12

Cognitive Pharmacology of Stimulants in Children with ADHD

COLIN B. DENNEY AND MARK D. RAPPORT

Psychostimulant effects on higher order cognitive processes in children with Attention-Deficit Hyperactivity Disorder (ADHD) are of central interest to researchers in the field of clinical child psychopharmacology. This interest reflects the importance of studying drug effects for understanding the etiology and treatment of the disorder. Several lines of evidence indicate that drugs affecting different neurochemical systems show generalized and in some cases selective effects on cognitive functioning, but the mechanisms responsible have not yet been definitively elucidated. One potential reason for this lies in the conceptual focus that has typically been brought to bear on the problem. Some investigations have examined the impact of stimulants or other drugs used to treat children with ADHD at the neurobiological level while noting correlated behavior changes in core features of the disorder (Rapoport & Zametkin, 1988). Others (Douglas et al., 1986; Rapport et al., 1994) have adopted a more clinical approach and examined the impact of different drugs within and across individual measures of interest. Although these approaches differ in their levels of analysis, they share a common emphasis on reducing the question of how a drug "works" to what the drug "does" at the pharmacological level or how its pharmacological effects are expressed on individual psychological measures. It is the thesis of the current chapter that the central issue must be conceptualized in broader terms. Comprehensive understanding of drug effects on cognitive processes among children with ADHD requires a program of research that links pharmacological effects of particular drugs to expected intra- and inter-individual variations within psychological domains. This can be accomplished by explaining how drugs influence hypothesized core cognitive and behavioral domains such as attention and/or behavioral inhibition, while accounting for the extent to which variations on particular measures propagate across the disorder's broader symptomatic spectrum.

We begin by outlining a conceptual model of ADHD that attempts to make the assumptions underlying pharmacological intervention explicit. Implications of the model for understanding relationships among core psychological processes and secondary features of ADHD are discussed. The ensuing section reviews studies relevant to understanding the effects of psychostimulants on learning and cognitive function, and emphasizes differences in methodological approaches for studying drug effects. The consistency with which neurocognitive tests, tasks, and experimental paradigms are able to differentiate children with ADHD from normal children is reviewed subsequently. This is complemented by task analysis to explore possible mechanisms responsible for core cognitive deficits associated with ADHD. An alternative conceptual model is presented in the final section to promote debate concerning core deficits of ADHD.

Theoretical and Conceptual Issues in Studying Cognitive Effects of Drugs in Children with ADHD

Assumptions

A basic conceptual model based on extant literature is illustrated in Figure 12–1 and elucidates the implicit and explicit causal assumptions underlying pharmacotherapeutic interventions for children with ADHD. It suggests that biological influences (e.g., genetics, prenatal insults) give rise to individual differences in the functional properties of neurobiological systems (e.g., dopaminergic/noradrenergic neurotransmission) that are etiologically responsible for the core psychological (i.e., cognitive and behavioral) features of ADHD. Secondary features of ADHD (e.g., those listed as associated features in DSM-IV; American Psychiatric Association, 1994) are conceptualized as causal by-products of the core features. For example, the academic underachievement observed in many children with ADHD is presumed to be

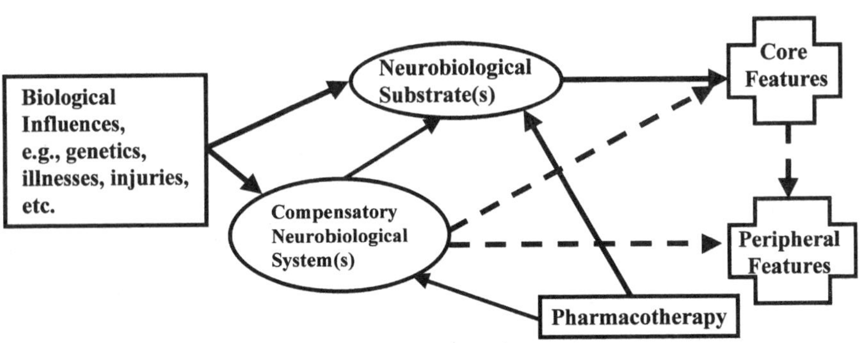

Figure 12–1. A conceptual causal model of underlying pharmacological treatments in ADHD. Biological influences give rise to individual differences in the functional properties of neurobiological systems that are etiologically responsible for core psychological features (e.g., attentional dysregulation, behavioral inhibition) of ADHD. Secondary features are conceptualized as causal by-products of core features.

a causal consequence of broader, more primary features of the disorder such as chronic inattention to and lack of persistence on classroom tasks. Other secondary features common to ADHD, such as inadequate social skills and peer relationships, low frustration tolerance, and strained family relationships, are similarly presumed to be due to core behavioral and cognitive influences. The conceptual framework can accommodate either categorical or dimensional views of psychopathology. Thus, neurobiological substrates may be either present or absent and give rise to correspondingly categorical symptomatic presentations, or they may vary incrementally across individual children, giving rise to continuously distributed psychological features. Comprehensive models of the disorder will need to reconcile which view of symptom distribution (i.e., categorical or dimensional) best describe the disorder (Gjone, et al., 1996; Levy, et al., 1997) with properties of the neurobiological substrates presumed to explain them (see Castellanos, Chapter 10).

Implications

Several implications for understanding pharmacotherapeutic effects in ADHD can be derived from the model. Drugs that effectively ameliorate the symptoms of ADHD are presumed to derive their therapeutic benefits from either of two sources. They may directly alter the functioning of the primary etiological substrate responsible for the disorder's core features, or alternatively, they may indirectly influence compensatory systems that ameliorate either the primary or secondary features. It is also plausible that the impact of a single drug may reflect contributions mediated by both pathways because many drugs (including dextroamphetamine and methylphenidate) affect multiple neurotransmitter systems (see Pliszka, Chapter 14). The specific means by which alteration of substrate systems influence core psychological processes in AD/HD must also be specified, and may involve hypothesized attentional regulation processes (Douglas, 1988) and/or those governing executive functions such as behavioral inhibition (Barkley, 1977). Corresponding change in secondary behaviors influenced by core processes are expected to the extent that secondary features rely on hypothesized core processes (i.e., the degree to which a secondary feature requires core processes for successful execution).

Psychostimulant Effects on Learning and Cognitive Function in Children with ADHD

There is abundant evidence documenting the effects of psychostimulants at the neurobiological level, albeit the precise mechanism(s) linking them to core processes remains speculative (see Taylor, Chapter 4). Complementary studies examining psychostimulant effects on diverse areas of functioning are thus needed to illuminate which processes are affected by drug, and whether similar dose-response patterns emerge across domains of cognitive function. Several avenues of research address these conceptual issues and are reviewed here.

Psychostimulant Effects on Task Performance

Rapport and Kelly (1991) reviewed 98 studies published between 1964 and 1991 that examined the effects of psychostimulants in children with ADHD using some form of neurocognitive test, task or experimental paradigm. Of these, 50 included a placebo control, investigated MPH in 6- to 12-year-old children with ADHD (or hyperkinesis prior to 1980), and included sufficient statistical information to permit determination of overall and/or between-dose effects. The types of tasks used in these studies and corresponding overall and between-dose (low vs high) effects are summarized in Table 12–1.

Two supraordinate categories of tasks were examined in the review. The first consists of tasks requiring convergent thinking skills, whereas the second includes tasks requiring divergent thinking skills. Those in the convergent category require a specific, pre-designated response (i.e., only one answer or response is correct). In contrast, divergent tasks require children to generate multiple responses representing different conceptual classes (e.g., Solanto & Wender, 1989). Convergent tasks were in turn listed under the core, secondary, or construct domains they are assumed to measure. For example, the attention domain consists of vigilance (continuous performance task or CPT) tasks that require children to identify single or combination letters (AX) or digits (1–9) that appear sequentially over an extended time interval. Test duration is typically between 9 and 15 minutes with both target and nontarget letters or digits (interspersed) appearing once per second on a computer monitor. Two tasks are listed under the impulsivity domain. Matching to sample tasks require children to study a stimulus picture or set of visual stimuli and identify which of several competing visual stimuli are exactly the same as the sample (i.e., children must withhold making a response and make careful comparisons prior to responding). The reaction time tasks typically involve withholding a response until receiving a designated verbal or visual signal (e.g., go/no-go paradigms).

Five types of tasks are listed under the learning construct heading and are considered measures of secondary features owing to their partial reliance on core variables for successful execution (i.e., attention and reflective thought processes). A broad range of learning tasks have been studied, and can be categorized as short/long-term recall and picture recognition tasks, paired associate learning (recalling picture or word combinations taught earlier), stimulus equivalence learning (child learns both directly taught and indirectly taught relationships among visual stimuli), and simple spelling tasks. The final construct shown in the table (labeled "other") contains various tasks that rely minimally on core features and thus would not be expected to be particularly sensitive to stimulant treatment. These tasks provide measures of children's ability to perceive and/or identify particular visual stimuli or to engage in instructed motor responses (e.g., tracing a dot-to-dot diagram, tapping one's finger in a particular sequence).

Scrutiny of the table reveals several findings relevant to issues concerning the overall and between-dose effects of psychostimulants on tasks presumed to represent hypothesized core and secondary domains of ADHD. The preponderance of studies examined cognitive performance under unitary dosage conditions. MPH

Table 12–1. Task × Dose Effects of MPH on Children's Cognitive Performance

Convergent Tasks	(*n*)	% of Tasks Using Doses 1	2	3	% Sign Med Effects	% Comparing Low/High-Dose Conditions	% Sign Between-Dose Difference: Low	NS/UK	High
Attention (core)									
Vigilance tasks		73	20	7	80	7	0	100/0	0
	(15)	(11)	(3)	(1)	(12)	(1)			
Impulsivity (core)									
Matching to sample		25	67	8	83	25	0	33/0	67
	(12)	(3)	(8)	(1)	(10)	(3)			
Reaction time		71	29	0	100	14	0	0/0	100
	(7)	(5)	(2)	(0)	(7)	(1)			
Learning (secondary)									
Short/long-term recall		33	67	0	50	0	—	—	—
	(6)	(2)	(4)	(0)	(3)	(0)			
Picture recognition		63	25	12	50	25	0	50/50	0
	(8)	(5)	(2)	(1)	(4)	(2)			
Paired associates		62	15	23	85	23	0	67/0	33
	(13)	(8)	(2)	(3)	(11)	(3)			
Stimulus equivalence		0	0	100	100	100	0	0/0	100
	(1)	(0)	(0)	(1)	(1)	(1)			
Spelling		100	0	0	100	0	—	—	—
	(2)	(2)	(0)	(0)	(2)	(0)			
Other (non-central)									
Visual/perceptual		50	50	0	0	0	—	—	—
	(2)	(1)	(1)	(0)	(0)	(0)			
Stimulus identification		83	17	0	33	0	—	—	—
	(6)	(5)	(1)	(0)	(2)	(0)			
Perceptual/motor		67	33	0	33	0	—	—	—
	(3)	(2)	(1)	(0)	(1)	(0)			
Divergent Tasks	(7)	28	29	43	43	43	0	0/67	33
	(7)	(2)	(2)	(3)	(3)	(3)			

The total number of tasks reviewed is listed under the "*n*" column. Some investigators examined performance under a single dose of methylphenidate (MPH) to discern overall drug effect (drug vs no-drug or placebo), whereas others used two or three doses to enable both overall and between-dose comparisons. The percentage of studies using 1, 2, or 3 doses is listed under the column titled (% of Tasks Using Doses: 1, 2, 3), with actual number of studies shown below in parentheses. Percentages of studies reporting significant overall MPH effects are shown in the column (titled % Sign Med Effects). Percentages and actual numbers (shown in parentheses) of studies comparing low- and high-dose conditions are shown in the column titled (% Comparing Low/High-dose Conditions). Low-dose conditions involved administering up to 10 mg MPH (<0.50 mg/kg), whereas high-dose conditions involved 20 mg or more (>.74 mg/kg) administered as a single oral dose prior to assessment. Percentage of studies reporting statistical information concerning low/high-dose contrasts is shown in the last column. None of the studies reported that children performed significantly better under low-dose MPH conditions compared to high-dose conditions. Several reported nonsignificant between-dose differences or failed to include statistical comparisons (middle row comparisons under NS/UK; NS = non-significant; UK = unknown). For those examining low/high-dose differences, between 33% and 100% of the studies reported that high doses resulted in significantly better performance.

resulted in improved performance for 85% (29 of 34) of the studies employing tasks considered central to core constructs (attention, impulsivity), whereas 70% (21 of 30) reported significant improvement on secondary measures of learning. In contrast, only 27% (3 of 11) found significant improvement on tasks considered noncentral to ADHD (visual/perceptual, stimulus identification, perceptual/motor). These findings support the premise that attention and impulsivity are either core features of ADHD or represent constructs that are associated with or governed by a higher-order construct. The finding that MPH improves performance on most learning tasks yet exerts relatively little influence on non-central tasks is also consistent with predictions stemming from the general model.

Low/high-dose contrasts were infrequently reported, but the limited information available yielded an interesting pattern of results. Performance on vigilance tasks is highly sensitive (12 of 15 studies) to overall drug effects (drug vs placebo), but higher doses of MPH consistently failed to result in additional improvement over low doses. Tasks tapping the impulsivity domain, in contrast, were found to be both highly sensitive in detecting overall (17 of 19 or 89% of studies) and between-dose effects, with higher doses yielding significantly improved performance over low doses (3 of 4 or 75% of studies). A majority of studies examining low/high-dose differences on learning tasks in which between-dose statistical analysis was reported favored high dose conditions (67% of studies), whereas none of the studies involving non-central tasks examined between-dose differences.

Tasks measuring divergent thinking were developed to address concerns that psychostimulants may result in a phenomenon described as "cognitive constriction" or "overfocused" cognitive states. This phenomenon refers to a person's inability to shift attention (or mental set) from one aspect of a stimulus situation to another when appropriate (vs staring or focusing too long on a particular stimulus with a loss of efficiency). Cognitive assessment of divergent thinking may take many forms. Traditional measures involve generating alternative responses such as naming animals within a category (Sebrechts et al., 1986), verbalizing alternative uses for a particular object (Solanto & Wender, 1989), or finding as many words as possible within the letters of one long word (Douglas et al., 1995). A summary of overall and between-dose MPH effects for divergent tasks is shown in Table 12–1 and include one recently published study (Douglas et al., 1995) in addition to the six previously reviewed by Rapport and Kelly (1991). Significant overall medication effects were reported in three of the seven studies (43%); three involved low/high-dose contrasts, but only one reported statistical findings and favored the high-dose condition.

Collectively, these results suggest that (at a group level) lower doses are required to optimize children's attention, whereas performance improves in a linear fashion with increasing dose on measures of impulsivity and some of the more complex learning (paired associate learning, stimulus equivalence learning) and divergent thinking tasks. Although the ability to attend is clearly central to all tasks reviewed, the results suggest that it may represent but one component of a higher-order core process inherent in children with ADHD. Consideration of this possibility necessitates the use of different avenues of research methodology that enable us to determine whether relationships among attention, impulsivity and expected secondary features are consistent with patterns anticipated by the general model.

Inter-individual Differences Within Domains and Intra-individual Differences Across Domains

Studying inter-individual differences within domains and intra-individual differences in drug response across domains can provide valuable insight concerning the psychological processes inherent to ADHD and how drugs affect these processes. Numerous studies and reviews are available (e.g., Rapport, 1990; Rapport & Kelly, 1993; Solanto, 1998). Select studies are highlighted in this section to elucidate possible reasons why inter- and intra-individual differences within and across different behavioral and cognitive domains occur, and the relevance of these findings for understanding dose-response differences in children.

Sprague and Sleator's (1977) seminal study served as an impetus for examining potential dissociated effects of MPH in children undergoing stimulant treatment and for studying drug-behavior mechanisms. Their results indicated that picture recognition learning was optimized at a low dose (0.3 mg/kg) and declined (albeit nonsignificantly) under the higher (1.0 mg/kg) dose, whereas teacher ratings of classroom behavior showed a linear dose-response pattern with increasing dose. Subsequent studies have not validated the implication that cognition and behavior follow different dose-response patterns. The effect of MPH on performance on a variety of cognitive measures and academic tasks has been either linear improvement with increasing dose up to 1.0 mg/kg, or curvilinear improvement with an asymptote at the highest doses tested (for a review, see Solanto, 1999).

Observing differences in dose-response curves between and among outcome domains, albeit informative, reveals little about the underlying processes involved and why change in one domain does or does not correspond with changes in other domains at a particular level of drug. For example, clear differences in dose-response patterns are apparent when comparing children's performance on different clinic or laboratory-based instruments under identical dose parameters. Performance on vigilance tasks nearly always shows initial improvement at low dose (5, 10mg MPH) with no further improvement at the higher (15, 20mg) dose range (Rapport et al. 1986). In contrast, paired associate learning and similar measures tend to show linear dose-response trends up through the intermediate dose range before reaching an asymptote (Rapport et al., 1985, 1989). And children's performance on tasks (Berman et al., 1999) or paradigms (Vyse & Rapport, 1989) involving more complex processing demands show continued improvement with increasing dose up to 1.0 mg/kg MPH.

Differences in dose-response effects within one domain (e.g., attention) but measured in different settings (e.g., clinic-based CPT vs direct observations of attention in school) also require explication. For example, whereas MPH dose effects on vigilance tasks such as the CPT nearly always asymptote at low doses, as described, observations of children's attention in the classroom consistently reveal linear dose-response patterns characterized by increased attention as a function of increasing dose (Rapport et al., 1986).

Understanding inter- and intra-individual differences in dose-response curves within and across outcome measures requires closer scrutiny of the particular tasks and outcome measures employed in these investigations and their underlying task parameters. For example, the CPT places minimal demands on working memory. A child merely has to remember the basic instructions and respond to the same target stimulus (e.g., "A" followed by "X") whenever it occurs over an extended time period. In contrast, tasks such as the Matching Familiar Figures Test (MFFT), Paired Associate Learning (PAL), complex visual memory search tasks, and stimulus equivalence learning paradigm (SEP; Vyse & Rapport, 1989) involve more complex information processing demands, and also require continuous evaluation of new information to be processed over extended time intervals. Similar arguments hold in comparing simple clinic-based assessment tasks and direct observations of children's attention and academic productivity in school settings. Paying attention and completing assignments in a classroom involve multiple, complex, interacting processes in an environment characterized by extraneous, distracting stimuli (which must be ignored). Instructions must be remembered, recall of past learning invoked, and new decisions continuously executed in accord with changing situational and task demands. Studies comparing clinic-assessed CPT performance and direct observations of children's attention in the classroom corroborate this explanation by demonstrating significant but weak correlations between the two settings (Kupietz & Richardson, 1978).

In summary, children's performance on tasks that place heavier demands on working memory and relevant executive functions involving attention and long-term recall show linear dose-response patterns, whereas performance is quickly brought to optimal levels under low doses of stimulant for those that place minimal demands on these processes. Additional research is needed to elucidate these relationships, and in particular, the influence and interaction of task parameter manipulations and stimulant response.

Polypharmacological Approaches to Understanding Core Deficits

This avenue of inquiry, limited in terms of published studies, examines the effects of drugs that affect multiple neurobiological substrates or that antagonize systems required for successful performance on particular cognitive tasks. Examination of relevant studies provides some insight concerning possible mechanisms of drug action.

Levy and Hobbes (1988) examined CPT performance in ADHD children under MPH alone and MPH preceded by 0.04 mg/kg haloperidol, a dopamine-blocking major tranquilizer. Consistent with previous studies, children showed significantly fewer errors on the CPT under MPH compared to placebo, whereas this effect was reversed when MPH was preceded by haloperidol. This finding suggests involvement of the dopaminergic system in tasks requiring sustained vigilance. Neuroleptics have been reported to enhance the positive effects of MPH on behavior ratings in some studies (e.g., Gittelman et al. 1976). This seeming contradiction, however, can be explained by examining differences in outcome measures among studies. Global improvement in children's behavior is observed with many drugs (including major tranquilizers), and suggests that the mechanism of action for reducing obstreperous behavior may be obtained by affecting several different neurotransmitter systems that may or may not be related to core processes. Blocking a demonstrated stimulant effect with a known antagonist in contrast provides reasonably strong evidence for dopaminergic involvement in vigilance.

Controlled outcome studies comparing TCAs with psychostimulants have been equally informative. Imipramine (IMI), desipramine (DMI), amitriptyline (AMI), and clomipramine (CMI) have been compared directly with psychostimulants in seven studies (for a review, see Rapport et al., 1993). Six of these studies incorporated one or more measures of cognitive function (including motor and perceptual skills). MPH is usually superior to the tricyclics in reducing CPT omission errors, whereas commission errors are either unaffected or reduced to a similar degree by MPH and IMI. Drug effects on motor and perceptual task performance are equivocal, with three studies reporting no drug effect, and three favoring MPH over IMI, DMI, or clomipramine. Separate and combined effects of MPH and DMI were examined by Rapport et al. (1993). MPH alone improved vigilance (CPT), both drugs alone improved short-term memory and visual problem solving, and combined drugs produced the most benefit for learning higher-order relationships.

Collectively, these results indicate that facilitation of dopaminergic neurotransmission may be responsible for improved vigilance, whereas either dopaminergic *or* noradrenergic neurotransmission may facilitate performance on paired-associate and visual problem-solving tasks. Moreover, noradrenergic agents contribute to performance on tasks requiring recognition of complex relationships over and above the effects of stimulant medications.

Conceptual Approaches to Understanding Core Deficits

Attention and impulsivity/hyperactivity are currently viewed as core features of AD/HD. The conceptual model implies that measures of core features should demonstrate moderate to high intercorrelations, and that treatment-induced gains in core domains should be necessary for attainment of broad-spectrum clinical improvement. The degree to which changes in core domains are sufficient to induce improvement on the more peripheral correlates of the disorder depends on the extent to which the latter are causally dependent on influences other than the core features.

A structural equation model is shown in Figure 12–2 as a means to illustrate this conceptual approach for examining expected patterns of relationships among core variables, and among core and secondary measures. The data and measures were collected as part of an outcome study reported by Rapport and colleagues (1994, 1999). Variables shown on the left side of the figure represent slopes of dose-response curves (based on placebo, 5 mg, 10 mg, 15 mg, and 20 mg MPH) for three classroom measures: (*a*) observed attention to task, (*b*) teacher ratings of self-control (Teacher Self-Control Rating Scale; Humphrey, 1982), and (*c*) hyperactivity (Abbreviated Conner's Teacher Rating Scale; Werry et al., 1975). The outcome variable to the right represents dose-response curve slopes of children's weekly academic performance (percent of academic assignments completed correctly). Intercorrelations among these slopes are shown by the coefficients adjacent to the double-headed arrows (i.e., .37, −.71, −.41). Coefficients on the single-headed arrows pointing toward the academic performance variable represent correlations between rates of change in academic performance and each of the three predictors after accounting for intercorrelations among the latter. These coefficients indicate that gains in attention were related to improved academic functioning (.37, $p < .05$), whereas teacher rated self-control and hyperactivity were not. The R^2 value beneath the academic performance variable indicates that the combination of gains in hypothesized core domains of attention, self-control, and hyperactivity account for 20% of the variation in the secondary variable academic improvement.

Several explanations may account for the above pattern of findings. These are briefly reviewed to highlight the complexity of conceptual issues involved in identifying core deficits. First, core and secondary variables were estimated in the model using single measures of construct domains. Given the inherent measurement error and relative impurity of any single measure, incorporating several measures of attention, impulsivity–hyperactivity, and academic functioning with demonstrated high internal consistency would be advisable to provide more accurate assessment of core and secondary constructs. Second, mediator variables or pathways may be required to correctly model the relationships or relevant processes bridging core (attention, impulsivity–hyperactivity) and secondary variables. For example, Rapport and colleagues (1999) demonstrated that the relationship between attention and long-term scholastic underachievement in ADHD is mediated by working memory processes. Finally, the relatively high inter-

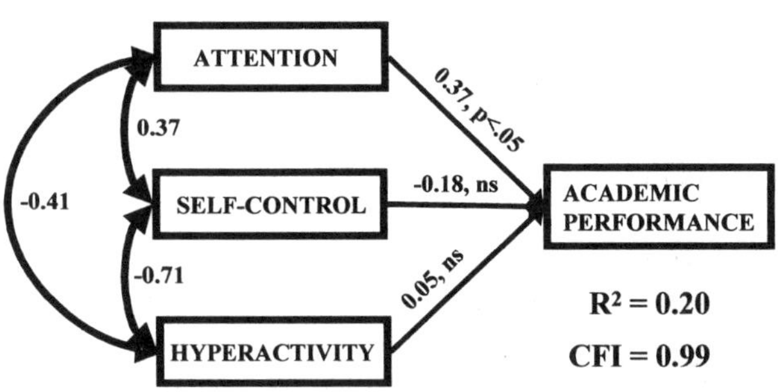

Figure 12–2. A structural equation model of expected patterns of relationship among hypothesized core (attention, self-control, hyperactivity) and secondary (academic performance) variables as a function of MPH therapy. R^2 = the percentage of variation in academic performance accounted for by the core features after controlling for their intercorrelations. CFI = Comparative Fit Index (values above .90 indicate adequate model fit).

correlations among core variables raises the possibility that they may represent complementary operations and behaviors underlying a more central, higher-order psychological process. This possibility is addressed in an ensuing section wherein an alternative model of ADHD is presented.

Summary and Implications of Stimulant Effects on Cognitive Processes

The methodological approaches used to review drug effects on cognitive and learning processes in children with ADHD highlight the complex nature of the phenomenon and tentativeness with which conclusions can be drawn. Extant patterns of psychostimulant response and dose-response curves nevertheless permit some inferences to be drawn concerning drug-behavior relationships and possible underlying mechanisms. For example, measures of hypothesized core (attention, impulsivity–hyperactivity) and secondary (learning, academic performance) variables are generally quite sensitive to psychostimulant effects, whereas measures of non-central features are not. Among the core and secondary variables, measures of impulsivity–hyperactivity and learning tend to be highly sensitive to between-dose differences and evince a linear pattern of improvement as a function of increasing dose. This observed pattern of response, however, appears to reflect a 3-way interaction involving dose, measure, and level of cognitive processing. That is, children with ADHD evince continued benefit on more complex processing tasks as a function of increasing dose. It is instructive to note, however, that the somewhat crude outcome measures used in clinical psychopharmacology merely indicate decreasing error rates (e.g., omission/commission errors) rather than informing us about underlying processes and the manner by which they become more efficient. Task analysis and manipulation of task parameters and how they affect performance may provide a useful means by which to further our understanding of core deficits and related processes underlying ADHD.

The rather consistent finding that measures of vigilance respond robustly under low doses of stimulant with no further improvement at higher doses suggests that maintenance of attention is an essential but complementary process required of learning and cognitive function. This role as a complementary rather than core process is also consistent with the linear dose-response curves obtained for directly observed attention in classroom settings and limited data from polypharmacological studies. Studies in the latter two areas indicate that classroom measures of attention probably reflect more complex cognitive processes over and above simple maintenance of attention, and that simple dopaminergic transmission represents an incomplete accounting for these effects.

The last area reviewed in this section was intended to provoke more questions than it suggests answers, acknowledging the methodological limitations outlined. The high intercorrelations among hypothesized core variables but moderate to nonsignificant relationships with secondary measures such as academic performance suggests that attention and impulsivity–hyperactivity are clearly related and may represent complementary operations and behaviors underlying a higher-order, more central process.

Differences in Cognitive Performance Between Children with ADHD and Normal Controls

A total of 98 studies published between 1980 and 1999 were located that compared the performance of children with ADHD and normal controls on neurocognitive tests, tasks, and experimental paradigms. Studies were categorized according to the type of task employed, then compared to determine whether independent research reports confirmed their reliability for differentiating between groups based on reported statistical significance and overall number of validating studies. Potentially reliable, moderately reliable and unreliable instruments are shown in Table 12–2.

Scrutiny of the table illuminates two relevant issues. First, a plethora of instruments has been employed in an attempt to differentiate children with ADHD from normal controls with little emphasis on conceptual or theoretical analysis of the constructs underlying them. Second, as illustrated in Table 12–2, performance on all of these measures requires comparative judgements of similarities or differences, either between pairs of stimuli or between concretely presented objects (e.g., letters, numbers, or visual figures) and cognitive representations of target stimuli. These judgments culminate in classification of stimuli as either targets or non-targets and are operationally measured in terms of either omission or commission errors. For example, stop-signal and go/no-go tasks require children to terminate repetitive behaviors in response to particular signals. Although the mathematical model underlying these tasks emphasizes measurement of response time to termination of a repetitive behavior rather than errors, the inference that the decision process at work involves a comparison of visually presented stimuli to a cognitive representation of the stop-signal is inescapable. Moreover, reaction time to termination of a behavior is only a continuous (as opposed to categorical) expression of a commission error. Parallel arguments can be made for the majority of instruments employed in the assessment and treatment of children with ADHD.

Central features and characteristics of reviewed tests, tasks, and experimental paradigms used to differentiate children with ADHD from normal controls are shown in Table 12–3 to facilitate comparisons between reliable and unreliable instruments. Most tasks rely on recognition rather than recall memory. Recognition tasks involve providing children with a copy of the information they need to find in memory. That is, one compares the representation of the perceptual input with representations stored in memory. Conversely, recall tasks require the child to access representations different from that of the question. Tasks such as Paired Associate Learning (PAL) are listed under both categories to reflect differences in procedures used in published studies.

Scrutiny of the table reveals several differences between reliable and unreliable tasks. Reliable tasks tend to involve more speeded processing than do unreliable

Table 12–2. Neuropsychological Tests, Tasks, and Experimental Paradigms: ADHD versus Normal Comparisons

Task Name or Description	Total Studies (n)	Significant Number (n)	Findings (%)
Highly Reliable Between-Group Differences			
Attention			
CPT, visual			
X-only	11	9	81
AX	26	24	92
Combined	8	8	100
Inhibition			
Go/no-go	3	3	100
Stop-Signal	6	5	83
Memory for digits	4	4	100
Paired Associate Learning	3	3	100
Stroop Color-Word	11	9	81
Totals	72	65	90
Moderately Reliable (or Tentative) Between-Group Differences			
Match-to-Sample			
MFFT	13	9	69
Planning/Motor Control			
Porteus mazes	3	2	67
Progressive mazes	2	2	100
Unspecified mazes	3	2	67
Visual Motor/Perceptual			
Rey-Osterrieth	8	5	63
WISC-R Block	6	4	67
Verbal Memory			
Brown Peterson	2	2	100
Rey AVLT	3	2	67
Tentative or Unreliable Between-Group Differences			
Cancellation Letters	3	0	0
Language miscellaneous	2	0	0
Motor			
Finger tapping	2	0	0
Pegboard	5	1	20
Trail Making			
Trails A	5	1	20
Verbal Fluency			
Boston Naming	2	0	0
Visual Motor			
Miscellaneous	5	1	20
Totals	24	3	12

Potentially reliable, moderately reliable and unreliable neurocognitive tests, tasks, and experimental paradigms used to differentiate children with ADHD from normal children. Reliable instruments were those studied by at least three independent investigators that yielded statistically significant between-group differences in at least 80% of the studies. Moderately reliable or tentative instruments showed statistically significant between-group differences in 60% or more independent investigations or reported higher rates but in fewer than three independent research reports. Unreliable instruments failed to show significant between-group differences in a majority of independent studies.

Table 12–3. Characteristics of Tests, Tasks, and Experimental Paradigms Used to Differentiate Children with ADHD from Normal Controls

Task	Recognition	Recall	Time Parameter	Working Memory: Subvocal Speech	Working Memory: Buffer	Response Stimulus Present
Reliable Tests, Tasks, and Experimental Paradigms						
CPT	yes	no	sec or msec	yes	yes	no
Go/no-go	yes	no	msec	yes	yes	no
Stop Signal	yes	no	msec	yes	yes	no
Digit Span	no	yes	sec	yes	yes	no
PAL-T	yes	yes	sec	yes	yes	no
Stroop Color Word	yes (priming)	no	sec	yes	yes	no
Unreliable tests, tasks, and experimental paradigms						
Cancellation Tasks	yes	no	sec-min	yes (min)	yes	yes
Finger Tapping	no	no	sec-min	no	no	yes
Pegboard	yes	no	sec-min	no	no	yes
Trails	yes	no	min	no	no	yes
Boston Naming	no	yes	sec-min	no	no	yes
Visual Motor	no	no	min	no	no	yes

tasks (i.e., milliseconds to seconds versus seconds to minutes, respectively). Clear differences also emerge with respect to demands placed on working memory. For example, reliable instruments place clear demands on the subvocal speech and buffer components of the articulatory control loop, whereas unreliable tasks do not. Subvocal speech and the buffer are conceptualized as the primary mechanisms of the articulatory control loop, the latter of which enables any information that can be phonologically encoded to be rehearsed subvocally and briefly stored in the buffer for immediate or long-term storage (for a thorough discussion, see Baddeley, 1998). A final distinction between the reliable and unreliable instruments is that on the latter the answer or stimulus required to generate a correct response is available to the child throughout the trial.

The foregoing observations suggest that manipulations of task parameters related to comparative judgments and speeded recognition processes are likely to yield insights into etiologically relevant and treatment-related cognitive processes implicated in ADHD. A rich body of literature exists pertaining to these processes and their mediation by working-memory. The processes subsumed by the working-memory construct have been implicated repeatedly in ADHD although typically they have been viewed as correlates of the disorder rather than a set of core causal cognitive processes. A theoretical model of ADHD based on the premise that working-memory constitutes a core feature of the disorder is outlined in the next section.

A Conceptual (Working-Memory) Model of ADHD

Overview

The model outlined below is rooted in cognitive models of recognition and recall processes and working-memory in particular. It also incorporates the behavioristic principle of negative reinforcement. In brief, working memory describes a set of memory processes that serve to construct, maintain, and manipulate cognitive representations of incoming stimuli. As such, it is also responsible for recognition of externally presented stimuli. In this context, recognition refers to four related processes (Glass & Holyoak, 1986). First, recognition requires that a representation of the input stimulus be constructed on the basis of its features and patterns of organization. In addition, the representation of the input guides a search of memory traces for representations of stimulus configurations that match or resemble it. Stored representations are "activated" (i.e., made available for further use) and maintained in working-memory in proportion to their degree of similarity to the input representation. Finally, a comparative judgment is rendered regarding the degree to which the input stimulus matches a particular cognitive representation or set thereof. The latter process enables access to a broad network of memory traces of data related to the input stimulus, including actions appropriate to it. This facilitates organized execution of situationally relevant behavior. Evidence has accrued to suggest that these inter-related components of recognition occur in parallel rather than in a temporal sequence (Allport, 1989).

The model posits that working memory plays a pivotal role in the organization of behavior. Specifically, it proposes that organized responding is functionally dependent upon the capacity of working memory to (*a*) generate and maintain representations of input stimuli, (*b*) search memory traces for matches thereto, and, (*c*) access and maintain representations of behavioral responses appropriate to input stimuli. If these assumptions hold, disruption of any of the outlined working memory processes should result in haphazard or tangential responding to environmental stimuli.

As stated, this model may help account for the poorly structured (i.e., disorganized) behavior characteristic of children with ADHD. It does not, however, imply anything about the rate at which behavior is executed. Consequently, additional assumptions are necessary to explain why hyperactivity accompanies disorganized behavior.

The basic components of the model are shown in Figure 12–3. The hypothesis implied in the figure is that failure of working memory not only leads to disorganized behavior but also motivates children to redirect their attention to other stimuli in the environment (i.e., a phenomenon described in the literature as "stimulation seeking"). Specifically, the inability to maintain working memory representations leads to behavior that increases the rate at which input is delivered to working memory so as to compensate for the rapid rate at which representations fade. This is a purely inductive inference based on the observations that children with ADHD demonstrate frequent, rapid shifts in activity, especially under conditions that can be characterized

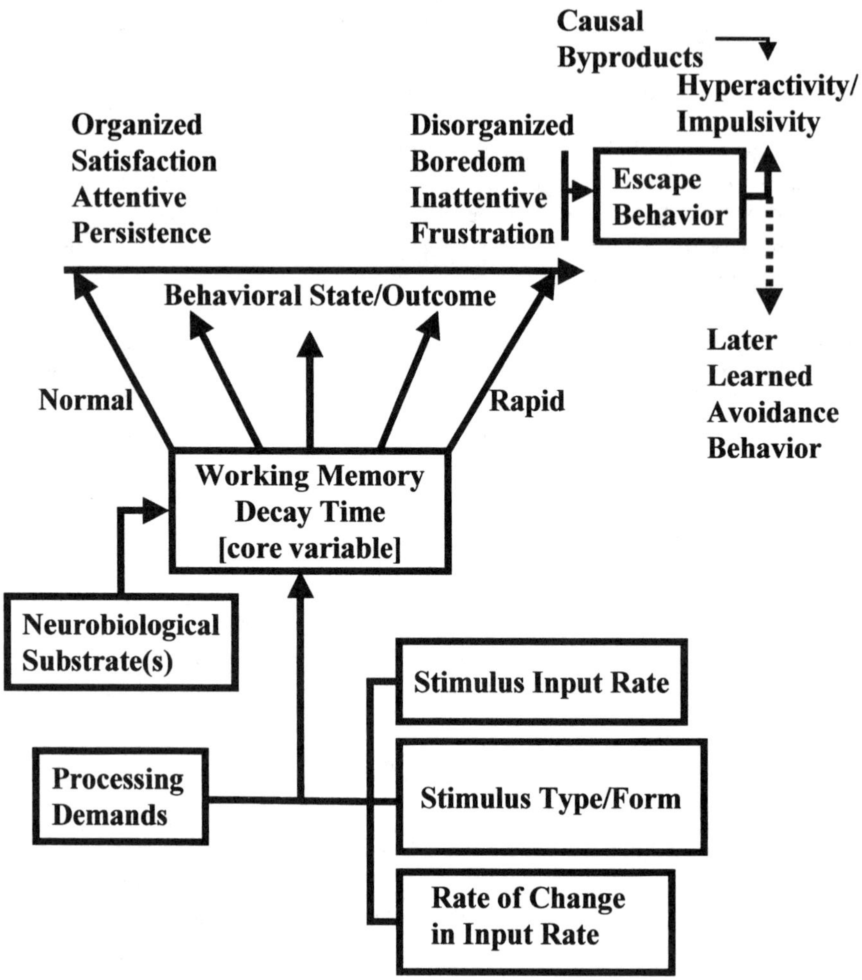

Figure 12–3. Core components of a conceptual model of ADHD. Failure of working memory leads to disorganized behavior and motivates children to redirect their attention to other stimuli in the environment (i.e., a phenomenon viewed by others as hyperactivity–impulsivity and described in the literature as "stimulation seeking"). Inability to maintain working memory representations leads to behavior that increases the rate at which input is delivered to working memory so as to compensate for the rapid rate at which representations fade.

as "monotonous" or too complex to enable thorough processing of stimuli. Alternatively, redirection of attention can be conceptualized as a form of escape from "monotonous" or high task demand conditions and observed by others as hyperactivity and impulsivity.

Note that the model provides for precise conceptual definitions of several terms and constructs frequently invoked in descriptions or analysis of ADHD. First, the term "monotonous" is defined in this context as the property of producing low rates of working memory activity. Low rates of working memory activity may derive from either low rates of stimulus input or perceptual constancy (i.e., minimal variation in the sensory properties of stimuli). Second, the concept of "attention" is defined in terms of working memory representations. Specifically, the target(s) of attention at a given moment are defined by current working memory content. Thus, the model implies that the term "inattention" is without meaning except in instances where conscious activity is precluded. This view reflects the old psychological adage that "one cannot not think." Third, the imposition of difficult task demands on a working memory system incapable of holding information to permit sufficient processing results in escape behavior owing to its aversive nature. Working memory is subsequently directed to alternative environmental stimuli to maintain a continual flow of input and observed by others as impulsive–hyperactive behavior. Finally, the current model argues that consideration of consequences prior to execution of a behavioral sequence is fundamentally dependent on working memory since the latter is the conduit through which learned behaviors find expression. Thus, "impulsive" acts are viewed as disorganized patterns of behavior that reflect an inability to maintain working memory representations of either the stimulus context, memory traces relevant thereto, or both.

The relationship between this model and the broader conceptual framework outlined previously is shown in Figure 12–4. First, this model emphasizes the

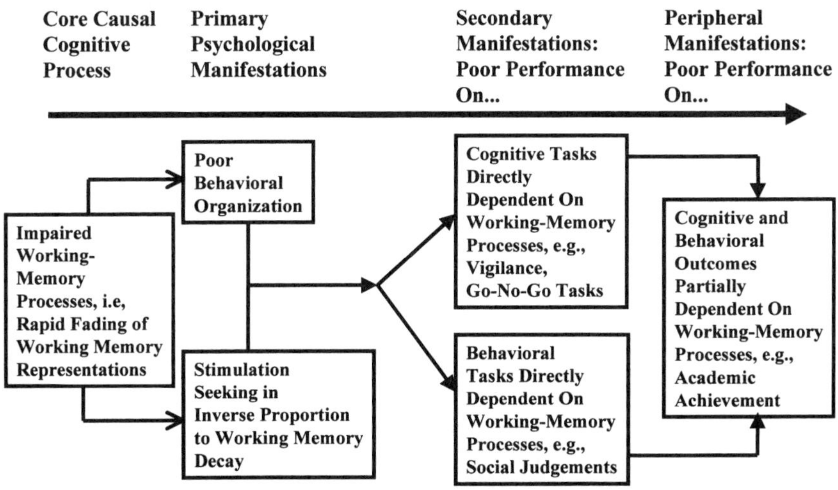

Figure 12–4. Implications for understanding the relationships among hypothesized core components of ADHD and related psychological processes. Core working memory processes are presumed to account for a variety of secondary characteristics in children with ADHD, such as poor performance on cognitive and behavioral tasks dependent and partially dependent on working memory processes.

psychological rather than the biological features of ADHD. No specific biological substrate is directly linked to working memory deficits in the model formulated here. Working memory is simply viewed as a core causal cognitive process that is the direct consequence of one or more neurobiological substrates. This core feature is presumed to account for several peripheral characteristics of ADHD. These include disorganization, boredom, inattentiveness, and low frustration tolerance, with impulsivity–hyperactivity representing the causal by-products of these states by means of negative reinforcement principles (see Figure 12–3). These features are further assumed to exert causal influences to varying degrees across a variety of other domains. Some of these, such as performance on tests of cognitive functioning, are likely to be directly affected by working memory processes. Others, such as academic achievement, reflect the cumulative impact of working memory failure on component academic skills (e.g., computational ability) combined with a variety of other influences (e.g., availability of tutoring or other compensatory resources).

Summary and Conclusions

A conceptual model was presented to illustrate the implicit and explicit causal assumptions underlying pharmacotherapeutic interventions for children with ADHD. Consideration of the model asseverates that the mechanisms by which neurobiological substrates govern core and peripheral psychological features of the disorder must be elucidated. To do so, it is necessary to (*a*) demonstrate that the neurobiological substrates of interest bear broad, puissant, causal relationships to the disorder's central features, and (*b*) model the relationships among the psychological features of the disorder and explicate the means by which they impact peripheral features of the disorder. Extant stimulant-treatment literature precludes firm conclusions regarding mechanisms of action of drug-effects on higher-order cognitive processes. Available data implicate DA mechanisms in vigilance performance whereas either NE or DA systems may influence performance in other domains. Both transmitters appear to participate in mediating stimulant effects on complex learning tasks, but additional studies are needed to dissect the individual contributions of noradrenergic and dopaminergic systems. The various methodological approaches for studying drug-behavior relationships reviewed here suggest that hypothesized core features of ADHD (attention, impulsivity–hyperactivity) may represent complementary operations underlying a higher-order, more central process. This possibility was explored by examining differences between children with ADHD and normal children on neuropsychological tests, tasks, and experimental paradigms, and complemented by theoretical analysis. A model citing working memory as a core feature of ADHD was offered as a plausible, testable explanation of the disorder. It is hoped that this view serves a heuristic purpose in furthering conceptualization and analysis of ADHD and individual differences in response to stimulant-treatment.

References

Allport, A (1989). Visual attention. In MI Posner (Ed.), *Foundations of cognitive science*. Cambridge, MA: The MIT Press.

American Psychiatric Association (1994). *Diagnostic and statistical manual of mental disorders, 4th edition (DSM-IV)*. Washington, DC: American Psychiatric Association.

Baddeley, A (1998). *Human memory: Theory and practice*. Boston: Allyn and Bacon.

Barkley, RA (1997). Behavioral inhibition, sustained attention, and executive functions: Constructing a unifying theory of ADHD. *Psychol Bull, 121*(1), 65–94.

Berman, T, Douglas, VI, & Barr, RG (1999). Effects of methylphenidate on complex cognitive processing in attention-deficit hyperactivity disorder. *J Abnorm Psychol, 108*, 90–105.

Denney, CB & Rapport, MD (1999). Predicting methylphenidate response in children with ADHD: Theoretical, empirical, and conceptual models. *J Am Acad Child Adolesc Psychiatry, 38*, 393–401.

Douglas, VI (1988). Cognitive deficits in children with attention deficit disorder with hyperactivity. In LM Bloomingdale & JA Sergeant (Eds.), *Attention deficit disorder: Criteria, cognition, intervention* (pp. 65–82). London: Pergamon.

Douglas, VI, Barr, RG, Desilets, J, & Sherman, E (1995). Do high doses of methylphenidate impair flexible thinking in attention-deficit hyperactivity disorder? *J Am Acad Child Adolesc Psychiatry, 34*, 877–885.

Douglas, VI, Barr, RG, O'Neill, ME, & Britton, BG (1986). Short term effects of methylphenidate on the cognitive, learning, and academic performance of children with attention deficit disorder in the laboratory and the classroom. *J Child Psychol Psychiatry, 27*(2), 191–211.

Gittelman, R, Klein, D, Kartz, S, Sarf, K, & Pollack, E (1976). Comparative effects of methylphenidate and thioridazine in hyperkinetic children. *Archives of General Psychiatry, 33*, 1217–1231.

Gjone, H, Stevenson, J, & Sundet, JM (1996). Genetic influences on parent reported attention related problems in a Norwegian general population twin sample. *J Amer Acad Child Adolesc Psychiatry, 35*, 588–596.

Glass, AL & Holyoak, KJ (1986). *Cognition* (2nd edition). New York: Random House.

Humphrey, LL (1982). Children's and teachers' perspectives on children's self-control: The development of two rating scales. *J Consult Clin Psychol, 50*, 624–633.

Kupietz, SS & Richardson, E (1978). Children's vigilance performance and inattentiveness in the classroom. *J Child Psychol Psychiatry, 19*, 145–154.

Levy, F, Hay, DA, McStephen, M, Wood, C, & Waldman, I (1997). Attention deficit hyperactivity disorder: A category or continuum? Genetic analysis of a large twin study. *J Am Acad Child Adolesc Psychiatry, 36*, 737–744.

Levy, F & Hobbes, G (1998). The action of stimulant medication in attention deficit disorder with hyperactivity: Dopaminergic, noradrenergic, or both? *J Am Acad Child Adolesc Psychiatry, 27*, 802–805.

Rapoport, JL, & Zametkin, A (1988). Drug treatment of attention deficit disorder. In LM Bloomingdale & JA Sergeant (Eds.), *Attention deficit disorder: Criteria, cognition, intervention* (pp. 161–182). New York: Pergamon Press.

Rapport, MD (1990). Controlled studies of the effects of psychostimulants on children's functioning in clinic and classroom settings. In CK Conners & M Kinsbourne (Eds.), *Attention Deficit Hyperactivity Disorder* (pp. 77–111) Germany: Medizin Verlag Munchen.

Rapport, MD, Carlson, GA, Kelly, KL, & Pataki, C (1993). Methylphenidate and desipramine in hospitalized children. II. Separate and combined effects on cognitive function. *J Am Acad Child Adolesc Psychiatry, 32,* 333–342.

Rapport, MD, Denney, C, DuPaul, GJ, & Gardner, MJ (1994). Attention deficit disorder and methylphenidate: Normalization rates, clinical effectiveness, and response prediction in 76 children. *J Am Acad Child Adolesc Psychiatry, 33,* 882–893.

Rapport, MD, DuPaul, GJ, Stoner, G, & Jones, JT (1986). Comparing classroom and clinic measures of attention deficit disorder: Differential, idiosyncratic, and dose-response effects of methylphenidate. *J Consult Clin Psychol, 54,* 334–341.

Rapport, MD & Kelly, KL (1991). Psychostimulant effects on learning and cognitive function: Findings and implications for children with attention deficit hyperactivity disorder. *Clin Psychol Rev, 11,* 61–92.

Rapport, MD & Kelly, KL (1993). Psychostimulant effects on learning and cognitive function in children with attention deficit hyperactivity disorder: Findings and implications. In J Matson (Ed.), *Handbook of Hyperactivity in Children* (pp. 97–136). New York: Pergamon Press.

Rapport, MD, Quinn, SO, DuPaul, GJ, Quinn, EP, & Kelly, KL (1989). Attention deficit disorder with hyperactivity and methylphenidate: The effects of dose and mastery level on children's learning performance. *J Abnorm Child Psychol, 17,* 669–689.

Rapport, MD, Scanlan, S, & Denney, CB (1999). Attention-deficit/hyperactivity disorder and scholastic achievement: A model of dual developmental pathways. *J Child Psychol Psychiatry, 40,* 1169–1184.

Rapport, MD, Stoner, G, DuPaul, GJ, Birmingham, BK, & Tucker, S (1985). Methylphenidate in hyperactive children: Differential effects of dose on academic, learning, and social behavior. *J Abnorm Child Psychol, 13,* 227–244.

Sebrechts, MM, Shaywitz, SE, Shaywitz, BA, Jatlow, P, Anderson, GM, & Cohen, DJ (1986). Components of attention, methylphenidate dosage, and blood levels in children with attention deficit disorder. *Pediatrics, 77,* 222–228.

Solanto, MV (1998). Neuropsychopharmacological mechanisms of stimulant drug action in attention-deficit hyperactivity disorder: a review and integration. *Behav Brain Res, 94,* 127–152.

Solanto, MV (1999). Dosage effects of ritalin on cognitive self-regulation, learning and memory, and academic performance. In: L Greenhill & B Osman (eds.). *Ritalin: Theory and patient management* (2nd edition). New York: Mary Ann Liebert.

Solanto, MV & Wender, EH (1989). Does methylphenidate constrict cognitive functioning? *J Am Acad Child Adolesc Psychiatry, 26,* 897–902.

Sprague, RL & Sleator, EK (1977). Methylphenidate in hyperkinetic children: Differences in dose effects on learning and social behavior. *Science, 198,* 1274–1276.

Vyse, SA, & Rapport, MD (1989). The effects of methylphenidate on learning in children with ADDH: The stimulus equivalence paradigm. *J Clin Consult Psychol, 57,* 425–435.

Werry, J, Sprague, R, & Cohen, M (1975). Conners' Teacher Rating Scale for use in drug studies with children—an empirical study. *J Applied Child Psychol, 3,* 217–229.

13

Comparative Psychopharmacology of Methylphenidate and Related Drugs in Human Volunteers, Patients with ADHD, and Experimental Animals

MITUL A. MEHTA, BARBARA J. SAHAKIAN AND TREVOR W. ROBBINS

Attention-Deficit/Hyperactivity Disorder (ADHD) is associated with deficits in functions commonly seen following frontal lobe damage (Pennington & Ozonoff, 1996; Kempton et al., 1999). Many of these impairments are improved after administration of stimulant medication (see Chapter 11 by Denney & Rapport,). Psychomotor stimulants boost catecholaminergic function by indirectly increasing synaptic levels of dopamine (DA) and norepinephrine (NE). The precise mechanisms and substrates of action of psychomotor stimulants, such as methylphenidate, on cognitive functions are, however, little understood. It is the aim of this chapter to discuss investigations in humans in which specific catecholaminergic (dopaminergic and noradrenergic) manipulations have been made and how subsequent behavioral changes have given insights into the involvement of these systems in cognitive functions, in particular those subserved by fronto-striatal systems. These systems are known to be connected by anatomically defined, segregated cortico-striatal circuits (Fig. 13–1). The key theoretical questions that will be addressed in this chapter, are:

1. Is methylphenidate acting, neurally and psychologically, in a similar manner in persons with ADHD and in normal volunteers?
2. Can the cognitive action of methylphenidate be explained by its action on the DA or NE systems, or some combination of both?
3. What are the neural loci of the effects of methylphenidate on cognitive performance?

In addressing these questions, the influences of psychomotor stimulants, particularly methylphenidate, on cognitive functions will be discussed in comparison

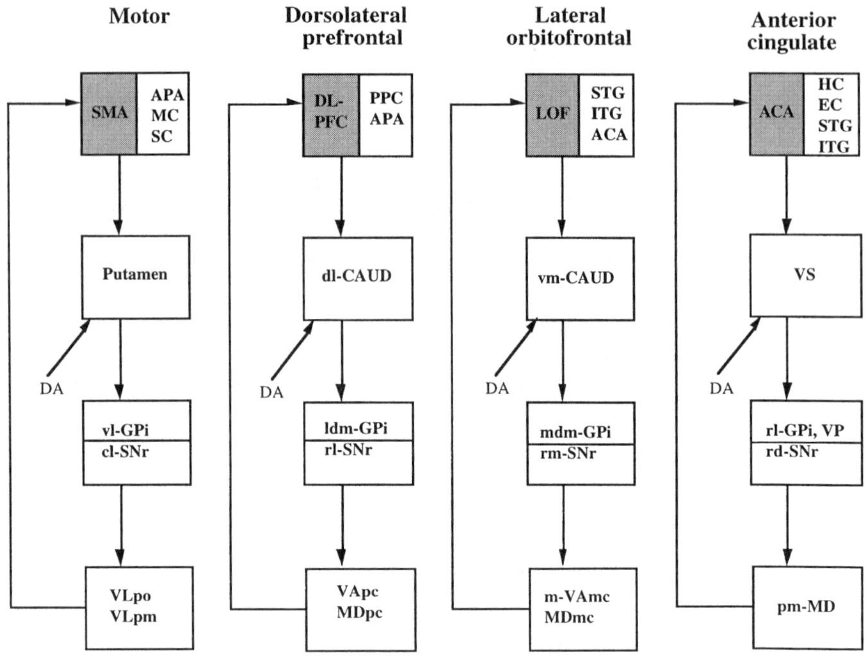

Figure 13–1. Basal ganglia-thalamocortical circuits based on the original scheme proposed by Alexander et al. (1986). Four of the putative segregated parallel "loops" are shown with the area of cortex that receives thalamic projections labeled (Motor, Dorsolateral prefrontal, etc.) and also shown in the left of the bisected boxes. Abbreviations by regions of brain are as follows: Areas of the cerebral cortex: SMA, supplementary motor area; APA, arcuate premotor area; MC, motor cortex; SC, somatosensory cortex; DLPFC, dorsolateral prefrontal cortex; PPC, posterior parietal cortex; LOF, lateral orbitofrontal cortex; STG, superior temporal gyrus; ITG, inferior temporal gyrus; ACA, anterior cingulate area; HC, hippocampal cortex; EC, entorhinal cortex. Areas of the striatum: dl-CAUD, dorsolateral caudate; vm-CAUD, ventromedial caudate; VS, ventral striatum. Areas of the pallidum: Gpi, internal segment of the globus pallidus; v1, ventrolateral; mdm, medial dorsomedial; rl, rostrolateral; VP, ventral pallidum. Areas of the substantia nigra: SNr, substantia nigra pars reticulata; c1, caudolateral; rl, rostrolateral; rd, rostrodorsal. Areas of the thalamus: VLpo, ventralis lateralis pars oralis; VLpm, ventralis lateralis pars medialis; VApc, ventralis anterior pars parvocellularis; MDpc, medialis dorsalis pars parvocellularis; m-VAmc, medial ventralis anterior pars magnocellularis; MDmc, medialis dorsalis magnocellularis; pm-MD, posteriomedial medialis dorsalis; dopamine (DA), ascending dopamine projections to the striatum.

to studies using specific dopaminergic and noradrenergic agents, with particular reference to studies in normal volunteers.

Methylphenidate is the most widely prescribed stimulant medication for the most common psychiatric disorder of childhood, with an estimated 2% of American schoolchildren receiving treatment (Barkley, 1990). However, children with ADHD have also been reported to respond to a range of treatments including antidepressants, neuroleptics and DA and NE agonists (see Pliszka, Chapter 14). The monoaminergic neurochemical systems upon which all of these drugs are known to act include both the dopaminergic and noradrenergic systems, which have both been implicated in the modulation of cognitive functions of the prefrontal cortex (PFC)-functions which are impaired in ADHD (Shue & Douglas, 1992; Pennington & Ozonoff, 1996; Kempton et al., 1999). The influence of such drug classes on cognitive abilities can be examined in a number of ways. One approach has been to determine the neuropsychological impairments exhibited by patient groups with specific neuropathologies. An example of this is the broad range of cognitive deficits that have been demonstrated in Parkinson's disease (PD) (see Robbins et al., 1998b for discussion) where the core neuropathology is the degeneration of the nigrostriatal DA system. The inherent limitation of this approach is that deficits seen in patients rarely implicate a single neurotransmitter system. Another approach is to investigate behavioral and cognitive effects of selective drugs with well-designed and validated paradigms in human volunteers. However, a major problem for human psychopharmacology is the necessity for the systemic administration of agents, so their site of action can only be inferred from knowledge of the neural networks involved in the performance of the tasks. This can be obviated to some extent by using functional neuroimaging techniques [e.g. positron emission tomography (PET) and functional magnetic resonance imaging (fMRI)] and observing interactions between drug and task in distinct brain areas (Grasby et al., 1996). Alternatively, experimental animals such as rats and monkeys can be used to make anatomically selective interventions, for example, using specific neurotoxins or intracerebral infusions. This approach is limited by the degree of neuropsychological comparability to humans, although a degree of behavioral homology across species can be utilized (Robbins, 1998). Despite the limitations of the numerous approaches described, convergence of results across the different techniques provides a powerful account of the roles of the different neurochemical systems in the cognitive processes being investigated.

Cognitive Effects of Dopaminergic Manipulations

Effects of Dopamine Manipulations in Animals

Although the presence of DA receptors and the innervation of DA neurons in the brain (see Lewis, Chapter 3; Taylor & Jentsch, Chapter 4) is widespread, it is in the modulation of the fronto-striatally mediated functions that DA plays a major role. The cognitive paradigms used most extensively in the assessment of prefrontal function in nonhuman primate are visuospatial delay tasks where the animal is required to hold 'on-line', for a short period of time, a stimulus attribute prior to making a response. These delayed response tasks (which include oculomotor delayed response tasks and delayed alternation tasks) are commonly used to assess working memory function (see Arnsten, Chapter 7). To summarize, the DA system appears to have a triphasic effect on working memory performance: antagonists producing impairment (Sawaguchi & Goldman-Rakic, 1991), low-dose agonists enhancing performance (Arnsten et al., 1994; Cai & Arnsten, 1997; Granon et al., 1999) and higher doses of agonists disrupting performance (Zahrt et al., 1997) (often referred to as the inverted-U function). Moreover, there is evidence that rats with higher indices of cortical DA turnover exhibit more efficient delayed alternation performance (Sahakian et al., 1985).

From animal studies (see Arnsten, Chapter 7), D1 receptors (presumably localized in the PFC) have been implicated in spatial working memory, with some evidence implying a possible role for D2 receptor systems (Arnsten et al., 1995; Arnsten, this volume). However, the precise role that D2 receptors play in spatial working memory in the PFC is not clear from the nonhuman primate research. By considering the greater density of D1 and D2 receptors in the striatum relative to the prefrontal cortex (Cortes et al., 1989; Camps et al., 1989) and the anatomical connectivity between these two brain areas (see Fig. 13–1), it is possible that striatal action of D2 receptor agents could explain the effects of systemic administration of the D2 agonist, quinpirole (Arnsten et al., 1995) on a working memory task sensitive to prefrontal cortex damage in the monkey.

Dopamine and Parkinson's Disease

Despite the vast literature describing the role of DA in the modulation of working memory functions of the PFC in animals (see Dunnett and Robbins, 1992; Arnsten, 1997; Goldman-Rakic, 1998, for reviews), studies in normal humans have been limited. By demonstrating impaired performance on a delayed response task in PD patients, Lueck and colleagues (1990) showed the relevance, to human patients, of the findings from the animal studies. Interpretation of this study is however, problematic, since all of the patients were receiving L-Dopa medication at the time of testing. Various studies on nonmedicated patients with PD have described a broad range of cognitive impairments on numerous tasks sensitive to frontal lobe damage (e.g., Lees & Smith, 1983; Morris et al., 1988; Owen et al., 1992, 1993, 1995a, 1997). More direct evidence for the influence of DA on cognition in PD comes from the deterioration of performance on tasks sensitive to frontal damage in PD patients when taken off L-Dopa medication (Lange et al., 1992; Kulisevsky et al., 1998). Specifically, in a study by Lange and colleagues (1992) performance decrement following L-Dopa withdrawal was seen on tests of spatial working memory and planning in the absence of effects on tests of visuospatial memory and cognitive flexibility (as assessed by the attentional-set shifting task, see Fig. 13–2). However, in a subgroup of patients exhibiting a "wearing-off" phenomenon in their motor response to L-Dopa medication, performance on a test of cognitive flexibility, the Wisconsin Card Sorting Test (WCST), was worsened following L-Dopa administration (Kulisevsky et al., 1996).

In addition, Swainson and colleagues (2000) have recently demonstrated specific impairments relative to control volunteers in both unmedicated and medicated patients with PD on different tasks. A visuospatial recognition memory test known to be sensitive to frontal lobe damage (Owen et al., 1997), and thought to involve

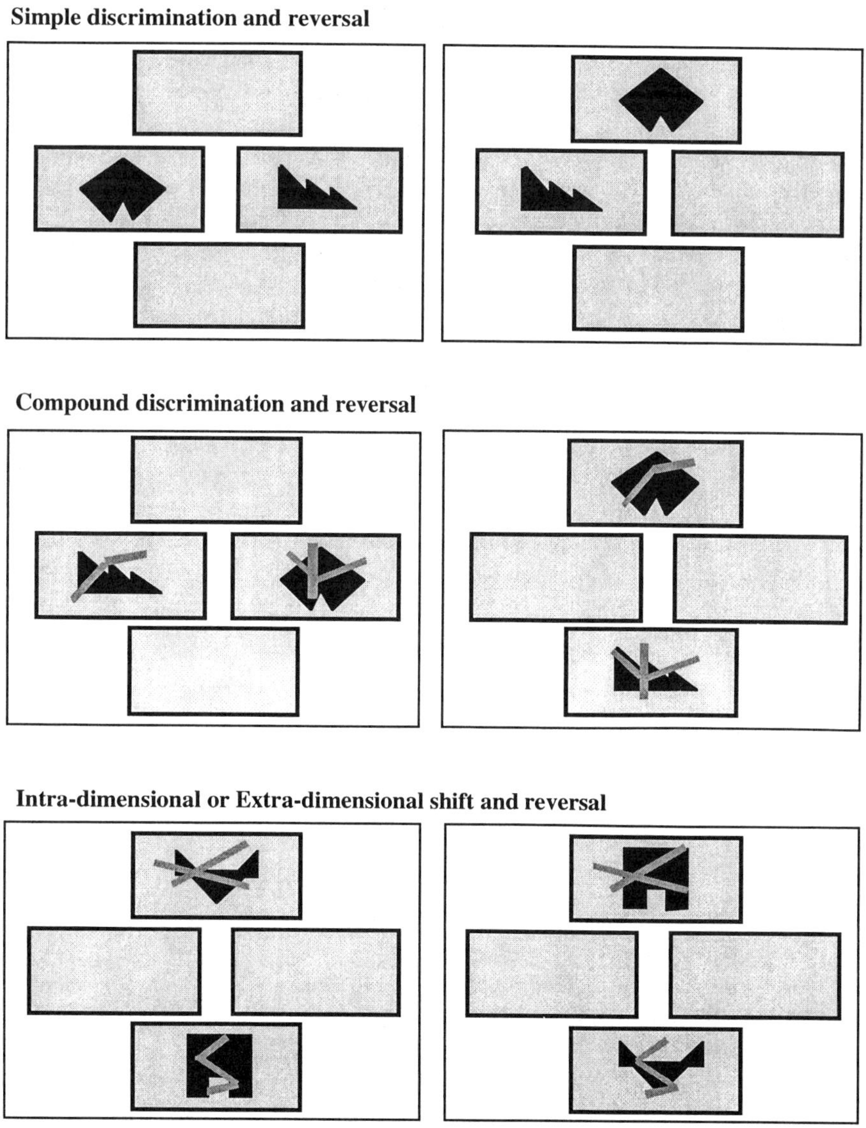

Figure 13–2. Attentional-set shifting task. Subjects initially learn a simple visual discrimination (e.g., respond to one of two presented shapes) and then learn to reverse this discrimination. (The location of stimuli is unimportant and varied.) After this, another dimension is introduced (e.g., lines) and after a set of compound discriminations and reversal have been successfully performed, new exemplars are introduced. Subjects are then required to continue responding to the previously reinforced dimension (e.g., shapes). This is termed an intra-dimensional shift and demonstrates an ability to generalize a learned rule. After another reversal of reward contingencies new exemplars are again introduced. For this, the critical stage of the test, subjects are required to respond to the previously *un*reinforced dimension (e.g., lines). This is termed an extra-dimensional shift and is formally akin to a category shift in the WCST. Finally, the stimulus-reward contingencies are reversed.

activation of a neural network including regions of the lateral prefrontal cortex (Jonides et al., 1993), was impaired in the unmedicated group, but *not* impaired in the medicated group. In contrast, on a reversal learning test thought to involve recruitment of orbital areas of the prefrontal cortex (Dias et al., 1996), the medicated group, but *not* the unmedicated group, exhibited impairments. These results were interpreted in terms of "normalization" of DA levels, in the medicated patients, in the dorsal striatum (which has anatomical connections to the lateral prefrontal cortex, see Fig. 13–1) and an "overdosing" of DA levels, in the medicated patients, in the ventral striatum (which has anatomical connections to the orbitofrontal cortex). These studies in PD argue for a specific influence of DA on the modulation of performance on tasks sensitive to frontal lobe damage, but highlight the necessity to understand the state of the patients being studied in terms of both clinical response and neurochemical pathology.

Studies in normal human volunteers allow investigation of the role of specific neurotransmitter systems in contrast to conditions such as Parkinson's disease, in which, at the very least, other neurotransmitter systems are involved in the neuropathology (Agid et al., 1989).

Effects of Dopamine Manipulations in Normal Human Volunteers

The effects of dopamine in normal volunteers have been assessed on measures of *executive function*, a term encompassing control processes that act to optimize performance on tasks requiring multiple elemental operations (Baddeley, 1986; Robbins, 1996). Cognitive abilities such as working memory, planning, attentional set-shifting, sustained and divided attention, and goal integration are examples of tasks that assess aspects of executive function.

Visuospatial delay tasks employed in much of the animal research into monoaminergic influences on cognition have been modified for use in psychopharmacological studies in normal volunteers. Cognitive effects mediated by specific DA receptors have been investigated using systemic administration of agents such as the D2 receptor agonist bromocriptine and D2 receptor antagonist sulpiride. The effects of bromocriptine on an analogue of the spatial-delayed response task from animal investigations has been assessed in a number of published studies to date. Of these, two have demonstrated enhanced spatial working memory performance using 1.25 mg and 2.5 mg bromocriptine in normal volunteers (Luciana et al., 1992, 1997). The latter study also found detrimental effects of the D1/D2 antagonist haloperidol on spatial working memory performance (a nonspatial working memory task was unaffected by either drug). The changes in performance were specific to the memory task, with tests of nonspecific arousal, attention and motor function being unaffected by the drugs. In contrast, however, one study with normal volunteers and one with a heterogeneous group of patients with traumatic brain injury (TBI) showed no beneficial effects of bromocriptine (2.5 mg) on the same delayed-response task used by Luciana et al. (1992, 1997) (Kimberg et al., 1997; McDowell et al., 1998), although both studies showed improved performance in other cognitive tasks. There is some indication that the D1 receptor system in humans is also important in delayed memory function. Pergolide (a D1/D2 agonist), but not bromocriptine, was reported to enhance performance of a visuospatial delayed-matching paradigm in normal volunteers (Müller et al., 1998).

In addition to effects on spatial working memory, the DA system of the primate brain has also been implicated in a number of other tasks known to be dependent on an intact PFC. Among these are the attentional set-shifting paradigm illustrated in Figure 13–2, analogous in concept to the Wisconsin Cart Sorting Test (WCST). Roberts et al. (1994) showed that monkeys with compensatory increases in striatal DA levels (after prefrontal DA depletion) performed better than control monkeys, specifically at the extra-dimensional (ED) shift stage of the test. (Importantly, the same monkeys also showed the expected impairment on a spatial delayed response task.) This is consistent with the findings from our own group in normal volunteers (Mehta et al., 1999) showing that a single dose of sulpiride impaired performance at the ED shift stages. Attentional set-shifting can also be impaired following frontal lobe damage and in patients with striatal dysfunction such as those with PD or Huntington disease (Owen et al., 1990, 1992; Lawrence et al., 1996). Validating the attentional set-shifting task, PET neuroimaging has shown that the neural correlates of the critical ED shift stage of the task are in specific regions within the anterior frontal lobe (Rogers et al., 2000).

Deficits following sulpiride (Mehta et al., 1999) were also seen on difficult tests of spatial working memory and spatial planning, both of which are known to be dependent on intact frontal lobes (Owen et al., 1995a). Bromocriptine (2.5 mg) enhanced executive function performance in normal volunteers with low verbal memory capacity, and improved impaired executive function performance in patients with TBI (Kimberg et al., 1997; McDowell et al., 1998). Using a lower dose of bromocriptine (1.25 mg) a computerized version of the Corsi block tapping task (Milner, 1971) was improved in normal young volunteers when compared with placebo administration (Mehta et al., 1998). The essential requirements of this task are similar to those engaged (storage, rehearsal and recall) by the delayed-response task employed by Luciana et al. (1992; 1997), Kimberg et al. (1997), and McDowell et al. (1998). However, the subject is required to remember a sequence of spatial locations on the former spatial span task, compared to only a single spatial location on the latter. It has been proposed that the former task is, *predominantly*, a probe for assessing rehearsal processes and the latter a probe for assessing storage processes (Smith & Jonides, 1998; Awh et al., 1996; D'Esposito & Postle, 1999).

The additional sequencing requirements of the spatial span task may be considered as reflecting greater executive components in the performance of this task compared to the single-location delayed response task. It is important to note that a recent study by Collins et al. (1998) demonstrated that DA depletion from the prefrontal cortex of marmosets that impaired spatial delayed response had no effect on a task of self-ordered spatial sequencing, which differed from the task employed by Mehta et al. (1999), by having a much reduced working memory load. The results of Collins et al. (1998) demonstrate that the effects of lesioning ascending DA projections are not always equivalent to those of lesions of the prefrontal cortex itself.

In summary, for humans, there is currently evidence that DA manipulations influence both storage and rehearsal processes in certain working memory tasks. More complex working memory tasks that require not only short-term storage but also manipulation (including executive control processes) of spatial mnemonic information (as in Fig. 13–3) have also been shown to be influenced by administration of dopaminergic agents (e.g., Lange et al., 1992; Mehta et al., 1999).

Studies using bromocriptine can be summarized as demonstrating modulation of certain executive and mnemonic functions (Luciana et al., 1992, 1997; Kimberg et al., 1997; McDowell et al., 1998; Mehta et al., 1998, 1999). However, the results on delayed response tests of working memory are mixed with some studies showing improvement (e.g., Luciana et al., 1992) and other finding no effect in either normal volunteers (Kimberg et al., 1997) or patients (McDowell et al., 1998).

A complex spatial working memory paradigm that requires manipulation of spatial information (the two-back task, which requires subjects to remember two-back in a spatial sequence) has recently been performed in the neuroimaging environment

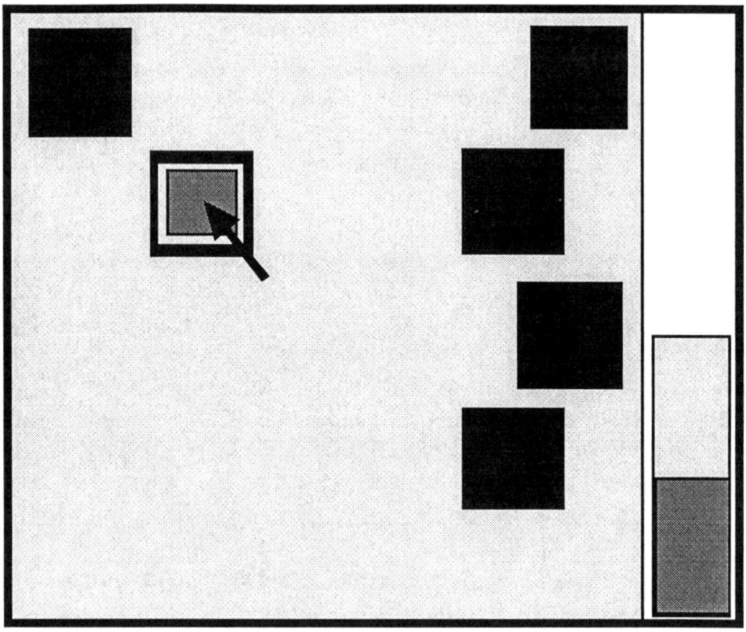

Figure 13–3. The self-ordered spatial working memory task. Subjects are required to search through an array of colored boxes for blue tokens, avoiding returns to locations where tokens previously have been found. Once a token is found (for example, as indicated by the arrow) subjects place it in the column at the side of the screen before beginning the search for the next token. Each box is only used once to hide a token and therefore two types of error are possible. A between-search error occurs when subjects return to a box in which they had previously found a blue token and a within-search error occurs when subjects return to a box within the same search.

using fMRI with subjects having taken either bromocriptine or placebo. Drug-induced attenuation of task-related increases in cerebral blood flow was observed in the inferior parietal lobe (Zarahn et al., 1998). It is noteworthy that the performance of the two-back working memory task in subjects after administration of bromocriptine was not significantly different to performance following placebo (Zarahn et al., 1998). Nevertheless, these suggest that working memory task-related neural activity may be modulated by bromocriptine via changes in parietal cortex activity.

In contrast, using PET, a reduction of task-induced frontal cortex activation and an augmentation of posterior cingulate cortex activation were observed following administration of the nonspecific DA agonist, apomorphine, where normal volunteer subjects were performing an auditory-verbal word list memory task (Grasby et al., 1992). The doses of apomorphine used (5 μg/kg and 10 μg/kg subcutaneous) impaired performance when compared to placebo (saline) injection. Importantly, Grasby and colleagues (1992) also demonstrated that these changes were specific, since the partial agonist of the serotonin$_{1A}$ receptor, buspirone, induced different patterns of task-related changes in cerebral blood flow. Taken together with the behavioral studies discussed earlier, these neuroimaging studies provide convergent evidence for a specific locus in the frontal cortex for the task-related effects of dopaminergic drugs and imply that more posterior areas may also be involved in this modulation of function. However, the differences between the loci of modulation of task-related activation in the studies by Grasby and colleagues (1992) and Zarahn and colleagues (1998) could be the result of a variety of factors, including the mode of action of the drugs, the drugs employed, and the nature of cognitive tasks used.

Effects of DA manipulations on selective (or focused) attention, sustained attention and covert orienting of attention have also been investigated in humans. The effect of D2 receptor blockade using sulpiride (Mehta et al., 1999) had no effect on the rapid visual information processing task (RVIP) of sustained attention (Wesnes & Warburton, 1984; Sahakian et al., 1989). The functional anatomy of this continuous performance task was examined by Coull et al. (1996a). Their findings were consistent with a right fronto-parietal network of sustained attention. These results provide preliminary evidence that D2 receptor modulation may have little influence in certain aspects of attentional control. This has recently been demonstrated in rats performing a version of the continuous performance task: D1 receptor agents influenced performance as measured by both latency and accuracy, but the D2 receptor antagonist, sulpiride, had no effects on any performance measure (Granon et al., 2000).

Other aspects of attentional function, specifically focused and divided attention, were compromised in normal subjects having taken the D1/D2 receptor antagonist droperidol (Clark et al., 1986a). These impairments could be reversed following administration of methylphenidate in the same volunteer subjects. Droperidol also reduced the cost of switching to invalid cues (i.e., facilitated the disengagement of spatial attention) and reduced the activational state of subjects, impairing their readiness to respond (Clark et al., 1987, 1989). This latter effect, highlighting the importance of the mesostriatal DA system in motor (response) readiness, is consistent with evidence from humans and experimental animals (Brown & Robbins, 1991; Clark et al., 1987). The role of the mesostriatal DA system innervating the dorsal striatum has previously been described in terms of the "activation" or energisation of behavior (see Dunnett & Robbins, 1992 for review). In fact it has been suggested that any apparent effects on attentional function produced by enhancing DA activity (in the nucleus accumbens) are secondary to behaviorally activating effects of the drug (Robbins, 1997).

More recently, neural network ensembles have been used to model the influence of DA on cognitive processes such as working memory (Durstewitz et al., 1999) and selective attention (Servan-Schreiber et al., 1998a). Explanations of numerous experimental effects have been embodied in these models, although their clear utility will be in the prediction of specific behavioral effects of drug manipulations (see Servan-Schreiber et al., 1998b).

Summary

In summary, it is clear that DA manipulations influence certain cognitive processes in human volunteers. The more complex working memory tasks that require information to be held in memory and sequentially recalled or manipulated can be enhanced by low doses of DA agonists (e.g., 1.25 mg or 2.5 mg bromocriptine) and impaired by DA antagonists. DA also appears to influence the holding "on-line" of remembered information in a similar manner. In certain situations in experimental animals, DA agonists can impair working memory function, particularly following intra-prefrontal cortical administration (Zahrt et al., 1997). Executive functions, which are assessed by various tasks, including tests of planning and attention-set shifting, are also influenced by DA manipulations, with antagonists impairing performance and agonists generally improving performance. Evidence for dopaminergic modulation of attentional processes is also present from empirical data and has also been implicated in neural network models. Knowledge of the specific receptor systems that mediate the effects of DA manipulations is currently limited, partly because of a lack of specific agents, particularly for human research. However, research in experimental animals has demonstrated interesting dissociations between D1 and D2 receptor modulation (e.g., Sawaguchi & Goldman-Rakic 1991; Granon et al., 2000). The precise mechanisms by which DA modulates cognitive functions are only beginning to be understood. Although neuroimaging has been able to define the neural networks involved in the performance of some tasks, the direct approach of assessing cognitive functions after administration of DA agents in the imaging environment is relatively new and is a promising tool for understanding the mechanisms by which centrally acting agents can modulate performance.

Cognitive Effects of Noradrenergic Manipulations

Effects of Norepinephrine Manipulations in Animals

The primary pathway of the noradrenergic system of the human brain originates in the locus coeruleus, from which axons project rostrally as the dorsal noradrenergic bundle (DNAB), dorsally into the cerebellum and caudally into the spinal cord.

The locus coeruleus innervates virtually all parts of the telencephalon and diencephalon, including the neocortex, hippocampus, amygdala, septum, thalamus, and hypothalamus (see Foote et al., 1983 and Berridge, Chapter 6, for reviews).

Cortical NE can be reliably reduced to less than 10% of control values in rats by injecting a neurotoxin, 6-hydroxydopamine, into the DNAB. Following this procedure there are no obvious changes in spontaneous locomotor activity, eating and drinking, or general behavior. In an adaptation of the continuous performance test of sustained attention from the human literature, rats with such lesions perform normally except: (*1*) when brief bursts of white noise precede the presentation of each stimulus (Carli et al., 1983); (*2*) when the timing of stimulus presentation is unpredictable (Cole & Robbins, 1992); (*3*) when the stimulant D-amphetamine is administered (Cole & Robbins, 1987). These three findings have two main implications. First is that impairments on the attentional task become apparent in the coeruleo-cortical NE depleted animals, under conditions of high arousal (e.g., 1 & 3 as mentioned previously). Thus DNAB lesions appear to affect controlled, rather automatic responding, perhaps via effects on attentional function (Robbins, 1997). Second, as demonstrated by situation (1), distracting stimuli, such as white noise, may be more potent in disrupting performance after DNAB lesions. Indeed Roberts et al. (1976) showed increased distractibility in DNAB lesioned rats performing a maze learning test. These ideas have proved useful in the interpretation of the effects of noradrenergic lesions and treatments on spatial working memory performance, where NA may have particular importance in preventing distraction from irrelevant stimuli (see Arnsten, Chapter 7).

Arnsten (Chapter 7) has described the possible role of noradrenergic manipulations in animals in the processes of "on-line" working memory functions. In keeping with the suggestions from studies in rats (previously cited), the influences of noradrenergic agents on working memory processes may be mediated by effects on attentional processes. Early theoretical (Robbins, 1984) and experimental findings (Arnsten & Constant, 1992) have described catecholamine modulation as "protective" against the disruptive effects of distractors presented during the delay period prior to a response being made on delayed response tasks. In the monkey, both clonidine and guanfacine (alpha-2 agonists) [see Berridge, Chapter 6 for a description of noradrenergic (and adrenergic) receptor subtypes] ameliorate the deleterious effects of including distractors within the delay period of the delayed response task (Arnsten & Constant, 1992). These results have been interpreted as reflecting predominantly postsynaptic actions by the alpha-2 agonists within the prefrontal cortex (Arnsten & Goldman-Rakic, 1985; Arnsten, 1997) and are consistent with the cognitive influence of these manipulations being mediated by changes in attentional functions. For example, guanfacine improves working memory performance without sedative or hypotensive effects that often accompany clonidine administration (Arnsten, 1997). Indeed both guanfacine and clonidine have been investigated for treatment efficacy for ADHD. Their efficacy has been noted in terms of improved scores for children with ADHD on parent rating scales and, for guanfacine, improved performance on a continuous performance task or sustained attention (Chappell et al., 1995; see Pliska, Chapter 14). The cognitive mechanism of this enhancement may be mediated by noradrenergic modulation altering inhibitory abilities of the PFC (Arnsten et al., 1996). In monkeys, infusion of the alpha-2 adrenergic antagonist, yohimbine, into the PFC impaired performance on a delayed go/no-go task (with increased errors of commission) (Li & Mei, 1994), a similar impairment to that seen in children with ADHD (e.g., Vaidya et al., 1998). In this context, it is noteworthy that in the study by Vaidya and colleagues (1998) the methylphenidate-induced effect on a version of the go/no-go task was similar in children with ADHD (whose impairment was reduced) and in normal control children (whose performance was enhanced).

Effects of Norepinephrine Manipulations in Human Volunteers

In adult humans, systemic administration of clonidine alters working memory performance in a dose-dependent manner. A lower dose of clonidine (1.5 μg/kg i.v.) had no effect on a computerized self-ordered spatial working memory task (Coull et al., 1995b), whereas a higher dose (2.5 μg/kg i.v.) enhanced performance, but only when the task was familiar to subjects. These two doses are thought to have different actions in the brain, with the lower dose acting presynaptically to reduce coeruleo-cortical activity (Frith et al., 1985; Charney & Heninger, 1986). Higher doses of alpha-2 agonists are thought to have greater action at post-synaptic (and post-junctional) receptors (Arnsten, 1997). Using oral doses of clonidine, Jäkälä et al. (1999b) demonstrated disruption of self-ordered spatial working memory performance following doses of 0.5 μg/kg or 5 μg/kg clonidine (with no effects following a dose of 2 μg/kg), in keeping with an inverted-U response curve.

The site of action of clonidine in the brain during cognitive performance has been investigated directly using PET neuroimaging (Coull et al., 1997). Following intravenous administration of 1.5 μg/kg clonidine, normal volunteer subjects performed a task that placed demands on working memory and sustained attention (the RVIP test). Performance of this task has previously been shown to be mediated by a neural network including the right frontal and parietal lobe regions (Coull et al., 1996a). In the study using clonidine, a significant drug × task interaction was found in the right thalamus, such that blood flow in the thalamus was reduced after clonidine in the resting condition with no difference in the task condition. This is consistent with clonidine having a greater detrimental influence in states of low arousal. Indeed, clonidine (1.5 μg/kg) impaired performance on the RVIP task of sustained attention, but only when it was familiar to subjects (Coull et al, 1995a). This produces a putatively lower state of arousal than when the task is novel to subjects and the deleterious effects of the drug might be counterbalanced by the stressful or arousing aspects of the novel test situation (Coull et al., 1995a).

It is important to note that this pattern of findings contrasts with the effects of DNAB lesions in rats, where continuous performance task decrements are seen in situations of heightened arousal (Carli et al., 1983; Cole & Robbins, 1987; 1992). However, the reduction in locus coeruleus activity following low doses of clonidine may have a much more modest functional effect than that produced by cortical noradrenergic depletion of over 90%.

The specific impairment induced by clonidine on the RVIP task in humans (Coull et al., 1995a) was a reduction in stimulus sensitivity with no effect on response bias. The impairment was not influenced by the amount of time on task. In children with ADHD deficits in sustained attention on continuous performance tasks with fast presentation rates appear similar to the effect of 1.5 μg/kg clonidine in normal adults, both showing reduced performance levels with, in general, no decline in performance with increased time on task (van de Meere, 1996). In normal adults, Coull et al. (1995a) showed that when the stimulus presentation rate on the RVIP task was halved (and the task length was doubled) 1.5 μg/kg clonidine had no significant effects on performance. With such task parameters (i.e., stimuli widely spaced in time) children with ADHD, however, demonstrate a decline in performance over time to a greater extent than normal controls (van de Meere, 1996). A higher dose of clonidine (2.5 μg/kg), which is thought to act at postsynaptic alpha-2-adrenoreceptors, produced a decrement in performance with increased time on task, which was not seen on placebo (Coull et al., 1995a). This effect was observed on the task with the faster presentation rate. Unfortunately, the influence of 2.5 μg/kg clonidine on the task with stimuli more widely spaced in time was not assessed. These findings indicate that a reduction in noradrenergic activity in healthy controls can simulate the general impairment seen on continuous performance tasks in children with ADHD via a reduction in level of arousal or vigilance. The possible influences of postsynaptic effects of clonidine in relation to ADHD are less clear.

Impairments in tests of sustained attention induced by clonidine are contrasted by the lack of detrimental effects on a visuospatial paired associates learning task equally sensitive to frontal and temporal lobe damage (Owen et al., 1995b), which was actually enhanced following 2.5 μg/kg i.v. or 2 μg/kg oral clonidine or 29 μg/kg guanfacine in normal volunteers (Coull et al., 1995a; Jäkälä et al., 1999a). These effects may be mediated by postsynaptic receptors whereas the impairment observed by Frith et al. (1985) on a difficult paired associates learning task following a lower dose of clonidine may be mediated by a presynaptic action. Moreover accuracy of performance on the attentional-set shifting (Fig. 13–2) task shown to be sensitive to frontal, but not temporal, lobe damage (Owen et al., 1991) is not influenced by either 1.5 μg/kg or 2.5 μg/kg i.v. clonidine or 7 μg/kg or 29 μg/kg oral guanfacine (Coull et al., 1995a; Rogers et al., 1999; Jäkälä et al., 1999b). Taken together, these findings highlight that not all tasks sensitive to frontal lobe damage are also sensitive to noradrenergic manipulations.

Clark et al. (1986b) have investigated the effects of clonidine administration on a test of visuospatial orienting of attention sensitive to parietal lobe damage (Posner et al., 1984). A dose of 200 μg i.v. clonidine (suggested by Clark et al. to inhibit firing of locus coeruleus neurons) apparently reduced the cost of switching, suggesting a facilitation in the disengagement of attention (Clark et al., 1989), but having no effect on reaction times. The effect on switching is similar to the findings with droperidol (as noted previously). Preliminary results from studies in monkeys show a discrepancy, with clonidine having no effect on the cost of switching, but instead reducing the "alerting" effect of the external warning cue to reduce reaction times (Witte et al., 1997). However, Clark and colleagues (1989) used a task with central cues and Witte and colleagues (1997) used a task with peripheral cues, both of which are sensitive to parietal lobe damage (Posner et al., 1984, 1987), but believed to engage different attentional systems (see Rosen et al., 1999). Children with ADHD are impaired on visual orienting tasks with central cues (Swanson et al., 1991; Wood et al., 1999), but not peripheral cues (Aman et al., 1998). Centrally cued tasks selectively activate the dorsolateral prefrontal cortex in addition to more posterior areas activated by the peripherally cued tasks in normal volunteers (Rosen et al., 1999). Thus, the effects of clonidine (and possibly dopamine agents) on orienting may be related to its modulation of prefrontally mediated processes, and the impairments in ADHD may be related to dysfunction of frontal lobe systems: a possible interpretation in keeping with the effects of alpha-2 adrenoreceptor modulation of tasks of sustained attention and working memory (Coull et al., 1995a; Jäkälä et al., 1999b).

The effects of idazoxan are compatible with the selective effects of noradrenergic manipulations on prefrontal cortex functions. The alpha-2 adrenoreceptor antagonist idazoxan is thought to act presynaptically to elevate synaptic levels of NE (Dennis et al., 1987), although it is important to note that idazoxan also influences imidazoline receptors as well as serotonin systems (Marsden et al., 1986) and Middleton and colleagues (1999) have shown that the psychological effects of idazoxan and clonidine are not necessarily mutually antagonistic. Coull et al. (1996b) used idazoxan to assess alterations in performance in 3 patients diagnosed with dementia of the frontal type. Dose-dependent improvements were found on tests of planning, sustained attention, verbal fluency, and episodic memory. In contrast, idazoxan administration led to impairment on a test of spatial working memory. Since working memory performance was relatively unimpaired in the patients in the Coull et al. (1996b) study, idazoxan may in fact be producing a disruptive effect via its action at the alpha-2 adrenoreceptor in keeping with the inverted-U hypothesis of catecholamine modulation of function (Robbins & Everitt, 1987). The inverted-U shaped function essentially conceptualizes an optimum level of arousal for the performance of a particular task (or a performance measure within a task) (Yerkes & Dodson 1908; Hockey 1979). In the current context of monoamine manipulations, the studies by Coull et al. (1995a; 1995b; 1996b) exemplify the existence of complex interactions between the nature of task requirements, arousal and the manipulation of neurochemical systems (for a fuller discussion of arousal constructs, see Broadbent, 1971; Robbins, 1984; Robbins & Everitt, 1987).

Summary

In summary, it appears that noradrenergic manipulation in normal human volunteers influences certain cognitive processes, in particular those dependent on an intact prefrontal cortex. These include tests of sustained attention, working memory, planning, and certain forms of covert orienting of attention. However, there are important differences between the effects of noradrenergic manipulations and dopaminergic manipulations. Tests of working memory indicate that DA is predominantly involved with maintaining information on-line, whereas NE reduces the disruptive effects of distracting stimuli. Articles by Clark et al. (1987,1989) and Robbins (1984) highlight the role of NE in regulating the capacity for conscious registration of external stimuli, and the role of DA in regulating the capacity to respond to external stimuli. In addition, preliminary evidence exists to suggest a possible dissociation of DA and NE in other tests of attention. Specifically, D2 receptor manipulations have been shown to have *no* influence on tasks of paired associates learning and sustained attention (Mehta et al., 1999), whereas NE manipulations clearly modulate performance of both types of task (Frith et al., 1985; Coull et al., 1995b; Jäkälä et al. 1999a). An important consideration is the effect of novelty in the interpretation of drug effects. The marked effect of clonidine in states of familiarity (or lower arousal) contrasts with the influence of DA manipulations in novel situations on some tasks. The validity of this distinction will become more apparent when comparative investigations are performed that carefully control parameters, for subjects, of task novelty and familiarity.

Cognitive Effects of Stimulant Drug Manipulations

The amphetamines and methylphenidate (as well as caffeine) are classified as stimulant drugs. For purposes of comparison with previous sections and drug treatment in ADHD the cognitive effects of caffeine [which have a very different mode of action from amphetamines and methylphenidate and have not been shown to improve hyperactive symptomatology (Arnold et al., 1978)] will not be discussed here. From the evidence discussed thus far we are in a position to make preliminary predictions regarding the influence of stimulant administration on certain cognitive processes in normal volunteers. Stimulant drugs such as methylphenidate block the reuptake of DA and NE (although to differing extents). The first prediction is that the cognitive effects of stimulant administration will probably reflect both DA and NE actions, although there are differences between the "presynaptic" effects of stimulants and the pre- and postsynaptic actions of agents such as clonidine and bromocriptine. The nature of these cognitive effects would be predicted to be improvements; however, many of the studies of separate DA and NE effects discussed involved down-regulation of these neurotransmitter systems leading to specific impairments. It has long been recognized that it might be much easier to demonstrate impaired rather than enhanced performance in normal volunteers because, in normal circumstances, they may already be performing at their optimal levels (Fagan et al., 1988). However, using sensitive neuropsychological tasks, improvements in performance have been demonstrated with selective DA and NE agents, as well as with stimulant drugs in normal volunteers.

Weiss and Laties (1962), in a classic article, reviewed the effects of stimulant drugs on various aspects of "human performance." They concluded that stimulant drugs (particularly D-amphetamine, as there was a paucity of studies using methylphenidate at the time) produced an absolute improvement in vigilance. However, their conclusions were based largely on the work of Hauty and Payne (1955) and Payne and Hauty (1954; 1955). In these studies the cognitive task used was a pursuit task, in which vigilance (or monitoring) was only one component of the test and in which concurrent responses had to be made using joysticks, levers, and pedals.

Koelega (1993) has more recently reviewed the influence of stimulant drugs on vigilance performance and concluded that (*1*) D-amphetamine can produce absolute improvements on tasks requiring sustained attention, exerting effects similar to those of other drugs (namely nicotine & caffeine) on measures of reaction time (decreased) and hits (increased), (*2*) these improvements are not restricted to fatigued subjects and protracted sessions, and (*3*) simple neurotransmitter models of attention and information processing may be untenable (drugs that affect different neurotransmitter systems can all influence, e.g., vigilance performance). Koelega (1993) also commented on the wide use of tasks that may be insensitive to stimulant administration or unreliable, or produce uninterpretable data. Included among his "good" tests of sustained attention was the RVIP continuous performance test. Elliott et al. (1997) used this test in normal volunteers after they had been given methylphenidate (40 mg) and, consistent with Koelega (1993), observed enhanced performance in terms of the speeding of responses. Camp-Bruno and Herting (1994) also demonstrated speeded responses to targets on a similar sustained attention task following methylphenidate (20 mg) when compared with placebo. Tasks of divided attention and covert orienting of attention have also been used in normal volunteers after methylphenidate administration. Clark et al. (1986a) demonstrated that methylphenidate (0.65 mg/kg) reversed the impairments in a dichotic auditory attention task produced by administration of the DA antagonist droperidol. However, methylphenidate administration alone had no effects on any of the performance measures, while producing subjective increases in feelings of elation, alertness, and energy (Clark et al., 1986a). These results indicate that in a test of divided attention, methylphenidate can reverse drug-induced impairments but is ineffective in producing absolute enhancements in auditory divided attention in normal subjects.

In addition, Elliott et al. (1997) assessed the effects of methylphenidate on a number of tests taken from CANTAB (Sahakian & Owen 1992; Robbins et al., 1994; Robbins et al., 1998a). Normal volunteers were assessed on these tasks in a crossover design in which drug and placebo sessions were counterbalanced. Methylphenidate was found to improve performance on a test of working memory (spatial span task) that was also improved following the D2 agonist bromocriptine (Mehta et al., 1998) and is known to increase blood flow in the ventrolateral prefrontal cortex (Owen et al., 1996). On a more complex self-ordered working memory task that activates both dorso- and ventrolateral prefrontal cortex (Owen et al., 1996), Elliott et al. (1997) demonstrated enhanced performance after methylphenidate. A test of planning, which also activates the prefrontal cortex (Baker et al., 1996), showed improved performance on methylphenidate compared with placebo, consistent with a general hypothesis of methylphenidate enhancing prefrontally mediated cognitive functions. The improvements in working memory tests and the test of planning following methylphenidate (Elliott et al., 1997) were all limited to the first test session, when the tasks and testing situation were novel to subjects.

In order to directly determine the locus of the methylphenidate improvements in cognitive performance we (Mehta et al., 2000) have recently performed positron emission tomography (PET) scans on normal volunteers after the administration of methylphenidate or placebo while they were performing the test of spatial working memory used by Elliott et al. (1997). Figure 13–4 shows that defined regions within the dorsolateral prefrontal cortex and posterior parietal cortex were revealed as loci of significant changes in blood flow in the interaction analysis of task and drug. This is consistent with other studies using D-amphetamine, demonstrating that the dorsolateral prefrontal cortex is the region of the brain showing a change in blood flow during performance of a test of cognitive flexibility (WCST) (Daniel et al., 1991; Mattay et al., 1996). However, the performance of normal volunteers on the WCST was not significantly different on D-amphetamine compared with placebo (Mattay et al., 1996) and, therefore, it was unclear which aspects of cognitive flexibility were modulated by stimulant administration.

In normal volunteers, Rogers and colleagues (1999) have performed a detailed analysis of the effects of methylphenidate (40 mg) on performance on a version of the attentional-set shifting test of cognitive flexibility (Fig. 13–2). Volunteers on methylphenidate demonstrated enhanced learning of the extra-dimensional shift compared to the intra-dimensional shift (irrespective of the novelty or familiarity of the task to them). Thus DA and NE stimulation appeared to facilitate the shifting of attention toward newly relevant stimuli. Additionally, however, the mean response latencies at both of the shift stages was lengthened after methylphenidate, highlighting that enhancements in task performance may sometimes be accompanied by

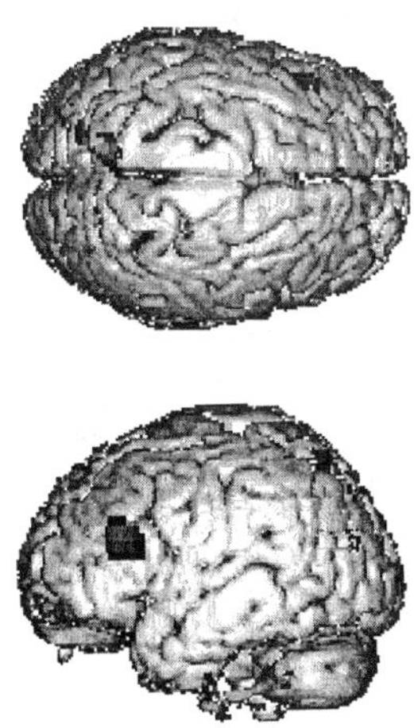

Figure 13–4. Cortical regions that are responsible for the neuromodulatory influence of methylphenidate on self-ordered spatial working memory in normal adults. These regions appear at the front of the brain to the right in the upper image and the left in the lower image.

performance costs on other measures. Elliott and colleagues (1997) also demonstrated impairments, as well as improvements, in performance after methylphenidate administration to normal volunteers: the impairments were observed on a version of the Tower of London task (Owen et al., 1990), but only when it was familiar to subjects.

Summary

In summary, studies to date have evidenced advantageous effects of stimulant administration on tests of sustained attention, working memory, planning, and cognitive flexibility (Koelega, 1993; Elliott et al., 1997; Rogers et al., 1999). Impairments have also been shown after methylphenidate administration on a test of planning when it was familiar to subjects and a version on the ID/ED task in terms of response latencies. Some of these apparently conflicting effects of methylphenidate may be related to the complex interaction between the cognitive requirements of tasks used, their novelty to subjects, and the drug administration (see Robbins & Everitt, 1987; Elliott et al., 1997). Additionally, methylphenidate can reverse the impairments on tests of divided attention and covert orienting of attention induced by the dopamine D1/D2 receptor antagonist droperidol (Clark et al., 1986a).

Discussion

Is Methylphenidate Acting, Psychologically, in a Qualitatively Similar Manner in ADHD and Normal Volunteers?

Earlier in this chapter three theoretical questions were raised regarding the influence of stimulant drugs, particularly methylphenidate, on cognitive performance. The first question was whether the psychological effects of methylphenidate were qualitatively similar in ADHD and normal volunteers. The importance of this question is that it helps to define whether the drug is remediating a specific form of brain dysfunction caused by ADHD, or if instead the drug is acting on a relatively normal catecholaminergic substrate. Robbins and Sahakian (1979) suggested that the "calming" effects of methylphenidate treatment of ADHD were not paradoxical, but were in fact part of the normal response based on a "rate-dependent" analysis of activity level affected by the drug in normal and hyperactive children and normal adults (see also Swanson & Volkow, Chapter 11). Attempts have been made to relate the rate-dependency hypothesis to the cognitive effects of methylphenidate in situations involving "attentional" processes (Lyon & Robbins, 1975; Sahakian & Robbins, 1977). More recent cognitive studies have supported the rate-dependency suggestion in the cognitive domain, because normal volunteers and patients with ADHD generally demonstrate improvements in similar task domains. For example, Kempton and colleagues (1999) observed better performance in a self-ordered working memory task (Fig. 13–3) in children with ADHD treated with methylphenidate compared to untreated patients which parallels the improvements found by Elliott and colleagues (1997) in normal volunteers on the same task. Recent data from our research group has demonstrated a similar improvement on the self-ordered working memory task in an adult diagnosed with ADHD (Mehta, MA, Calloway, P, & Sahakian, BJ, in press). Kempton and colleagues (1999) also demonstrated better performance in children with ADHD taking their normal medication (compared to an unmedicated group) on the attentional-set shifting task (Fig. 13–2), in keeping with previous findings of improved cognitive flexibility in children with ADHD after various acute doses (0.3–0.9mg/kg) of stimulant treatment (Tannock et al., 1995; Douglas et al., 1995). Using the same task as Kempton and colleagues (1999), Elliott and colleagues (1997) did not find improved attentional-set shifting performance in normal volunteers. However, using a more difficult version of the same task, Rogers and colleagues (1999) did demonstrate reduced errors at the extra-dimensional shift stage in normal volunteers who had taken methylphenidate compared with placebo administration. Tests of sustained attention are, similarly, improved after methylphenidate treatment in children (e.g., Klorman et al., 1991; de Soneville et al., 1991) and adults (Riordan et al., 1999) with ADHD, although a composite performance measure was used in the study by Riordan et al. (1999). Elliott et al. (1997) also demonstrated that normal adults administered an acute dose of methylphenidate improved on a test of sustained attention. However, future studies of sustained attention must note that task parameters are known to affect performance in ADHD (van de Meere, 1996; Denney & Rapport, Chapter 12) so that direct comparisons with normal volunteers can be made.

On a test of planning (Owen et al., 1990), which—as with the tests of spatial working memory, attentional-set shifting and sustained attention thus far discussed—is known to be sensitive to frontal lobe damage, children with ADHD treated with methylphenidate perform better than untreated patients (Kempton et al., 1999). However, normal adults who had taken methylphenidate were *impaired* on the same task, but only when it was familiar to them, and they also made fewer errors on a similar task with reduced motoric requirements after drug administration (Elliott et al., 1997). This finding highlights that, although the effects of methylphenidate in patients with ADHD and normal volunteers on certain cognitive measures are broadly similar, there might be some important differences that may be related to a distinct neuropathology in ADHD, or an interaction between drug response, and the novelty of the tasks to the subjects and the specific requirements of the task.

Baseline performance of subjects may also be a crucial factor in determining the effects of methylphenidate on cognitive performance. Indeed, Kimberg and colleagues (1997) demonstrated that after an acute dose of the dopamine D2 agonist bromocriptine, performance on various tasks of executive function was enhanced in subjects with low baseline verbal memory capacity and impaired in subjects with higher baseline verbal memory capacity. In our group we have demonstrated a similar interaction between baseline performance and the cognitive effects of methylphenidate administration in normal volunteers (Mehta et al., 2000). Specifically, subjects' baseline memory span correlated with the methylphenidate-induced improved performance on a version of the self-ordered spatial working memory task shown in Fig. 13–3: the subjects with lower memory capacity demonstrated a greater reduction in errors on the spatial working memory task. Similar relationships have not been examined in ADHD.

Can the Cognitive Action of Methylphenidate Be Explained by Its Action on the DA or NE Systems, or Some Combination of Both?

The specific mediation of the cognitive effects of methylphenidate by the separate DA and NE systems was the subject of the second theoretical question posed in the introduction. The cognitive tasks that have been shown to be sensitive to methylphenidate administration are broadly similar to the tests that are influenced by both DA manipulations and NE manipulation. For example, a sustained attention task that was enhanced after methylphenidate administration has been shown to be sensitive to NE manipulations using clonidine (Coull et al., 1995a), but *in*sensitive to dopamine D2 receptor antagonist administration (Mehta et al., 1999). Certain tests of working memory have also been shown to be influenced by methylphenidate administration (e.g., Elliott et al., 1997) with separate proposed contributions from the DA and NE systems: the DA system being involved in the maintenance of information held in working memory and the NE system being involved in the preventing distraction from irrelevant stimuli (possibly by focusing attention). However, the pharmacological mechanisms that mediate the effects of methylphenidate on cognitive performance may differ from the pre- and postsynaptic manipulations of single neurotransmitter systems. Studies that attempt to antagonize the effects of methylphenidate with specific DA or NE receptor agents can assist in deciphering the specific contributions of these systems to the beneficial effects of methylphenidate. Levy and Hobbes (1988, 1996) used such an approach to examine, on a continuous performance task of sustained attention, the effect of preceding methylphenidate administration with the D1/D2 (and nonadrenergic) antagonist haloperidol in children with ADHD. The significant improvement in reaction times produced by methylphenidate on the sustained attention task was blocked by haloperidol (Levy & Hobbes 1988, 1996). In normal volunteer adults, Clark and colleagues (1986a) demonstrated a reversal of the negative effects of droperidol administration using methylphenidate on a dichotic auditory attention task. In summary, the studies discussed in this chapter suggest that dopaminergic and noradrenergic mechanisms both contribute to the cognitive effects of methylphenidate.

Interactions between separate catecholamine systems (e.g., Thierry et al., 1992) may also be important in understanding the relative efficacy of methylphenidate, compared to medications that influence single neurotransmitter systems, in the clinical treatment of ADHD (see Pliszka, Chapter 14) and cognitive performance (Rapport et al., 1993). The recent finding of reduced motor activity in an animal model of ADHD using a serotonin agonist indirectly implies a role for serotonin in the neurochemical pathology of ADHD (e.g., Gainetdinov et al., 1999). It is unclear if alterations in serotonergic activity are responsible for the cognitive effects of methylphenidate administration at clinically effective doses, although possible interactions exist between the dopamine and serotonin systems (Kapur & Remington, 1996) and opposing roles of these systems have been demonstrated on a test of working memory in human volunteers (Luciana et al., 1998).

What Are the Neural Loci of the Effects of Methylphenidate on Cognitive Performance?

The final theoretical question posed in the introduction of this chapter concerned the neural loci of the effects of methylphenidate on cognitive performance. In general, the tasks that are influenced by methylphenidate are sensitive to fronto-striatal damage, although some tasks are also sensitive to damage to, or are known to activate, more posterior areas of the brain. However, only three studies to date have directly assessed the functional loci of the action of stimulant drugs in normal volunteers using neuroimaging (Mattay et al., 1996; Vaidya et al., 1998; Mehta et al., 2000). The studies by Mattay et al. (1996) and Mehta et al. (2000) highlighted the left dorsolateral prefrontal cortex as an important region in the task-related modulation of regional cerebral blood flow by stimulant administration. Of particular relevance to ADHD is the study by Mehta et al. (2000) that examined the effect of methylphenidate on a self-ordered, spatial working memory task that is known to be impaired in children with ADHD (Kempton et al., 1999). The application of this neuroimaging methodology (Mehta et al., 2000) to adults with ADHD, in order to compare mechanisms of stimulant drug action during working memory performance, is currently being planned. The feasibility of such an approach has been demonstrated by Vaidya and colleagues (1998) using a version of the go/no-go response inhibition task, which is also sensitive to ADHD. Vaidya et al. (1998) showed a differential task-related, blood flow response (using fMRI) to methylphenidate administration in normal volunteers compared with children with ADHD. Specifically, in the basal ganglia, methylphenidate attenuated blood flow in normal children but increased blood flow in children with ADHD. However, it should be noted that methylphenidate improved performance in both patients and normal volunteers, and that blood flow was also increased in the frontal lobes in both groups.

Conclusions

Despite advances in psychiatric genetics, the root causes of ADHD are still not understood and stimulants remain the best current therapeutic strategy. The neuropsychological effects of specific noradrenergic and dopaminergic agents have provided valuable insights into the pharmacological systems that may mediate the cognitive response to stimulant drugs such as methylphenidate. In addition, comparisons between normal volunteers and patients with ADHD have demonstrated broadly similar patterns of cognitive response to methylphenidate that are related to its effects on both the DA and NE systems of the brain. A convergence of different methodological approaches, including recent neuroimaging studies, have defined the frontal lobes as an important region in the modulation of the cognitive effects of stimulant drugs. It is recommended that future research in these areas focus on (*1*) the continued definition of the effects of methylphenidate on specific cognitive processes, (*2*) developing an understanding of the mechanisms by which methylphenidate modulates performance, using techniques such as functional neuroimaging and sophisticated task and study designs, and (*3*) an examination of the cognitive psychopharmacology of stimulant treatment in relation to possible subgroups of ADHD (e.g., Zametkin & Rapoport, 1987; Halperin et al., 1993; Pennington & Ozonoff, 1996; Pennington et al., 1993).

We thank Dr. Robert Rogers for discussion. Much of the work described in this chapter was supported by a programme grant from the Wellcome Trust to Professors T. W. Robbins and B. J. Everitt and Drs. B. J. Sahakian and A. C. Roberts. Mitul Mehta was funded by a Medical Research Council studentship.

References

Agid, Y, Cervera, P, Hirch, E, Javoy-Agid F, Lehericy, S, & Raisman, R, et al. (1989). Biochemistry of Parkinson's disease 28 years later: A critical review. *Mov Disorf, 4*(Suppl. 1), S126–S144.

Alexander, GE, DeLong, MR, & Strick, PL (1986). Parallel organization of functionally segregated circuits linking the basal ganglia and cortex. *Ann Rev Neurosci, 9*, 357–381.

Aman, CJ, Roberts, RJ, & Pennington, BF (1998). A neuropsychological examination of the underlying deficit in Attention Deficit Hyperactivity Disorder: Frontal lobe versus right parietal lobe theories. *Dev Psychol, 34*, 956–969.

Arnold, LE, Christopher, J, Huestis, R, & Smeltzer, DJ (1978). Methylphenidate vs dextroamphetamine vs caffeine in minimal brain dysfunction: Controlled comparison by placebo washout design with Bayes' analysis. *Arch Gen Psychiatry, 35*, 463–473.

Arnsten, AF, & Goldman-Rakic, PS (1985). Alpha 2-adrenergic mechanisms in prefrontal cortex associated with cognitive decline in aged non-human primates. *Science, 230*, 1273–1276.

Arnsten, AFT, & Constant, TA (1992). Alpha-2 adrenergic agonists decrease distractibility in aged monkeys performing the delayed response task. *Psychopharmacology, 108*, 159–169.

Arnsten, AFT, Cai, JX, Murphy, BL, & Goldman-Rakic, PS (1994). Dopamine D1 receptor mechanisms in the cognitive performance of young adult aged monkeys. *Psychopharmacology, 116*, 143–151.

Arnsten, AFT, Cai, JX, Steere, JC, & Goldman-Rakic, PS (1995). Dopamine D2 receptor mechanisms contribute to age-related cognitive decline: The effects of quinpirole on memory and motor performance in monkeys. *J Neurosci, 15*, 3429–3439.

Arnsten, AFT, Steere, JC, & Hunt, RD (1996). The contribution of alpha 2-noradrenergic mechanisms to prefrontal cortical cognitive function: potential significance for Attention-Deficit Hyperactivity disorder. *Arch Gen Psychiatry, 53*, 448–455.

Arnsten, AFT (1997). Catecholamine regulation of the prefrontal cortex. *J Psychopharm, 11*, 151–162.

Awh, E, Jonides, J, Smith, EE, Schumacher, EH, Koeppe, RA, & Katz, S (1996). Dissociation of storage and rehearsal in verbal working memory: Evidence from PET. *Psychol Sci, 7*, 25–31.

Baddeley, AD (1986). *Working memory.* Oxford: Clarendon Press.

Baker, SC, Rogers, RD, Owen, AM, Frith, CD, Dolan, RJ, & Frackowiak, RS et al. (1996). Neural systems engaged by planning: A PET study of the Tower of London task. *Neuropsychologia, 34*, 515–526.

Barkley, RA (1990). Attention-deficit hyperactivity disorder: A handbook for diagnosis and treatment. New York: Guilford Press.

Broadbent, DE (1971). *Decision and stress.* London: Academic Press.

Brown, VJ & Robbins, TW (1991). Simple and choice reaction time performance following unilateral striatal dopamine depletion in the rat. *Brain, 114*, 513–525.

Cai, JX & Arnsten, AFT (1997). Dose-dependent effects of the dopamine D1 receptor agonists A77636 or SKF81297 on spatial working memory in aged monkeys. *J Pharmacol Exp Ther, 283*, 183–189.

Camp-Bruno, JA, & Herting, RL (1994). Cognitive effects of milacemide and methylphenidate in healthy young adults. *Psychopharmacology, 115*, 46–52.

Camps, M, Cortes, R, Gueye, B, Probst, A, & Palacios, JM (1989). Dopamine receptors in the human brain: Autoradiographic distribution of D2 sites. *Neuroscience, 28*, 275–290.

Carli, M, Robbins, TW, Evenden, JL, & Everitt, BJ (1983). Effects of lesions to ascending noradrenergic neurons on performance of a 5-choice serial reaction time task: Implications for theories of dorsal noradrenergic bundle function based on selective attention and arousal. *Behav Brain Res, 9*, 361–380.

Chappell, PB, Riddle, MA, Scahill, L, Lynch, KA, Schultz, R, & Arnsten, A, et al. (1995). Guanfacine treatment of comorbid attention-deficit hyperactivity disorder and Tourette's syndrome: Preliminary clinical experience. *J Am Acad Child Adolesc Psychiatry, 34*, 1140–1146.

Charney, DS, & Heninger, GR (1986). Abnormal regulation of adrenergic function in panic disorder. *Am J Psychiatry, 43*, 1042–1054.

Clark, CR, Geffen, GM, & Geffen, LB (1986a). Role of monoamine pathways in the control of attention: Effects of droperidol and methylphenidate in normal adult humans. *Psychopharmacology, 90*, 28–34.

Clark, CR, Geffen, GM, & Geffen, LB (1986b). Role of monoamine pathways in attention and effort: Effects of clonidine and methylphenidate in normal adult humans. *Psychopharmacology, 90*, 35–39.

Clark, CR, Geffen, GM, & Geffen, LB (1987). Catecholamines and attention II: Pharmacological studies in normal volunteers. *Neurosci Biobehav Rev, 11*, 353–364.

Clark, CR, Geffen, GM, & Geffen, LB (1989). Catecholamines and the covert orientation of attention in humans. *Neuropsychologia, 27*, 131–139.

Cole, BJ, & Robbins, TW (1987). Amphetamine impairs the discriminative performance of rats with dorsal noradrenergic bundle lesions on a 5-choice serial reaction time task: new evidence for central dopamine-noradrenergic interactions. *Psychopharmacology, 91*, 458–466.

Cole, BJ, & Robbins, TW (1992). Forebrain norepinephrine: Role in controlled information processing in the rat. *Neuropsychopharmacology, 7*, 129–141.

Collins, P, Roberts, AC, Dias, R, Everitt, BJ, & Robbins, TW (1998). Perseveration and strategy in a novel spatial self-ordered sequencing task for nonhuman primates: Effects of exitotoxic lesions and dopamine depletions of the prefrontal cortex. *J Cog Neurosci, 10*, 332–354.

Corbett, B, & Stanczak, DE (1999). Neuropsychological performance of adults evidencing Attention-Deficit Hyperactivity Disorder. *Arch Clin Neuropsych, 14*, 373–387.

Cortes, R, Gueye, B, Pazos, A, Probst, A, & Palacios, JM (1989). Dopamine receptors in the human brain: Autoradiographic distribution of dopamine D1 sites. *Neuroscience, 28*, 263–273.

Coull, JT, Middleton, HC, Robbins, TW, & Sahakian, BJ (1995a). Clonidine and diazepam have differential effects on tests of attention and learning. *Psychopharmacology, 120*, 322–332.

Coull, J, Middleton, HC, Robbins, TW, Sahakian, BJ (1995b). Contrasting effects of clonidine and diazepam on tests of working memory and planning. *Psychopharmacology, 120,* 311–321.

Coull, JT, Frith, CD, Frackowiack, RSJ, & Grasby, PM (1996a). A fronto-parietal network for rapid visual information processing: A PET study of sustained attention and working memory. *Neuropsychologia, 34,* 1085–1095.

Coull, JT, Sahakian, BJ, & Hodges, JR (1996b). The alpha-2 antagonist idazoxan remediates certain attentional and executive dysfunction in patients with dementia of the frontal type. *Psychopharmacology, 123,* 239–249.

Coull, JT, Frith, CD, Dolan, RJ, Frackowiack, RSJ, & Grasby, PM (1997). The neural correlates of the noradrenergic modulation of human attention, arousal and learning. *Eur J Neurosci, 9,* 589–598.

Daniel, DG, Weinberger, DR, Jones, DW, Zigun, JR, Coppola, R, & Handel, S, et al. (1991). The effect of amphetamine on regional cerebral blood flow during cognitive activation in schizophrenia. *J Neurosci, 11,* 1907–1917.

de Sonneville, LM, Njiokiktjien, C, & Hilhorst, RC (1991). Methylphenidate-induced changes in ADHD information processors. *J Child Adolesc Psychiatry, 32,* 285–295.

D'Esposito, M, & Postle, BR (1999). The dependence of span and delayed-response performance on prefrontal cortex. *Neuropsychologia, 37,* 1303–1315.

Dennis, T, L'Heureux, R, Carter, C, & Scatton, B (1987). Presynaptic alpha-2 adrenoreceptors play a role in effects of idazoxan on cortical noradrenaline release (as measured by in vivo dialysis) in the rat. *J Pharmacol Exp Ther, 24,* 642–649.

Dias, R, Robbins, TW, & Roberts, AC (1996). Dissociation in prefrontal cortex of affective and attentional shifts. *Nature, 380,* 68–72.

Douglas, VI, Barr, RG, Desilets, J, & Sherman, E (1995). Do stimulants impair flexible thinking in attention-deficit hyperactivity disorder? *J Am Acad Child Adolesc Psychiatry, 34,* 877–885.

Dunnett, SB, & Robbins, TW (1992). The functional role of mesotelencephalic dopamine systems. *Biol Rev Camb Philos Soc, 67,* 491–518.

Durstewitz, D, Kelc, M, & Güntürkün, O (1999). A neurocomputational theory of the dopaminergic modulation of working memory functions. *J Neurosci, 19,* 2807–2822.

Elliott, R, Sahakian, BJ, Matthews, K, Bannerjea, A, Rimmer, J, & Robbins, TW (1997). Effects of methylphenidate on spatial working memory and planning in healthy young adults. *Psychopharmacology, 131,* 196–206.

Fagan, D, Swift, CG, & Tiplady, B (1988). Effects of caffeine on vigilance and other performance tests in normal volunteers. *J Psychopharm, 2,* 19–25.

Foote, SL, Bloom, FE, & Aston-Jones, G (1983). Nucleus locus ceruleus: New evidence of anatomical and physiological specificity. *Physiol Rev, 63,* 844–914.

Frith, CD, Dowdy, J, Ferrier, IN, & Crow, TJ (1985). Selective impairment of paired associated learning after administration of a centrally-acting adrenergic agonist (clonidine). *Psychopharmacology, 87,* 490–493.

Gainetdinov, RR, Wetsel, WC, Jones, SR, Levin, ED, Jaber, M, & Caron, MC (1999). Role of serotonin in the paradoxical calming effect of psychostimulants on hyperactivity. *Science, 283,* 397–401.

Goldman-Rakic, PS (1998). The cortical dopamine system: role in memory and cognition. *Adv Pharmacol, 42,* 707–711.

Granon, S, Passetti, F, Thomas, KL, Dalley, JW, Everitt, BJ, & Robbins, TW (1999). Enhanced and impaired attentional performance following infusion of D1 dopaminergic receptor agents into rat prefrontal cortex. *J Neurosci, 20,* 1208–1215.

Grant, DA, & Berg, EA (1948). A behavioural analysis of degree of reinforcement and ease of shifting to new responses in a Weigl-type card-sorting problem. *J Exp Psychol, 38*, 404–411.

Grasby, PM, Friston, KJ, Bench, CJ, Frith, CD, Paulesu, E, & Cowen, PJ, et al. (1992). The effect of apomorphine and buspirone on regional cerebral blood flow during the performance of a cognitive task—measuring neuromodulatory effects of psychotropic drugs in man. *Eur J Neurosci, 4*, 1203–1212.

Grasby, P, Malizia, A, & Bench, C (1996). Psychopharmacology—*in vivo* neurochemistry and pharmacology. *Biol Psychiatry, 52*, 513–526.

Halperin, JM, Newcorn, JH, Matier, K, Sharma, V, McKay, KE, & Schwartz, S (1993). Discriminant validity of attention-deficit hyperactivity disorder. *J Am Acad Child Adolesc Psychiatry, 32*, 1038–1043.

Hauty, GT & Payne, RB (1955). Mitigation of work decrement. *J Exp Psychol, 49*, 60–67.

Hockey, R (1979). Stress and the cognitive components of skilled performance. In V Hamilton & D Warburton (Eds.), *Human stress and cognition.* Chichester: Wiley.

Jäkälä, P, Sirviö, J, Riekkenen, M, Koivisto, E, Kejonen, K, & Vanhanen, M, et al. (1999a). Guanfacine and clonidine, alpha-2-agonists, improve paired associates learning, but not delayed matching to sample, in humans. *Neuropsychopharmacology, 20*, 119–130.

Jäkälä, P, Riekkinen, M, Sirviö, J, Koivisto, E, Kejonen, K, & Vanhanen, M, et al. (1999b). Guanfacine, but not clonidine, improves planning and working memory performance in humans. *Neuropsychopharmacology, 20*, 460–470.

Jonides, J, Smith, EE, Koeppe, RA, Awh, E, Minoshima, S, & Mintun, MA (1993). Spatial working memory in humans as revealed by PET. *Nature, 363*, 623–625.

Kapur, S, & Remington, G (1996). Serotonin-dopamine interaction and its relevance to schizophrenia. *Am J Psychiatry, 153*, 466–476.

Kebabian, JW, & Calne, DB (1979). Multiple receptors for dopamine. *Nature, 277*, 93–99.

Kempton, S, Vance, A, Maruff, P, Luk, E, Costin, J, & Pantelis, C (1999). Executive function and attention deficit hyperactivity disorder: Stimulant medication and better executive function performance in children. *Psych Med, 29*, 527–538.

Kimberg, DY, D'Esposito, M, & Farah, MJ (1997). Effects of bromocriptine on human subjects depends on working memory capacity. *Neuroreport, 8*, 3581–3585.

Klorman, R, Brumaghim, JT, Fitzpatrick, PA, & Borgstedt, AD (1991). Methylphenidate speeds evaluation processes of attention deficit disorder adolescents during a continuous performance test. *J Abnorm Child Psychol, 19*, 263–283.

Koelega, HS (1993). Stimulant drugs and vigilance performance: A review. *Psychopharmacology, 111*, 1–16.

Kulisevsky, J, Avila, A, Barbanoj, M, Antonijoan, R, Berthier, ML, & Gironell, A (1996). Acute effects of levodopa on neuropsychological performance in stable and fluctuating Parkinson's disease patients at different levodopa plasma levels. *Brain, 119*, 2121–2132.

Lange, KW, Robbins, TW, Marsden, CD, James, M, Owen, AM, & Paul, GM (1992). L-Dopa withdrawal in Parkinson's disease selectively impairs cognitive performance in tests of frontal lobe function. *Psychopharmacology, 107*, 394–404.

Lawrence, AD, Sahakian, BJ, Hodges, JR, Rosser, AE, Lange, KW, & Robbins, TW (1996). Executive and mnemonic functions in early Huntington's disease. *Brain, 119*, 1633–1645.

Lees, AJ, & Smith, E (1983). Cognitive deficits in the early stages of Parkinson's disease. *Brain, 106*, 257–270.

Levy, F, & Hobbes, G (1988). The action of stimulant medication in attention deficit disorder with hyperactivity: Dopaminergic, noradrenergic, or both? *J Am Acad Child Adolesc Psychiatry, 27*, 802–805.

Levy, F, & Hobbes, G (1996). Does haloperidol block methylphenidate? Motivation or attention? *Psychopharmacology, 126*, 70–74.

Li, B-M, & Mei, Z-T (1994). Prefrontal injection of the alpha-2-adrenergic receptor antagonist yohimbine impairs performance of a delayed go/no-go task in monkeys. *Acta Physiologica Sinica, 46*, 249–257.

Luciana, M, Depue, RA, Arbisi, P, & Leon, A (1992). Facilitation of working memory in humans by a D2 dopamine receptor agonist. *J Cog Neurosci, 4*, 58–67.

Luciana, M, & Collins, PF (1997). Dopaminergic modulation of working memory for spatial but not object cues in normal volunteers. *J Cog Neurosci, 9*, 330–347.

Luciana, M, Collins, PF, & Depue, RA (1998). Opposing roles for dopamine and serotonin in the modulation of human spatial working memory functions. *Cereb Cortex, 8*, 218–226.

Lueck, CJ, Tanyeri, S, Crawford, TJ, Henderson, L, & Kennard, C (1990). Antisaccades and remembered saccades in Parkinson's disease. *J Neurol Neurosur Psychiatry, 53*, 284–288.

Lyon, M, & Robbins, TW (1975). The action of central nervous system stimulant drugs: A general theory concerning amphetamine effects. In W Essman & L Valzelli (Eds.) *Current developments in psychopharmacology*, vol. 2. New York: Spectrum-Publications.

Marsden, CA & Martin, KF (1986). Involvement of 5-HT1A- and alpha 2-receptors in the decreased 5-hydroxytryptamine release and metabolism in rat suprachiasmatic nucleus after intravenous 8-hydroxy-2-(n-dipropylamino) tetralin. *Brit J Pharmacol, 89*, 277–286.

Mattay, VS, Berman, KF, Ostrem, JL, Esposito, G, Van Horn, JD, & Bigelow, LB, et al. (1996). Dextroamphetamine enhances "neural network-specific" physiological signals: a positron-emission tomography rCBF study. *J Neurosci, 16*, 4816–4822.

McDowell, S, Whyte, J, & D'Esposito, M (1998). Differential effect of a dopaminergic agonist on prefrontal function in traumatic brain injury patients. *Brain, 121*, 1155–1164.

Mehta, MA, Swainson, R, Sahakian, BJ, Ogilvie, AD, & Robbins, TW (1998). The dopamine D2 receptor agonist bromocriptine enhances spatial memory span in human volunteers. *J Psychopharm, 12*, Supp. A. 171.

Mehta, MA, Sahakian, BJ, McKenna, PJ, & Robbins, TW (1999). Systemic sulpiride in young healthy volunteers simulates the profile of cognitive deficits in Parkinson's disease. *Psychopharmacology, 146*, 162–174.

Mehta, MA, Owen, AM, Sahakian, BJ, Mavaddat, N, Pickard, JD & Robbins, TW (2000). Methylphenidate enhances working memory by modulating discrete & frontal and parietal lobe regions in the human brain. *J Neurosci, 20,* RC65 (1–6).

Middleton, HC, Sharma, A, Agouzoul, D, Sahakian, BJ, Robbins, TW (1999). Idazoxan potentiates rather than antagonizes some of the cognitive effects of clonidine. *Psychopharmacology, 145*, 401–411.

Milner, B (1971). Inter-hemispheric differences in the localisation of psychological processes in man. *Br Med Bull, 27*, 272–277.

Morris, RG, Downes, JJ, Sahakian, BJ, Evenden, JL, Heald, A, & Robbins, TW (1988). Planning and spatial working memory in Parkinson's disease. *J Neurol Neurosur Psychiatry, 51*, 757–766.

Müller, U, Yves von Cramon, D, & Pollman, S (1998). D1- versus D2-receptor modulation of visuospatial working memory in humans. *J Neurosci, 18*, 2720–2728.

Owen, AM, Downes, JJ, Sahakian, BJ, Polkey, CE, & Robbins, TW (1990). Planning and spatial working memory following frontal lobe lesions in man. *Neuropsychologia, 28*, 1021–1034.

Owen, AM, Roberts, AC, Polkey, CE, Sahakian, BJ, & Robbins, TW (1991). Extra-dimensional versus intra-dimensional set shifting performance following frontal lobe excisions, temporal lobe excisions or amygdalo-hippocampectomy in man. *Neuropsychologia, 29*, 993–1006.

Owen, AM, James, M, Leigh, PN, Summers, BA, Marsden, CD, & Quinn, NP, et al. (1992). Fronto-striatal cognitive deficits at different stages of Parkinson's disease. *Brain, 115*, 1727–1751.

Owen, AM, Beksinska, M, James, M, Leigh, PN, Summers, BA, & Marsden, CD, et al. (1993). Visuo-spatial memory deficits at different stages of Parkinson's disease. *Neuropsychologia, 31*, 627–644.

Owen, AM, Sahakian, BJ, Hodges, JR, Polkey, CE, Summers, BA, & Robbins, TW (1995a). Dopamine-dependent fronto-striatal planning deficits in early Parkinson's disease *Neuropsychology, 9*, 126–140

Owen, AM, Sahakian, BJ, Semple, J, Polkey, CE, & Robbins, TW (1995b). Visuo-spatial short term recognition memory and learning after temporal lobe excisions, frontal lobe excisions or amygdalo-hippocampectomy in man. *Neuropsychologia, 33*, 1–24.

Owen, AM, Evans, AC, & Petrides, M (1996). Evidence for a two-stage model of spatial working memory processing within lateral frontal cortex: A positron emission tomography study. *Cereb Cortex, 6,* 31–38.

Owen, AM, Iddon, JL, Hodges, JR, Summers, BA, & Robbins, TW (1997). Spatial and non-spatial working memory at different stages of Parkinson's disease. *Neuropsychologia, 35*, 519–532.

Payne, RB, & Hauty, GT (1954). The effects of experimentally induced attitudes on task proficiency. *J Exp Psychol, 47,* 267–273.

Payne, RB, & Hauty, GT (1955). Factors affecting the endurance of psychomotor skill. *J Aviat Med, 25*, 382–389.

Pennington, BF, Groisser, D, & Welsh, MC (1993). Contrasting cognitive deficits in attention deficit hyperactivity disorder versus reading disability. *Dev Psychol, 29*, 511–523.

Pliszka, SR, McCracken, JT, & Maas, JW (1996). Catecholamines in attention-deficit hyperactivity disorder: Current perspectives. *J Am Acad Child Adolesc Psychiatry, 35*, 264–272.

Pennington, BF, & Ozonoff, S (1996). Executive functions and developmental psychopathology. *J Child Psychol Psychiatry, 17*, 51–87.

Posner, M, Walker, JA, Friedrich, FJ, & Rafal, RD (1984). Effects of parietal injury on the covert orienting of visual attention. *J Neurosci, 4*, 1863–1874.

Posner, MI, Inhoff, AW, Friedrich, FJ, & Cohen, A (1987). Isolating attentional systems: a cognitive-anatomical analysis. *Psychobiology, 15*, 107–121.

Rapport, MD, Carlson, GA, Kelly, KL, & Pataki, C (1993). Methylphenidate and desipramine in hospitalized children: I. Separate and combined effects on cognitive function. *J Am Acad Child Adolesc Psychiatry, 32*, 333–342.

Riordan, HJ, Flashman, LA, Saykin, AJ, Frutiger, SA, Carroll, KE, & Huey, L (1999). Nueropsychological correlates of methylphenidate treatment in adult ADHD with and without depression. *Arch Clin Neuropsychol, 14*, 217–233.

Robbins, TW (1984). Cortical noradrenaline, attention and arousal. *Psych Med, 14*, 13–21.

Robbins, TW (1996). Dissociating executive functions of the prefrontal cortex. *Phil Trans R Soc Lond B Biol Sci, 351*, 1463–1471.

Robbins, TW (1997). Arousal systems and attentional processes. *Biol Psychol, 45,* 57–71.

Robbins, TW (1998). Homology in behavioural pharmacology: An approach to animal models of human cognition. *Behav Pharmacol, 9*, 509–519.

Robbins, TW & Everitt, BJ (1987). Psychopharmacological studies of arousal and attention. In SM Stahl, SD Iversen, & EC Gilman (eds.) *Cognitive neurochemistry.* Oxford: Oxford University Press.

Robbins, TW, James, M, Owen, AM, Sahakian, BJ, Lawrence, AD, & McInnes, L, et al. (1998a). A study of performance on tests from the CANTAB battery sensitive to frontal lobe dysfunction in a large sample of normal volunteers: Implications for theories of executive functioning and cognitive aging. *J Int Neuropsych Soc, 4*, 474–490.

Robbins, TW, James, M, Owen, AM, Sahakian, BJ, McInnes, L, & Rabbitt, P (1994). Cambridge Neuropsychological Test Automated Battery (CANTAB): A factor analytical study of a large sample of normal elderly volunteers. *Dementia, 5*, 266–281.

Robbins, TW, Owen, AM, & Sahakian, BJ (1998b). The neuropsychology of basal ganglia disorders: An integrative cognitive and comparative approach. In MA Ron & AS David (Eds.) *Disorders of brain and mind*. Cambridge: Cambridge University Press.

Robbins, TW & Sahakian, BJ (1979). 'Paradoxical' effects of psychomotor stimulant drugs in hyperactive children from the standpoint of behavioural pharmacology. *Neuropharmacology, 18*, 931–950.

Roberts, DCS, Price, MT, & Fibiger, HC. (1976). The dorsal noradrenergic projection: An analysis of its role in maze learning. *J Comp Physiol Psychol, 90*, 363–372.

Roberts, AC, De Salvia, MA, Wilkinson, LS, Collins, P, Muir, JL, Everitt, BJ, & Robbins, TW (1994). 6-Hydroxydopamine lesions of the prefrontal cortex in monkeys enhance performance on an analog of the Wisconsin card sort test: Possible interactions with subcortical dopamine. *J Neurosci, 14*, 2531–2544.

Rogers, RD, Blackshaw, AJ, Middleton, HC, Matthews, K, Deakin, JFW, & Sahakian, BJ, et al. (1999). Tryptophan depletion impairs stimulus-reward learning while methylphenidate disrupts attention control in healthy young adults: Implications for the monoaminergic basis of impulsive behaviour. *Psychopharmacology, 146*, 482–491.

Rogers RD, Andrews TC, Grasby, PM, Brooks D, & Robbins, TW (2000). Contrasting cortical and sub-cortical PET activations produced by reversal learning and attentional-set shifting in humans. *J Cog Neurosci, 12*, 142–162.

Rosen, AC, Rao, SM, Caffarra, P, Scaglioni, A, Bobholz, JA, & Woodley, SJ et al. (1999). Neural basis of endogenous and exogenous spatial orienting: A functional MRI study. *J Cog Neurosci, 11*, 135–152.

Sahakian, BJ, Jones, GMM, Levy, R, Gray, JA, & Warburton, DM (1989). The effects of nicotine on attention, information processing and short-term memory in patients with dementia of the Alzheimer type. *Br J Psychiatry, 154*, 797–800.

Sahakian, BJ, & Owen, AM (1992). Computerised assessment in neuropsychiatry using CANTAB: Discussion paper. *J R Soc Med, 85*, 399–402.

Sahakian, BJ & Robbins, TW (1977). Are the effects of psychomotor stimulant drugs on hyperactive children really paradoxical? *Med Hypotheses, 3*, 154–158.

Sahakian, BJ, Sarna, GS, Kantamaneni, BD, Jackson, A, Hutson, PH, & Curzon, G (1985). Association between learning and cortical catecholamines in non-drug-treated rats. *Psychopharmacology, 86*, 339–343.

Sawaguchi, T & Goldman-Rakic, PS (1991). D1 dopamine receptors in prefrontal cortex: Involvement in working memory. *Science, 251*, 947–950.

Seamans, JK, Floresco, SB, & Phillips, AG (1998). D1 receptor modulation of hippocampal-prefrontal cortical circuits integrating spatial memory with executive functions in the rat. *J Neurosci, 18*, 1613–1621.

Serven-Schreiber, D, Bruno, RM, Carter, CS, & Cohen, JD (1998a). Dopamine and the mechanisms of cognition: Part I. A neural network model predicting dopamine effects on selective attention. *Biol Psychiatry, 43*, 713–722.

Servan-Schreiber, D, Carter, CS, Bruno, RM, & Cohen, JD (1998b). Dopamine and the mechanisms of cognition: Part II. D-Amphetamine effects in human subjects performing a selective attention task. *Biol Psychiatry, 43*, 723–729.

Shue, KL, & Douglas, VI (1992). Attention deficit hyperactivity disorder and the frontal lobe syndrome. *Brain Cog, 20*, 104–124.

Smith, EE, & Jonides, J (1998). Neuroimaging analysis of human working memory. *Proc Natl Acad Sci USA, 95*, 12061–12068.

Solanto, MV (1998). Neuropsychopharmacological mechanisms of stimulant drug action in attention-deficit hyperactivity disorder: A review and synthesis. *Behav Brain Res, 94*, 127–152.

Swainson, R, Rogers, RD, Sahakian, BJ, Summers, BA, Polkey, CE & Robbins, TW (1999). Probabilistic learning and reversal deficits in patients with Parkinson's disease or frontal or temporal lobe lesions: possible adverse effects of dopaminergic medication. *Neuropsychologia, 38*, 596–612.

Swanson, JM, Posner, M, Potkin, S, Bonforte, S, Youpa, D, & Fiore, C et al. (1991). Activating tasks for the study of visual-spatial attention in ADHD children: A cognitive anatomic approach. *J Child Neurol, 6*, S119–S127.

Tannock, R, Schacher, R, & Logan, G (1995). Methylphenidate and cognitive flexibility: Dissociated dose effects in hyperactive children. *J Abnorm Child Psychol, 23*, 235–266.

Thierry, AM, Mantz, J, & Glowinski, J (1992). Influence of dopaminergic and noradrenergic afferents on their target cells in rat medial prefrontal cortex. *Adv Neurol, 57*, 545–554.

Vaidya, CJ, Austin, G, Kirkorian, G, Ridlehuber, HW, Desmond, JE, & Glover, GH et al. (1998). Selective effects of methylphenidate in attention deficit hyperactivity disorder: A function magnetic resonance study. *Proc Natl Acad Sci U S A, 95*, 14494–14499.

Van de Meere, J (1996) The role of attention. In: Sandberg S (Ed.) *Hyperactivity disorders of childhood.* Cambridge: Cambridge University Press.

Weiss, B, & Laties, VG (1962). Enhancement of human performance by caffeine and the amphetamines. *Pharmacol Rev 14*, 1–36.

Wesnes, K, & Warburton, DM (1984). Effects of scopolamine and nicotine on human rapid visual information processing performance. *Psychopharmacology, 82*, 147–150.

Witte, EA, & Marrocco, RT (1997). Alteration of brain noradrenergic activity in rhesus monkeys affects the alerting component of covert orienting. *Psychopharmacology, 132*, 315–323.

Wood, C, Maruff, P, Levy, F, Farrow, M, & Hay, D (1999). Covert orienting of visual spatial attention in Attention Deficit Hyperactivity Disorder: Does comorbidity make a difference? *Arch Clin Neuropsychol, 14*, 179–189.

Yerkes, RM, & Dodson, JP (1908). The relation of strength of stimulus to rapidity of habit formation. *J Comp Neurol Psychol, 19*, 459–482.

Zahrt, J, Taylor, JR, Mathew, RG, & Arnsten, AFT (1997). Supranormal stimulation of D_1 dopamine receptors in the rodent prefrontal cortex impairs spatial working memory performance. *J. Neurosci, 17*, 8528–8535.

Zametkin, AJ & Rapoport, JL (1987). Neurobiology of attention deficit disorder with hyperactivity: Where have we come in 50 years? *J Am Acad Child Adolesc Psychiatry, 26*, 676–686.

Zarahn, E, D'Esposito, M, Aguirre, GK, Armstrong, M, Lease, J, & Kimberg, DY (1998). Cortical effects of bromocriptine, a D-2 dopamine receptor agonist, on working memory and prefrontal function revealed by fMRI. *Soc Neurosci Abstracts*, 494.2.

14

Comparing the Effects of Stimulant and Non-Stimulant Agents on Catecholamine Function: Implications for Theories of ADHD

STEVEN R. PLISZKA

The catecholamines norepinephrine (NE) and dopamine (DA) have been hypothesized to be involved in a number of psychiatric disorders, primarily because a wide range of psychotropic medications affect these neurotransmitters. Previous chapters have reviewed the anatomy and physiology of these pathways, as well as their extensive roles in cognition, motor behavior, and mood. Effects of psychotropic medications on catecholamine systems may provide insight into the role catecholamines play in the pathophysiology of Attention-Deficit/Hyperactivity Disorder (ADHD) and the mechanism of stimulant treatment. This chapter will review the effects of the stimulant medications on NE, DA, and their metabolites in humans. It will then proceed to examine the effects of several major classes of psychotropic agents on the catecholamines: tricyclic antidepressants, novel noradrenergic antidepressants, dopamine agonists, neuroleptics, alpha-agonists, and serotonergic (5-HT) agents. The 5-HT agents are of interest because of increasing information about extensive interaction between the 5-HT and catecholamine systems (Kapur & Remington, 1996), as well as the recent finding that 5-HT agonists may reduce motor activity in an animal model of ADHD (Gainetdinov et al., 1999).

Brief Overview of Stimulant Effects on Catecholamines

In humans, catecholamines may be measured in either cerebrospinal fluid (CSF), plasma, or urine. The relationship of peripheral (urine, plasma) to central catecholamine measures has been reviewed elsewhere (Pliszka et al. 1996). Briefly, the locus coeruleus (LC) receives excitatory input from the paragigantocellularis (PGi), which also innervates the interiomediolateral (IML) cell column of the spinal cord. The IML is the source of the preganglionic sympathetic neurons. The

PGi receives a complex input from the nucleus tractus solitarius, which conveys information about the vegetative state of the body (blood pressure, volume, blood pH, etc.) and multiple sensory inputs from the spinal cord and brain stem sensory nuclei (Aston-Jones et al., 1990). It is by these routes that LC responds to sensory events. The PGi not only has connections with the LC but strongly activates the peripheral sympathetic nervous system through excitatory input into the IML (Aston-Jones et al., 1991). Thus various types of stimuli will simultaneously activate the LC and sympathetic nervous system (Maas & Leckman, 1983; Elam et al., 1986; Elam et al., 1984). Aston-Jones and colleagues (1991) summarized the relationship of these two systems as follows:

> the activation of the peripheral sympathetic nervous system prepares the animal physically for adaptive phasic responses to urgent stimuli, while parallel activation of LC increases vigilance and attentiveness, preparing the animal cognitively for adaptive responsiveness to such stimuli. It is proposed that the LC serves as the cognitive limb of the global sympathetic nervous system, and the cognitive and peripheral sympathetic responses are integrated and coordinated through PGi. (p. 516)

NE, after it is released, undergoes extraneuronal metabolism via catechol-O-methyl transferase (COMT) into normetanephrine (NM). When NE is taken back up into the neuron, monoamine oxidase (MAO) plays a more predominant role and formation of 3-methoxy, 5 hydroxyphenylglycol (MHPG) is more likely. Thus, higher amounts of NM reflect the fact that more release of NE has occurred, whereas higher amounts of MHPG reflect a lower rate of NE release, as most of the NE remains within the neuron to be metabolized by MAO. NE is also metabolized by aldehyde dehydrogenase to vanillylmandelic acid (VMA) , as shown in Figure 14–1. An analogous pattern is found for epinephrine; COMT metabolizes EPI to metanephrine (MN).

Figure 14–1 also shows how the ratios of these metabolites to each other can be used to make inferences regarding the activity of the NE system. In the upper panel of the figure, the action of NE is terminated by reuptake (which leads to metabolism to MHPG and VMA), while that remaining in the cleft is metabolized predominately to NM and VMA. Thus, the ratios of NE to MHPG and NM are both 1.0. In a disease state with abnormally high release, the reuptake mechanism is saturated and more NE remains in the synaptic cleft where it is metabolized to NM. Thus, the NE/NM ratio falls while the NE/MHPG ratio rises (i.e., proportionately less intraneuronal metabolism). When the reuptake site is blocked, a similar situation emerges, except that little NE can enter the neuron and MHPG production falls. The NE/MHPG ratio rises even more than in the high-release disease state.

Dopamine is metabolized via both COMT and MAO into its principal metabolite, homovanillic acid (HVA). Only about half of the HVA in plasma is derived from the brain; the other half is principally derived from the sympathetic nervous system and the adrenal medulla. It is not clear that there is a strong functional link between the central and peripheral DA systems. Maas et al. (1985) found only a non-significant ($r = .39$) correlation between plasma and CSF HVA in seven human subjects until they blocked the production of peripheral HVA by administering debrisoquin, an MAO inhibitor that does not cross the blood-brain barrier.

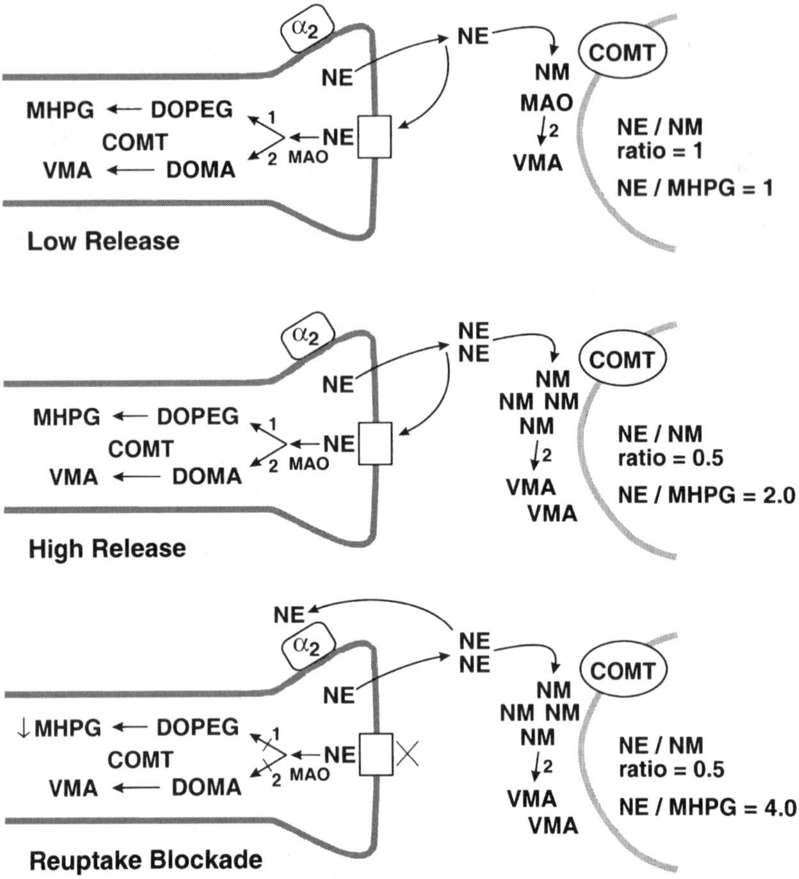

Figure 14–1. Norepinephrine and its metabolism. NE—norepinephrine; NM—Normetanephrine; DOPEG—3, 4-dihydroxyphenylglycol; DOMA—3,4 dihydroxymandelic acid; MHPG—3-methoxy, 4-hydroxy phenylglycol; VMA—vanillylmandelic acid; COMT—catechol-O-methyl transferase; 1—Aldehyde reductase; 2—Aldehyde dehydrogenase.

Debrisoquin lowered plasma, but not CSF HVA. After debrisoquin treatment there was a robust correlation between CSF and plasma HVA ($r = .95$). Without debrisoquin pretreatment (which was never done in pediatric studies) the relationship between plasma/urinary and central dopamine functioning remains unclear.

Methylphenidate (MPH) blocks the reuptake of both NE and DA at their respective transporters. Amphetamine not only blocks the reuptake of these neurotransmitters but can enhance their release even in the absence of a neuronal impulse (Seeman & Madras, 1998) (see Grace, Chapter 5). In the case of DA, stimulants raise the baseline level of extracellular dopamine, which activates presynaptic autoreceptors. This has the effect of lowering the relative rise in the pulsatile release of dopamine with each neuronal impulse.

The results of studies of the effects of psychotropic medications on urinary catecholamines are shown in Table 14–1. Arrows indicate a statistically significant increase or decrease in that metabolite on active drug relative to placebo, while a dash indicates that the metabolite was not assessed during that study. Acute administration of dextroamphetamine leads to increases in plasma NE in ADHD children (Mikkelsen et al., 1981). However, when dextroamphetamine is consistently administered over several weeks, it decreases urinary MHPG, and such decreases in this metabolite correlate with improvement in teacher ratings. Dextroamphetamine blocks the reuptake of NE and the extra NE in the synaptic cleft stimulates alpha-2 autoreceptors (Zametkin et al., 1985). This may lead to long-term decreases in LC firing. This effect of dextroamphetamine on MHPG may not be necessary or sufficient for a therapeutic effect in ADHD, however, as methylphenidate (MPH) does not lower MHPG (Zametkin et al., 1985; Zametkin & Hamburger, 1988; Elia et al., 1990). NM declined after amphetamine treatment in two studies (Shekim, Dekirmenjian, & Chapel, 1977; Zametkin, et al., 1985b), while methylphenidate increased NM (Zametkin et al., 1985). Because dextroamphetamine increases the release of NE as well as blocking its reuptake, the sharp decline in MHPG is most likely due to decreased intraneuronal metabolism via MAO. Because NM levels were also decreased; this also suggests that the chronic effects of amphetamine are to reduce NE output. In contrast, methylphenidate increased NE and NM in only one study (Zametkin et al., 1985), while all other studies failed to show any significant effects of methylphenidate on NE or its metabolites.

Table 14–1. Urinary Catecholamines Before and After Treatment with Psychotropic Agents in ADHD Children

Investigator	Drug	NE	EPI	NMN	MN	VMA	MHPG	HVA
Shekim et al., 1977	AMP	—	—	↓	↓	—	↓	ns
Shekim, et al., 1983	AMP	—	—	—	—	—	↓	↑
Brown, et al., 1981	AMP	—	—	—	—	—	↓	ns
Zametkin et al., 1985	AMP	ns	—	ns	—	ns	↓	ns
Zametkin et al., 1985b	AMP	↓	—	↓	—	↓	↓	ns
Donnelly et al., 1989	AMP	↓	↑	—	—	↓	↓	ns
Elia et al., 1990	AMP	—	↑	—	↑	↓	↓	ns
Donnelly et al., 1989	FEN	↓	↓	—	—	↓	↓	ns
Zametkin et al., 1985	METH	↑	—	↑	—	ns	ns	ns
Zametkin et al., 1988	METH	ns	—	ns	—	ns	ns	—
Elia et al., 1990	METH	ns	↑	ns	↑	ns	ns	ns
McCracken et al., 1990	METH	ns	↑	—	—	—	—	ns
Zametkin et al., 1985b	MAOI	ns	—	↑	—	↓	↓	↓
Donnelly et al., 1986	DMI	↓	ns	—	—	↓	↓	—

↓ = metabolite significant decreased post treatment; ↑ = metabolite significantly increased post-treatment; METH = methylphenidate; AMP = dextroamphetamine; MAOI = monoamine oxidase inhibitor; DMI = desipramine, FEN = fenfluramine.

Although drugs that are effective in ADHD do not seem to have any consistent effect on urinary NE, three studies (Donnelly et al., 1989; Elia et al., 1990; McCracken et al., 1990) have shown that stimulants increase urinary EPI. Indeed, increasing urinary EPI is the only biochemical effect that MPH and dextroamphetamine share (Elia et al., 1990). Mefford and Potter (1989) proposed that central EPI may be deficient in ADHD. Consistent with this hypothesis, urinary EPI has been found to be decreased in ADHD boys relative to controls (Hanna et al., 1996a); lower EPI in urine correlated with increased ratings of restlessness and inattention (Hanna, et al., 1996b). McCracken (1991) has hypothesized that stimulant-induced increases in EPI may be part of the mechanism of drug response in ADHD. McCracken (McCracken, 1991) suggested two possible mechanisms. First, the stimulant may increase, through reuptake blockade, the availability of EPI at central EPI synapses. Central EPI neurons might inhibit the LC leading to decreased baseline firing of the LC. Alternatively, peripheral effects of EPI in the bloodstream (increased pulse, blood pressure, etc.) are transmitted to the LC via sensory afferents.

As noted in Table 14–1, stimulant drugs have few effects on urinary HVA, and studies of the dopamine system have focused on the CSF levels of HVA. Unfortunately, CSF is nearly impossible to obtain in normal controls due to ethical and logistical issues. Riddle et al. (1986) collected CSF from 50 children who had treatment for acute lymphoblastic leukemia but were in remission and without evidence of neurological disease. Levels of HVA declined with age, and no effect of gender was noted. Another metabolite of DA—2,4, dihydroxyphenylacetic acid—-also declines from infancy to childhood (Anderson et al., 1988).

Shaywitz et al. (1977) found decreased CSF HVA in a small group of boys with minimal brain dysfunction relative to controls, but only after probenecid treatment. Levels of probenecid differed between the groups, however. A small study ($n = 10$) assessed CSF HVA before and after dextroamphetamine treatment and showed a significant decline in this metabolite post treatment (Shetty & Chase, 1976). Castellanos and colleagues (1996) measured CSF HVA in 45 boys with ADHD prior to a stimulant trial. All the patients underwent a double-blind placebo-controlled crossover trial of dextroamphetamine and methylphenidate, 16 of the subjects received a subsequent open trial of pemoline. CSF HVA correlated positively with baseline teacher and parent ratings of hyperactivity, both in the 29 subjects who had been part of an earlier study (Castellanos et al., 1994) and in the subset of additional subjects ($n = 20$) recruited for this study. In the sample of 45 who underwent the drug trial, higher baseline CSF HVA predicted better response of ADHD symptoms to all three stimulants; lower HVA was associated with worsening of some symptoms during stimulant treatment.

There was no relationship of baseline CSF MHPG to treatment response, although there were methodological difficulties with the MHPG assays. The study clearly showed a role for the dopaminergic system in the pathophysiology of ADHD, but suggested that stimulants worked by down-regulating the DA system, possibly through presynaptic autoreceptor stimulation. Higher CSF HVA correlating positively with severity of ADHD symptoms is also consistent with theories of an excessively active DA system in producing impulsive, reward-driven behavior in these children (Solanto, 1998). However, as Taylor and Jentsch point out in Chapter 4, blocking DA reuptake is expected to decrease CSF HVA by shifting dopamine metabolism into the extracellular compartment. Thus, results from CSF studies in children are fundamentally ambiguous. They are no longer continuing because of their limitations, and also because of ethical constraints that make it impossible to enroll normal controls.

Monoamine Oxidase Inhibitors

Zametkin et al. (1985a) treated 14 ADHD boys with MAO inhibitors (MAOI). Six received clorgyline, a MAO-A selective inhibitor. MAO-A has EPI, NE, DA, and 5-HT as its substrates and is found mainly in DA and NE neurons. The remaining eight boys received tranylcypromine, a nonselective MAO A+B inhibitor. MAOB is present in 5-HT neurons; DA and 5-HT are its main metabolites, but not NE. Both MAOI were equally efficacious relative to dextroamphetamine in improving the symptoms of ADHD. Similar to stimulants, the MAOIs decreased the excretion of NE and its metabolites, particularly MHPG and VMA, although MAOI increased NM while dextroamphetamine decreased this particular metabolite. Clorgyline and tranylcypromine are the only nonstimulant medications ever shown to have efficacy equal to the stimulants. Subsequently, Donnelly (as reported by Rapoport et al. (1985)) found deprenyl (a selective MAO-B inhibitor) to be minimally effective in the treatment of ADHD. Because a MAO-B inhibitor would block the metabolism of DA and 5-HT but not NE, such results would suggest a strong role for both the NE and DA systems in the pathophysiology of ADHD, given that the MAO-A inhibitors are effective.

Tricyclic Antidepressants

Since the early 1970s, tricyclic antidepressants (TCAs) have been used as second-line drugs for the treatment of ADHD after at least two different stimulants have failed. Clomipramine (CLO), imipramine (IMI), and desipramine (DMI) have all been shown to be effective in the treatment of ADHD, though studies that directly compare TCAs to stimulants find the stimulants to be superior (Pliszka, 1987). Garfinkel and colleagues (1983) compared CLO and DMI at doses of 3.5 mg/kg day and found both drugs equal to each other, superior to placebo, but not as efficacious as methylphenidate in treating ADHD. Using doses of DMI of 4.6 mg/kg/day, 68% of DMI-treated patients were very much improved versus only 10% of placebo patients (Biederman, et al., 1989). No study has directly compared this higher dose of DMI, that seemed to yield better results than the lower dose used in the Garfinkel et al. (1983) study, to stimulants. However, concern about possible TCA cardiac toxicity (Werry et al., 1995) will likely prevent such a study in the future. Similar to stimulants, DMI reduces VMA and MHPG but does not increase EPI (Donnelly et al., 1986).

All tricyclic antidepressants block the reuptake of NE and 5-HT, and although the TCAs may vary in their relative in vitro potency of NE/5HT reuptake blockade, such small differences are unlikely to be relevant in vivo (Frazer, 1997). Growth hormone rises after an acute dose of desipramine (DMI), which is thought to be mediated by postsynaptic alpha-2 receptors (Laakman et al., 1986). Esler and colleagues (1981) administered an acute dose of 125 mg of DMI to normal adult volunteers. They used a tritiated NE infusion technique to study NE kinetics. Post-DMI, reuptake of NE from plasma dropped by 20%, but the spillover rate (rate of release of NE from neurons) also fell, resulting in no net change in plasma NE levels. It appears that as with stimulants, an acute dose of TCA produces a blockade of NE reuptake, but very quickly autoreceptors are stimulated, which leads to decreased neuronal firing.

Much of the research on the chronic effects of tricyclic antidepressants on the catecholamine system in humans was carried out by the NIMH-Clinical Research Branch Collaborative study (Bowden et al., 1987). In this multicenter study, 132 patients with major depressive disorder (both unipolar and bipolar) underwent a 14-day drug-free baseline period, at the end of which 24-hour urinary and CSF measures of NE, DA and 5-HT were obtained. Patients were then randomized to receive either AMI or IMI. One hundred and three patients completed all four weeks of treatment, and the neurochemical measures were reassessed. There were no changes in the urinary excretion of NE or EPI, but there were significant reductions in urinary MHPG (–30%). AMI and IMI did not differ in their effects on urinary catecholamine in any important way.

Effects of IMI and AMI were noted on CSF measures (Bowden et al., 1985). Both tricyclics reduced CSF MHPG (–49%) and 5-HIAA (–75%), with no changes in CSF HVA. Bowden and colleagues (1985) reviewed the many studies examining the effects of tricyclics on these CSF metabolites, and the results have been highly consistent: four studies showed that TCAs reduce CSF MHPG, and nine studies have shown declines in CSF 5-HIAA after TCA treatment of depression. No study has shown that TCAs decrease CSF HVA.

TCAs acutely block the reuptake of NE and 5-HT, but stimulation of autoreceptors results in decreased neuronal firing in the locus coeruleus and raphe. Thus the TCAs reduce NE and 5-HT output over the long term. Paradoxically, this reduction in NE/5HT output may lead to more functional NE/5-HT transmission (Frazer, 1997). The TCAs may serve to "tune" an excessively noisy 5-HT or NE system. TCAs do not appear to have any measurable effects on the DA system. Again, the fact that these agents have little impact on the dopaminergic system and are effective in ADHD presents difficulties for any theory of ADHD that rests solely on a hypothesized dysfunction in the dopaminergic system.

Novel Noradrenergic Antidepressants

Tomoxetine is a highly selective noradrenergic reuptake inhibitor with little affinity for serotonergic or dopaminergic systems. Using a crossover design, tomoxetine (80 mg/day) was recently shown to be superior to placebo in 11 of 12 adults with

ADHD (Spencer et al., 1998). The response rate was less than that seen in trials of methylphenidate in adults (Spencer et al., 1995), although further dose-ranging studies need to be done. The authors also felt a longer trial of the agent might have increased the effectiveness; such trials in both adults and children are currently under way.

Open trials of venlafaxine, a combined noradrenergic/serotonergic reuptake blocker, suggest that it may be effective in the treatment of ADHD (Findling et al., 1996; Olvera et al., 1996; Reimherr et al., 1995), but no controlled trials have been performed.

Bupropion is a compound with weak affinity for the NE and DA (but not the 5-HT) transporter. Its main metabolite, hydroxybupropion may reach levels in humans that permit NE transporter blockade, while the parent compound does not block reuptake of NE at clinical doses (Ascher et al., 1995). Bupropion has been shown to be superior to placebo in the treatment of ADHD, though the effect size is smaller than that seen with stimulants (Conners et al., 1996). In a series of studies, Golden and colleagues (Golden, et al., 1988a; Golden et al., 1988b) examined the effects of bupropion on the NE and DA systems in adults with major depression. Plasma HVA did not change with bupropion treatment for the group as a whole, but the four nonresponders (in terms of depression) had an increase in plasma HVA levels. CSF MHPG, HVA, and HIAA did not change, in contrast to the effects of TCAs which lower MHPG and HIAA. Twenty-four hour urine assessment of NE and its metabolites before and after treatment showed that whole-body NE turnover was reduced with bupropion treatment, falling to 82% of baseline. There was no change in the ratio of NM/MHPG +VMA. As discussed, if bupropion (or its metabolite) had blocked the reuptake of NE, one would expect this ratio to rise. Because this ratio did not change, its appears bupropion achieved this reduction in NE turnover through some other mechanism than increasing NE in the synaptic cleft and thus stimulating autoreceptors. Despite decreasing NE turnover, bupropion treatment nonsignificantly increased 6-hydroxymelatonin output (Golden et al., 1988). The pineal secretes melatonin, and it is innervated by the sympathetic NE system—stimulation of beta receptors on pinealocyctes leads to melatonin production. Since desipramine, MAOIs and bupropion all reduced total NE body turnover but maintained melatonin production, this again suggests that the therapeutic effect of antidepressants is to “tune down” a noisy NE system, making it more efficient.

Alpha-Agonists—Clinical Issues

Hunt et al. (1985) first performed a double-blind crossover study of clonidine in the treatment of ADHD. The study period was 12 weeks; all subjects ($N = 10$) received eight continuous weeks of clonidine. Some subjects received placebo for 2 weeks before the clonidine treatment and again for 2 weeks after the clonidine phase, while others received clonidine for 8 weeks followed by 4 weeks of placebo. Relative to placebo, the investigators found significant decreases on both the parent (34%) and teacher (44%) Conners Hyperactivity Index ratings. In the next study, 10 children with ADHD underwent a double-blind placebo-controlled crossover study of two doses of methylphenidate (Hunt, 1987). After this study, all the subjects were then entered into an open trial of 8 weeks of oral clonidine. After the open trial of oral clonidine, eight of the subjects then underwent a trial of transdermal clonidine. The length of time the subjects were on the patch was not specified. Clonidine (both oral and transdermal) and methylphenidate both produced an equivalent reduction in Conners Parent and Teacher Ratings relative to placebo, but the stimulant and clonidine groups were not compared to each other in a double-blind fashion.

Gunning (1992) evaluated the effectiveness of clonidine and methylphenidate in two groups of ADHD children: 32 with ADHD and tic disorders, and 72 with ADHD alone. A double-blind placebo-controlled parallel groups design was used. Curiously, clonidine was not effective relative to placebo in ADHD children with comorbid tics. For the children with ADHD without tics, methylphenidate and clonidine were equally effective (50% response rate in each active drug group), whereas only 13% responded to placebo. It is unusual that the methylphenidate response rate was smaller in this study than in most clinical trials of stimulants. Leckman et al. (1991) found that clonidine significantly reduced both clinician and parent ratings of ADHD symptoms relative to placebo in comorbid ADHD/Tourette's syndrome children. Singer and colleagues (1995), however, did not detect any significant effects of fixed dose clonidine when compared with placebo for either ADHD or tic symptoms in a double-blind controlled trial.

The go/no-go task was obtained on in ADHD subjects before and after drug treatment, but the task was obtained at three different inter-stimulus intervals (van der Meere et al., 1999). The methylphenidate group had a faster reaction time than the clonidine group, but neither drug was different from placebo on this measure. Both drugs reduced the number of errors relative to placebo, although this difference was not statistically significant.

The theoretical basis for the role of guanfacine in the treatment of ADHD has been reviewed (Arnsten et al., 1996) (see Chapter 7 by Arnsten). Clonidine is equally effective at stimulating all three subtypes (A, B, C) of the alpha-2 receptor. Alpha-2B receptors are found exclusively in thalamus; clonidine's action in this area accounts for its sedative effect. Alpha-2C receptors are found in the brain stem and are responsible for the hypotensive effect of clonidine. Alpha-2A receptors are found predominantly in the prefrontal cortex and are important in cognitive function. Guanfacine has a greater affinity for alpha-2A receptors and improves attention and impulse control in monkeys with much less sedation than clonidine. Thus, the preclinical data suggest that guanfacine might have a more direct effect on attention with less sedative effect in the treatment of ADHD.

Hunt et al. (1995) first explored the use of guanfacine in ADHD. The 13 subjects underwent an open trial of guanfacine at a mean dose of 3.2 mg day (0.091 mg/kg/day). The medication was dosed in the morning, at 12 noon, at 4 p.m., and at bedtime. The subjects showed a significant decline in parent ratings of inattention and hyperactivity, although teacher ratings were not obtained. Sedative and hypotensive side effects were minimal. Chappell et al. (1995) reported on the use of guanfacine in 10 children with comorbid ADHD and Tourette's syndrome. Guanfacine total daily doses ranged from 1.5 to 3.0 mg/day; most children were dosed two or three times daily. The length of treatment ranged from 4 to 20 weeks. The authors obtained Continuous Performance Test (CPT) scores and Conners parent ratings before and after treatment, but no teacher data were available. Guanfacine produced a significant reduction in the number of errors on the CPT, but no overall improvement was found on the parent ratings. On examining individual data, the authors found that four of the 10 subjects markedly improved in parent ratings of hyperactive behavior, three showed no change, and three showed worsening of symptoms.

Thirteen children with severe ADHD and various comorbid conditions (predominantly ODD, Tourette's, and learning disorders) underwent an open trial of guanfacine (Horrigan & Barnhill, 1995). The subjects' ages ranged from 7 to 17 years; all had failed an average of two other medications before entering the study. The dose of guanfacine ranged from 0.5 to 3 mg/day (mean 1.27 mg). Parent and teacher ratings of ADHD symptoms were obtained at baseline and after 4 to 8 weeks of guanfacine treatment. There were no statistically significant effects of guanfacine on either blood pressure or pulse. Both parents and teacher rating scales showed significant declines in ratings of inattention and impulsivity.

Alpha Agonists—Effects on Catecholamines

There is a large body of literature examining the impact of alpha-2 agonists and antagonists on catecholamine measures, primarily looking at adults with hypertension. Hansson and Hokfelt (1981) administered clonidine and tiamentiadine (a related alpha-2 agonist) to adult males with newly diagnosed essential hypertension. After a baseline assessment of plasma and 24-hour urine NE and EPI, patients were randomized to receive either clonidine or tiamenidine for 4 to 6 weeks. The medication was then discontinued, and subjects were crossed over to the other alpha agonist. Plasma NE levels fell by half at the end of the six week treatment period for both drugs, plasma EPI levels also fell. Twenty-four hour urine NE and EPI showed a similar pattern. Interestingly, both NE and EPI rebounded to *higher* than baseline levels that persisted 3 days after drug withdrawal.

Taittonen et al. (1998) administered clonidine at low and high doses to six normal volunteers who then performed the cold pressor test. Relative to placebo, clonidine reduced plasma NE and EPI before, during, and after the cold pressor test. On placebo, the cold pressor test induces an increase in NE and EPI, which was blunted by the higher clonidine dose. Goldstein et al. (1985) compared the effect of clonidine on both normotensive ($n = 41$) and hypertensive ($n = 44$) subjects. The hypertensive subjects had higher plasma NE levels before receiving a single 0.3 mg dose of clonidine. NE fell by 157% in controls and 218% in the hypertensive subjects. In patients treated with alpha-2 agonists for hypertension, urinary NE levels remain depressed even after chronic treatment (>100 days) (Hokfelt, Hedeland, & Dymling, 1970). This suggests there are no neuronal compensatory mechanisms that return NE release to baseline levels when individuals are treated with alpha agonists.

Only a few studies have examined the effects of clonidine on neurochemical variables in patients with ADHD. Hunt et al. (1984) administered a 0.1 mg dose of clonidine to eight boys with ADHD before and after a 12-week trial of methylphenidate.

Before treatment, the ADHD boys showed a robust response of growth hormone to clonidine. The ADHD boys were not compared with a control group, but the authors noted that their mean response of growth hormone to clonidine was greater than that shown by a group of boys with constitutional short statue. After 12 weeks of methylphenidate treatment at a dose of 10 to 20 mg a day, the growth hormone response to clonidine was significantly blunted relative to baseline. It was still blunted, although clearly returning to baseline, one day after the last methylphenidate dose. These studies suggest chronic stimulant treatment down-regulates alpha-2 receptors. Garfinkel and colleagues (1986) found that young adults with a history of ADHD had an increased growth hormone response to an acute dose (10 mg) of dextroamphetamine relative to controls. This would be consistent with alpha-2 receptor hypersensitivity, usually seen when a neurotransmitter system is less active. However, mianserin, an alpha-2 antagonist that increases NE plasma levels, was completely ineffective in the treatment of ADHD in a small pilot trial (Langer et al., 1984). Thus increasing activity of the noradrenergic system does not appear to ameliorate the symptoms of ADHD, inconsistent with a hypothesis of NE underactivity in ADHD.

Further double-blind, parallel-group studies of the effectiveness of alpha agonists in ADHD are needed. Such studies should compare the alpha agonists not only to placebo but also to stimulants. In addition to measures of behavior, the studies should also obtain measures of cognition, as well as precise measures of sedation. It is critical to establish that the alpha agonists do not exert their therapeutic effect via sedation rather than increased cognition. If improvements in behavior correlate with side effects (i.e., sedation) and improvements on cognitive measures are not obtained, then the effect of the alpha-agonists are probably not relevant to the pathophysiology of ADHD. In contrast, if behavioral and cognitive improvements are obtained without sedation, then it would be important to establish which mechanism (pre- or postsynaptic) is most important. Studies could be done examining growth hormone output and plasma NE during alpha-agonist treatment. If treatment response correlated with decreased plasma NE, this would suggest that alpha-agonist efficacy in ADHD was related to alpha-2 presynaptic receptor stimulation and reduction of NE activity. In contrast, if growth hormone increases correlate with efficacy, then alpha-agonists might be exerting their effects through stimulation of alpha-2 postsynaptic receptors in the brain.

Dopamine Agonists

Drugs with primarily dopaminergic effects have not been shown to be effective in the treatment of ADHD. The dopamine agonists piribidel (Brown et al., 1979) and amantadine (Mattes & Gittelman, 1979) were ineffective in small pilot trials. Eight boys with ADHD were treated in a 3-week double-blind crossover trial of carbidopa/levodopa (Langer et al., 1982). The children showed modest, but statistically significant, decreases in teacher ratings of hyperactivity, but not in ratings of inattention or conduct, which normally show robust changes to stimulants. Motor activity as assessed by actometer did not differ between placebo and carbidopa/levodopa phases.

Neuroleptics

Today, neuroleptics are never used for treatment of ADHD per se, but in the 1970s they were often viewed as an alternative to stimulants. If ADHD were associated with a hypodopaminergic state, one would expect neuroleptics to worsen ADHD behavior, but this is not what clinical studies have shown. Gittelman-Klein and colleagues (1976) randomized hyperactive children to placebo ($n = 41$), methylphenidate ($n = 41$), thioridazine ($n = 41$) and a combination of thioridazine and methylphenidate ($n = 42$) for 12 weeks. The range of thioridazine doses was 50 to 300 mg/day, with a mean dose of 160 mg (4.5 mg/kg/day). All three active treatments were superior to placebo, with the two methylphenidate conditions being superior to thioridazine alone. The combination of methylphenidate and thioridazine was superior to methylphenidate alone at 4 weeks, but at 12 weeks these two regimens were equally efficacious. Thus, a dopamine-blocking agent was partially therapeutic in treating ADHD, and combining a neuroleptic with methylphenidate did not block the therapeutic effects of the stimulant.

Werry and Aman (1975) found that low doses of haloperidol improved both behavior and cognition in ADHD children (similar to methylphenidate), whereas higher doses of haloperidol impaired performance on a laboratory measure of attention despite behavior improvement. Using a crossover design, Levy and Hobbes (1988) administered either placebo or a single dose of haloperidol (0.04 mg/kg) in the morning. Subjects then took an 11:00 a.m. dose of methylphenidate (10 mg) and performed the Continuous Performance Test (CPT) an hour later. The subjects also performed the CPT on placebo alone (9:00 a.m. and 11:00 a.m.) and haloperidol alone. Haloperidol alone caused a decline in CPT performance relative to placebo, while methylphenidate improved CPT performance. Haloperidol given before methylphenidate blocked the stimulant-induced improvement on CPT.

The positive clinical effects of neuroleptics on ADHD behaviors are consistent with the hypothesis of a hyperdopaminergic state in this condition, rather than a dopamine deficit (Solanto, 1998). The effects of haloperidol on the CPT are more difficult to interpret, since they do not correlate with behavioral effects of the drugs. In any case, neuroleptics affect other neurotransmitter systems, particularly at alpha-adrenergic and cholinergic receptors. Atypical neuroleptics also affect the 5-HT system. Thus neuroleptics are not ideal clinical probes for exploring neurotransmitter involvement in ADHD.

Serotoninergic Agents

Is there a role for the serotoninergic (5-HT) agents in the treatment of ADHD? Interactions between the catecholamine systems and the serotoninergic systems are complex. The locus coeruleus projects to the cell bodies of the raphe neurons; NE released at this site stimulate alpha-1 receptors, which appear to enhance raphe activity (Baraban & Aghajanian, 1980). In contrast, the locus coeruleus also projects to the terminal fields of the raphe neurons, where NE is released onto alpha-2 heteroreceptors; these decrease the release of 5-HT (Frazer, 1997). Raphe neurons, in turn, project to the cell bodies of neurons in both the substantia nigra and ventral tegmental area, as well as to heteroreceptors in the terminal fields of these neurons. There is considerable debate as to the specific role of 5-HT in modulating DA transmission. Does the serotoninergic system inhibit or facilitate DA transmission? Animal and human studies are very much in conflict with each other. Raphe projections to the substantia nigra inhibit the firing of dopamine neurons (Jacobs & Azmitia, 1992; Kelland, et al., 1990), possibly modulated by 5HT2 receptors (Kapur & Remington, 1996). Stimulation of the raphe-nigral projections also leads to inhibition of striatal neuronal firing (Waldmier & Delini-Stula, 1979). It has been speculated that atypical neuroleptics have a lower incidence of extrapyramidal symptoms (EPS) due to their 5HT2 receptor antagonism (Huttunen, 1995). By opposing the effect of 5-HT in the striatum and frontal cortex, atypicals increase the release of DA, lessening the impact of D2 postsynaptic receptor blockade. Among patients treated with SSRIs, about one in a thousand will experience EPS or an akathisia-like syndrome (Kapur & Remington, 1996). This would be consistent with an antagonist effect of 5-HT on DA, but EPS is such a rare syndrome in SSRI-treated patients one must question how important 5-HT antagonism of DA is in vivo.

Animal models have long postulated a role for the 5-HT system in inhibition of behavior (Soubrie, 1986; Spoont, 1992). Spoont (1992) reviewed the effects of 5-HT on a range of behaviors facilitated by amphetamine in rodents. Administration of 5-HT agonists generally inhibits amphetamine-induced increases in locomotion and exploratory behavior; drugs that enhance 5-HT transmission generally decrease intracranial self-administration of amphetamine. Spoont (1992) concluded that the 5-HT system served to "constrain" the behaviors facilitated by the DA system, that is the 5-HT opposed the actions of the DA system. Recently, a role for the 5-HT system has been shown in the reduction of hyperactivity in DA transporter "knockout" mice (Gainetdinov et al., 1999). These mice lack a functional DA transporter and thus have very high levels of DA in extracellular fluid; knockout mice are extremely active relative to controls both in their home cages and in a novel environment. When given high doses of methylphenidate or amphetamine, motor activity in control mice increases, whereas it declines in the knockout mice. NE reuptake inhibitors do not affect activity in either the wild or knockout mice. Methylphenidate increased extracellular DA levels in the wild-type mice, whereas DA levels were not affected by stimulant in the knockout mice. Fluoxetine and other 5-HT agonists decreased activity in the knockout mice but did not affect DA levels (inconsistent with the earlier animal studies suggesting an inhibitory effect of 5-HT on DA).

Based on their observations in these knockout mice, the authors (Gainetdinov et al., 1999) suggested that 5-HT agonists might have a role in the treatment of ADHD. There are, of course, substantial differences between ADHD and the DA transporter knockout mice. While a DNA polymorphism has been associated with the DA transporter in ADHD subjects (see Chapter 10 by Castellanos), there is no evidence that ADHD children have nonfunctional DA transporters. Furthermore, very high doses of stimulant were used, high enough to induce increases in motor activity in the wild-type mice. Clinical doses of stimulants used in normal children do not cause increases in motor activity; there is nothing paradoxical about their effect in ADHD children (Rapoport et al., 1980). Finally, it is clear that knockout mice demonstrate an extreme sort of phenotype that entails numerous other compensatory effects, any of which could explain the observed serotoninergic effects in these mice when treated with stimulants. Before embarking on new trials of 5-HT agonists in ADHD, we should note (*1*) the interactions of 5-HT and DA remain unclear, and in humans, 5-HT may facilitate, rather than inhibit DA, and (*2*) trials of 5-HT agonists to date have not shown these agents to be effective in the treatment of ADHD.

Benloucif et al. (1992) infused a variety of 5-HT agonists directly into the neostriatum of rats, and assessed DA levels via microdialysis. All the agonists increased the release of DA in a dose-dependent manner. The 5-HT1B receptors were most potent in inducing this release, followed by 5-HT4 receptors. The 5-HT2 and 5-HT1A agonists also induced release, but were far less potent than the 5-HT1B and 5-HT4 agonists. Studies show that administering 5-HT agonists systemically leads to decreased ventral tegmental area firing (Di Mascio & Esposito, 1997; Di Mascio et al., 1998; Prisco & Esposito, 1995); studies in which 5-HT agonists are infused directly into the neostriatum consistently show increased release of DA (Benloucif & Galloway, 1991; Chen & Reith, 1995; Li, Perry, & Fuller, 1996; Yadid et al., 1994). In baboons, administration of intravenous 5-HT agonists appears to oppose DA release (Dewey et al., 1995), but in humans the reverse has been found (Smith et al., 1997). When oral fenfluramine is administered to humans, 11C-raclopride binding as measured by PET is decreased. Raclopride binds to D2 DA receptors, thus the decrease in binding post-fenfluramine can be interpreted as an increase in DA release, because more raclopride is being displaced by newly released DA. A review (Agren et al., 1986) of 14 human studies of CSF HVA and 5-HIAA have all shown robust positive correlations (range, $r = .47–.93$), a finding recently confirmed in children ($r = .63$) (Castellanos & et al., 1996). A statistical analysis of adult CSF studies (Agren et al., 1986) was highly consistent with a strong facilitatory effect of 5-HT on DA in humans.

If ADHD is associated with a DA deficiency, it might be possible to ameliorate the symptoms with 5-HT agonists. The 5-HT agonist might cause an increase in DA release, with subsequent improvements in motor activity and cognition. Three well-controlled studies of 5-HT agonists have not shown these agents to be effective in ADHD, however. Tryptophan, the amino-acid 5-HT precursor, was tested along with tyrosine and placebo in a double-blind crossover design in 1-week trials (Nemzer et al., 1986). Tyrosine did not differ from placebo on any of the outcome measures; and while tryptophan did not differ from placebo on teacher ratings, a small but significant effect of tryptophan was found on parent ratings.

Fenfluramine, a drug that both blocks the reuptake and enhances the release of 5-HT, was compared to dextroamphetamine in a double-blind, placebo-controlled fashion (Donnelly et al., 1989). Fenfluramine was given to 16 ADHD boys at doses ranging from 0.6 to 2.0 mg/kg/day. After 3 weeks, fenfluramine significantly reduced platelet 5-HT and increased plasma prolactin. These effects were not seen with dextroamphetamine, showing that fenfluramine had specific effects on the 5-HT system. Fenfluramine was not different from placebo on any of the measures, including teacher ratings and the Clinical Global Impressions. Motor activity as measured by actometer did not differ between placebo and fenfluramine, whereas dextroamphetamine induced very significant reductions. Recently, a trial of buspirone, a 5-HT1A agonist, was found to be ineffective in the treatment of ADHD (L. Greenhill, personal communication, May 1999).

On the basis of these studies, it is difficult to be enthusiastic that other SSRIs might be effective in ADHD. The 5-HT system does not appear to play a major role in ADHD, though it may have a significant role in the modulation of aggressive behavior, which is often comorbid with ADHD. Animal studies have recently shown that SSRIs can increase amphetamine concentration in brain tissue, most likely via actions on the cytochrome P450 system (Sills et al., 1999). In a substantial case series, it was reported that adding an SSRI to methylphenidate converted a number of "partial" stimulant responders to "full responders"(Gammon & Brown, 1993). It is possible that the observed benefits from SSRI increased stimulant levels in the brain—not that 5-HT agonism had direct positive effect on ADHD behavior.

Conclusions

Both older TCA antidepressants and more novel noradrenergic antidepressants appear to "dampen" the activity of the NE system but do not affect the DA system. Alpha-agonists suppress NE activity and tolerance does not appear to develop to this effect. Alpha agonists could still be exerting a noradrenergic effect by directly stimulating postsynaptic alpha-2 receptors. The efficacy of alpha-agonists in treating ADHD, particularly on cognitive measures, remains to be established. Increasing the activity of the NE system does not appear to be helpful in the treatment of ADHD, nor are agents that exclusively affect the DA or 5-HT system.

Stimulants and MAOIs that affect the NE and DA systems remain the most effective agents in the treatment of ADHD. These agents also increase the activity of noradrenergic systems. It seems likely that both the NE and DA systems are involved in stimulant response. One hypothesis is that both these systems are excessively noisy in ADHD. Agents that are most effective in ADHD ultimately down-regulate these systems, allowing for both more efficient information processing via the NE system, and more efficient executive function via the DA system (Pliszka et al., 1996).

References

Agren, H, Mefford, IN, Rudorfer, MV, Linnoila, M, & Potter, WZ (1986). Interacting neurotransmitter systems. A non-experimental approach to the 5-HIAA-HVA correlation in human CSF. *J Psychiatr Res*, *20*, 175–193.

Anderson, GM, Riddle, MA, Hoder, EL, Feibel, FC, Shaywitz, BA, & Cohen, DJ (1988). The ontogeny of 3,4-dhydroxyphenylacetic acid in human cerebral spinal fluid. *J Neurol Neurosurg Psychiatry*, *51*, 1100–1102.

Arnsten, AF, Steere, JC, & Hunt, RD (1996). The contribution of alpha 2-noradrenergic mechanisms of prefrontal cortical cognitive function. Potential significance for attention-deficit hyperactivity disorder. [Review]. *Arch Gen Psychiatry, 53*, 448–455.

Ascher, JA, Cole, JO, Colin, JN, Feighner, JP, Ferris, RM, & Fibiger, HC et al. (1995). Bupropion: A review of its mechanism of antidepressant activity. *J Clin Psychiatry, 56*, 395–401.

Aston-Jones, G, Chiang, C, & Alexinsky, T (1991). Discharge of noradrenergic locus coeruleus neurons in behaving rats and monkeys suggests a role in vigilance. *Prog Brain Res, 88*, 501–520.

Aston-Jones, G, Shipley, MT, Ennis, M, Williams, JT, & Pierbone, VA (1990). Restricted afferent control of locus coeruleus neurones revealed by anatomical, physiological, and pharmacological studies. In DJ Heal & CA Marsden (Eds.), *The Pharmacology of noradrenaline in the central nervous system* (pp. 187–247). Oxford: Oxford Medical Publications.

Baraban, JM & Aghajanian, GK (1980). Suppression of firing activity of 5-HT neurons in the dorsal raphe by alpha-adrenoreptor antagonists. *Neuropharmacology, 19*, 355-363.

Benloucif, S & Galloway, MP (1991). Facilitation of dopamine release in vivo by serotonin agonists: studies with microdialysis. *Eur J Pharmacol, 200*, 1–8.

Benloucif, S, Keegan, MJ, & Galloway, MP (1992). Serotonin-facilitated dopamine release in vivo: pharmacological characterization. *J Pharmacol Exp Ther, 265*, 373–377.

Biederman, J, Baldessarini, RJ, Wright, V, Knee, D, & Harmatz, JS (1989). A double-blind placebo controlled study of desipramine in the treatment of ADD: I. Efficacy. *J Am Acad Child Adolesc Psychiatry, 28*, 777–784.

Bowden, CL, Koslow, S, Maas, JW, Davis, J, Garver, DL, & Hanin, I (1987). Changes in urinary catecholamines and their metabolites in depressed patients treated with amitriptyline or imipramine. *J Psychiatr Res, 21*, 111–128.

Bowden, CL, Koslow, SH, Hanin, I, Maas, JW, & Davis, JM (1985). Effects of amitriptyline and imipramine on brain amine neurotransmitter metabolites in cerebrospinal fluid. *Clin Pharmacol Ther, 37*, 316–324.

Brown, GL, Ebert, MH, Hunt, RD, & Rapoport, JL (1981). Urinary 3-methyoxy-4-hydroxyphenylglycol and homovanillic acid response to d-amphetamine in hyperactive children. *Biol Psychiatry, 16*, 779–787.

Brown, GL, Ebert, MH, Mikkelsen, E, Bushsbaum, MS, & Bunney, WE (1979) Dopamine agonist piribedel in hyperactive children. Paper presented at the annual meeting of the American Psychiatric Association, Chicago, IL.

Castellanos, FX, Elia, J, Kruesi, MJP, Gulotta, CS, Mefford, IN, & Potter, WZ et al. (1994). Cerebrospinal fluid monoamine metabolites in boys with attention deficit hyperactivity disorder. *Psychiatry Res, 52*, 305–316.

Castellanos, FX, Elia, J, Kruesi, MJP, Marsh, WL, Gulotta, CS, & Potter, WZ et al. (1996). Cerebrospinal homovanillic acid predicts behavioral response to stimulants in 45 boys with attention-deficit hyperactivity disorder. *Neuropsychopharmacology, 14*, 125–137.

Chappell, PB, Riddle, MA, Scahill, L, Lynch, KA, Schultz, R, & Arnsten, A et al. (1995). Guanfacine treatment of comorbid attention-deficit hyperactivity disorder and Tourette's syndrome: Preliminary clinical experience. *J Am Acad Child Adolesc Psychiatry, 34*, 1140–1146.

Chen, NH & Reith, MEA (1995). Monoamine interactions measured by microdialysis in the ventral tegmental area of rats treated systemically with -8-hydroxy-2-(Di-n-Propylamino) tetralin. *J Neurochem, 64*, 1585–1597.

Conners, CK, Casat, CD, Gualtieri, CT, Weller, E, Reader, M, & Reiss, A et al. (1996). Bupropion hydrochloride in attention deficit disorder with hyperactivity. *J Am Acad Child Adolesc Psychiatry, 35*, 1314–1321.

Dewey, SL, Smith, GS, Logan, J, Alexoff, D, Ding, Y, & King, P et al. (1995). Serotonergic modulation of striatal measured with positron emission tomography (PET) and in vivo microdialysis. *J Neurosci, 15*, 821–829.

Di Mascio, M, Di Giovanni, G, Di Matteo, V, Prisco, S, & Esposito, E (1998). Selective serotonin reuptake inhibitors reduce the spontaneous activity of dopaminergic neurons in the ventral tegmental area. *Brain Res Bull, 46*, 547–554.

Di Mascio, M & Esposito, E (1997). The degree of inhibition of dopaminergic neurons in the ventral tegmental area induced by selective serotonin reuptake inhibitors is a function of the density-power-spectrum of the interspike interval. *Neuroscience, 79*, 957–961.

Donnelly, M, Rapoport, JL, Potter, WZ, Oliver, J, Keysor, CS, & Murphy, DL (1989). Fenfluramine and dextroamphetamine treatment of childhood hyperactivity. Clinical and biochemical findings. *Arch Gen Psychiatry, 46*, 205–212.

Donnelly, M, Zametkin, AJ, Rapoport, JL, Ismond, DR, Weingartner, H, & Lane, E et al. (1986). Treatment of childhood hyperactivity with desipramine: Plasma drug concentration, cardiovascular effects, plasma and urinary catecholamine levels and clinical response. *Clin Pharmacol Ther, 39*, 72–81.

Elam, M, Svensson, TH, & Thoren, P (1986). Locus coeruleus neurons and sympathetic nerves: Activation by cutaneous sensory afferents. *Brain Res, 366*, 254–261.

Elam, M, Yao, T, Svensson, TH, & Thoren, P. (1984). Regulation of locus coeruleus neurons and splanchnic sympathetic nerves by cardiovascular afferents. *Brain Res, 290*, 281–287.

Elia, J, Borcherding, BG, Potter, WZ, Mefford, IN, Rapoport, JL, & Keysor, CS (1990). Stimulant drug treatment of hyperactivity: Biochemical correlates. *Clin Pharmacol Ther, 48*, 57–66.

Esler, M, Jackman, G, Leonard, P, Skews, H, Bobik, A, & Korner, P (1981). Effect of norepinephrine uptake blockers on norepinephrine kinetics. *Clin Pharmacol Ther, 29*, 12–20.

Findling, RL, Schwartz, MA, Flannery, DJ, & Manos, MJ (1996). Venlafaxine in adults with attention-deficit/hyperactivity disorder: An open clinical trial. *J Clin Psychiatry, 57*, 184–189.

Frazer, A (1997). Pharmacology of antidepressants. *J Clin Psychopharmacol, 17*, 2S–18S.

Gainetdinov, RR, Wetsel, WC, Jones, SR, Levin, ED, Jaber, M, & Caron, MC (1999). Role of serotonin in the paradoxical calming effect of psychostimulants on hyperactivity. *Science, 283*, 397–401.

Gammon, GD & Brown, TE (1993). Fluoxetine and methylphenidate in combination for treatment of attention deficit and comorbid depressive disorder. *J Child Adolesc Psychopharmacol, 3*, 1–10.

Garfinkel, BD, Brown, WA, Klee, SH, Braden, W, Beauchesne, H, & Shapiro, SK (1986). Neuroendocrine and cognitive responses to amphetamine in adolescents with a history of attention deficit disorder. *J Am Acad Child Psychiatry, 25*, 503–508.

Garfinkel, BD, Wender, PH, Sloman, L, & O'Neill, I (1983). Tricyclic antidepressant and methylphenidate treatment of attention deficit disorder in children. *J Am Acad Child Psychiatry, 22*, 343–348.

Gittelman-Klein, R, Klein, DF, Katz, S, Saraf, K, & Pollack, E (1976). Comparative effects of methylphenidate and thioridazine in hyperkinetic children. I. Clinical results. *Arch Gen Psychiatry, 33*, 1217–1231.

Golden, RN, Markey, SP, Risby, ED, Rudorfer, MV, Cowdry, RW, & Potter, WZ (1988a). Antidepressants reduce whole-body norepinephrine turnover while enhancing 6-Hydroxymelatonin output. *Arch Gen Psychiatry, 45*, 150–154.

Golden, RN, Rudorfer, MV, Sherer, MA, Linnoila, M, & Potter, WZ (1988b). Bupropion in depression I. Biochemical effects and clinical response. *Arch Gen Psychiatry, 45*, 139–143.

Goldstein, DS, Levinson, PD, Zimlichman, R, Pitterman, A, Stull, R, & Keiser, HR (1985). Clonidine suppression testing in essential hypertension. *Ann Intern Med, 102*, 42–48.
Gunning, B (1992). A controlled trial of clonidine in hyperkinetic children. Department of Child/Adolescent Psychiatry, Academic Hospital, Rotterdam, Netherlands.
Hanna, GL, Ornitz, EM, & Hariharan, M (1996a). Urinary catecholamine excretion and behavioral differences in ADHD and normal boys. *J Child Adolesc Psychopharmacol, 6*, 63–73.
Hanna, GL, Ornitz, EM, & Hariharan, M (1996b). Urinary epinephrine excretion during intelligence testing in attention-deficit hyperactivity and normal boys. *Biol Psychiatry, 40*, 553–555.
Hansson, BG & Hokfelt, B (1981). Changes in blood pressure plasma catecholamines and plasma renin activity during and after treatment with tiamenidine and clonodine. *Br J Clin Pharmacol, 1*, 73–77.
Hokfelt, B, Hedeland, H, & Dymling, JF (1970). Studies on catecholamines renin and aldosterone following catapresan (2,6-dichlor-phenylamine)-2-imidazoline hydrochloride) in hypertensive patients. *Eur J Pharmacol, 10*, 389–397.
Horrigan, JP & Barnhill, LJ (1995). Guanfacine for the treatment of attention deficit hyperactivity disorder in boys. *J Child Adolesc Psychopharmacol, 5*, 215–223.
Hunt, RD (1987). Treatment effects of oral and transdermal clonidine in relation to methylphenidate: An open pilot study in ADD-H. *Psychopharmacol Bull, 23*, 111–114.
Hunt, RD, Arnsten, AFT, & Asbell, MD (1995). An open trial of guanfacine in the treatment of attention deficit hyperactivity disorder. *J Am Acad Child Adolesc Psychiatry, 34*, 50–54.
Hunt, RD, Cohen, DJ, Anderson, G, & Clark, L (1984). Possible change in noradrenergic receptor sensitivity following methylphenidate treatment: Growth hormone and MHPG response to clonidine challenge in children with attention deficit disorder with hyperactivity. *Life Sci, 35*, 885–897.
Hunt, RD, Minderaa, RB, & Cohen, DJ (1985). Clonidine benefits children with attention deficit disorder and hyperactivity: Report of a double-blind placebo-crossover therapeutic trial. *J Am Acad Child Psychiatry, 24*, 617–629.
Huttunen, M (1995). The evolution of the serotonin-dopamine antagonist concept. *J Clin Psychopharmacol, 15*, 4S–10S.
Jacobs, BL & Azmitia, EC (1992). Structure and function of the brain serotonin system. *Physiol Rev, 72*, 165–229.
Kapur, S & Remington, G (1996). Serotonin-dopamine interaction and its relevance to schizophrenia. *Am J Psychiatry, 153*, 466–476.
Kelland, MD, Freeman, AS, & Chiodo, LA (1990). Serotonergic afferent regulation of the basic physiology and pharmacological responsiveness of nigrostriatal dopamine neurons. *J Pharmacol Exp Ther, 253*, 803–811.
Laakman, G, Zygan, K, Shoen, H-W, Weiss, A, Wittman, M, & Meisner, R (1986). Effect of receptor blockers (methysergide, propranolol, phentolamine, yohimbine, and prazosin) on desipramine induced pituitary hormone stimulation in humans- I. Growth hormone. *Psychoneuroendocrinology, 11*, 447–461.
Langer, DH, Ebert, MH, Lake, CR, & Nee, LE (1984). Pilot trial of mianserin hydrochloride for childhood hyperactivity. In B Shopsin & L Greenhill (Eds.), *Psychobiology of childhood: Profiles of current issues* (pp. 197–210). New York: Spectrum Publications.
Langer, DH, Rapoport, JL, Brown, GL, Ebert, MH, & Bunney, WE (1982). Behavioral effects of carbidopa/levodopa in hyperactive boys. *J Am Acad Child Psychiatry, 21*, 10–18.
Leckman, JF, Hardin, MT, Riddle, MA, Stevenson, J, Ort, SI, & Cohen, DJ (1991). Clonidine treatment of Gilles De La Tourette's syndrome. *Arch Gen Psychiatry, 48*, 324–328.

Levy, F & Hobbes, G (1988). The action of stimulant medication in attention deficit disorder with hyperactivity: Dopaminergic, noradrenergic, or both? *J Am Acad Child Adolesc Psychiatry, 27*, 802–805.

Li, XM, Perry, KW, & Fuller, RW (1996). On the in-vivo modulation of neostriatal dopamine release by fluoxetine and 5-hydroxy-L-tryptophan in conscious rats. *J Pharm Pharmacol, 48*, 825–828.

Maas, JW, Contreras, SA, Bowden, CL, & Weintraub, SE (1985). Effects of debrisoquin on CSF and plasma HVA concentrations in man. *Life Sci, 36*, 2163–2170.

Maas, JW, & Leckman, JF (1983). Relationship between central nervous system noradrenergic function and plasma and urinary MHPG and other norepinephrine metabolites. In JW Maas (Ed.), *MHPG: Basic mechanisms and psychopathology* (pp. 33–44). London: Academic Press, Inc.

Mattes, JA & Gittelman, R (1979). A pilot trial of amantadine in hyperactive children. Presented at the New Clinical Drug Evaulation Unit (NCDEU), Key Biscayne, FL.

McCracken, JT (1991). A two-part model of stimulant action on attention-deficit hyperactivity disorder in children. *J Neuropsychiatry Clin Neurosci, 3*, 201–209.

McCracken, JT, Hinshaw, SP, Henker, B, Whalen, CK, & Zupan, B (1990). Urinary catecholamine differences in ADHD versus normal children and effects of methylphenidate. Presented at the annual meeting of the Society for Research in Child and Adolescent Psychopathology, Los Angeles.

Mefford, IN & Potter, WZ (1989). A neuroanatomical and biochemical basis for attention deficit disorder with hyperactivity in children: A defect in tonic adrenaline mediated inhibition of locus coeruleus stimulation. *Med Hypotheses, 29*, 33–42.

Mikkelsen, E, Lake, RC, Brown, GL, Ziegler, MG, & Ebert, MH (1981). The hyperactive child syndrome: Peripheral sympathetic nervous system function and the effect of d-amphetamine. *Psychiatry Res, 4*, 157–169.

Nemzer, ED, Arnold, LE, Vololato, NA, & McConnell, H (1986). Amino acid supplementation as therapy for attention deficit disorder. *J Am Acad Child Adolesc Psychiatry, 25*, 509–513.

Olvera, RL, Pliszka, SR, Luh, J, & Tatum, R (1996). An open trial of venlafaxine in children and adolescents with ADHD. *J Child Adolesc Psychopharmacol, 6*, 241–250.

Pliszka, SR (1987). Tricyclic antidepressants in the treatment of children with attention deficit disorder. *J Am Acad Child Adolesc Psychiatry, 26*, 127–132.

Pliszka, SR, McCracken, JT, & Maas, JW (1996). Catecholamines in attention deficit hyperactivity disorder: Current perspectives. *J Am Acad Child Adolesc Psychiatry, 35*, 264–272.

Prisco, S & Esposito, E (1995). Differential affects of acute and chronic fluoxetine administration on the spontaneous activity of dopaminergic neurones in the ventral tegmental area. *Br J Pharmacol, 116*, 1923–1931.

Rapoport, JL, Buchsbaum, MS, Weingartner, H, Zahn, TP, Ludlow, C, & Mikkelson, EJ (1980). Dextroamphetamine: Its cognitive and behavioral effects in normal and hyperactive boys and normal men. *Arch Gen Psychiatry, 37*, 933–943.

Rapoport, JL, Zametkin, AJ, Donnelly, M, & Ismond, MA (1985). New drug trials in attention deficit disorder. *Psychopharmacol Bull, 21*, 232–236.

Reimherr, FW, Hedges, DW, Strong, RE, & Wender, PH (1995). An open trial of venlafaxine in adult patients with attention deficit hyperactivity disorder. Presented at the annual meeting of the New Clinical Drug Evaluation Unit Program (NCDEU), Orlando, FL.

Riddle, MA, Anderson, GM, McIntosh, S, Harcherik, DF, Shaywitz, BA, & Cohen, DJ (1986). Cerebrospinal fluid monoamine precursor and metabolite levels in children treated for leukemia: Age and sex effects and individual variability. *Biol Psychiatry, 21*, 69–83.

Seeman, P & Madras, BK (1998). Anti-hyperactivity medication: Methyphenidate and amphetamine. *Mol Psychiatry, 3*, 386–396.

Shaywitz, BA, Cohen, DJ, & Bowers, MB (1977). CSF monoamine metabolites in children with minimal brain dysfunction: Evidence for alteration of brain dopamine. *J Pediatr, 90*, 67–71.

Shekim, WO, Dekirmenjian, H, & Chapel, JL (1977). Urinary catecholamine metabolites in hyperkinetic boys treated with d-amphetamine. *Am J Psychiatry, 134*, 1276–1279.

Shekim, WO, Javaid, J, Davis, JM, & Bylund, DB (1983). Urinary MHPG and HVA excretion in boys with attention deficit disorder and hyperactivity treated with d-amphetamine. *Biol Psychiatry 18*, 707–714.

Shetty, T & Chase, TN (1976). Central monoamines and hyperkinesis of childhood. *Neurology, 26*, 1000–1002.

Sills, TL, Greenshaw, AJ, Baker, GB, & Fletcher, PJ (1999). The potentiating effect of sertraline and fluoxetine on amphetamine-induced locomotor activity is not mediated by serotonin. *Psychopharmacology, 143*, 426–432.

Singer, HS, Brown, J, Quaskey, S, Rosenberg, LA, Mellits, ED, & Denckla, MD (1995). The treatment of attention-deficit hyperactivity disorder in Tourette's Syndrome: A double blind placebo controlled study with clonidine and desipramine. *Pediatrics, 95*, 74–81.

Smith, GS, Dewey, SL, Brodie, JD, Logan, J, Vitkun, SA, & Simkowitz, P et al. (1997). Serotonergic modulation of dopamine measured with (^{11}C) raclopride and positron emission tomography (PET) in normal human subjects. *Am J Psychiatry, 154*, 490–496.

Solanto, MV (1998). Neuropsychopharmacological mechanisms of stimulant drug action in attention-deficit hyperactivity disorder: A review and integration. *Behav Brain Res, 94*, 127–152.

Soubrie, P (1986). Reconciling the role of central serotonin neurons in human and animal behavior. *BehavBrain Sci, 9*, 319–364.

Spencer, T, Biederman, J, Wilens, T, Prince, J, Hatch, M, & Jones, et al. (1998). Effectiveness and tolerability of tomoxetine in adults with attention deficit hyperactivity disorder. *Am J Psychiatry, 155*, 693–695.

Spencer, T, Wilens, T, Biederman, J, Faraone, SV, Ablon, JS, & Lapey, K (1995). A double-blind, crossover comparison of methylphenidate and placebo in adults with childhood-onset attention-deficit hyperactivity disorder. *Arch Gen Psychiatry, 52*, 434–443.

Spoont, MR (1992). Modulatory role of serotonin in neural information processing: Implications for human psychopathology. *Psychol Bull, 112*, 330–350.

Taittonen, MT, Aantaa, R, & Kirvela, OA (1998). Effect of clonidine on changes in plasma catecholamine concentrations and oxygen consumption caused by the cold pressor test. *Br J Anaesth, 81*, 144.

van der Meere, J, Gunning, B, & Stemerdink, N (1999). The effect of methylphenidate and clonidine on response inhibition and state regulation in children with ADHD. *J Child Psychol Psychiatry, 40*, 291–298.

Waldmier, PC & Delini-Stula, AA (1979). Serotonin-dopamine interactions in the nigrostriatal system. *Eur J Pharmacol, 55*, 363–373.

Werry, JS & Aman, MG (1975). Methylphenidate and haloperidol in children. *Arch Gen Psychiatry, 32*, 790–795.

Werry, JS, Biederman, J, Thisted, RA, Greenhill, LL, & Ryan, ND (1995). Resolved: Cardiac arrhythmias make desipramine an unacceptable choice in children. *J Am Acad Child Adolesc Psychiatry, 34*, 1239–1245.

Yadid, G, Pacak, K, Kopin, IJ, & Goldstein, DS (1994). Endogenous serotonin stimulates striatal dopamine release in conscious rats. *J Pharmacol Exp Ther, 270*, 1158–1165.

Zametkin, A, Rapoport, JL, Murphy, DL, Linnoila, M, & Ismond, D (1985a). Treatment of hyperactive children with monoamine oxidase inhibitors. I. Clinical efficacy. *Arch Gen Psychiatry*, *42*, 962–966.

Zametkin, A, Rapoport, JL, Murphy, DL, Linnoila, M, Karoum, F, & Potter, WZ et al. (1985b). Treatment of hyperactive children with monoamine oxidase inhibitors. II. Plasma and urinary monoamine findings after treatment. *Arch Gen Psychiatry, 42*, 969–973.

Zametkin, AJ & Hamburger, SD (1988). The effect of methylphenidate on urinary catecholamine excretion in hyperactivity: A partial replication. *Biol Psychiatry, 23*, 350–356.

Zametkin, AJ, Karoum, F, Linnoila, M, Rapoport, JL, Brown, GL, & Chuang, LW et al. (1985). Stimulants, urinary catecholamines, and indoleamines in hyperactivity: A comparison of methylphenidate and dextroamphetamine. *Arch Gen Psychiatry, 42*, 251–255.

IV

INTEGRATION

15

The Neuroscience of Stimulant Drug Action in ADHD

MARY V. SOLANTO, AMY F. T. ARNSTEN, AND F. XAVIER CASTELLANOS

The purpose of this book has been to deepen our understanding of the mechanisms of therapeutic drug action in ADHD by integrating knowledge across the relevant disciplines of basic and clinical neuroscience, presented in the foregoing chapters. The goals of this final chapter are to present the results of that integration thus far and to chart a course for further research. It represents the efforts and insights not only of the named authors but of all of the contributors who shared their thoughts, ideas, and knowledge in the process of producing this volume.

ADHD: Phenomenology

Attention-Deficit/Hyperactivity Disorder (ADHD) is a common disorder, occurring, conservatively, in 3% to 5% of the school-aged population. It is characterized by developmentally inappropriate and functionally impairing levels of inattentiveness and/or hyperactivity-impulsivity. Diagnostic criteria further require that symptoms be manifest in at least two settings (usually home and school), have an onset by the age of 7 years, and a duration of at least 6 months. As of the most recent *Diagnostic and Statistical Manual* (DSM-IV) (American Psychiatric Association, 1994), three subtypes are recognized: Predominantly Inattentive type (IN), Predominantly Hyperactive–impulsive (HI), and the Combined type (CB), which is characterized by the presence of *both* symptom types. The Combined type corresponds most closely to ICD-10 "Hyperkinetic Disorder" and has been by far the type most widely recognized, researched, and treated with stimulant medications. As such, it has been the focus of this book.

Behavioral and Cognitive Deficits

The symptoms of ADHD comprehend both behavioral and cognitive deficits (Solanto, Chapter 1). Technologically sophisticated studies of activity level in children with ADHD show that they are motorically more active than non-ADHD children across a variety of in-school academic and after-school play situations, as well as during sleep (Porrino et al., 1983b). These findings strongly suggest that overactivity is a primary feature of the combined type, and not simply secondary to reduced attention or increased impulsivity (although these functions may also influence activity). The symptoms of inattention and impulsivity have proven more difficult to quantify and study in the laboratory (see Solanto, Chapter 1). Denney and Rapport (Chapter 12) have pointed out the relatively low correlations among measures of these constructs in children with ADHD. Their analysis of cognitive test results suggests that those laboratory measures that reliably differentiate ADHD and normal children are likely to place demands on working memory and to involve speeded processing. These findings have led them to postulate that an impairment in working memory—which is critical to the self-regulation of behavior by means of internally represented stimuli, schema, and rules—is the central deficit in ADHD. They further propose that inattentive and impulsive behaviors serve to increase the rate of stimulus input and thereby compensate for weak working memory representations, or may facilitate escape from monotonous conditions. Barkley (1997) has alternatively proposed that disinhibition is the central deficit in ADHD and that dysfunctions in working memory, attention, and impulse-control all proceed from a primary inability to appropriately delay a behavioral response. Potentially reconciling these alternate points of view are recent neuroimaging data indicating that working memory and behavioral inhibitory tasks activate the *same* area of dorsolateral prefrontal cortex (Diamond et al., 1998) and may represent facets of the same construct (Goldman-Rakic, 1996). Researchers agree that the search for a pivotal behavioral deficit, and its corresponding neurobiological substrate, is particularly challenging in ADHD because the population of individuals encompassed is heterogeneous, even within subtype, and is characterized by variations in severity and relative prominence of core symptoms, as well as presence or absence of aggression, learning disabilities, and comorbid disruptive and internalizing disorders.

Genetics

A longstanding issue is whether ADHD is a qualitatively distinct syndrome or, alternatively, whether the symptoms of inattention, overactivity, and impulsivity represent extreme points on normal distributions of these behaviors. The results of large-scale twin studies (Gjone et al., 1996; Levy et al., 1997) provide support for the latter interpretation, and further indicate that the behavioral traits of attentiveness, and hyperactivity–impulsivity have different genetic determinants (Neuman et al., 1999). Heritability estimates for ADHD in nine twin studies ranged from .80 to .98, whereas the contribution of shared environment was negligible. Genetic polymorphisms that have been linked to ADHD are the 10-repeat allele for the dopamine transporter (DAT1), and the 7-repeat allele for the DRD4 receptor.

Most recently, polymorphisms at three noradrenergic genes (alpha-2a receptor, alpha-2c receptor, and dopamine beta-hydroxylase) were additively associated with increased risk for ADHD in a large sample of individuals with Tourette syndrome (Comings et al., 1999). However, each of these candidate dopamine and norepinephrine gene polymorphisms, which are widely distributed in the population, increases the relative risk for ADHD only slightly, suggesting that other genetic factors are contributory.

Neuroimaging

Neuroimaging studies (reviewed by Castellanos, Chapter 10) have contributed to the characterization of the ADHD phenotype. Anatomic MRI has revealed smaller volume in the right prefrontal brain regions, caudate nucleus, globus pallidus, and a subregion of the cerebellar vermis; furthermore, volume of these regions (with the exception of cerebellum) was negatively correlated with inhibitory control on laboratory measures in children with ADHD. However, unsettled methodological issues remain in anatomic studies, and even more so in functional studies. Not surprisingly, results of studies employing functional MRI (fMRI) and positron emission tomography (PET) have not been entirely consistent. One study found greater frontal activation on an easy response-inhibition task, and lesser striatal activation on a harder version of the task in children with ADHD compared with normal children (Vaidya et al., 1998). Another study showed lesser activation in the right medial and right inferior prefrontal regions during a relatively difficult stopping task in ADHD adolescents (Rubia et al., 1999). Differences in age groups and small sample sizes (10 or fewer in the ADHD group in each study) may have contributed to the different results. PET neuroimaging has shown increased [^{18}F] F-DOPA uptake in right midbrain (presumably stemming from substantia nigra and ventral tegmentum) in adolescents with ADHD with no differences in prefrontal cortex (PFC) or striatum (Ernst et al., 1999) but revealed decreased [^{18}F] F-DOPA uptake in left and medial PFC of adults with ADHD (Ernst et al., 1998). Small samples and the very low signal obtainable from [^{18}F] F-DOPA outside the striatum make it hard to interpret these results. Furthermore, it is important to remember that [^{18}F] F-DOPA binding is a measure of dopa decarboxylase activity, and thus would reflect *both* norepinephrine (NE) and DA terminals in cortex.

In the midst of these conflicting findings, there has been one point of convergence. Two independent groups found that while stimulants administered intravenously produced mixed effects in most brain regions, regional glucose metabolism in the cerebellum was consistently and significantly increased by either methylphenidate (Volkow et al., 1997) or D-amphetamine (Ernst et al., 1994).

The Neurobiology of ADHD and Stimulant Drug Action

The neurobiological basis of ADHD has become the focus of both clinical and basic research. There is increasing agreement that dysregulation of fronto-striatal circuits likely underlies many of the symptoms of ADHD, and that many of the therapeutic effects of stimulants may arise from actions on these circuits. The catecholamine innervation of these circuits has been beautifully characterized in both rodents and primates (see Chapter 3 by Lewis), and we have begun to appreciate the profound effect of catecholamines on the functioning of these circuits (see Chapter 6 by Berridge; Chapter 4 by Taylor & Jentsch; Chapter 5 by Grace; Chapter 7 by Arnsten; and Chapter 13 by Mehta, Sahakian, & Robbins). However, basic research on ADHD has lagged behind that on many other neuropsychiatric disorders.

PFC Dysfunction

The PFC plays a critical role in regulating attention, inhibiting inappropriate or impulsive behaviors, and using working memory to plan, organize, and guide behavior effectively. Convergent evidence from a variety of fields, including the neuroimaging data described, suggests that the PFC is dysfunctional in patients with ADHD, with particular emphasis on the right hemisphere. Research in rodents and monkeys indicates that both dopamine and norepinephrine have profound effects on the cognitive functioning of the PFC: low levels of dopamine D1 and norepinephrine alpha-2 receptor stimulation are essential for PFC function, while very high levels of dopamine D1 and norepinephrine alpha-1 receptor stimulation impair PFC function. Research in normal humans, reviewed by Mehta and colleagues (Chapter 13) has attempted to specify, by administration of receptor-specific drugs, the catecholaminergic mediation of specific cognitive functions. These studies showed that D2 agonists (bromocriptine) enhanced spatial (but not nonspatial) working memory performance, whereas D2 antagonists (sulpiride, pergolide) impaired these functions. The effects of alpha-2 noradrenergic agents, such as clonidine, however, appeared to be highly dependent on the dose administered, the novelty of the task, and/or the baseline state of arousal of the individual. For example, a high dose of clonidine (2.5 μg/kg) enhanced performance on a spatial working memory task when the task was familiar, but impaired it when it was novel. A low dose (1.5 μg/kg), by contrast, impaired performance when the task was familiar but had no effect when novel. These results suggested the possibility that the combination of novelty and high dose induced an excessive and maladaptive level of arousal. The lower doses, possibly by acting presynaptically, may have reduced noradrenergic "arousal" to suboptimal levels. This would impair performance on practiced tasks but might have no effect on tasks that are more challenging/stressful and therefore inherently "arousing." Possibly parallel effects of improvement at high doses and impairment at low doses were also reported on a visuospatial paired-associate learning task. More consistent beneficial effects on cognitive performance have been observed with guanfacine, a selective alpha-2A adrenergic agonist with fewer side effects and less potent presynaptic actions (see Arnsten, Chapter 7; Jakala et al., 1999). In this regard, it is of interest that guanfacine is now being developed as a treatment for ADHD.

Because stimulants have marked effects on catecholamine transmission, the PFC is a likely site for their therapeutic actions. In the studies described by Mehta and colleagues (Chapter 13), stimulant medication improved performance on measures of spatial working memory and sustained attention in normal adults, as has also been shown for children with ADHD. Other research has shown improvement in word list learning in adults and improved paired-associate learning in children following stimulant administration. In addition, a recent fMRI study showed that stimulant treatment increased blood flow in the PFC and simultaneously improved performance of a response inhibition task in ADHD subjects and in age-matched controls (Vaidya et al., 1998). Increased regional glucose metabolism in the PFC in normal adults following MPH in normal adults has also been shown (Volkow et al., 1997).

How might stimulants enhance PFC function, and how might these actions relate to possible etiologies of ADHD symptomatology? There are several possibilities. The prevailing view is that many ADHD symptoms may arise from insufficient catecholamine receptor stimulation in the PFC, which could in turn arise from a number of factors—for example, genetic alterations in the synthetic enzymes for the catecholamines, altered postsynaptic receptors that are less responsive to NE or DA stimulation, or presynaptic receptors that are more responsive to catecholamine stimulation, reduced DA or NE innervation of PFC, and so forth. This view of ADHD further assumes that stimulants ameliorate ADHD symptoms by increasing catecholamine release and normalizing function. This assumption is rarely challenged, but there have been so few studies of oral, *low* dose stimulant treatment in awake animals that there are few direct data to support or refute this assumption.

Biochemical studies in animals suggest that catecholamine-depletion in the PFC actually produces an exaggerated dopamine release during stress (e.g., see Deutch et al., 1990). Thus, one could have both insufficient and excessive catecholamine activity that is state-dependent. Normalization of catecholamine release might be particularly important under these conditions.

Changes in PFC catecholamines may also be therapeutic even if catecholamine activity in PFC is completely normal. For example, ADHD might arise simply from having an immature PFC (e.g., a smaller PFC, reduced neuropil). Data from both animal and human studies indicate that one can produce supranormal performance on PFC tasks by small increases in catecholamine stimulation in the PFC of normal young subjects (see Arnsten, Chapter 7, and Mehta et al., Chapter 13). Thus, small increases in catecholamine release induced by stimulants might enhance the functioning of an immature or small PFC and help to normalize function. Strengthening PFC function would also enhance PFC regulation of subcortical structures that may be affected in ADHD (see what follows).

None of these ideas are mutually exclusive, and it is very likely that many different etiologies can lead to similar symptoms. The evidence from animal studies suggests that the PFC has to have everything "just right"to function optimally, and that small changes in the neurochemical environment can have marked effects on PFC function. Thus, there could be a wide variety of alterations that could result in similar PFC dysfunction and ADHD symptomatology. It is widely recognized that ADHD is a heterogeneous disorder, and it is likely that many genetic and environmental factors contribute to the expression of these symptoms.

It is tempting to speculate that the predominantly inattentive subtype of ADHD might not involve PFC dysfunction but instead impairment of posterior cortical areas that devote attentional resources to stimulus processing (in contrast to the PFC, which inhibits the processing of irrelevant stimuli). The attentional abilities of the parietal cortex do not seem to be enhanced by manipulation of NE alpha-2 receptors (Witte & Marrocco, 1997), but are very sensitive to cholinergic manipulations (Voytko et al., 1994). As the neurochemical needs of posterior cortices may differ from those of the PFC, children with these symptoms may have different responses to medication and perhaps require alternative medication strategies.

Subcortical Dysfunction

Basal Ganglia

The PFC projects to many subcortical targets such as the striatum (both dorsal and ventral—i.e., the caudate and nucleus accumbens, respectively), the thalamus, the amygdala, the dopaminergic cells of the substantia nigra/ventral tegmental area, the noradrenergic cells of the locus coeruleus (LC), and more distant targets such as the cerebellum via the pons. Imaging data indicate that several of these structures may be altered in ADHD (see Castellanos, Chapter 10). Notably, several studies have found reduced size of the caudate and the floccunodular lobe of the cerebellum. (Small or poorly delineated structures such as the nucleus accumbens or LC are difficult to quantify, and thus it is not known whether these structures are altered in ADHD). The dorsal striatum (caudate and putamen) is thought to be essential for the selection, initiation, and execution of responses; the putamen may perform these functions for movement but some have speculated that the caudate may serve these same functions in the cognitive domain (Alexander et al., 1986).

The ventral striatum (i.e. nucleus accumbens) plays a critical role in reward and incentive behavior (reviewed in Taylor & Jentsch, Chapter 4). High levels of phasic dopamine activity in these structures promote movement and can induce locomotor hyperactivity. Grace (Chapter 5) has developed an elegant model of tonic/phasic dopamine activity in the nucleus accumbens which succeeds in reconciling contradictory observations and hypotheses concerning a hypo- versus hyper-dopaminergic state in ADHD, and presynaptic inhibitory versus postsynaptic excitatory effects of stimulant drug treatment. This model proposes that individuals with ADHD have low tonic dopamine levels (possibly due to insufficient stimulation from an underactive or underdeveloped PFC), and consequently elevated phasic dopamine release. By blocking dopamine reuptake, methylphenidate, administered orally in low, therapeutic doses, produces accumulation of dopamine in the synapse, which diffuses into the extrasynaptic space. Increased dopamine in the extrasynaptic space stimulates impulse-modulating autoreceptors, producing a decrease in spike-dependent dopamine release. The net effect is a reduction in phasic dopamine, and increase in tonic dopamine, to normal levels. Interestingly, Grace's model is consistent with PET findings of Ernst and colleagues (1999) showing elevated [^{18}F] DOPA binding in the right midbrain (substantia nigra and ventral tegmental area) of adolescents with ADHD. Also relevant in this context may be Castellanos' finding in children with ADHD of increased CSF HVA, which was positively correlated with severity and also with response to MPH (Castellanos et al., 1996). Finally, it is noteworthy that although PET showed MPH *increased* metabolism in the cerebellum, as described, it *reduced* metabolism in the basal ganglia (Volkow et al., 1997).

Cerebellum

The cerebellum has traditionally been thought of as a structure that is important for the on-line guidance of movement, permitting smooth, coordinated responses. The flocculonodular lobe has been thought to be particularly involved in vestibular coordination. However, recent imaging studies indicate that the cerebellum is also activated during demanding cognitive situations. Therefore, the finding of reduced cerebellar size in the area of the floccunodular lobe in ADHD patients is intriguing. Might this reduced size relate to symptoms of motor clumsiness? Or might there be cognitive correlates of altered floccunodular volume? Recent research in animals suggests that cerebellar dysfunction may relate to the hyperactivity observed in ADHD (see Ferguson, Chapter 8). Thus, the cerebellum is an important focus for future research.

Locus coeruleus

The noradrenergic nucleus, locus coeruleus (LC), may be another subcortical structure that is highly relevant to ADHD. As reviewed in Chapter 6 by Berridge, recordings from the LC in monkeys demonstrate a fascinating relationship between behavioral state and the response of the LC to targets versus distractors (Rajkowski et al., 1998). The LC showed a progressive increase in baseline firing from drowsy to alert to hyperaroused states. In contrast to baseline firing, the phasic responses of the LC to target stimuli were most pronounced during an optimal alert state, and diminished during either drowsy or hyperaroused conditions, while the responses to distractors were suppressed during alert attentive states and evident during either drowsy or hyperaroused states. It is possible that the PFC regulates the appropriate phasic response of the LC, for the PFC provides one of the few "intelligent" inputs to the LC. During alert waking, the LC exhibits moderate baseline firing, which may lead to optimal levels of NE release in the PFC, engaging postsynaptic alpha-2A receptors and enhancing PFC modulation of the LC (LC responses to targets but not distractors). A recent computer model of LC cell interactions suggests that coupling between LC cells may be increased during improved cognitive performance, and PFC inputs may enhance this coupling (Usher et al., 1999). In contrast, the LC may exhibit inappropriate phasic responses (responses to distractors but not targets) during drowsiness or hyperarousal due to poor PFC regulation of the LC. The PFC may be "off-line" during periods of low arousal due to very low levels of NE release (insufficient postsynaptic alpha-2 receptor stimulation in the PFC), and "off-line" during periods of hyperarousal due to excessive NE release (stimulation of alpha-1 receptors in the PFC, impairing PFC function). Interestingly, studies in anesthetized rats suggest that exposure to chronic stress increases basal LC firing as well as increasing LC phasic responses to salient stimuli (electric shocks; Mana & Grace,

1997). Thus, stress may contribute to the high rate of LC firing in hyperaroused, distractible monkeys. Alternatively, excessive firing of the locus coeruleus, and an associated hyperactive and distractible behavioral state may result from low levels of tonic inhibition, perhaps due to insufficient NE stimulation of presynaptic, down-modulatory autoreceptors (analogous to the model described by Grace for tonic/phasic dopamine regulation in nucleus accumbens). Stimulants have also been shown to potently inhibit firing of locus coeruleus in anesthetized rats (Graham & Aghajanian, 1971). In addition, the adverse cognitive effects of clonidine at low (0.5 μg/kg) and high (5.0 μg/kg), but not intermediate (2.0 μg/kg) doses in normal human subjects have been explained as possibly due to stimulation of presynaptic autoreceptors (Jakala et al., 1999) at low doses, which might be facilitative in an individual with a "hyperaroused" locus coeruleus (Hunt et al., 1994; Pliszka et al., 1996). It would be interesting to see how stimulants might alter LC firing in monkeys during differing attentional states. As both the PFC and the LC project to the DA cell bodies of the ventral tegmental area, a similar regulatory relationship may be paralleled in the DA system as well. Abnormal activity of either the NE or DA cells could have widespread effects on the functioning of cortico- striatal circuits that regulate behavior.

Catecholamine Dysregulation

An obviously primary question with respect to mechanisms of action of stimulants concerns the extent to which their therapeutic effects are differentially mediated by norepinephrine and dopamine. Research at the cellular level (see Taylor & Jentsch, Chapter 4, Grace, Chapter 5, and Berridge, Chapter 6) indicate that, at higher doses, methylphenidate and D-amphetamine potentiate the effects of both NE and DA by blocking their reuptake and, in the case of D-AMP, facilitating their release. PET studies of labeled MPH administered orally to primates at doses in the clinical range reveal occupancy of over 50% of total striatal DA transporter sites (see Chapter 11 by Swanson & Volkow). Furthermore, functional MRI has shown increased activation of striatum during performance of a challenging motor inhibition task in children with ADHD (Vaidya et al., 1998), contrasting with reduced activation in comparison children. These data, as well as the abundant evidence in animals and humans linking striatum to regulation of motor function, strongly suggest that the effects of stimulants on motor activity in ADHD are at least partly mediated by effects of stimulants on dopamine in striatum. To the extent that activity level reflects more general behavioral disinhibition, however, norepinephrine is likely to be involved as well. Furthermore, it is possible that NE can indirectly modulate striatum by enhancing cortical regulation thereof. It is also important to consider that output from striatum back to cortex is noradrenergically regulated via the thalamus. Dopamine may be primarily involved in mediating any effects of stimulants on incentive motivation and response to reinforcement as discussed by Taylor & Jentsch (Chapter 4).

The best current evidence indicates that effects of the stimulants on cognitive self-regulatory processes, including attention and impulse-control, are likely to be mediated by *both* dopamine and norepinephrine. As reviewed by Arnsten and also by Mehta and colleagues, delayed responding and working memory, which appear to be fundamental to cognitive self-regulatory processes, are sensitive to both dopaminergic (D1 and D2) and noradrenergic (alpha-1 and alpha-2) manipulations. Investigation of effects of receptor-specific drugs on well-validated cognitive tasks in humans has the potential to delineate the different roles of the catecholamines in mediating such processes as sustained and focused attention, spatial orienting, attentional set-shifting, interference control, and paired-associate learning, as well as verbal and nonverbal working memory and planning. Thus far, such studies, reviewed by Mehta and colleagues (Chapter 13) have revealed that in normal adults sustained attention and paired associate learning were influenced by noradrenergic manipulations but not by D2 receptor agonists and antagonists. However, Rapport and colleagues (1993) reviewed studies comparing the cognitive effects of stimulants and tricyclic antidepressants (NE reuptake inhibitors) on cognitive functions in children with ADHD, and reached a different conclusion—that sustained attention was more likely dopaminergically mediated whereas tasks requiring both inhibitory control and higher-order learning processes were more likely noradrenergically mediated. These differences underscore the fact that, given a dearth of receptor-specific agents that can be administered to humans—notably the lack of D1 and alpha-1 agonists and antagonists—any inferences at present concerning mediation of discrete cognitive processes by specific catecholamines are premature.

Studies of nonstimulant medications in clinical use to treat ADHD, comprehensively reviewed by Pliszka (Chapter 14), yield the conclusion that the effects of stimulants on overt behavior, as observed and rated by parents and teachers, are also mediated by effects on both norepinephrine and dopamine. An earlier review by Zametkin and Rapoport (1987) of clinical trials of drugs with predominantly dopaminergic, noradrenergic, or combined effects reached a similar conclusion. Of particular interest is that, in contrast to the minimal or moderate effects of more selective agents, the effects of MAO-A and MAO A + B inhibitors (that inhibit catabolism of both NE and DA) were "clinically indistinguishable" from those of the stimulants in improving behavior ratings and enhancing performance on the CPT (Zametkin et al., 1985). Unfortunately, the dietary restrictions necessitated by the effects of these drugs limit their clinical use.

While it is not possible at present to differentiate the contributions of norepinephrine and dopamine to the clinical effects of stimulants, it is possible to exclude any direct role of serotonin in mediating stimulant drug effects (see Berridge, Chapter 6). Kuczenski and colleagues demonstrated that MPH failed to increase levels of extracellular serotonin at any dose and that D-amphetamine did so only at doses much greater than those used clinically (Kuczenski & Segal, 1997). This finding is entirely congruent with the lack of therapeutic effectiveness of fenfluramine and the minimal benefit of serotonin reuptake inhibitors in the treatment of ADHD (Pliszka, Chapter 14). Despite the apparent lack of a direct effect, there remains the possibility of indirect effects of stimulants on serotonin, mediated by primary effects on NE and DA, as elaborated by Pliszka.

Stimulants as Therapeutic Agents

Domains of Demonstrated Efficacy

Many well-controlled studies have demonstrated that stimulants produce improvements in manifestations of the three core symptoms of ADHD on the basis of parent and teacher ratings on questionnaires and direct behavioral observations (Greenhill, Chapter 2). Significant reductions in motor activity across school and weekend play situations have been shown using truncal actometers (Porrino et al., 1983a). Improvement has also been objectively demonstrated on laboratory measures of reaction time, vigilance, inhibitory control, and paired-associate learning (Denney & Rapport, Chapter 12). Contrary to an early and influential study suggesting optimization of cognition at lower doses than optimization of behavior (Sprague & Sleator, 1977), dose-response effects for both teacher-rated behavior and performance on cognitive tasks have generally been linear at doses up to 1.0 mg/kg. In fact, studies reviewed by Denney and Rapport (Chapter 12) have suggested that tasks of higher-order "controlled" cognitive processing require *higher* doses for maximum improvement than do simpler "automatic" tasks, a conclusion also reached by Douglas in a recent review (Douglas, 1999). In one important and interesting study, however, normalization of academic performance occurred at lower doses, and at a lower overall rate, than did normalization of attention and behavior in the classroom (Rapport et al., 1994). The results indicated that drug-induced improvements in attention and behavior were "necessary but not sufficient" to optimize academic functioning. A review of the voluminous stimulant drug literature indicates that effect sizes have generally been larger for behavioral (0.8–1.0) than for cognitive (0.6–0.8) changes (Spencer et al., 1996). However, adverse effects on cognition at higher doses, relative to placebo, have not been reported in any study. Although the positive response rate to stimulants is 75% to 95%, and the adverse response rate is small (less than 10%), there is wide individual variation in dose-response functions both within and across domains of effectiveness. At present, there are no clinically useful behavioral or physiological predictors of individual drug response.

Disjunctions between the effects of stimulants on cognitive processes and activity may have significance for understanding their mechanisms of action. In addition to the differences in dose-response relationships described above and also observed at a subclinical dose (Solanto, 1986), differences in time-action curves for drug effects on attention and overt behavior have been described. In several studies, the effects of methylphenidate (Pelham et al., 1987; Swanson & Volkow, Figure 11–1, Chapter 11) and of D, L-amphetamine (Adderall) (Pelham et al. 1999) on activity and classroom behavior lasted as long as 7 to 8 hours, compared to only 2 to 3 hours for measures of attention. The same divergence between time–action curves for activity and attention was previously reported in studies employing laboratory measures of motor activity and cognition, including vigilance (Porrino et al., 1983a) and reaction time (Solanto & Conners, 1982). Taken together, these findings of stimulant drug effects on activity at lower doses and for longer periods than those on cognition suggest that the striatum and the prefrontal cortex may differ in their sensitivity to stimulant effects and in their time courses. Alternatively, it may be more difficult to optimize neurochemical conditions for cognitive processing than for motor control. Regional differences in dopamine transporter distribution (see Taylor & Jentsch, Chapter 4) as well as in distribution and function of presynaptic autoreceptors (see Grace, Chapter 5) may play a role in these differences in drug response.

Nonspecificity of Stimulant Effects

A long-standing enigma in the field of stimulant drug research has been the issue of why a presumably activating drug should *decrease* activity in children with ADHD. For many years, this effect was described as "paradoxical" and unique to children with ADHD; in normal children it was assumed that the drug would cause the expected *increase* in activity. It was not until 1978 that a pivotal study at NIMH by Rapoport and colleagues disproved this notion by demonstrating qualitatively similar changes in ADHD children and normal children and adults—that is, D-AMP enhanced attention and reduced activity level and impulsivity in all three groups of subjects (Rapoport et al., 1978, 1980). Recent studies have also shown that methylphenidate reduces symptoms in adults with ADHD as well (Spencer et al., 1995).

The qualitatively similar effects of stimulants on behavior in persons with ADHD and in normal individuals suggest that stimulants do not target a specific deficit in ADHD. This conclusion is also supported by the failure to find interactions between methylphenidate effects and specific attentional or information-processing deficits in ADHD (Solanto, 1998), and also by the effectiveness of the stimulants in children of varying ages with varying comorbidities (Greenhill, Chapter 2), as well as in those who do not meet full criteria for the disorder. However, the response to stimulants may be along a continuum. This hypothesis would be consistent with results to date from "rate-dependency" research, described below, as well as with the reports in some clinical studies that increasing severity is a positive predictor of clinical response to stimulants (Buitelaar et al., 1995; Taylor et al., 1987). It is also supported by recent familial genetic studies indicating that the symptoms of inattention, hyperactivity, and impulsivity are continuously distributed. Grace (Chapter 5) has proposed a neurobiological mechanism that is able to account for both a continuous distribution of symptoms and a continuously distributed response to stimulant medication.

Rate-Dependent Effects

The activity-reducing effects of stimulants have been viewed in the context of "rate-dependency," which refers to the observation, documented in extensive previous studies in animals (Dews & Wenger, 1977), that drug-induced change in frequency of behavioral response is an inverse function of baseline response frequency, such that low rates of responding are increased whereas higher baseline rates are increased to a lesser extent or are decreased. Robbins and Sahakian (1979) re-analyzed data from several studies of stimulant effects in children and adults, relating change in activity level on drug (as a percentage of baseline) to baseline level of activity and demonstrated a negative linear regression slope. This documentation that points for both normal children and adults did not significantly deviate from the regression curve for children with ADHD indicated that the reduction in activity was inherent in the pharmacodynamics of the drug effect and was in no way "paradoxical."

Mathematical and statistical challenges to the rate-dependency effect have been posed by Gonzalez and Byrd (1977) and by Swanson (1988). Gonzalez and Byrd pointed out the inherent dependency in relating drug response rate, expressed as a percentage of baseline rate, to baseline rate, and the likelihood that this can produce spuriously high correlations. Swanson demonstrated that the *chance* correlation between baseline response rate and percent change in response on drug is 0.7, indicating that any study purporting to show rate-dependency would have to demonstrate a correlation significantly higher than this value. Although higher correlations have been reported—for example 0.94, $n = 12$, reported by Solanto (1986)—most studies have not explicitly fulfilled this criterion. A second and related statistical concern is that rate-dependent reduction in response rate reflects the phenomenon of regression to the mean. Although analysis of placebo effects by Robbins and Sahakian (1979) did not support this interpretation, repeated testing may be necessary to fully resolve this matter.

Examination of earlier studies in primates and rodents, not previously reviewed within this context, yields support both for activity-reducing effects of stimulants at low doses and also for rate-dependent effects. D-amphetamine (0.1–0.8 mg/kg p.o.) dose-dependently *decreased* locomotor activity in squirrel monkeys (Isaac & Kallman, 1979; Lowther & Isaac, 1976), and rhesus macaque monkeys (Alexander & Isaac, 1965). Fixed-interval responding was also reduced at these doses (Goethe & Isaac, 1977; Stinnette & Isaac, 1975). Of particular interest, the reduction in activity was especially pronounced in monkeys who were spontaneously hyperactive, decreasing to levels below that of normal monkeys when treated with comparable doses (Davis, 1957). In addition, species differences were noted such that the nocturnal owl monkey, normally less active in light, showed an *increase* in daytime activity following low-dose D-AMP, whereas the diurnal squirrel monkey showed the decrease described (Isaac & Troelstrup, 1969). Similar strain-dependent differences in locomotor response to D-amphetamine (0.5–1.0 mg/kg) were also seen in mice (Kitahama & Valatx, 1979; Oliverio et al., 1973); strains responding with a decrease in locomotor activity were found to have higher D_2 receptor density than those showing an increase (Helmeste & Seeman, 1982). Rats (Cherokee-SD male albino) showed *increased* activity following either low-dose D-AMP (0.2–1.6 mg/kg) or low-dose methylphenidate (0.4–6.4 mg/kg) (Kallman & Isaac, 1975). The possibility that the decrease in locomotor activity in these studies was due to a subtle form of stereotypy has not been completely ruled out.

An important theoretical question that may be posed concerning rate-dependent effects is whether the drug effect is a function specifically of the *rate* of behavior (or, more generally, the corresponding behavioral or neurobiological "state") or, alternatively, a function of the neurobiological substrate or "trait" that generates the rate as a stable feature of the individual's baseline behavior. This question might be addressed by generating different levels of activity level *within* a given individual or animal and observing drug effects as a function of those differences. Demonstration of rate-dependency under these circumstances would provide support for the first alternative hypothesis. Several studies have employed this approach by generating different response frequencies on operant tasks by manipulating reinforcement schedules. Two such studies in animals (Heyman, 1992; Sagvolden et al., 1988) and two in children with ADHD (Rapport et al., 1985; Weber, 1985) failed to yield unequivocal support for the rate-dependency model. Changes in response rate were better explained as a result of MPH-induced changes in reinforcement gradient or reinforcement efficacy, which introduced a confound. Thus far, then, the only support for rate-dependent effects of stimulants in ADHD has come from studies in which the baseline rate of behavior was that which was *spontaneously* generated by the normal or hyperactive child, adult, or animal subject, and that may therefore be, at least in part, a function of a neurobiological substrate (e.g., D_2 receptor density) characteristic of that individual or group. In order to fully resolve these questions, further studies are needed of stimulant effects in individuals with ADHD and in normal subjects in whom baseline levels of activity or response are varied by environmental manipulations (ideally not involving reinforcement). In addition, the statistical concerns described above must be addressed.

Acute versus Chronic Drug Effects

Concern has been raised that long-term use of stimulants, which are typically administered for months or years, might produce tolerance or sensitization. Sensitization, or "reverse tolerance," refers to the phenomenon, observed in animals and in two studies from one laboratory in humans, wherein after repeated use, a given dose of a stimulant produces *greater* effects on behavior than it did initially. Sensitization might be of concern if it conferred a greater risk for drug abuse later in life. Studies, however, have not found evidence of tolerance to stimulants in clinical use (Greenhill, Chapter 2). In addition, recent analysis of data from the titration trial portion of the Multimodal Treatment Study of ADHD (MTA), during which placebo and three different doses of MPH were switched daily and randomly, yielded no evidence of sensitization of behavioral response over the course of the five weekly administrations of any of the three doses (Swanson et al., unpublished). The mean daily dose at the end of the titration trial was 30.5 mg/day; 14 months later, the subjects treated with medication alone were taking 37.7 mg/day, and the group treated with medication and intensive psychosocial interventions took an average of 31.2 mg/day (Vitiello et al., in review). In a naturalistic study extending 3 to 10 years, stimulant dose corrected for body surface area was nearly constant, with no loss of effect (Safer & Allen, 1989). Thus, there is no evidence that the long-term clinical response to stimulants varies significantly from the acute response. This further suggests that long-term changes in receptor sensitivity do *not* occur as a function of continued stimulant drug treatment. However, the possibility that long-term treatment with stimulants might represent an independent risk factor for later substance abuse has been raised (Lambert & Hartsough, 1998). Subsequently, two independent, albeit flawed, longitudinal studies have reported the reverse: that stimulant treatment appears to diminish the risk of later substance abuse (Biederman et al., 1999; Paternite et al., 1999). The absence of euphorigenic effects of stimulants at clinically effective doses in children (Greenhill, Chapter 2) as well as the use of much lower doses and an oral route of administration, also suggest that mechanisms of stimulant drug action in ADHD differ in important ways from those that occur in drug abusers. These differences would concomitantly appear to reduce the abuse potential of these drugs in ADHD.

Methylphenidate Versus D-Amphetamine

Neurochemical studies of the effects of D-amphetamine and methylphenidate have revealed different mechanisms of action of these compounds at the cellular level. Functionally, D-amphetamine inhibits reuptake of dopamine and norepinephrine and also facilitates their release, whereas methylphenidate only inhibits catecholamine reuptake (Chapter 5 by Grace, and Taylor & Jentsch, Chapter 4). Furthermore, these two stimulants differ in the relative potency of their effects on norepinephrine and dopamine (Berridge, Chapter 6). It is somewhat surprising, therefore, that no qualitative differences have been found in the therapeutic response or side effects profiles of these compounds (Elia et al., 1991), perhaps because both norepinephrine and dopamine can exert facilitative effects on prefrontal cortical function. In group comparisons, the few distinctions appear to be primarily related to the much faster rate of excretion of methylphenidate relative to D-amphetamine. Thus, at comparable doses, the behavioral half-life of amphetamines such as Adderall (Pelham et al., 1999; Swanson et al., 1998) is longer than that of MPH by 1 to 2 hours. D-amphetamine produced greater insomnia and greater diminished appetite than methylphenidate in a recent study (Efron, 1999). Of particular interest is that Elia et al. (1991) reported that 25% of their sample of 48 improved on only one of the two stimulants when both were tried. Refinement of our understanding of the core features of ADHD and the corresponding effects of the psychostimulants (Denney & Rapport, Chapter 12), and parallel elucidation of the cognitive and behavioral functions mediated by norepinephrine (Arnsten, Chapter 7, and Berridge, Chapter 6) and dopamine (Arnsten, Chapter 7; Taylor & Jentsch, Chapter 4; and Grace, Chapter 5) may help investigators to uncover clinically relevant differences in the psychopharmacology of these drugs.

Animal Models

Animal models can be helpful in understanding the etiology of neuropsychiatric disorders and in identifying novel pharmacotherapies. However, it has not been easy to develop animal models of ADHD which truly (*1*) mimic behavioral symptoms of ADHD, (*2*) show calming responses to stimulant medication, and (*3*) share some of the same neural changes observed in patients.

It is very easy to create animal models that *superficially* express the symptoms of ADHD. For example, a large variety of perturbations can produce hyperactive locomotor activity in animals, particularly in rodents. However, it is likely that many of these perturbations would be irrelevant to the hyperactivity that occurs in ADHD patients. Furthermore, "attention" is often so poorly defined that one can readily induce "attention deficits" that have no similarity to the types of cognitive changes that occur in ADHD. "Inattention" can be caused by either improper attention regulation—paying attention to irrelevant external or internal stimuli—or by insufficient delegation of attentional resources, as seen in its most exaggerated form in patients with neglect. The former type of attention problem often arises from dysfunction of the PFC and is common in ADHD (Satterfield et al., 1988), while the latter deficit can be produced by lesions to the right parietal cortex and has generally not been associated with ADHD (Swanson et al., 1991). Modeling these symptoms in animals can be problematic, especially in rodents where very indirect and poorly validated measures of "attention" (e.g., speed of extinction, rearing duration) are often used. The widespread use of rodents as animal models can also be problematic because rodents, (particularly mice), have very little PFC. Thus the brain region most likely affected in patients is minimally functional in many animal models. Nonhuman primates have much more developed cortices and can be very useful for studying the neural mechanisms underlying higher cognitive functions. Indeed, some labs have succeeded in having monkeys perform human cognitive tasks. However, even in nonhuman primate models performing the "same" tasks as humans, one must be cautious that the animals are not using different cognitive strategies to solve the problem, and thus challenging differing neural circuits.

Most animal models of ADHD would require an amelioration of symptoms with stimulant treatment; however, this too is an area where the researcher must remain cautious. Stimulants can "reduce" locomotor activity by producing stereotyped behavior rather than truly ameliorating symptoms. Stereotypy occurs in all animals with increasing dose of drug; thus one must be cautious that reduced locomotor activity is not simply an artifact of increased drug sensitivity. Many automated activity monitors are unable to detect perseverative or stereotyped behaviors, and thus these data are often unavailable to researchers. As described previously, animal models should observe amelioration of symptoms with low dose, oral stimulant administration if they are to truly mimic the disorder. ADHD is so often misunderstood by both the scientific and lay communities, that researchers developing animal models all too often rely on myths about the disorder rather than real data. For example, a common myth is that ADHD patients show reduced activity, while normal individuals show increased activity, following stimulant administration (the so-called "paradoxical" effect of stimulants in ADHD). This myth has falsely guided several animal models (e.g., Gainetdinov et al., 1999), and it is not surprising that such models often lead to erroneous conclusions (e.g., that serotoninergic mechanisms underlie the beneficial effects of stimulants in ADHD (Gainetdinov et al., 1999).

Imaging studies and genetic analyses have begun to indicate important brain structures that could serve as intelligent targets for ADHD models. For example,

ADHD patients often show reduced size of the PFC, caudate, and floccunodular lobe, suggesting significant alterations in these structures. Indications of polymorphisms in the dopamine transporter, D4 DA receptor, and NE alpha-2 receptors in ADHD patients suggest possible molecular targets. However, it is important to note that polymorphisms are not equivalent to knockouts, and likely have qualitatively different effects on receptor function. For example, the D4 knockout mouse demonstrates *improved* coordination and is less active than wild-type controls (Rubinstein et al., 1997).

Animal models can either begin with a subject that shows symptoms similar to ADHD, and then proceed to try to identify neural changes that may underlie the behavioral abnormalities (e.g., the spontaneously hypertensive rat [SHR] model; Sagvolden, Chapter 9), or, conversely, can produce a known neural change and then observe whether any of the ensuing behavioral changes are relevant to ADHD (e.g., rats with selective depletion of DA). As described by Ferguson (Chapter 8), Shaywitz and colleagues showed more than 20 years ago that global DA depletion in rat pups induced hyperactivity that could be reduced by amphetamine administration. A few years later, adult rats with more selective lesions of the VTA were shown to be hyperactive, impulsive, and impaired on a test of working memory (reviewed in Le Moal & Simon, 1991). These same behavioral changes can be created by more selective depletions of the PFC. 6-OHDA lesions that deplete both DA and NE in the PFC induce hyperactivity and working memory deficits (Bubser & Schmidt, 1990; Simon, 1981) and impulsivity (Sokolowski & Salamone, 1994). Many of these changes have been observed in monkeys with catecholamine depletion restricted to the PFC (Brozoski et al., 1979). PFC cognitive deficits can also be seen in monkeys treated with low doses of MPTP, using a regimen that depletes striatal DA and PFC NE (Roeltgen & Schneider, 1991). MPTP- treated monkeys exhibit errors of omission that can be ameliorated by D2 agonists (Schneider et al., 1994). The effects of low dose, orally administered stimulants have not been studied in these lesioned rats or monkeys, although VTA-lesioned rats show reduced activity following stimulant administration (Le Moal & Simon, 1991). The SHR model has also been found to have reduced DA release in the PFC in some studies (Russell et al., 1995); norepinephrine was not examined. These data suggest that altered catecholamine transmission in the PFC may underlie ADHD symptoms in at least a subgroup of patients. It is noteworthy that most of this research has focused on DA, even though NE is often just as likely to have been depleted (Brozoski et al., 1979; Bubser & Schmidt, 1990; Roberts et al., 1994; Simon, 1981) or contributed to the indices measured (Ernst et al., 1998). Because changes in NE are known to induce distractibility and poor attention regulation in rats (Carli et al., 1983) and monkeys (Rajkowski et al., 1998), and because NE compounds are effective in treating ADHD (Spencer et al., 1998), greater focus on NE in future research seems warranted. Indeed, very recent biochemical studies in rats suggest that low doses of methylphenidate increase extrasynaptic norepinephrine levels while having no effect on dopamine (Berridge, Chapter 6). Thus, there is a great need for more research on NE mechanisms relevant to ADHD.

Future Research

Basic Neuroscience

One of the most important areas for future research is to examine how low-dose, *orally* administered stimulants alter behavior and neurochemistry in animal studies, particularly in young animals with developing brains. To date, almost all research on stimulants in animals has used injection as the route of administration. As discussed in the chapter by Swanson and Volkow, oral administration results in very different kinetics. Indeed, it is possible that orally administered stimulants may have effects opposite to that of injected drug—for example, primarily increasing tonic rather than phasic release (see Chapter 5 by Grace). Most studies on stimulants in animals have used higher amphetamine doses, examining mechanisms relevant to drug abuse or schizophrenia. There has been much less research directed specifically at the etiology of ADHD. However, we must be cautious in extending data from rodents to humans, particularly as there are potentially significant differences in stimulant metabolism between rats and humans. For example, the half-life for amphetamine in rats is about one hour (Lewander, 1971; Melega et al., 1995); it is 6 to 10 hours in humans (Brown et al., 1979; Matin et al., 1977). This faster metabolism may render rodents more susceptible to the addictive or sensitizing properties of stimulants.

Another methodological concern is that conditions for testing locomotor activity in rodents are often not conducive to ADHD research, as they are usually aimed at detecting increases rather than decreases in locomotor activity. For example, animals are often adapted to the activity cages before drug treatment, thus exhibiting very low levels of baseline activity. Thus, "floor" effects can often preclude observation of reductions in locomotor activity. Using more appropriate methods, we may be able to observe whether low-dose, orally administered stimulants can reduce activity in normal animals as well as in animal models of ADHD. It may be helpful to conduct these studies in an apparatus where the opportunity for focused investigatory behavior may provide a measure of attention regulation as well. With these methods in hand, one could then perform a detailed examination of the receptor mechanisms and neurochemical changes underlying low-dose, orally administered stimulant actions. Changes in both DA and NE systems would be relevant, in brain regions such as the PFC, dorsal and ventral striatum, and cerebellum (NE innervates the cerebellum). In primates, there is a dense DA innervation of the premotor and motor cortices; it would be of interest to observe whether these cortices are affected in ADHD and/or are a site of action for stimulant therapeutic effects. Given that the DA innervation of cortex appears to be less tightly regulated by autoreceptor mechanisms than is the projection to striatum (Deutch & Roth, 1990), it is possible that low-dose, orally administered stimulants could produce a small increase in phasic DA release in cortex while decreasing phasic DA release in striatum. It would also be important to characterize the effects of low-dose, orally administered stimulants on incentive behavior, associative processes, impulse inhibition, working memory, and attention regulation. Without these studies we may be misled by conclusions reached from research aimed at understanding brain changes in drug abuse and schizophrenia. Basic research focused specifically on ADHD is greatly needed to guide policies and strategies for pharmacological treatment of ADHD.

Translational Research

Goals of translational research in ADHD are to render the behavioral phenotype more precise and to identify neurobiologically meaningful cognitive/behavioral subtypes that then can be used to develop drug and other treatments specifically targeted to the symptoms of these subtypes. This could be accomplished in a number of ways. There is a need to first, develop an improved taxonomy of the cognitive and behavioral characteristics of ADHD. These data can then be viewed in the context of the neural circuits that subserve these functions, an area of burgeoning research in both humans and animals. Second, characterize how stimulants influence these differing cognitive abilities in normal adults and animals, as well as in ADHD patients, taking care to systematically assess the interaction between drug effects and arousal and task parameters. Parallel studies using functional MRI may further help to identify the cortical regions influenced by stimulant treatment. Third, elucidate the distinct neurochemical needs of different parts of cortex that subserve distinct cognitive functions. Recent research suggests that even *within* a cortical region (e.g., the dorsolateral PFC), neurochemical needs may alter depending upon the operations carried out by that cortex (e.g., working memory vs. attention shifting is optimally regulated by moderate vs. low levels of dopamine (Roberts et al., 1994). Most research in this area has focused on the dorsolateral PFC in monkeys and ventromedial PFC in rats. It is possible that other subregions of the PFC are modulated differently, but there has been little or no research in this area. Finally, use the more detailed understanding of how cognitive operations are modulated in brain to develop and test catecholaminergic drugs that target specific cognitive and behavioral symptoms in ADHD subtypes. This endeavor may be facilitated by the currently active development of neurotransmitter- and receptor-specific drugs (Spencer et al., 1998; Wilens et al., 1999).

Genetic and neuroimaging data play an important role in validating ADHD subtypes with differing drug treatment needs. Genetic studies of ADHD, proceeding on two independent tracks, are expected to eventually yield important insights into the neurodevelopmental mechanisms that result in the various forms of ADHD and related disorders. The results of candidate gene studies in ADHD have been unexpectedly fruitful, yielding preliminary positive reports of a number of genes of moderate effect in several dopaminergic and noradrenergic genes. Despite these encouraging reports, each of the positive findings has been sufficiently modest that there has been substantial support for the formation of confederations of investigators so that sufficiently large data sets can be accumulated to test the statistical robustness of candidate genes with adequate statistical power. Given current trends in psychiatric genetics, most catecholamine-related genes are likely to be tested in ADHD in the near future. However, even if a central role of catecholamine dysregulation in ADHD were to be confirmed, the genetic roots of such dysregulation may lie in distant gene families. Thus, work is also continuing on the second track of genetic research, leading to the collection of large numbers of families with multiple affected individuals, and particularly with affected siblings. Such families, and larger extended pedigrees, can be used to search for genetic loci that may be unrelated to existing pathophysiologic theories of ADHD. Findings from this so-called positional approach are likely to expand our insights into ADHD dramatically, and to lead to more targeted candidate gene studies in an iterative manner.

Imaging data have also outlined important areas for further research. A review of potential future technologies and directions for this research can be found in the chapter by Castellanos). Why are structures such as the frontal lobes, caudate, and flocculonodular lobe smaller in ADHD subjects? Can MR spectroscopy provide us with any clues? Might neuropathological data be available to help guide this research? And what are the functional consequences of these regions being smaller? Are there changes in neurotransmitter systems that can be observed in vivo using neuroreceptor imaging? Neuroreceptor imaging may be particularly helpful not only in identifying differences in individuals with ADHD but also in examining whether chronic stimulant usage produces long-term changes in catecholamine systems. Imaging data have helped scientists and the lay public alike to take ADHD more seriously as a neurobiological disorder, and it is likely that this arena of research will continue to provide some of our most important clues regarding this disorder.

Conclusion

In 1970, Conrad Kornetsky (1970) wrote, "If we make the assumption that in the hyperkinetic child there is an increase in the synthesis or turnover rate of norepinephrine, and this is instrumental in the hyperkinetic syndrome, we can entertain the idea that amphetamine blocks this norepinephrine turnover and this block in turnover rate is primarily responsible for the therapeutic effect of amphetamine" (p. 130). This prescient prediction is still viable and yet fundamentally untested. In the intervening three decades, the pendulum of pathophysiologic theories of ADHD has swung from norepinephrine to dopamine, with the occasional inclusion of epinephrine. If, as this chapter and this volume have demonstrated, we have not yet resolved this basic question, the chief reason has been that basic scientists shifted their focus to paradigms related to substance abuse, particularly of cocaine and amphetamines. This volume was motivated by the notion that the time is ripe for a new dialogue between neuroscientists and clinical investigators so that the scientific questions regarding ADHD and the therapeutic mechanisms of psychostimulant drugs in ADHD can be approached with the panoply of techniques that are now available to us. In this new decade and century, we still cannot expect "breakthrough" discoveries, whether in genetics, neuroimaging, neuropharmacology, cognitive neuroscience, or animal models, to make sense of the puzzle that is

ADHD. Rather, each of these fields, and others not yet glimpsed, has the potential to contribute to an integrated multidimensional perspective that will be required and that is within our reach.

References

Alexander, CE, DeLong, MR, & Strick, PL (1986). Parallel organization of functionally segregated circuits linking the basal ganglia and cortex. *Ann Rev Neurosci, 9,* 357–381.

Alexander, M, & Isaac, W (1965). Effect of illumination and D-amphetamine on the activity of the rhesus macaque. *Psychol Rep, 16,* 311–313.

American Psychiatric Association (1994). *Diagnostic and statistical manual of mental disorders* (4th ed.). Washington, DC: Author.

Barkley, RA (1997). Behavioral inhibition, sustained attention, and executive functions: Constructing a unifying theory of ADHD. *Psychol Bull, 121,* 65–94.

Biederman, J, Wilens, T, Mick, E, Spencer, T, & Faraone, SV (1999). Pharmacotherapy of attention-deficit/hyperactivity disorder reduces risk for substance use disorder. *Pediatrics, 104,* E201–E205.

Brown, GL, Hunt, RD, Ebert, MH, Bunney, WE, & Kopin, IJ (1979). Plasma levels of D-amphetamine in hyperactive children. *Psychopharmacology (Berl), 62,* 133–140.

Brozoski, TJ, Brown, RM, Rosvold, HE, & Goldman, PS (1979). Cognitive deficit caused by regional depletion of dopamine in prefrontal cortex of Rhesus monkey. *Science, 205,* 929–932.

Bubser, M, & Schmidt, W (1990). 6-OHDA lesions of the rat prefrontal cortex increases locomotor activity, impairs acquisition of delayed alternation tasks, but does not affect uninterupted tasks in the radial maze. *Beh Brain Res, 37,* 157–168.

Buitelaar, JK, Van der Gaag, J, Swaab-Barneveld, H, & Kuiper, M (1995). Prediction of clinical response to methylphenidate in children with attention-deficit hyperactivity disorder. *J Am Acad Child Adolesc Psychiatry, 34,* 1025–1032.

Carli, M, Robbins, TW, Evenden, JL, & Everitt, BJ (1983). Effects of lesions to ascending noradrenergic neurons on performance of a 5-choice serial reaction task in rats: implications for theories of dorsal noradrenergic bundle function based on selective attention and arousal. *Behav Brain Res, 9,* 361–380.

Castellanos, FX, Elia, J, Kruesi, MJP, Marsh, WL, Gulotta, CS, & Potter, WZ, et al. (1996). Cerebrospinal homovanillic acid predicts behavioral response to stimulants in 45 boys with attention-deficit/hyperactivity disorder. *Neuropsychopharmacology, 14,* 125–137.

Comings, DE, Gade-Andavolu, R, Gonzalez, N, Blake, H, Wu, L, & MacMurray, JP (1999). Additive effect of three noradrenergic genes (ADRA2a, ADRA2C, DBH) on attention-deficit hyperactivity disorder and learning disabilities in Tourette syndrome subjects. *Clin Genet, 55,* 160–172.

Davis, GD (1957). Effects of central excitant and depressant drugs on locomotor activity in the monkey. *Am J Physiol, 188,* 619–623.

Deutch, AY, Clark, WA, & Roth, RH (1990). Prefrontal cortical dopamine depletion enhances the responsiveness of mesolimbic dopamine neurons to stress. *Brain Res, 521,* 311–315.

Deutch, AY, & Roth, RH (1990). The determinants of stress-induced activation of the prefrontal cortical dopamine system. *Prog Brain Res, 85,* 367–403.

Dews, PB, & Wenger, GR (1977). Rate-dependency of the behavioral effects of amphetamine. In T Thompson & PB Dews (Eds.), *Advances in behavioral pharmacology* (Vol. 1, pp. 167–227). New York: Academic Press.

Diamond, A, O'Craven, KM, & Savoy, RL (1998). Dorsolateral prefrontal cortex contributions to working memory and inhibition as revealed by fMRI. *Society for Neuroscience Abstracts, 24,* 1251.

Douglas, VI (1999). Cognitive control processes in Attention-Deficit/Hyperactivity Disorder. In HC Quay & AE Hogan (Eds.), *Handbook of disruptive behavior disorders* (pp. 105–138). New York: Kluwer Academic/Plenum.

Efron, D (1999). Methylphenidate versus dextroamphetamine in ADHD. *J Am Acad Child Adolesc Psychiatry, 38,* 500.

Elia, J, Borcherding, BG, Rapoport, JL, & Keysor, CS (1991). Methylphenidate and dextroamphetamine treatments of hyperactivity: Are there true nonresponders. *Psychiatry Res, 36,* 141–155.

Ernst, M, Zametkin, AJ, Matochik, JA, Jons, PH, & Cohen, RM (1998). DOPA decarboxylase activity in attention deficit disorder adults. A [fluorine-18] fluorodopa positron emission tomographic study. *J Neurosci, 15,* 5901–5907.

Ernst, M, Zametkin, AJ, Matochik, JA, Liebenauer, L, Fitzgerald, GA, & Cohen, RM (1994). Effects of intravenous dextroamphetamine on brain metabolism in adults with attention-deficit hyperactivity disorder (ADHD). Preliminary findings. *Psychopharmacol Bull, 30,* 219–225.

Ernst, M, Zametkin, AJ, Matochik, JA, Pascualvaca, D, Jons, PH, & Cohen, RM (1999). High midbrain 18F-DOPA accumulation in children with ADHD. *Am J Psychiatry, 156,* 1209–1215.

Gainetdinov, RR, Wetsel, WC, Jones, SR, Levin, ED, Jaber, M, & Caron, MG (1999). Role of serotonin in the paradoxical calming effect of psychostimulants on hyperactivity. *Science, 283,* 397–401.

Gjone, H, Stevenson, J, & Sundet, JM (1996). Genetic influence on parent-reported attention-related problems in a Norwegian general population twin sample. *J Am Acad Child Adolesc Psychiatry, 35,* 588–596.

Goethe, KE, & Isaac, W (1977). Effects of amphetamine and methylphenidate on fixed-interval responding in the squirrel monkey. *Pharmacol, Biochem, and Behav, 7,* 79–82.

Goldman-Rakic, PS (1996). The prefrontal landscape: implications of functional architecture for understanding human mentation and the central executive. *Philos Trans R Soc Lond B Biol Sci, 351,* 1445–1453.

Gonzalez, FA, & Byrd, LD (1977). Mathematics underlying the rate-dependency hypothesis. *Science, 195,* 546–550.

Graham, AW, & Aghajanian, GK (1971). Effects of amphetamine on single cell activity in a catecholamine nucleus, the locus coeruleus. *Nature, 234,* 100.

Halperin, JM, Newcorn, JH, Koda, VH, Pick, L, McKay, KE, & Knott, P (1997). Noradrenergic mechanisms in ADHD children with and without reading disabilities: A replication and extension. *J Am Acad Child Adolesc Psychiatry, 36,* 1688–1696.

Helmeste, DM, & Seeman, P (1982). Amphetamine-induced hypolocomotion in mice with more brain D_2 dopamine receptors. *Psychiatry Res, 7,* 351–359.

Heyman, GM (1992). Effects of methylphenidate on response rate and measures of motor performance and reinforcement efficacy. *Psychopharmacology (Berl), 109,* 145–152.

Hunt, RD, Hoehn, R, Stephens, KJ, Riley, W, & Osten, C (1994). Clinical patterns of ADHD: A treatment model based on brain functioning. *Compr Ther, 20,* 106–112.

Isaac, W & Kallman, MD (1979). Locomotor effects of D-amphetamine and methyphenidate in young squirrel monkeys. *Bulletin of the Psychonomic Society, 14*, 315–317.

Isaac, W & Troelstrup, R (1969). Opposite effect of illumination and D-amphetamine upon activity in the squirrel monkey (Saimiri) and owl monkey (Aotes). *Psychopharmacology (Berl), 15*, 260–264.

Jakala, P, Riekkinen, M, Sirvio, J, Koivisto, E, Kejonen, K, & Vanhanen, M, et al. (1999). Guanfacine, but not clonidine, improves planning and working memory performance in humans. *Neuropsychopharmacology, 20*, 460–470.

Kallman, WM & Isaac, W (1975). The effects of age and illumination on the dose- response curves for three stimulants. *Psychopharmacology (Berl), 40*, 313–318.

Kitahama, K & Valatx, J (1979). Strain differences in amphetamine sensitivity in mice. *Psychopharmacology (Berl), 66*, 189–194.

Kornetsky, C (1970). Psychoactive drugs in the immature organism. *Psychopharmacology (Berl), 17*, 105–136.

Kuczenski, R & Segal, DS (1997). Effects of methylphenidate on extracellular dopamine, serotonin, and norepinephrine: Comparison with amphetamine. *J Neurochem, 1997, 68*, 2032–2037.

Lambert, NM & Hartsough, CS (1998). Prospective study of tobacco smoking and substance dependencies among samples of ADHD and non-ADHD participants. *J Learn Disabil, 31*, 533–544.

Le Moal, M & Simon, H (1991). Mesocorticolimbic dopaminergic network: functional and regulatory roles. *Physiol Rev, 71*, 155–234.

Levy, F, Hay, DA, McStephen, M, Wood, C, & Waldman, I (1997). Attention-deficit hyperactivity disorder: a category or a continuum? *J Am Acad Child Adolesc Psychiatry, 36*, 737–744.

Lewander, T (1971). On the presence of p-hydroxynorephedrine in the rat brain and heart in relation to changes in catecholamine levels after administration of amphetamine. *Acta Pharmacol Toxicol, 29*, 33–48.

Lowther, WR & Isaac, W (1976). The effects of D-amphetamine and illumination on behaviors of the squirrel monkey. *Psychopharmacology (Berl), 50*, 231–235.

Mana, MJ & Grace, AA (1997). Chronic cold stress alters the basal and evoked electrophysiological activity of rat locus coeruleus neurons. *Neuroscience, 81*, 1055–1064.

Matin, SB, Wan, SH, & Knight, JB (1977). Quantitative determination of enantiomeric compounds. I. Simultaneous measurement of the optical isomers of amphetamine in human plasma and saliva using chemical ionization mass spectrometry. *Biomed Mass Spectrom, 4*, 118–121.

Melega, WP, Williams, AE, Schmitz, DA, DiStefano, EW, & Cho, AK (1995). Pharmacokinetic and pharmacodynamic analysis of the actions of D- amphetamine and D-methamphetamine on the dopamine terminal. *J Pharmacol Exp Ther, 274*, 90–96.

Neuman, RJ, Todd, RD, Heath, AC, Reich, W, Hudziak, JJ, & Bucholz, KK, et al. (1999). Evaluation of ADHD typology in three contrasting samples: A latent class approach. *J Am Acad Child Adolesc Psychiatry, 38*, 25–33.

Oliverio, A, Eleftheriou, BE, & Bailey, DW (1973). Exploratory activity: genetic analysis of its modification by scopolamine and amphetamine. *Physiol Behav, 10*, 893–899.

Paternite, CE, Loney, J, Salisbury, H, & Whaley, MA (1999). Childhood inattention-overactivity, aggression, and stimulant medication history as predictors of young adult outcomes. *J Child Adolesc Psychopharmacol, 9*, 169–184.

Pelham, WE, Gnagy, EM, Chronis, AM, Burrows-MacLean, L, Fabiano, GA, & Onyango, AN, et al. (1999). A comparison of morning-only and morning/late afternoon Adderall to morning-only, twice-daily, and three times-daily methylphenidate in children with attention deficit hyperactivity disorder. *Pediatrics, 104*, 1300–1311.

Pelham, WE, Jr., Sturges, J, Hoza, J, Schmidt, C, Bijlsma, JJ, Milich, R, & Moorer, S (1987). Sustained release and standard methylphenidate effects on cognitive and social behavior in children with attention deficit disorder. *Pediatrics, 80*, 491–501.

Pliszka, SR, McCracken, JT, & Maas, JW (1996). Catecholamines in attention deficit hyperactivity disorder: Current perspectives. *J Am Acad Child Adolesc Psychiatry, 35*, 264–272.

Porrino, LJ, Rapoport, JL, Behar, D, Ismond, DR, & Bunney, WE (1983a). A naturalistic assessment of the motor activity of hyperactive boys II. Stimulant drug effects. *Arch Gen Psychiatry, 40*, 688–693.

Porrino, LJ, Rapoport, JL, Behar, D, Sceery, W, Ismond, DR, & Bunney, WE (1983b). A naturalistic assessment of the motor activity of hyperactive boys I. Comparison with normal controls. *Arch Gen Psychiatry, 40*, 681–687.

Rajkowski, J, Kubiak, P, Ivanova, S, & Aston-Jones, G (1998). State-related activity and reactivity of locus coeruleus neurons in behaving monkeys. *Adv Pharmacol, 42,* 740–744.

Rapoport, JL, Buchsbaum, MS, Weingartner, H, Zahn, TP, & Ludlow, C (1978). Dextroamphetamine: Cognitive and behavioral effects in normal prepubertal boys. *Science, 199*, 560.

Rapoport, JL, Buchsbaum, MS, Weingartner, H, Zahn, TP, & Ludlow, C (1980). Dextroamphetamine: Cognitive and Behavioral effects in normal and hyperactive boys and normal men. *Archives of General Psychiatry, 37*, 933.

Rapport, MD, Carlson, GA, Kelly, KL, & Pataki, C (1993). Methylphenidate and desipramine in hospitalized children: I. Separate and combined effects on cognitive function. *J Am Acad Child Adolesc Psychiatry, 32*, 333–342.

Rapport, MD, Denney, C, DuPaul, GJ, & Gardner, MJ (1994). Attention deficit disorder and methylphenidate: Normalization rates, clinical effectiveness, and response prediction in 76 children. *J Am Acad Child Adolesc Psychiatry, 33*, 882–893.

Rapport, MD, DuPaul, GJ, & Smith, NF (1985). Rate-dependency and hyperactivity: Methylphenidate effects on operant responding. *Pharmacol Biochem Behav, 42, 23*, 77–83.

Robbins, TW & Sahakian, BJ (1979). "Paradoxical" effects of psychomotor stimulant drugs in hyperactive children from the standpoint of behavioural pharmacology. *Neuropharmacology, 18*, 931–950.

Roberts, AC, De Salvia, MA, Wilkinson, LS, Collins, P, Muir, JL, Everitt, BJ, & Robbins, TW (1994). 6-hydroxydopamine lesions of the prefrontal cortex in monkeys enhance performance on an analog of the Wisconsin card sort test: Possible interactions with subcortical dopamine. *J Neurosci, 14,* 2531–2544.

Roeltgen, DP & Schneider, JS (1991). Chronic low-dose MPTP in nonhuman primates: A possible model for attention deficit disorder. *J Child Neurol, 6,* S80–S87.

Rubia, K, Overmeyer, S, Taylor, E, Brammer, M, Williams, SCR, Simmons, A, & Bullmore, ET (1999). Hypofrontality in Attention Deficit Hyperactivity Disorder during higher-order motor control: A study with functional MRI. *Am J Psychiatry, 156,* 891–896.

Rubinstein, M, Phillips, TJ, Bunzow, JR, Falzone, TL, Dziewczapolski, G, & Zhang, G, et al. (1997). Mice lacking dopamine D4 receptors are supersensitive to ethanol, cocaine, and methamphetamine. *Cell, 90,* 991–1001.

Russell, V, de Villiers, A, Sagvolden, T, Lamm, M, & Taljaard, J (1995). Altered dopaminergic function in the prefrontal cortex, nucleus accumbens and caudate-putamen of an animal model of Attention Deficit Hyperactivity Disorder — the spontaneously hyperactive rat. *Brain Res, 676,* 343–351.

Safer, DJ & Allen, RP (1989). Absence of tolerance to the behavioral effects of methylphenidate in hyperactive and inattentive children. *J Pediatr, 115,* 1003–1008.

Sagvolden, T, Slatta, K, & Arntzen, E (1988). Low doses of methylphenidate (Ritalin) may alter the delay-of-reinforcement gradient. *Psychopharmacology (Berl), 95,* 303–312.

Satterfield, JH, Schell, AM, Nicholas, T, & Backs, RW (1988). Topographic study of auditory event-related potentials in normal boys and boys with attention deficit disorder with hyperactivity. *Psychophysiology, 25,* 591–606.

Schneider, JS, Sun, Z-Q, & Roeltgen, DP (1994). Effects of dopamine agonists on delayed response performance in chronic low dose MPTP-treated monkeys. *Pharmacol Biochem Behav, 48,* 235–240.

Simon, H (1981). Dopaminergic A10 neurons and the frontal system. *J. Physiol, 77,* 81–95.

Sokolowski, JD & Salamone, JD (1994). Effects of dopamine depletions in the medial prefrontal cortex on DRL performance and motor activity in the rat. *Brain Res, 642,* 20–28.

Solanto, MV (1986). Behavioral effects of low-dose methylphenidate in childhood attention deficit disorder: Implications for a mechanism of stimulant drug action. *J Am Acad Child Adolesc Psychiatry, 25,* 96–101.

Solanto, MV (1998). Neuropsychopharmacological mechanisms of stimulant drug action in attention deficit/hyperactivity disorder: A review and integration. *Behav Brain Res, 94,* 127–152.

Solanto, MV & Conners, CK (1982). A dose-response and time-action analysis of autonomic and behavioral effects of methylphenidate in attention deficit disorder with hyperactivity. *Psychophysiology, 19,* 658–667.

Spencer, T, Biederman, J, Wilens, T, Prince, J, Hatch, M, Jones, J, Harding, M, Faraone, SV, & Seidman, L (1998). Effectiveness and tolerability of tomoxetine in adults with attention deficit hyperactivity disorder. *Am J Psychiatry, 155,* 693–695.

Spencer, T, Biederman, J, Wilens, TE, Harding, M, O'Donnell, D, & Griffin, S (1996). Pharmacotherapy of attention-deficit hyperactivity disorder across the life cycle. *J Am Acad Child Adolesc Psychiatry, 35,* 409–432.

Spencer, T, Wilens, T, Biederman, J, Faraone, S, Ablon, J, & Lapey, K (1995). A double-blind, crossover comparison of methylphenidate and placebo in adults with childhood-onset attention deficit hyperactivity disorder. *Arch Gen Psychiatry, 52,* 434–443.

Sprague, RL & Sleator, EK (1977). Methylphenidate in hyperkinetic children: differences in dose effects on learning and social behavior. *Science, 198,* 1274–1276.

Stinnette, MJ & Isaac, W (1975). Behavioral effects of D-amphetamine and caffeine in the squirrel monkey. *Eur J Pharmacol, 30,* 268–271.

Swanson, J, Wigal, S, Greenhill, L, Browne, R, Waslik, B, & Lerner, M, et al. (1998). Analog classroom assessment of Adderall in children with ADHD. *J Am Acad Child Adolesc Psychiatry, 37,* 519–526.

Swanson, JM (1988). What do psychopharmacological studies tell us about information processing deficits in ADD/Hyperactive children? In J Sergeant & L Bloomingdale (Eds.), *Attention: Criteria, cognition, intervention* (Vol. 5). New York: Pergamon Press.

Swanson, JM, Greenhill, LL, Vitiello, B, Davies, M, Abikoff, H, & Pelham, WE et al. An evaluation of sensitization to clinical doses of methylphenidate. Unpublished manuscript.

Swanson, JM, Posner, M, Potkin, S, Bonforte, S, Youpa, D, Fiore, C, et al. (1991). Activating tasks for the study of visual-spatial attention in ADHD children: A cognitive anatomic approach. *J Child Neurol, 6(Suppl),* S117–S127.

Taylor, E, Schachar, R, Thorley, G, Weiselberg, HM, Everitt, B, & Rutter, M (1987). Which boys respond to stimulant medication? A controlled trial of methylphenidate in boys with disruptive behaviour. *Psychological Medicine, 17,* 121–143.

Usher, M, Cohen, JD, Servan-Schreiber, D, Rajkowski, J, & Aston-Jones, G (1999). The role of locus coeruleus in the regulation of cognitive performance. *Science, 283,* 549–554.

Vaidya, CJ, Austin, G, Kirkorian, G, Ridlehuber, HW, Desmond, JE, & Glover, GH, et al. (1998). Selective effects of methylphenidate in attention deficit hyperactivity disorder: A functional magnetic resonance study. *Proc Natl Acad Sci USA, 95,* 14494–14495.

Vitiello, B, Severe, J, Greenhill, L, Arnold, E, Abikoff, H, & Elliott, G, et al. (In press). Establishing the best maintenance dose of methylphenidate for children with attention deficit hyperactivity disorder: The MTA Experience. *J Am Acad Child Adolesc Psychiatry.*

Volkow, ND, Wang, GJ, Fowler, JS, Logan, J, Angrist, B, & Hitzemann, R, et al. (1997). Effects of methylphenidate on regional brain glucose metabolism in humans: relationship to dopamine D2 receptors. *Am J Psychiatry, 154,* 50–55.

Voytko, ML, Olton, DS, Richardson, RT, Gorman, LK, Tobin, JT, and Price, DL (1994). Basal forebrain lesions in monkeys disrupt attention but not learning and memory. *J Neurosci, 14,* 167–186.

Weber, K (1985). Methylphenidate: Rate-dependent drug effects in hyperactive children. *Psychopharmacology, 85,* 231–235.

Wilens, TE, Biederman, J, Spencer, TJ, Bostic, J, Prince, J, & Monuteaux, MC, et al. (1999). A pilot controlled clinical trial of ABT-418, a cholinergic agonist, in the treatment of adults with attention deficit hyperactivity disorder. *Am J Psychiatry, 156,* 1931–1937.

Witte, EA, & Marrocco, RT (1997). Alterations of brain noradrenergic activity in rhesus monkeys affects the alerting component of covert orienting. *Psychopharmacology, 132,* 315–323.

Zametkin, A, Rapoport, JL, Murphy, D, Linnoila, M, & Ismond, D (1985). Treatment of hyperactive children with monoamine oxidase inhibitors. *Arch Gen Psychiatry, 42,* 962–966.

Zametkin, AJ, & Rapoport, JL (1987). Neurobiology of attention deficit disorder with hyperactivity: Where have we come in 50 years? *J Am Acad Child Adolesc Psychiatry, 26,* 676–686.

APPENDIX

The Behavior of Children Receiving Benzedrine*, †

CHARLES BRADLEY

The use of benzedrine (beta-aminopropylbenzene; benzyl methyl carbinamine) as a central nervous system stimulant has recently aroused considerable comment. A number of reports are now available describing its effect upon the mood and other psychological reactions of adults,[1–7] but so far no one appears to have studied its effect upon the behavior of children. Since the psychological response of children to drug therapy of any sort appears to have been a generally neglected subject,[8] and since some of the changes produced by benzedrine in adults are of a type which might theoretically prove of clinical value in certain behavior problem children, the present studies were undertaken at the Emma Pendleton Bradley Home.

The facilities of this institution[9] are admirably adapted to the observation of children's behavior under controlled conditions without introducing the artificial distractions of a laboratory situation.The hospital deals only with resident patients of normal intelligence, the majority being children under 12 years of age presenting neurological and behavior disorders. Since such children require prolonged treatment the hospital setting has been designed to provide most of the desirable features of an outside community, including school, scouting activities, sports, etc. Although each child busies himself with a very natural sort of daily activity, he is under constant observation, and careful records of his behavior are kept as a matter of routine. In general, the patients are quite unaware of this close scrutiny.

To observe the effect of benzedrine upon the behavior of children, 30 patients, already resident in the hospital, suffering from a variety of behavior disorders, were

*** From the Emma Pendleton Bradley Home.**
Portions of this study were made possible by a grant from the Committee for Research in Dementia Præcox founded by the Supreme Council, 330, Northern Masonic Jurisdiction, U.S.A.

selected for study. Every child in this group had already been under observation for at least a month. Twenty-one boys and nine girls were studied, this ratio of sexes approximating that noted for all admissions to the hospital. The ages ranged from 5 to 14 years, two boys and one girl having passed their 13th birthday. The children's behavior disorders were severe enough to have warranted hospitalization, but varied considerably. They ranged from specific educational disabilities, with secondarily disturbed school behavior, to the retiring schizoid child on the one hand and the aggressive, egocentric epileptic child on the other. The patients' intelligence was in general quite within the so-called "normal" range.

In addition to routine observation and recording of behavior by ward nurses and teachers, each child was personally observed for a period of three weeks by a special psychiatric nurse. Because this nurse's customary duties had already involved daily close association with all of the children, she was able to observe carefully each patient without his recognition of her special function at this time. During the second week of observation a daily morning dose of benzedrine[10] was administered to each patient upon arising. Later the records of his behavior during this week were compared with those of the week prior and the week following, which were taken as control periods for this particular child.

School Performance

Possibly the most striking change in behavior during the week of benzedrine therapy occurred in the school activities of many of these patients. Fourteen children responded in a spectacular fashion. Different teachers, reporting on these patients, who varied in age and school accomplishment, agreed that a great increase of interest in school material was noted immediately. There appeared a definite "drive" to accomplish as much as possible during the school period, and often to spend extra time completing additional work. Speed of comprehension and accuracy of performance were increased in most cases. Insight into school improvement was generally present, though few of the children attributed it to the medication they had received earlier in the day. The improvement was noted in all school subjects. It appeared promptly the first day benzedrine was given and disappeared on the first day it was discontinued.

There were 15 children who did not show so striking a change and one child who was not of school age. Of the 15, eight did show some improvement, accompanying other temporary changes in their personal characteristics. Five others showed no change in school performance, although their behavior in other respects was distinctly altered. Two children showed no response to the drug in school or elsewhere.

Emotional Response

In discussing emotional responses it must be kept in mind that in many of these children the conduct disorder was essentially an external symptom of emotional maladjustment. Fifteen of the 30 children responded to benzedrine by becoming distinctly subdued in their emotional responses. Clinically in all cases this was an improvement from the social viewpoint. In many it was accompanied by a sense of well-being; but superficially the appearance was that of a child whose mood swings had become diminished. In this group were some children who had expressed their irritability in group activities by noisy, aggressive, domineering behavior. Such children under the influence of the drug became more placid and easy-going. Coincidentally, they seemed to become more interested in their surroundings. Such changes, however, by no means paralleled the improved school performance noted above as only seven children were common to both groups.

A variety of emotional responses were noted in the remaining 15 children. Seven of them gave evidence only of a sense of well-being, even to the point of mild euphoria. Although questioning the children in regard to their subjective feelings was studiously avoided in all instances, spontaneous remarks such as, "I have joy in my stomach," "I feel fine and can't seem to do things fast enough today," "I feel peppy," "I start to make my bed and before I know it, it is done," and "I feel fine," seemed significant. All of these patients showed a widening of interest in things around them, and superficially at least, a diminished tendency to be preoccupied with themselves.

Three other patients were observed to cry much more easily during the week that they received benzedrine, as though their emotions were more easily aroused. Two others were noted to wear worried expressions quite foreign to their usual appearance. Only one child of the entire 30 responded to the medication by becoming more hyperactive, aggressive, and irritable.

As was the case with alterations in school performance, all these changes appeared promptly on the first day of treatment and disappeared as soon as the drug was stopped.

Other Psychological Changes

A drug whose action is that of a stimulant may naturally be expected to produce changes in motor activity. In children, however, it is extremely difficult to measure motor activity *per se* apart from the individual's general behavior as shown by his mood and his intellectual activity. In our patients some decrease of motor activity was usually noted in the children who became emotionally subdued. A few of those who responded with elation possibly became a little more active, particularly in their daily routine about the hospital. Only three patients showed an increase in motor activity not explained by other changes. These three boys became active to the point of agitation, ran about rather than walked, talked more rapidly, and in general were so stimulated as to interfere with the efficiency of their daily work in school and on the ward. One of these was the child who became noisy and defiant.

Another boy, previously mentioned as having a worried expression, and also included among the three whose motor activity increased, expressed feelings of agitation and worry, was afraid to go out for fear he might run away, and was at times fearful of death. He had not manifested such symptoms before.

It will be noted that comparatively little mention has been made of the patients' subjective feelings under the influence of benzedrine. Personal questioning of any child was deliberately avoided so as not to influence his behavior artificially through suggestion or by indicating that he was the object of particular attention. Records of subjective responses are therefore limited to the occasional spontaneous remarks which the children made, and which have already been mentioned.

Physiological Observations

In view of the observations of others that certain physiological changes[1, 6, 11, 12] may follow benzedrine therapy, such routine examinations as would not artificially influence behavior were carried out during these studies. On the doses of benzedrine which were given blood pressure was noted to rise moderately but inconstantly during each day on about half the children. There was no significant change in the erythrocyte count or hemoglobin level as a result of the week's treatment. Body weight was not appreciably influenced. In six of the 30 children nocturnal sleep was mildly disturbed, as evidenced by a delay in the time of falling asleep for the first night or two. One patient remained awake to a late hour for four nights. Loss of appetite and nausea at the time of the morning meal were observed in a few of the children at the start of the experiment. Intolerance of the conventional dose when present was uniformly expressed by malaise, nausea and epigastric distress, persisting for an hour or two, and usually relieved by rest. Agitation and irritability, said to be characteristic of overdosage in adults, were observed in but one child.

Duration of Effect

Psychological effects were noted to appear in the majority of the children 30 to 45 minutes after oral ingestion of the drug. Behavior changes were at their height during the second and third hours, with gradual disappearance over a period of 6 to 12 hours. Occasional sleep disturbances 16 or more hours after the morning dose suggested that in a few individuals the stimulating effect was more persistent. In all instances the full effects appeared on the first day of administration, continued daily throughout the week and disappeared permanently as soon as the drug was discontinued.

Dosage

The 30 children who were the objects of this study received single morning doses of benzedrine sulfate upon arising. It was found, on the basis of giving gradually increasing trial doses to a few of the children, that 20 mg. was generally a satisfactory amount to produce changes in behavior without undesirable physiological consequences. All the children studied were then given a trial dose of 20 mg., but in eight instances this resulted in gastrointestinal distress sufficient to color all the child's reactions. These eight children were therefore given 10 mg. as their daily dose during the week. Two of them showed no response of any sort on this amount, suggesting that the largest dose they could tolerate was therapeutically ineffectual. The remaining patients all received daily doses of 20 mg., except one boy who tolerated 30 mg. very well and received that amount. There was no apparent relationship between the children's ages or weights and their reaction to benzedrine.

Comment

Possibly the most spectacular change in behavior brought about by the use of benzedrine was the remarkably improved school performance of approximately half the children. This is the more striking when we note that these patients were of good intelligence and that they were receiving adequate attention for any personality disorders which might affect their school progress. Moreover they were already in a school where specially trained, sympathetic teachers dealt with their pupils either on an individual basis or in very small groups. To see a single daily dose of benzedrine produce a greater improvement in school performance than the combined efforts of a capable staff working in a most favorable setting, would have been all but demoralizing to the teachers, had not the improvement been so gratifying from a practical viewpoint.

No entirely plausible reason for this acceleration of school performance is evident aside from the fact that a stimulating drug was used. However, at least half of the whole group of children did not show this striking response in school, in spite of the fact that most of them responded in other ways to the medication. Only half of the group showing favorable mood changes shared to a notable degree in the school improvement.

As far as improvement in individual school subjects was concerned the teachers were most impressed by changes in arithmetic performance, since speed of comprehension, degree of accuracy and quantity of output were all favorably affected.

It appears paradoxical that a drug known to be a stimulant should produce subdued behavior in half of the children. It should be borne in mind, however, that portions of the higher levels of the central nervous system have inhibition as their function, and that stimulation of these portions might indeed produce the clinical picture of reduced activity through increased voluntary control. Presumptive evidence that this may be the case is suggested by electroencephalographic studies,[13] which formed part of the clinical examination in 20 of our 30 patients. Definite abnormalities of the brain rhythms were noted in 11 of these children, eight of whose records showed the "spike and slow wave" form characteristic of the "seizure waves" seen in patients with petit mal epilepsy[14] and in some children whose erratic, irritable and aggressive behavior resembles that of epileptic patients.[15, 16] It may be significant that all but one of these children with "seizure waves" became distinctly subdued under the influence of benzedrine. Since it has also been noted that "slow waves" in the electroencephalogram are likewise present in a number of conditions where consciousness is impaired, it has been suggested that their appearance indicates impaired cortical functions.[17] One might postulate that a stimulating drug like benzedrine may tend to arouse this impaired cortical activity and alleviate any behavior disorder resulting therefrom. Further studies are necessary to confirm this.

There are, of course, a great many other possible factors entering into the genesis and clinical manifestation of any child's behavior problem. Complete understanding of the fact that children vary in their response to benzedrine awaits correlation of this variability with these many factors. From preliminary observations of the 30 children in this series there seems to be very little apparent relationship between a child's response to benzedrine and the conventional clinical characteristics of age, sex, past history, physical condition, reaction type, intelligence and special mental abilities or disabilities.

In spite of the attractive results obtained in this small group of patients, and the apparent low toxicity of the drug as evidenced by its prolonged use in the hands of others observers,[6, 7] it seems wise to await more complete knowledge of the action of benzedrine before recommending its clinical use in behavior problem children. Presumably the uncomfortable, though not necessarily dangerous, gastrointestinal symptoms which accompany overdosage indicate that the drug should be administered only by the experienced physician. The factors involved in children's behavior disorders are so many and varied that they can be competently evaluated only by physicians adequately trained in child psychiatry.

Any indiscriminate use of benzedrine to produce symptomatic relief might well mask reactions of etiological significance which should in every case receive adequate attention.

Summary

The psychological reactions of 30 behavior problem children who received benzedrine sulfate for one week were observed. There was a spectacular improvement in school performance in half of the children. A large proportion of the patients became emotionally subdued without, however, losing interest in their surroundings. A variety of other definite behavior changes were also noted. Dosage, unfavorable responses, and duration of effect of the drug are discussed, and it is suggested that its use in the treatment of children with behavior disorders be supervised by physicians trained in child psychiatry.

References

1. Peoples, S. A., and Guttmann, E.: Hypertension produced with benzedrine. Lancet, I:1107, May 16, 1936.
2. Myerson, A.: Effect of benzedrine sulfate on mood and fatigue in normal and in neurotic persons. Arch. Neurol. Psychiat., 36:816, Oct. 1936.

3. Davidoff, E. M.: A clinical study of the effect of benzedrine therapy on self-absorbed patients. Psychiat. Quart., 10:652, Oct., 1936.
4. Sargant, W., and Blackburn, J. M.: The effect of benzedrine on intelligence scores. Lancet, 2:1385, Dec. 12, 1936.
5. Carlisle, C. L.: Use of benzedrine sulphate in catatonic stupors. Med. Bull. Veterans' Admin., 13:224, Jan. 1937.
6. Nathanson, M. H.: The central action of beta-aminopropylbenzene. J. A. M. A., 108:528, Feb. 13, 1937.
7. Wilbur, D. L., MacLean, A. R., and Allen, E. V.: Clinical observations on the effect of benzedrine sulphate. Proc. Staff Meetings Mayo Clin., 12:97, Feb. 17, 1937.
8. Darrow, C. W.: Psychological effects of drugs. Psychol. Bull., 26:527, 1929.
9. Bradley, C.: A children's hospital for neurologic and behavior disorders. J. A. M. A., 107:650, Aug. 29, 1936.
10. Benzedrine sulfate for this study kindly supplied by Smith, Kline and French Laboratories, Philadelphia, Pa.
11. Myerson, A., Loman, J., and Dameshek, W.: Physiologic effects of benzedrine and its relationship to other drugs affecting the autonomic nervous system. Am. J. Med. Sc., 192:560, Oct. 1936.
12. Anderson, E. W., and Scott, W. C. M.: Cardiovascular effects of benzedrine. Lancet, 2:1461, Dec. 19, 1936.
13. Jasper, H. H., and Andrews, H. L.: Human Brain Rhythms. I. Recording Techniques and Preliminary Results. J. Gen. Psychol., 14:98, 1936.
14. Gibbs, F. A., Davis, H., and Lennox, W. G.: The electroencephalogram in epilepsy and conditions of impaired consciousness. Arch. Neurol. Psychiat., 34:1133, Dec. 1935.
15. Solomon, P., Jasper, H. H., and Bradley, C.: Studies in behavior problem children. Arch. Neurol. Psychiat. (In press.)
16. Kiene, H. E., and Solomon, P.: Why people misbehave. Rhode Island Med. J., 20:58, April 1937.
17. Davis, H.: Some aspects of the electrical activity of the cerebral cortex. Cold Spring Harbor Symp. Quant. Biol., 4:285, 1936.

Index